System Engineering for

Machine

Development

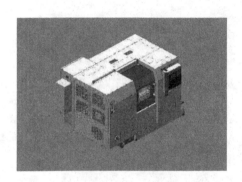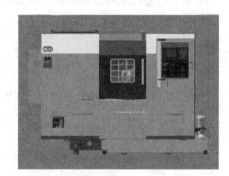

Jyoti Mukherjee

Licensed Professional Engineer

Ex-Chair Professor, IIT Kharagpur, India

iUniverse

SYSTEM ENGINEERING FOR MACHINE DEVELOPMENT

iUniverse books may be ordered through booksellers or by contacting:

iUniverse
1663 Liberty Drive
Bloomington, IN 47403
www.iuniverse.com
1-800-Authors (1-800-288-4677)

ISBN: 978-1-5320-7805-7 (sc)
ISBN: 978-1-5320-7807-1 (e)

Library of Congress Control Number: 2019909030

Print information available on the last page.

iUniverse rev. date: 07/03/2019

This book is dedicated to my students, well-wishers, mentors and professionals who have helped me to be what I am today. Hope this book will help them to design and develop best possible machines in future. Also, this work is dedicated to Lord Venkateshwara for granting me the energy to complete the task.

ABOUT THE AUTHOR

The author of this book, JYOTI MUKHERJEE, obtained his Bachelor of Mechanical Engineering (B.E.) from the Calcutta University, India, Master in Machine Design (M. Tech) from Indian Institute of Technology, Kharagpur, India, Post Graduate Diploma in Business Management (PGDBM) from Xavier's Labor Relations Institute (XLRI), Jamshedpur, India. He also received Doctor of Engineering in Mechanical Engineering from Cleveland State University, Cleveland, Ohio, USA. He also obtained Master of Science (S.M.) in System management from MIT, Cambridge, Massachusetts, USA and Master of Business Administration (MBA) in Operation Management from University of Syracuse, Syracuse, NY, USA. He is also a certified Professional Engineer (P.E.) from Ohio, USA.

His professional experience for more than 40 years includes working for Tata Motor Company in India as superintendent of Maintenance for production machines, in Warner and Swasey and Harding Brothers in USA as design engineer, in White Motor Consolidated, Cleveland, Ohio and in Ford Motor Company as a system reliability testing engineer and chassis design engineer. He also has worked for Makino Machine Tool, Japan. He also has experience as a consultant in small to medium machine tool companies designing CNC machines. In addition, he has partnered with various machine tool design engineers to design and manufacture CNC machines for indigenous and export markets. He holds several US and international patents to his credit in the area of machine tool design and development. He has obtained awards for his work from all the companies that he has worked for during his professional career.

P R E F A C E

Motivation for Writing This Book

First and foremost, I would like to express my motivation for writing this book, which outlines almost all the system engineering process details, help, and suggestions that I wished I'd had when I was a novice machine designer forty years back. Even now, I can't find a book of this nature for machine tool applications. There were enough engineering textbooks on the analysis of machine elements, but the process for managing a machine project to increase productivity and efficiency was very much lacking in the industry. I found my mentors focusing only on design, specifically on developing a design that would satisfy management's requirements but not the customers'. System engineering concepts were never applied in detail to make the project cost-effective and competitive in the marketplace. This is the single most important purpose of this book.

The contents of this book came from my 50 years of experience in the industries and teaching in several universities in the USA and overseas. As a chair professor in the systems and industrial engineering department at the Indian Institute of Technology, Kharagpur, I had an opportunity to teach several system engineering subjects and decided to write this book using my professional and academic experiences. I blended system engineering with product development topics to make it more interesting to various disciplines of engineering. My students there were industrial engineering, reliability engineering, and MBA students. My students at the University of Arizona in Tucson also helped me to develop the course content. This book was written to help the working professional in the fields of product development and reliability testing for machine tools and other commercial products. The idea is to apply system engineering principles along with product design development tools to machine tool design, development, and reliability enhancements.

In several instances in this book, I have commented on the topic of product design and development using system engineering applications. What is more important, design or manufacturing, to maintain the reliability and quality of a machine tool or any product in general. The single-word answer to this question in my mind is design, absolutely design. There is no question about it. A correct design is the core and quality, reliability, ease of manufacturing, and ease of assembly are all flavors around the design. Quality without proper design is a misnomer. To enhance the quality and reliability of a machine, one has to enhance the design. I have tried to emphasize this point throughout the book. I strongly believe in it and practiced the same throughout my career. The book is predominately written around this theme.

The first and foremost driving force behind this effort is a combination of my education, experience, and training in the design and development of machine tools in the United States and overseas over the last 50 years. During my tenure in the industry, I had an opportunity to work in the machine tool and automobile industries.

I also worked as a consultant in the machine tool industry for a decade or so. Moreover, I tried to start up a few machine manufacturing companies on my own or as a partner. I also had an opportunity to work in a machine tool company in Japan for a while, living in Japan and working with Japanese engineers for some time. For the last decade or so, I have taught as an adjunct professor in mechanical engineering and chair professor of industrial engineering.

I wanted to put down my observations and thoughts in a book dedicated to the design, development, and manufacturing of machinery. This book is primarily a reflection of my training and experiences in machine tool and automobile companies. This book answers the questions that students, business professionals, business owners, and engineers have asked me over the years and that my answers to them were not able to satisfy. I wanted to show the way machine tool design, development, and manufacturing could be done using system engineering principles. The idea came from one of my students, who asked me if a machine tool is a primary system or not. This book is an answer to that basic question. Also, the title of this book is dedicated to one of my students, who asked me to explain the reasons why a machine is a system.

While working in a very famous machine tool company in Japan, I noticed how their engineers designed and developed machines in a systemic way. I saw and learned how they designed and developed machines of the highest possible quality and reliability. The tremendous success of system engineering in the automobile and aerospace industries convinced me that system engineering is also the best tool for all kinds of machine design and development. A systemic approach is required to make machines more reliable and cost-effective. The basic principles of system engineering help design a product that makes the customers happy and proud to own the product. I came to realize that design is the source of all evils if not executed properly. The design has to satisfy all the customer requirements to make a successful product. This book is the answer to some of my thoughts about what has to be done to design, manufacture, and maintain machines in a very successful way using system engineering principles.

While working as a design consultant for very small and medium-sized companies, I realized the immense difficulty they face in managing machine design and development. Cost is the primary reason why customer requirements are not met. It is still a prevailing thought for some manufacturers that customers will like whatever the company designs and introduces to the market. Instead of designing what customers want, they design what the designers and company want and then justify the product for the market. Japanese products satisfy customers only, and that is why they are more successful.

I have made several attempts to write this book. This is the final version for this edition. This effort is not complete by any stretch of the imagination, but it lays down the basic steps of system engineering principles for all types of machine tools irrespective of volume and complexity. Nevertheless, the principles can also be used for any commercial products. Several times, I have visited factories and talked to designers and manufacturers about the application of system engineering in their day-to-day efforts. They talk about reliability, quality, cost, specifications, manufacturing methodologies, etc. in a very compartmentalized way without realizing that it is all about a total system that they are trying to produce.

System engineering allows engineers to think systematically while the product is being designed. I strongly hope that this book is a justification for my thoughts about the usefulness of system engineering applications for machine tool design and development. That is why I named this book *System Engineering Principles for Machines*. Time will tell if the book successfully explains my ideas to its target audience. I strongly believe it does since the text has been written about what I have practiced and what I have seen not being practiced in machine tool design and development.

Other very important and pertinent questions have always come to my mind, and students and some professionals have asked me some as well:

- What is more important to be a successful machine design engineer: in-depth training or in-depth education?
- Are system engineering principles really applicable to machine tool design and development?
- What is more important to enhance quality in machine tools: good design, good manufacturing, or quality slogans?
- What enhances the reliability of a machine tool: design or quality planning?
- Why don't Japanese machine tool companies brag about product quality as much as U.S. companies do?
- Does a well-designed product sell itself?
- Why don't quality consultants talk about design quality?

Obviously, one textbook cannot answer these and similar questions. But these are very common questions in the minds of professionals connected with machine design, development, and manufacturing. I have tried to answer some of these questions. I have witnessed academicians talking about machine design without any experience with designing a machine. I have also listened to consultants and professors talk about quality improvements and planning without ever having set foot in a machine tool factory. Consequently, many develop quality signals and banners but not a quality product.

I have tried to develop this book with these and other questions in mind. I have modified the general principles for application in machine tool and other similar commercial products, and I hope readers agree with my intentions. To answer the controversial question about what is more important, training or education, in the machine tool arena, without mincing words, I would say in-depth training with basic education. My exposure to machine tool industries says so.

Target Audience for This book

System Engineering Principles for Machines is primarily geared towards the following folks who are interested in designing, developing, and manufacturing reliable machine tools:

- **Professional System Engineers** practicing core system engineering for machinery and other commercial products

- **Professional Design and Development Engineers** responsible for designing and developing machinery to satisfy customers

- **Engineering Students** pursuing a machine design and system engineering professional career

- **Management Students** pursuing careers in the reliability and quality management profession

I also strongly believe and hope that this book will be useful for business managers and marketing professionals to understand the scope of system engineering and product development for various machinery and commercial products. This will help them to understand why customer satisfaction is the only goal of product design and development, and nothing else.

Organization of the Book

The introductory chapter very briefly explains the status of machine tool engineering in the United States. It also justifies the application of system engineering principles for machine tools. The next few chapters focus on system engineering principles specifically geared towards machine tool applications. These chapters basically set the stage for system development and system planning methodologies for machine tools. The underlying systemic principles for machine tool design and development are explained in detail. These chapters are not in serial in organization since it is very difficult to line up the contents in this way due to their descriptive nature. The first few chapters have general content on the following subjects:

- System engineering principles modified for machine tool

- Design and development for machine products in general

- Integrated system planning for machine tool development

- Customer needs and requirements development for machines

- Product planning steps and procedures

The next few chapters demonstrate how to design and develop machine tools step by step using system engineering principles. Examples to describe the principles are always given for several types of machines. So, the focus of these chapters is on a design that satisfies customer requirements. The theme is that a good system design that satisfies customer requirements always sells well. Everything starts with the design. The product development portion of this book concentrates on:

- Machine planning and development procedures
- Target machine system requirements development
- Machine concept development steps
- Machine concept selection procedures
- Design, analysis, and testing requirements and procedures
- Prototype building and testing criteria

Once the machine development procedures have been explained and a specific design has been selected, the question becomes: what are the requirements for quality, reliability, and robustness of the system? They are part and parcel of machine system requirements and have specific metrics. Hence, the next few chapters are dedicated to explaining machine tool quality, reliability, and robustness in a production or automatic setting. These chapters have been laid out to make the point that design requirements must include these metrics at the beginning and they should not be afterthoughts. In order to make sure design development time and cost are within controls, a project has to be managed properly and efficiently. Reliability has to be built into the design right from the beginning.

So, these chapters concentrate on:

- Reliability and robustness considerations for machines
- Quality management for machines
- Industrial design and safety considerations for the design of machines
- Design considerations for ease of manufacturing and assembly
- Project management guides and tools

The next few chapters deal with the applications of various system engineering procedures and tools in the area of machine tool design, development, and reliability management. The idea is to focus on the shortcomings that I witnessed during various stages of machine design and development. These chapters have been written to show how system engineering principles such as root-cause analysis (RCA), design and process failure mode, and effects analysis (DFMEA and PFMEA) should be applied during the design and prototype-development stages. Even if these procedures have been found to make a system more efficient and productive in the automobile and aerospace industries, they have not been used extensively and productively in machine tool industries.

The last few chapters concentrate on the following:

- Application of reliability principles for machine design and testing
- System engineering applications for machines
- Functional analysis and decomposition for system development

- Failure mode and effects analysis for design and process

In total, there are 20 separate chapters dedicated to individual system design and development for machines. The contents have been developed in a very general fashion so that the subject matter can be very easily modified and applied for systems other than machinery. The subject of system engineering is very wide and deep, and covering it in one textbook is simply impossible. Hence, I have focused on a few principles that are very pertinent and important for machine tool development.

Nevertheless, efforts toward applying systemic engineering principles to machine design, development, and manufacturing must be made very sincerely to ensure that future machines are reliable, productive, and able to handle increasingly complex and demanding requirements. The current highly automated production systems need unmanned and automated machines with more complex requirements and highly robust systems. There is no way around it if the goal is to be globally competitive.

A C K N O W L E D G M E N T S

Obviously, I could not have finished such a mammoth and complicated task without the help, suggestions, discussion, and guidance from an innumerable number of friends, teachers, and industry professionals. In particular, my training mentors, my students, and my professors from educational institutions around the globe have been instrumental in shaping my thoughts and logistics for this book. Nevertheless, I would be failing in my duties if I did not mention the special help I received from the following people.

First, I'd like to thank the key people who encouraged me to take this task to its completion: Joe Christensen and Kerry Ballard from Smithy Company in Ann Arbor, Michigan. Beyond my expectations, they gave me suggestions for improving this book after going through a few chapters. They are my examples to learn from. Some key people at Smithy Company also provided helpful suggestions from time to time. I sincerely appreciate their help in this regard. Next is my mentor Professor Parviz Nikravesh from the University of Arizona in Tucson. He always helped me while I was writing this book even though it is not his field of expertise. Professor Nikravesh guided me from time to time on the content and arrangement of the chapters of this book, and I could not have completed it without his timely help and suggestions.

I would also like to thank Professor Ara Arabian, the director of the UA Engineering Design Program, who gave me the incredible opportunity to learn from industry projects by offering me a mentoring position in his clinic. I learned a lot from the students and their projects, and also from teaching them system engineering and industrial reliability. I am thankful to these students for their encouragement and suggestions from time to time. Professor Ara also directly and indirectly influenced my thoughts on how to efficiently complete projects within a stipulated time. I thank him for the opportunity he provided me during the time in my life when I wanted it most.

I am also deeply indebted to the esteemed professors of the systems and industrial engineering department of my alma mater, the India Institute of Technology in Kharagpur. Each and every professor of the department always asked me to put down my thoughts on system engineering and its applications in industry, and they gave me input whenever possible. Professor P.K. Ray and Professor Mohanty from this department have been my examples for pursuing such an enormous task. At the same time, I am also indebted to my students at the institute for giving me the chance to start this book during my tenure there as a chair professor of the department. They helped me beyond my expectations. Their thoughts on system engineering, quality, and reliability helped me to write a significant part of this book. My sincere thanks go to them for giving me help when I needed it the most.

I also have been encouraged by the books and class notes written by my professors from MIT in Cambridge. I learned a lot from Professor Eppinger's book on product development, which really helped to order my thoughts on the product development chapters in this book. I remain obliged and thankful to him for encouraging me with his writing and class lectures at MIT.

Professor Whitney of MIT also set a good example for me, and I learned a lot from his lectures on product development. I would definitely be falling short in my duties if I did not also mention the immense amount that I learned from the system engineering lectures and projects of Prof C. Boop of MIT. He is an outstanding expert on system engineering in the field of aerospace, and he helped me to think of an assembly as a system. He is the primary motivator for this undertaking as well. His lectures influenced me a lot, and I have referenced his class notes throughout this book.

Most of my professional learning and training came from the Ford Design Institute (FDI) of Ford Motor Company, Dearborn, Michigan. My professional experience about systems, reliability, and testing would have been unfulfilled if I had not had the wonderful opportunity to work for Ford, which, I believe, is the best automobile company in the world. All my system engineering applications and reliability experience came from Ford. It is where I came to know how to blend system engineering and reliability principles with system applications, and it is where I practiced system engineering and product development concepts. I will never forget everything I learned from my fellow Ford engineers. I remain much obliged to all of them for their help and mentoring during my tenure at Ford. The training in reliability and system engineering that I received there influenced my thoughts and the content of this book immensely, and I have referenced their materials when necessary. I remain indebted to Ford and its training facilities. In addition, I remain indebted to various engineering professionals from Warner & Swasey Company, Motch & Merry Weather Company, Target machinery Company and Leblond Makino machine Tool Company etc.

Last but not least, I'd like to finally thank my family members for their extreme patience and support. Nanda, Anna, Sompura, Matt, and Chris showed me endless encouragement and patience during the time I was writing this book. Their love and encouragement are priceless. I also recognize timely help and guidance from Dr. Reetobroto Mookherjee, Vice President of Fandango, Los Angeles, CA, for his time to time advice and help while writing this book. I also recognize help and guidance from Dr. Reetobroto Mookherjee, Vice President of Fandango, Los Angeles, CA, for his time to time advice and help while writing this book.

I dedicate this book to my students and my fellow academicians and professionals who helped and encouraged me throughout my career to complete this project as best as I possibly could.

Jyoti Mukherjee

Tucson, Arizona, USA

CONTENTS

C H A P T E R 1

1 Introduction

1.1 Brief History of Machine Industry

Due to the enormity of the subject and the intricate details of various possible topics on production machines in a broader sense, this book only makes an attempt to nibble at the few aspects that the author thinks are very important for the systemic design and development of a business plan for such machines. When necessary, the subject matter and principles of system engineering and the development of such a business plan will be explained in light of a few production machines such as lathes, milling machines, machining centers, and grinding machines. The primary focus will be on general-purpose machinery for production purposes. This book is geared toward explaining the methodologies and principles used to design and develop a business plan for machines or machinery as a system for producing desired outputs. In this context, a machine is defined as a conglomerate of subsystems acting in unison to create a production part, whereas machinery is defined as a system of machines working together to create a part or family of parts.

The machine tool industry as a whole has come a long way, starting from a simple special-purpose machine consisting of a pulley, screw, and lever, developed by the Greek philosopher Archimedes around the third century BC, to the present-day era of computer numerically controlled (CNC) machines or machinery. The Industrial Revolution most probably was the single reason for the sporadic evolution and phenomenal growth of the machinery industries into its present form. It all started from mechanized cotton spinning, the power loom concept, in the textile industry for enhancement of productivity. The other predominant factors were the invention of steam engines and iron making. All of these events needed a higher production of components at a much lower cost. An era of automation emerged. Machine tools were invented to give rise to such demands with high speed and lower cost.

The first set of machine tools were most probably a screw-cutting lathe for round parts, a cylinder-boring machine for finishing operations, and milling machine for prismatic parts in the metalworking industry. The concept of migrating to automation from handmade parts came into being. The interchangeability of mass-produced parts without customization, to be used for the production of firearms and textile applications, demanded speed, quality, and cost-effectiveness. This program was promoted by the U.S. Department of War in the early 19th century.

Such a stellar, exemplary, and timely effort led to the development of the concept of accuracy for mass-produced machined components. This seems to be a major U.S. contribution to industrialization. This also gave birth to the development of jigs and fixtures, precision blocks, and gages to measure the accuracy of the part produced. For this process, Mr. E Whitney designed and developed the milling machine. The Blanchard lathe was produced by Thomas Blanchard. The standardized way of manufacturing mass-produced, interchangeable parts became known as the American system of manufacturing. The precision manufacturing methods required machinery that mechanized the shoe, textile, and watch industries, and the machine tool industry enjoyed sustainable growth in support of them. This innovation, which took the machine tool industry to a whole new level of significance, could be called a disruptive technology.

As can be inferred, development in the machine tool industry followed the requirements of other industries, so when production requirements went up substantially, the machine tool industry reacted accordingly with machines having higher speed and quality to enhance production requirements or improve productivity. World War II was another incentive for higher machine tool production and demand. The production of machine tools almost tripled during the war in order to cope up with the demand for weaponry. The substantial growth of the machine tool industry trend continued until the 70s, and the production of machines throughout the world for China, Japan, the USA, the UK, Spain, Italy, Germany, and other countries also showed phenomenal growth.

During the 80s, there was significant growth in machine tool production throughout the world. The USA and Germany shared the two top positions, followed by Japan. Germany's metalworking machine tool production almost doubled from 1977 to 1980, but it declined sharply to about $4 billion after 1981. The US had a similar rise and fall during the same period, and Japan's machine tool production peaked and surpassed that of the US and Germany. I think the primary reason for Japan's rise is the fact that they focused on manufacturing general-purpose machines with quality, reliability, and reduced cost. They took advantage of everybody's downfall and gained market share in the world.

Recently, the growth has slowed down consistently in spite of the use of high technology in machine tool control. First, NC, and then CNC, technology did bring about a major facelift for the machine tools. The digital control capability did empower the machine industry to move from general-purpose machines to high production machinery.

Machines became less dependent on human interference during production when higher productivity was demanded in the metal cutting industry. Analog and digital control converted machines almost into robots with higher production performance. Machines became way less dependent on mechanical elements, gears, cams, hydraulics, etc. The net result was much higher production and odd-shaped part production capability at the cost of price of the machine tools in general.

Once the digital control wave slowed down, the wave of "abilities" came into being. To attract buyers and survive, the machine tool industry has tried to enhance process capability, reliability, sustainability, machinability, maintainability, increased availability, and repairability. This is the modern trend. No such disruptive technology has arrived on the horizon yet to take another quantum leap for growth and profitability. In general, the machine tool industry lacks growth and innovation since survival mode has given precedence to excellence. Instead of the user industry taking the lead, the machine tool industry has had to produce machines with much higher capabilities and accuracy without increasing cost. In order to accomplish that objective, machines have to be designed based on system engineering principles.

There are many underlying questions that the machine tool industry has to answer. Some of these are mentioned below as thought questions. From the three-dimensional approach of "speed, quality, and cost," machine industries have to migrate to the era of "multi-dimensional universality" by embracing the answers to the following questions:

- How can we reduce the cost of machinery to reduce parts costs?

- How can we help the user industry reduce the cost of production without sacrificing quality, reliability, and availability of the machines?

- How can we design platform- or legacy-based machines to reduce the cost?

- How can we design a universal machine that can turn, mill, grind, shape, drill, and bore for all types of materials?

- How can we design a machine where accuracy does not depend on the work-holding capabilities or fixturing methods?

- How can we convert a general-purpose machine into a special-purpose machine, or vice versa, without a substantial investment of time and money?

Such questions remain unanswered in my opinion. In order to answer them systematically and methodically, management and design personnel must look within the industry for a disruptive technology. The beginning of such a hard path has to start by following system-based principles to understand what we are doing and what we are supposed to do in the future.

1.2 Strategic Approach for Growth

The primary strategic business approach of a company to making a sustainable profit is to identify and recognize customers' requirements and bake them into a product that will satisfy these needs. Moreover, to thrive, the business also has to offer quality products at a competitive price in the marketplace. Both of these objectives can be met if the company has a vision and plan to reach its goals as quickly as possible. In my opinion, the machine tool industry as a whole has to take the lead to guide the users to what they need. In order for the machine tool industry to survive, the developer must be innovative in its design development, manufacturing, and servicing. Otherwise, it will have a premature and undesirable end.

The design approach should also encompass innovation, manufacturing, quality, reliability, and other human factors to develop a machine that will satisfy the customer. However, manufacturing and quality control are not the answer. The design team has to take the lead in delivering a desired product and ensuring that a product is produced with all the constraints satisfied as best as possible. The product design strategy has to be ahead of other functional strategies, such as marketing, financial, and operational. The Japanese machine tool industry has proven beyond a doubt that this is a feasible and workable strategy to survive and grow again. In the beginning, the focus should be on design and design only.

I also believe that the U.S. machine tool industry has gone down the drain due to its primary focus on finance, sales, and marketing. A desirable product sells itself. Hence, a coherent product design strategy must be at the forefront to lead the industry out of its glut. The machine tool industry should be led by the innovators and designers and not by financiers, whose only goal is to be profitable.at any cost. I strongly believe that too much focus on the financial side of business and pseudo-cost reduction efforts without rhyme or reason were the two most important factors in the death of many machine tool companies, such as Warner and Swasey, Motch and Merry Weather, Gisholt, etc. We need a long-term focus instead of the short-term goal of making money at any cost. The U.S. machine tool industry needs to be innovative and bring a disruptive technology that will show the user industry how to be competitive and save by producing more, not less. We need a systemic approach to the business of machine tools.

Measuring the success of managers of machine tool companies in a very short period and the way to live on a day-to-day basis seems to be the way of life for Americans. These people are compensated on the basis of financial results of the companies over a very short duration. The financial managers and investors are too worried about the bottom line and balance sheet of machine tool companies on a daily basis. They fail to understand the fact this particular industry depends on innovation and design to satisfy customer demand and requirements. The financial managers and investors treat the companies as though they were consumer-oriented companies, where demand drives supplies. The performance of machine tool companies must be measured over a continuous and long period.

Machine tool companies are significantly different from the automobile and consumer-based industries. Their shortsighted approach in this regard has brought, in my opinion, the end of machine tool industries, in the US at least. On the other hand, Machine tool industries from Japan, China, Taiwan, and Korea have taken advantage of this approach and flooded the market with machines that are cost effective and productive. The financial managers have to understand the core business principles of machine development and marketing. The ROI for the investors has to be measured over a much longer period. Again, this approach of evaluating academic professionals, design engineers, and other directly involved professionals, unless significantly changed very soon, will make U.S. machine tool manufacturers much weaker, and consequently, the USA will have to depend on others for machines.

In the 70s and 80s, the U.S. machine tool industry exported its product to growing industrialized countries due to the fact that the machines were very highly technologically advanced and suitable for high productivity. Again, technology and innovation were the cause of such demand overseas, and not aggressive sales and marketing. The product was first, and everything else followed suit. Over time, the export market dried up since the US lost its edge, sharpness of design, and flexibility. Now the trend is to buy machines from Japan and Korea to replace machine tools from the USA. Since the indigenous market is shrinking at a very high rate, U.S. machine tool builders have to focus on being competitive and cost-effective. The machines have to be flexible, robust, and reliable enough to attract customers from inside and outside the country. We have to go back to the drawing board again.

In the recent past, the strategy in the USA was to survive by service industries alone. I fail to understand one point: if the products are not designed and produced in the USA, what are you going to service? We have to design and manufacture innovative products first, keeping quality in mind. We have to take a strategic and systemic approach to bring the machine tool industry back to the way it used to be in the 50s and 60s. The author strongly believes in this approach, and this textbook is focused in promoting this methodology.

Keeping our future of the nation in mind, the daily approach of performance evaluation has to become a decade-long evaluation of the machine tool industry. Design, manufacture, and quality of machines should be the performance indicators and not the financial ratios of the balance sheet. If our government could save the automobile industries, couldn't they also help the machine tool industry to thrive again? If other nations can impose additional tariffs on imports from the USA, why can't we use the same method with those countries that we are purchasing machines from? Can't we just get even with them to bring back our machine industry again? We need a fair approach in this regard to get back in line again

1.3 Systemic and Innovative Design Matters

The gradual decline of machine tool design and manufacturing in the USA and the increase in the importation of machine tools from Japan, China Taiwan, Korea, and other countries have been noticed over last 20–25 years. Well-established machine tool companies have moved their manufacturing and design functions overseas, and they have decided to only sell and service the products in the USA. The reason for such a decision is financial. The decline of design innovations and manufacturing in the USA have forced the engineers to shift their jobs outside the industry, basically becoming service-oriented industries.

At present, all castings and forgings come from overseas, and electronic controls and software are also imported, mostly from Japan, Germany, and China. We have lost the edge over others due to lack of investment and loss of manpower. We have a fake machine tool manufacturing industry where we bring most of the core components from overseas and do very little in the USA. The author has been involved several times with customizing overseas machine tools for U.S. consumers.

The million-dollar question is how to reduce labor cost without compromising quality and design innovation. Are a strong design and manufacturing community for machine tools important for the national economy? Can we compete cost-wise and excel again? Can we design and innovate machine tools that are the best in the world? The answers to all of these pertinent questions are yes. We have to take a systemic approach towards the machine tool business and development for such equipment. We have to design and produce first and then worry about profit later. I strongly believe that the product has to be created first and then produced and serviced when it is in use. So, design, manufacturing, and service are necessary for a successful product. Unit labor costs for some industrial sectors has risen much faster in Japan in recent years than that in the US. Still, we import many more times from Japan than we export to them.

Another argument has been made recently that innovation stays in the US and everything else goes overseas. My argument against that idea is the fact that proof of a disruptive innovation is in the product itself. So, the new products and manufacturing have to stay together to be successful to prove the technology. This will also support the high cost of research and development activities. We have to be the leader in bringing innovative products to the market to maintain our edge over other advanced countries.

It appears to me that Japan has taken a lead in many such industries, such as the automobile industry, electronic industry, machine tool industry, and the steel industry. They have integrated design, development, manufacturing, and quality into their products to gain ground in each field they have decided to enter. They have combined design, manufacturing, and servicing functions together, and they sell the product as a package. Thus, if manufacturing and design move overseas, service will follow suit very soon. The Japanese, in particular, have surpassed the US in patents and new innovations. The top patent winners are Hitachi, Canon, Mitsubishi, Fuji, Toshiba, GE, Kodak, Siemens, and IBM. Half of these companies are from Japan.

I agree that the U.S. economy is still thriving, healthy, and growing at a moderate rate. The notion that it is migrating towards a service industry and away from the design and manufacturing sectors is, in my humble opinion, totally wrong and baseless. U.S. economic growth cannot be sustained by only servicing products. Software applications in a cell phone are not the core technology; creating a cell phone design to accept software and communicate accordingly is the success. Again, design first, and everything else follows. In order to reduce the cost of a machine without compromising quality, we have to focus on disruptive and innovative design to satisfy customers' requirements. Due to the high level of customer acceptance, product volume will go up. This will reduce our fixed costs as well.

1.4 Status of the Machine Tool Industry in the USA

As per the federal report in September 1983 (1), the U.S. machine tool industry is still the second largest in the world, but we are behind Japan for sure. We have become specialized machine producers and given up our position of largest machine tool producer in the world to Japan and China. Capacity utilization is the lowest in the industry.

Many companies have gone under even after mergers and acquisitions. Expenditures on R&D have gone down substantially. More significantly, investors have moved out of machine tool industries and have permanently joined other investment avenues. To make the situation worse, machine tool imports have risen significantly. Imports have replaced indigenous machine tools in the last 20 years. Worst of all, U.S. machine tool consumers believe that U.S.-produced machine tools are worse in quality and higher in cost than foreign imports.

Also, U.S. producers take much longer to deliver products than foreign producers. U.S. machine tool builders have become better sales and service providers, as strategically intended. U.S.-produced machine tools are highly sophisticated and more for special purpose than general purpose. So, we have become SPM machine builders instead of GPM machine builders because SPMs bring in more money more quickly. The problem is, SPMs are only rarely required, so we have less volume of production and our costs go up substantially. Moreover, the exportation of machine tools has reached its lowest level in recent years. It also has been reported that Japan's machine tool productivity has doubled in the same time span.

It has been reported that U.S. machine tool industries are going to focus on flexible manufacturing systems, CAD, and CAM. I have not seen that trend yet. Historically, FMS has been very unproductive and cost prohibitive. The downtime is very high. Changeover timing and complexities from part to part are also comparatively very high. I have yet to see a very successful and productive FMS installed in the USA. The economists and finance folks have identified several prime factors, such as foreign exchange rate, high labor costs, and higher input costs, as fundamental reasons for such a consistent downturn. My argument is that we have lost touch with innovativeness in design and manufacturing. We have lost the leadership for this core technology. A perfect design has to encompass all these shortcomings.

We have to take a systemic approach to FMS concepts. We have to build the machine elements first, with high quality, reliability, and reduced cost, which can justify the installation of FMS. We have to design part handling systems to cooperate with the machines. We have to build tool and part storage systems. We have to build part transfer systems between machines. We have to design concurrent part inspection systems. We have to build machine maintenance and repair systems. We have to integrate the sub-systems using mechanics, electronics, and software to create a synergy for such systems. So, in a nutshell, we have to design, build, and fabricate the sub-systems first as per the system requirements.

Application of technology for technology's sake is absolutely unwanted and unproductive. We have to build machines with high flexibility and process capabilities. The process capability, in particular, has to be extremely high to make the system economically viable. That brings me to my point of the importance of knowing how to design reliable and quality machines in the first place for such applications. The mere use of controls and software will not make an economically viable product. Control technology is just a tool, and it is not a core technology. We have to make the whole system economically justifiable. Otherwise, we are better off staying in the general-purpose-machine arena for some time to come.

1.5 Realistic and Strategic Framework for Excellence

It is obvious that the U.S. machine tool industry has to step back and think about what went wrong and how to act strategically to put itself back on the track of success. It is going to be a monumental task, but it is definitely feasible. This country has pulled off such reversals time and again. Instead of using a two-dimensional approach, marketing and finance, we have to take a multi-dimensional systemic approach in the following directions:

- Design and innovation
- Robustness and reliability
- Cost competitiveness
- Flexibility and adaptability
- Manufacturability and process capability
- Delivery promptness
- Warranty and service
- Marketing and customer focus
- Finance and investment

First, the design has to embrace all other aspects of the business. It is a front-end process that starts with customer requirements. The design has to be totally customer-focused, and it should generate values for the customer. The design should be innovative and disruptive in nature to stay way ahead of the competition. The design has to open up a new dimension of making money. It has to change the culture of machining in the industry. The design of a product will bring brand awareness and new identification for the industry. The design has to blend customer requirements and the technology to produce the universal platform of machines. This is what I call design innovation.

It can be seen I have not mentioned quality as a new dimension. For any future product, quality is a given. Several measurable indices, such as robustness, reliability, cost of maintenance, maintainability, and availability, have to replace the term "quality." Quality is a very subjective term, whereas other indices are metric. We have to put metrics around the "quality circle". Customers will decide the quality of the machine, not the builder and designers. The product has to reflect quality to convince the customers; the company can't just talk about quality. Again, the product has to have reliability over the time of use, which is defined as robustness. The mean and variability of production from a machine have to be maintained at an almost constant level over time of use for the machine. A design has to embrace quality in objective terms.

I have also put marketing and finance on the back bench. The first few dimensions mentioned above are all the part of the design function of a company. We have to get to design first, which will encompass all of these aspects. The design team has to deliver a design that has quality indices, cost constraints, ease of manufacturing, minimum service requirements, and reduced life-cycle cost. No doubt, that the task is monumental, but we have to do it to turn the industry around.

By flexibility and adaptability, I mean the product platform should consist of a series of products to satisfy a wide range of desired production. It should also cater to repetitive, one-of-a-kind, and continuous production. The fixturing and workholding should be universal to adapt to several forms and shapes of parts, and the product has to change faster than these demands. For example, the HAAS machine has a similar design for the same platform of machinery. The customer should be able to adjust the product mix on a real-time basis as and when required by their customers with minimum change of cost.

Such a systemic approach will help us to bring uniqueness and singularity to the marketplace and will differentiate us from the rest of the builders. The product itself will dictate the cost that customers can bear. For example, George Fischer lathes were the costliest machine in the 80s for automotive production, but it commanded the highest market share for repetitive automotive production due to its capability, reliability, and high productivity. Gisholt lathes were also universally accepted for due to the company's high productivity for repetitive parts. K&T machining centers are another example.

Systemic thinking will position the firm strategically to be competitive and allow it to be the leader of the industry and not the follower. Once we start following, we lose independence of thinking. We cease to be leaders. Over time, we have handed our leadership baton to Japan and Korea, who learned the tools from us to use them against the U.S. machine tool industry. The primary function of business managers of the machine tool industry is to implement the multi-dimensional strategies to achieve world leadership again in the world marketplace. It has to create an environment where the design team can deliver desirable products without the fear of being laid off.

1.6 Business and Operations Strategy

Instead of taking a narrow approach of reviving only a section of the machine tool industry, such as metalworking or metal forming, I would like to expand the strategy proposed by the Skinner of Harvard Business School for the manufacturing industry to the whole machine tool industry. Against the classical view of business strategy, I would like to propose that the industry focus on the following areas:

- Customer orientation
- Innovation and design focus
- Productivity
- Robustness
- Less variability
- Long-term view

The management approach for the revival of the machine tool industry should be to promote activities geared toward customer satisfaction. A product has to satisfy customers' requirements, nothing less or more than what they want and at a cost they can afford. Moreover, customer feedback about the product has to be taken into account to modify it, if necessary, down the line.

Skinner also considered a very important business strategy for the industry. The focus should be on blending technology with a design philosophy to satisfy market demand at a cost that customers can afford. Just as a quality machine tool does not mean much unless some positive metrics are put around the machine to prove its quality, the focus also has to be given on enhancing durability, reliability, and robustness to reduce the part cost. Nevertheless, the value of these metrics must satisfy customers' expectations and not some arbitrary values. This will keep the cost competitive. We have to think about customized values for these metrics for different machines and not a common value for all machines.

Machine tool production volume and cost should be complementary to each other. The production output should be commensurate with the availability of plant capacity, tooling and fixtures, and process capability of the plant producing the machines. Here, the focus has to be on the high utilization rate of available resources to keep the cost down and enhance company profitability. The plant has to keep the work-in-progress inventory to a minimum. As suggested by Hayes and Wheelwright, focusing on the production of a few types of machines could be better than making everything possible, enhancing productivity and profitability. For example, Heald Company always focused on internal grinding machines and excelled in producing such machines, whereas Cincinnati Milacron started with external or universal grinding machines and then tried to expand into other machine types without much success.

Also, project management and the design team have to work concurrently to implement the focus functions for the product using a system engineering approach. A product will be the outcome of such a concerted effort. One of the major debating points is to how to evaluate a project's outcome. The project has to be evaluated against the metrics that were the requirements of the project, which would include the complete business, design, marketing, finance, legal aspects, and human resource objectives baked together. Customer satisfaction is at the core of this effort, and it will dictate price and profitability. Hence, the product development team, which consists of design, marketing, management, human resources, and finance has to lead the effort until the product is introduced into the market and customers are satisfied. We have to remember customers decide the quality and viability of the product, not anyone inside the company. The mission statement for the product puts all the constraints together to design, manufacture, and deliver the product into the market to satisfy customers.

1.7 Worldwide Competition

The machine tool market has expanded worldwide, creating an enormous opportunity for machine tool builders in the USA. Our balance of trade with Japan, China, and Korea has consistently gone up, indicating that these countries are able to produce and supply quality machine tools competitively in the world market, whereas U.S. machine builders have failed miserably to compete. Even if the U.S. labor rate in the machine tool industry was on par with that of Japan, or even less than Japan, we cannot compete. The other question to be considered is whether our quality, cost, and service have been out of alignment with these countries. A similar trend has been noticed in industries for steel, electronics, and heavy machinery for construction work.

As Porter suggested in the 90s, the answer may lie in the culture of the country, since some countries have been successful in certain industries, such as automobiles, electronics, video games, and machine tools in Japan; low-cost machinery from Korea; luxury cars in Germany, etc. I do not believe this is due to the culture of the country. I think the country as a whole finds a nitch market for its products and delivers products to satisfy that segment of the market. They satisfy the customers to start with instead of their own pockets. They buy the customers first, providing goods and services that the customers like. Once successful in that segment, they expand globally. This is the strategy of entering in a very small way and then growing big to gain market. It is also a fact that machines from Japan and Germany are far superior to the U.S.-built machines in every sense.

The historical fact that some industries are strong in some countries does not always hold water. The US used to be an innovative and cost-effective producer of quality machine tools. Cheap labor is also not an issue. There are many countries where labor is very cheap but they do not produce quality products. Cheap labor does not necessarily guarantee innovative products for the marketplace. Examples of such countries are too many to name here. The tax structure is also not an issue since you pay tax only when you make money. U.S. workers have not, all of a sudden, become lazy overnight. U.S. culture does not have the character of being lazy and unproductive. History also does not support this fact. Our natural resources have not gone down recently.

I think the primary reason is the fact that U.S. builders decided to stay out of this market because they failed to compete. The ROI on the investment has been comparatively less than that from service industries. This is absolutely a very shortsighted vision on the part of builders. As a result, we have lost the market and are struggling to compete globally. From my experience of working in the machine tool industry for a very long time, it would not be out of place to point the finger for such a dramatic downfall at the failure of management as a whole. Management in the machine tool industry lacked direction and visionary outlook and was too bullish to take care of their own personal needs at the cost of others. The rise of fall of the Warner and Swasey Company in Cleveland, Ohio, is a perfect example of such a failure. Management did not direct its people toward the path of victory. They did not recognize the imminent danger of competition from Japanese products.

A constant debate often arises as to how to be competitive or how to measure competitiveness in the market place. The clear answer is the acceptance of the product in the desired market place. If the product fails to satisfy customers' requirements, the company becomes less competitive and eventually dies. Once market share is lost, the marginal effort and resources required to gain it is exponentially higher than what could have been required before it was lost.

Such competitive loss shows up in every direction of the company, such as productivity, the balance sheet, the balance of trade, the creation of jobs, the loss of technical people, the morale of the workers, etc. These business metrics are the net results of the management's lack of performance and vision for the company. I do not support the argument that productivity is the measure of performance, as Porter suggested in the 90s. The reason is that productivity could also go down if the products are not desired in the marketplace. If the product is not what the customers want, you lose revenue, and eventually you lose the company. Productivity has nothing to do with it. Workers are still as productive as they were before, but they have nothing to produce since the company has lost the market to their competition.

It also has to be understood that the machine tool industry is an index of a nation's pride and success. It is often said that Japan and Germany produce quality machine tools. The machine tool industry is a core industry that gets attached to the nation's cultural index. Once upon a time in the past, the US was very well known for building quality but highly complex machine tools. For example, Warner and Swasey was known for turret lathes, Head and Cincinnati Milacron for grinding machines, Gisholt for automated lathes, K&T for machining centers, and Cincinnati for universal milling machines, to name a few. These were world-class machinery produced only by U.S. builders. The US used to excel in building quality machine tools.

I strongly believe that such a dramatic downturn could only happen due to management's poor short-term vision, strategy, and utter unwillingness to stay competitive in the marketplace because they were following others. When we follow, we cease to become leaders. The firms were not ready or willing to compete globally. Management style became balance-sheet oriented, without any regard for technicality. Financial managers, who had the least knowledge of machine tools, were at the helm of affairs, deciding on the fate of firms. Management graduates were running companies instead of engineering graduates.

In order to revive this core industry, the government has to play a much bigger role in bringing the industry as a whole to a level that the U.S. machine tool industry can compete again. For sick industries, such as steel and machine tools, the government can take certain steps to make competition fair. Why can't the federal government impose tariffs like other countries do when the US exports its products to them? This reciprocity of taxes must be done to level the field of competition for machine tool industries. Each country who wants to export its products to the US has the ultimate desire of earning foreign currency. They believe the foreign currencies have much higher values than their own. Some countries even adjust or deflate their own currencies to create a business condition for their internal industries so that their businesses have a much better chance to compete. Why can't the United States do the same? The U.S. machine tool industry needs fairness, not any special favors, to be competitive again. I think imposing a similar tax structure is fair game.

One another note, the U.S. machine tool industry cannot survive by only serving the domestic market, as suggested by many economists. Since most manufacturing activities have been shipped overseas to reduce cost, the domestic consumption of machine tools has been reduced significantly over the last 20–30 years.

In order to survive in the future, the U.S. machine tool industry has to market and compete their products globally. It is a known fact that machine tool Industries from Japan, Korea, India, and China get substantial subsidies and financial support from their governments to keep their costs down. Any nation's economic performance also depends on its exports and foreign currency reserves. With higher reserves, they gain financial strength and can use the gain to penetrate foreign markets. So, we have to depend on international competition for the machine tool sector.

It has also been argued that U.S. productivity is poor compared to that of other countries. This is basically false in my opinion. Productivity is reduced since we have ceased to be global machine tool producers. Expertise and technology have disappeared due to lack of interest from investors. Machine tools have to be treated like a base core industry upon which many other dependent industries thrive on, such as manufacturing and the automobile industry. In order to be more productive, we have to create a business environment that promotes innovative design and manufacturing. Only then will we be able to compete internationally and succeed as a global producer of machine tools. In a nutshell, we have no other choice than to be globally competitive, which will eventually make us profitable and productive.

1.8 Strategic Initiatives for Revival

Time and again, throughout my career in the machine tool industry, I have heard arguments that it does not bring money, cannot attract investors, and is not profitable by nature. All financial managers that I have come across have never given a counterproposal as to how to make it profitable or how to generate money internally instead of borrowing from financial institutions. If these allegations are right, why are Japanese, Korean, or Chinese machine tool companies thriving, and how have they taken over the global market? Are they losing money by selling products consistently year after year? How are these companies surviving? Has anybody explored that avenue? From my experience of working in some of these countries, let me pinpoint some of the differences that I have observed that might be crucial in bringing this core industry to a level where it can compete globally again.

Just-in-time (JIT) Concept: We have taken this concept on a piecemeal basis and have geared ourselves to apply this concept in the areas of supply chain and inventory management. In my opinion, this concept should be expanded to all elements of the business. We could apply this concept in managing finances, e.g., borrowing only when we need it and not before that.

Financial institutions should be treated just like other suppliers. We have to reduce the work-in-process inventory to a minimum level. For example, we can buy controls and electrical components just before shipment and final testing. Long lead items such as machine casting and forging components could be procured earlier.

The manufacturing process should dictate procurement procedure, using inventory management principles under uncertainty demand conditions. In other words, these avenues of application of JIT should be explored while designing the machine as a system. The principle of this system should be applied to every element of the business wherever applicable. The WIP inventory has to be optimized to keep the investment down to a bare minimum. The JIT principles should also be used in negotiating the supply of parts in small but frequent deliveries, development of suppliers as a single source of parts, developing long-term partnering relationships with the suppliers, and using local businesses as suppliers.

Nevertheless, we should be very judicious in applying JIT principles in this industry, and the application of such procedures should not be used as an excuse in case of failure of the system. JIT is a support system to produce parts, and parts should not be designed in alignment with JIT principles. If the application of JIT does not help reduce inventories, finances required, or throughput time for a particular production system, we should not use it. JIT is not a cure-all tool for all situations.

Push or Pull System: For machine tool applications, I think the pull system is more applicable than the push system due to uncertainties of demand and acceptability in the marketplace. Materials have to be pulled through the system rather than pushed, as recommended by MRP systems. For a low volume of production, we do not have the luxury of using costly MRP systems or any of their variants. Again, the focus is on system design and innovativeness and not on other support systems. We have to understand that the machine tool industry is nowhere near the high volume of automotive production, so we do not need an exhaustive MRP system, robots, and other applications for machine tool environments. Demand should be the criteria of materials and inventory pulled through the manufacturing systems.

Reduction of Set-up Time and Customization: I have also noticed that in the US, many times, parts are customized or manually adjusted to fit into the overall system. Such customization can only increase the set-up time. Set-up fixtures such as jigs should be used extensively to make production uniform.

This will also enhance the accuracy of the products, and variability effects will be minimized. I have seen the extensive use of such methods in Japan and China. Fixtures can be used to set up, for example, the cross-slide positioning with respect to the spindle centerline.

Reduction of Variability in Production: It is about time to apply quality standards to the machines produced. Since manufacturing and machining depend primarily on these products, I would strongly recommend using the three-sigma principle for the accuracy and reliability of these machines instead of six-sigma limits. The design team has to take the initiative to design these products in a way that these statistical quality control measures are feasible for the products without increasing the cost.

The design, manufacturing, assembly, and testing functions have to be coordinated to stick to this quality mandate. Another reason for such stricter quality control is to enhance uniformity and reduce the variability of the output. Accurate design and architecture are of the essence here.

Concept of Market Timing: Another severe limitation of the machine tool industry in the USA is the failure to deliver intended products on time. In my experience, the time it takes to go from concept to market has consistently been about three years. In three years, technology, design requirements, design trends, manpower, and financial requirements change to a point the product might not be feasible or viable anymore. In my opinion, the total time from conceptualization to market should be limited to 12–14 months. Most automobile projects now have about 18–24 months of gestation. This ensures the product will still satisfy the market conditions. This reduced time is also necessary to stay ahead of completion. Again, it is the design of the product that will mostly dictate this timing. If we design for easy manufacturing and assembly principles, this timing can be maintained to stay ahead of the competition.

In general, we have to take care of the following factors to be competitive again:

- Reliability consciousness
- Satisfaction of customers' requirements at a cost they can afford
- Innovative and flexible products
- Universality and applicability of the products for various applications
- Cost constraints
- On-time delivery
- Service and warranty
- Variability reduction
- Product flexibility

1.9 Summary

This chapter focused on the status of the machine tool industry in the US and future strategies to revive it. It also discussed the fact that we need a systemic strategy as a whole for this industry to be competitive and profitable again. When most of the machine tool industry is struggling to survive and global competition is on the rise to capture the U.S. market, the importance of refocusing on this core and basic industry is beyond any question. In the long run, the measure of success of this industry will depend on reliability, robustness, accuracy, flexibility, productivity, cost, and profitability.

We have to redefine the quality indices for machines. Instead of just producing quality machine tools, we have to define metrics of performance and productivity that will define the quality of these machines.

We have to reduce the variability of the machines produced. We also have to introduce three-sigma statistical quality instead of six-sigma quality. We have to design in the quality characteristic and metrics for the quality in the product starting from the design stage. We have to design machines with very high process capability for a series of parts, symmetrical and prismatic.

Our design has to be universal by nature to suit various requirements. For example, a turning machine should be consistently capable of creating a ground surface for hard metals. It should also be capable of milling operations. On the other hand, machining centers should be capable of producing round parts when required. Even if the process for milling and turning might not be the same, the capability of doing both should be there.

It also has been pointed out that the quality characteristic of any basic industry represents the capability of the workforce and its pride for the nation. It also shows the capacity for innovativeness of the people who work in the industry. The author thinks that the U.S. machine tool industry will be wiped out sooner rather than later unless federal and state governments intervene to be creative in making the environment more suitable for a level playing field for everybody.

In order to quickly turn things around, company management and engineers have to think in terms of systemic management. All possible design constraints need to be defined and spelled out very clearly in the beginning. For example, cost, service, profitability, reliability, robustness, manufacturing, and sales constraints have to be considered to create the machine as a total system that satisfies customer requirements. This is what I call a systemic approach to design.

I strongly believe that time has arrived for management and engineers to apply what we know to produce a global machine. Before we try to learn something new, we should apply what we already know.

1.10   References and Bibliography

Abernathy, W.J. and Townsend, P.L., 1975, "Technology, Productivity and Process Change," *Technological Forecasting and Social Change* 7(4), 379–96.

Blackburn, J.D., 1991, *Time Based Competition: The Next Battleground in American Manufacturing*, McGraw-Hill Erwin Companies, NY.

Devinney, T.M., 1987, "Entry and Learning," *Management Science* 33(6), 706–24.

Drucker, P.F., 1991, "Japan: New Strategies for a New Reality," *Wall Street Journal*, October 2.

Eckes, A.E., Stern, P., and Haggart, V.A., US International Trade Commission Commissioners, 1984, Report to United States International Trade Commission on Investigation No. 332-149 Under section 332 of Traffic Act of 1930, "Competitive Assessment of the U.S. Metal Working Machine Tool Industry."

Hill, T.J., 1999, *Manufacturing Strategy: Text and Cases*, McGraw-Hill Erwin Companies, NY.

Hopp, W.J. and Spearman, M.L., 2000, *Factory Physics*, second edition, McGraw-Hill Erwin Companies, NY.

Kidder, T., The Soul of a New Machine, Avon Books, NY.

Krugman, P., 1994, *Peddling Prosperity: Economic Sense and Nonsense in the Age of Diminished Expectations*, W.W. Norton and Company, NY.

Nahmias, S., 2000, *Production and Operations Management*, fourth edition, McGraw-Hill Erwin Companies, NY.

Tooze, A., 2009, *Machine Tools and the International Transfer of Industrial Technology*, "The Global History of Machine Tools, Knowledge, Narratives and Fiction," 30-31, King's College, Cambridge.

Ulrich, T.K., and Eppinger S.D., 2011, *Product Design and Development*, McGraw-Hill Companies, NY.

1.11 Review Questions

- Can we take up a machine-development project as a system-development task? If so, what are the new approaches that we have to take?

- What are the salient features of a sick industry? Is it necessary to revive the machines tool industry in the USA? What are the pros and cons of such a revival? What are your thoughts on the idea that a nation can be proud of its core industries?

- What are the actions that a sick industry has to take to excel, grow again, and stay ahead of the competition?

- Do we need a disruptive technology and process for the machine tool industry to improve its status?

C H A P T E R 2

2 Machine System Engineering Principles

2.1 Introduction to System Engineering

The universe consists of different functional elements working in unison to create desirable or undesirable system behavior. The emergent behavior or functions of a system can be substantially different from the functions of its constituents. In order to function properly and to generate a system's behavior, the sub-systems or elements have to follow a set of rules. A machine can also be defined as an integration of different basic elements or sub-assemblies. For example, for a CNC lathe, the spindle, cross slide, longitudinal slide, tailstock, base, index system, servo motors, spindle motors, controls, etc. work together to create an emergent behavior of the system. The machine system functions can be numerous and vary from machine to machine. The fundamental promise of system engineering is to manage and optimize the overall emergent system functions while following some specific interaction rules to have maximum or synergetic output out of the system. The output of the system also interacts with controlled and/or uncontrolled input variables to the system, and the system depends on the interrelated processes used to control it. The design and development of such a system will encompass both process and product. Nevertheless, the objective is to produce parts with maximum accuracy, maximum value addition, high productivity, and minimum scrap. The other objectives could be to have maximum process capability, flexibility, durability, robustness, and reliability. Compatibility of the system elements will determine system simplicity, accuracy, and productivity.

The fundamental system concepts have to consist of the following system behaviors:

- System must be adaptable and flexible with respect to time
- System design should have multiple solutions for the desired output
- System must be resistant to uncontrolled variables or noise
- System behavior and their interactions should be mathematically represented and simulated to predict behaviors for different system outputs
- System must satisfy system requirements and constraints

A well-designed system, whether it is a machine or system of machinery, must have predictable system performance over a long useful period. The system has to have the desired operating life with an acceptable performance degradation or robustness. The system also should be designed with the future trend in technology in mind. The machine must adapt easily to future control changes. The system design must support the enhancement of accuracy, capability, reliability, and ease of maintenance to keep the life-cycle cost to the minimum. The system must be designed to increase return on investment for the owner of the machine, and it should be designed to satisfy customer requirements, which must include life-cycle cost, profitability, and robustness. The system optimization process should include all the alternative methods of design and manufacturing methodologies. It also should focus on environmental friendliness and energy consciousness. Finally, the system design and development process should be a closed-loop process.

2.2 Why Follow a System Engineering Process?

The International Council on Systems Engineering (INCOSE) has defined system engineering as "an interdisciplinary approach and means to enable the realization of successful systems." In other words, system engineering is a tool to achieve system efficiency and project success. It is for the management of highly complex, technologically rich projects such as machine tool development. It brings robustness to the project and reduces the chance of common mistakes in a project. System engineering considers the whole system as an entity, from the mission statement to marketing a product. It manages complexity to reduce the risk of the project.

A complex machine system has various sub-systems, such as slides, spindles, index systems, servo motors, tooling systems, work holding systems, controls, instrumentations, work loading and unloading systems, safety shrouding, and computers. Other than the functional elements, there are business management, inspection and control mechanisms, inventory management, sales and marketing management, order control processes, and supply chain management to name a few. All of these activities need special skills and interaction to complete the project. To design and manage these critical elements of the business so that a project can be completed on time and without exceeding budget constraints, we need a total system engineering process. Moreover, the activities of each sub-system interact with each other to create emergent behavior. These functionalities need to be managed effectively and productively to create synergy.

A system engineering approach for the business as a whole is needed to drive the project, and it should be monitored and controlled by system engineers. Unfortunately, most machine tool projects do not provide for system engineers, and they are not as valued as other inter-disciplinary engineers or stakeholders on the project team. Using them makes it harder for powerful managers to take shortcuts according to their views to hasten the project. In the long run, such shortcuts lead to the project going over budget and also not being completed on time. The necessity of having system engineers has been exemplified in a recent government accountability report for defense projects.

It has been noted that most of the projects are overrun and do not deliver the proposed requirements and specifications. The report also suggests that a well-administered system engineering process for a large or medium complex project will help to deliver capabilities when promised and will help the project to be completed on time and more efficiently. A system management approach will enhance the probability of success and minimize risk for the project. The system approach will identify the mission of the project, its requirements, a statement of work, its concept of operation, and project enablers.

I have noticed time and again that many complex projects are run without any system concepts and system engineers to implement the system approaches. The primary reason for the absence could be attributed to the belief that the value added by the system engineers (SE) is not worth the cost of their employment. The system engineer's recommendation of the project was not considered important enough for the project's success. System engineers can help substantially in the following areas if implemented properly:

- Identify and mitigate risk
- Identify customer requirements and incorporate them into the product
- Convert the voice of the customer into the product portfolio
- Enhance customer satisfaction
- Eliminate product deficiencies
- Reduce amount of rework
- Ensure effective prototype building
- Ensure effective project management

It is true that these advantages and enhancements of the system engineering approach are hard to quantify, and usually, the system engineer's contribution to the project's success remains unrecognized. A quantitative evaluation of the project must be done with and without contributions made by the system engineers. Such evaluation and evidence will establish the value of system engineering and will show its impact on the program schedule, program efficiency, program productivity, program cost, and technical performance. When resources are limited and the time to complete the project is very critical for the survival of the company, a system engineering approach is especially useful. it can only help, justifying the cost associated with the implementation of such an effective approach.

During my involvement in the machine tool industry as an engineer and as a consultant, I experienced a shortfall of system-based approaches in the machine tool design and development process. I noticed the fact that system engineering was hardly deployed for complex product development activity and, as a result, program performance was compromised. The design was based on the management's requirements and ideas and not on what customers wanted. Instead of a market-pull system, we were using a market-push system. As a result, most of the products failed in the marketplace. The blame was placed on the engineers who were asked to design the product on the basis of management's likes and dislikes.

I also noticed, while working in Japan, a very strong relationship between the implementation of a systematic approach and the performance of the project and the product. When the nature of the program was very complex and the program incorporated a new technology, a system approach was always found to be the best at delivering project performance. During my tenure in the automobile industry, I saw how effectively system management principles were deployed for each project. Program complexity necessitates implementing a system engineering approach to ensure program performance.

To summarize, the implementation of a system engineering approach for highly complex projects will enhance the following important aspects of the project:

- Development of a mission statement
- Development of system requirements
- Benchmarking and tradeoff studies
- Product architecture development
- Product concept and screening
- Concept selection
- Product design and analysis
- System and sub-system integration
- System validation and verification
- Quality control activities
- Risk management and mitigation
- Program management and control
- Coordination of activities

It is the system engineer's (SE) sole responsibility to start the project with customer requirements in mind and deliver the product to the customers. SEs start with the customer and end with the customer. This is very critical for the machine tool industry if it wants to become more effective and productive again and deliver innovative products. SEs start by benchmarking the products, developing the process, and interacting with development engineers to help during design, building prototypes, and testing the prototypes. They identify the risks associated with each process and develop a plan for how to mitigate the risks and move the project forward. The system engineer also should help to build supply-chain management for the product. It is their responsibility to monitor and control the project till the end. A successful project and a profitable product are the two final benchmarks for the effectiveness of the system approach and system engineers.

2.3 Characteristics of a successful System Design

The question is, how can a company evaluate a system? What are the basic parameters for acceptability of a system's design? How can a company know a system is successful? There are various possible factors in evaluating a system. Some of the important ones, in my opinion, are listed below:

- **System Requirements Satisfaction:** The system must satisfy customers' requirements, which could include cost, durability, robustness, process capability, and profitability, at a cost that they can afford.

- **System Performance:** The system has to deliver a desired level of performance, which could include speed, quality, accuracy, efficiency, productivity, weight, process capability, variability of output, and robustness. The ultimate proof of a product's performance is the customer's acceptance, resulting in higher market share and profitability for the manufacturer.

- **System Cost:** This should represent the total system cost for the product. The cost should be commensurate with net cost per part over the life of the machine. The total cost should include amortization cost, energy consumption cost, maintenance cost, spare parts costs, operator's cost, inspection cost, and fixture cost over the life of the machine.

- **System Robustness:** Robustness is the reliability of a machine over time and also the variance of outputs over time. The reliability and variance of the system output have to be controlled over the expected life of the system. The accelerated key life test should be carried out to determine the performance variation over time under variable input conditions. The robustness is related to the cost since reduced performance over the life of the machine will increase the cost of maintenance and reduce the machine's availability for production. Higher robustness will ensure a higher rate of return for the equipment.

- **System Properties:** The system response as a whole has to be evaluated against controlled inputs and external uncontrolled variables (noise). The system has to be evaluated for sympathetic vibration, environmental temperature, sound, coolant effects, etc. The response has to also be evaluated against static and dynamic responses. The system stiffness, damping, and vibration attenuation have to be measured to enhance the accuracy of the system. The input conditions must consist of static and dynamic inputs. The system has to be tested against a real-world usage profile obtained from customer input.

- **System Development Time:** This is the total time from business planning initiation to the product's introduction into the market. This time has to be as short as possible in order to stay ahead of the competition. This planning-to-market timing also determines how quickly the company gets a return on its investment. A systemic approach helps the team get product feedback and make corrections, if necessary, for re-introduction of the modified product. Again, a system approach to product development should help keep this time to a minimum if administered efficiently.

- **System Insensitivity to Environmental Conditions:** One of the fundamental reasons for following the systemic approach to design and development of any product is to create a system that is unaffected by environmental factors for the business, both internal and external. This is a design constraint that a system must satisfy. The external environmental noise factors of the business could be competition, market conditions, economic conditions, acceptance or rejection of a product by customers, technology factors, and political factors of the country, to name a few. The internal factors could be financial and human resources, the intellectual capability of the firm, team capability, product legacy, cost and time pressure, service and warranty considerations, testing facilities, and others. The business and product development team have to interact and face these constantly changing and interacting factors during the product development period and beyond.

A systemic approach to product development activities for a business can only make it insensitive to these factors since most have been thought about and taken into account while developing the mission statement for the product. Moreover, preparedness against unknowns can only lead to the higher probability of economic success of the firm marketing the products or machines. I know of many companies, such as Warner and Swasey in Solon, Ohio, who undertook product development activities without any regard to system approach and ended up in bankruptcy. They used a push system instead of a more desirable pull system for the machines, which were way more complicated than what customers had asked for. Instead, OKUMA, Mori Seiki, and other Japanese companies flooded the market with the machines or system of machines that customers desired.

Another reason for the dismal failure of many companies to develop a winning design could be their top-down business approach for product development. The business office, without consulting the other team members of the company, decides on the product that the development team should consider developing. They also dictate the flow of design, such as for the sub-assemblies. Then they integrate the sub-assemblies into a system, manufacture the elements, and then assemble and test the system's performance metrics, which were not fixed earlier. This is a non-systemic approach to product development that has caused of failure of many machine tool companies in the USA.

This is what I call a classical and conventional compartment approach to developing a machine, or any system for that matter. In my opinion, most companies who adopted such an approach were forced into bankruptcy by management due to their shortsightedness and tunnel-vision strategy. Without such a failure, a company like Warner and Swasey could not go bankrupt. A systemic approach forced all these stakeholders to come to a common table and develop the system or product from start to finish.

Although I realize the fact that twenty to thirty years back, there were few strategic enterprise software applications and computers available to develop a product as a system. The systemic approach does need extensive use of established and proven software, such as design and drawing software, prototype development techniques and tools, finance software to keep track of finance on a regular basis, an integrated approach to human resources, etc., for completing a complex product. Very few statistical tools were available back then for testing purposes. In other words, resources for system development of any product were limited. Nevertheless, if management had thought to use the system approach for machine development, software industries would have responded to the demand with better products. For a particular capability to be created, someone has to ask that it be developed.

2.4 System Engineering for Machine Tool

As has been mentioned before, the system engineering approach to machine tool industries includes the total function of the business, i.e., business planning, marketing function, design and development, prototype development, manufacturing activities, purchasing, human resources, quality functions, sales, and service should be bundled together as a total system. It should also include the activities in the customer's facility too during installation and job tryout of the machine.

All these activities are interrelated and focused on satisfying the customer. So, activities can be external and internal to the machine tool supplier. This bundled approach puts the builder and user in one basket to live and act together throughout the life cycle of the product. The V life-cycle model shows the logical relationship among various system engineering processes. Such a relationship can be shown in a flow diagram, as in Figure 2.4. The complete V life-cycle functions can be detailed in two sections: pre-critical design audit and post-critical design audit. The success of a system project process highly depends on project planning and control, which will be dealt with in the design review section of this chapter.

The machine tool is a very complex system to start with. In order to cope with this complexity, the principles and methodologies of system engineering should be deployed during project execution. System engineering is particularly useful for this industry, helping to manage the cost and time of completion. The possible activities of a machine tool design project are detailed in Figure 2.1.

The detailed functions are given for example only. Further dissection of each activity is quite possible, as mandated by the nature of the machine tool.

Pre-audit activities are detailed in Figure 2.2, and post-audit activities in Figure 2.3. It is quite possible to have several other activities similar to what has been mentioned here.

For an efficient and productive system engineering process, it must be a closed-loop system with feedback loops for control purposes. The inputs of the system are processed to convert inputs into desired outputs and noise. Also, some of the inputs could be noise. For example, environmental inputs to the system, such as heat and sound generated during the cutting process, temperature, sympathetic vibration, and environmental conditions, are normally treated as uncontrolled. The system that emerges is a result of interactions among its elements, as described in the next section.

The closed-loop system process is shown in Figure 2.5. Two such design and process audits are recommended. One is the system input audit during the system review process, and the other is the system output audit. When the product is introduced into the market, customer response could be used as external feedback on the overall system, representing customer feedback about the system in general. This is the customer voice feedback about the product.

System engineering functions for machines can entail the following activities, which will be detailed in Chapter 3:

- Business planning
- Voice of the customer identification or customer need determination matrix
- Project mission and vision details
- Development of system requirements matrix (SRM)
- System decomposition: functional, non-functional, technical, and performance
- Risk management and mitigation plan
- Team development and integration
- Development of target system requirements matrix (SRM)
- Development of sub-system and component requirement details
- Requirements verification and validation
- System architecture development and interface development
- System, subsystem, and component design and details as per requirements
- System integration
- System validation for confirmation of requirements
- Customer feedback

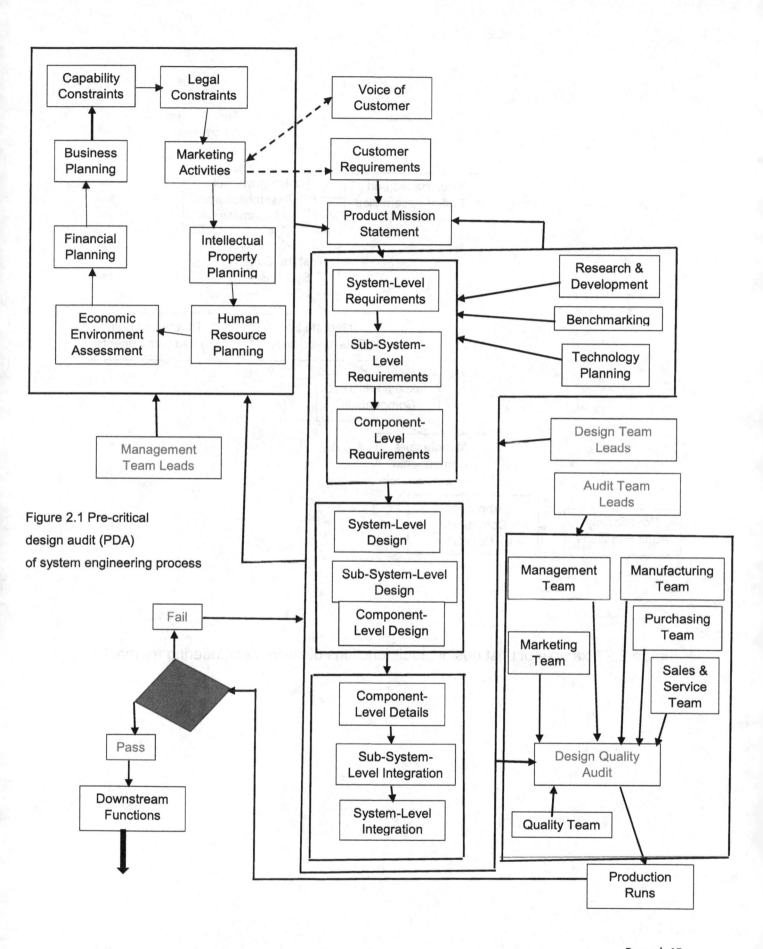

Figure 2.1 Pre-critical
design audit (PDA)
of system engineering process

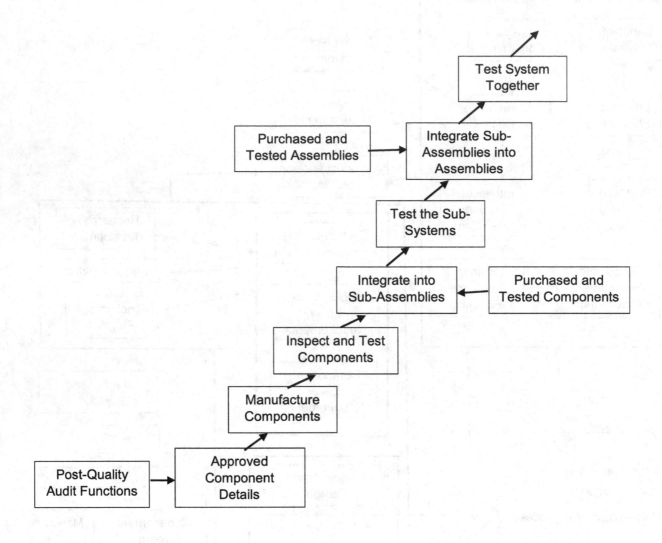

Figure 2.2 Example of post design audit functions of system engineering for machines

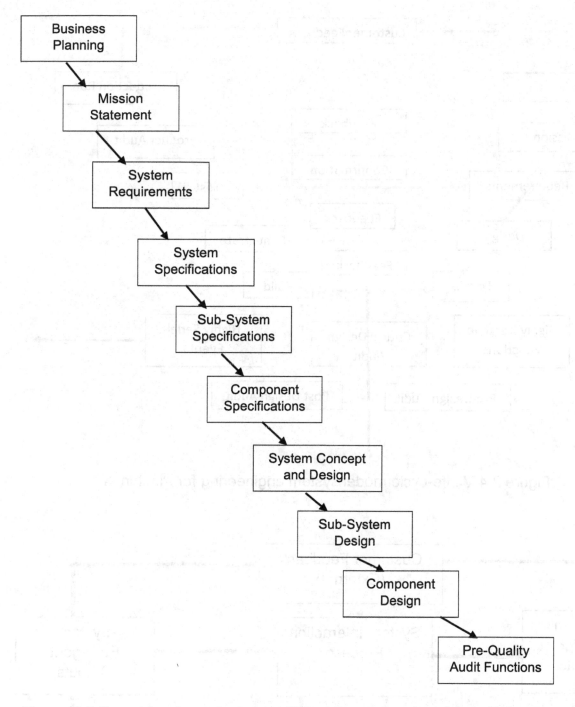

Figure 2.3 Example of pre-design audit functions of system engineering for machines

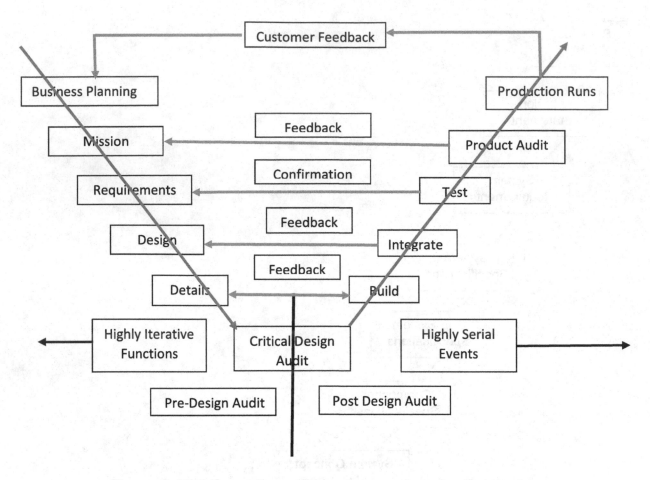

Figure 2.4 V Life-cycle model system engineering for machines

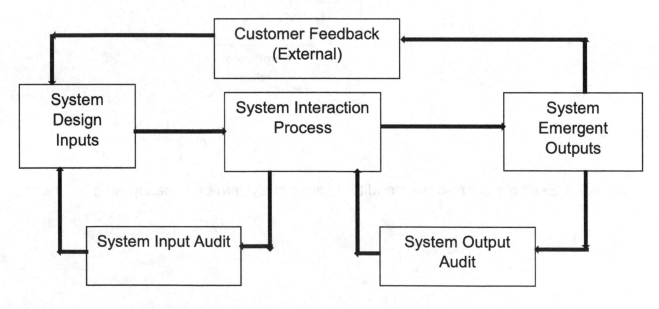

Figure 2.5 Closed-loop system of the engineering process

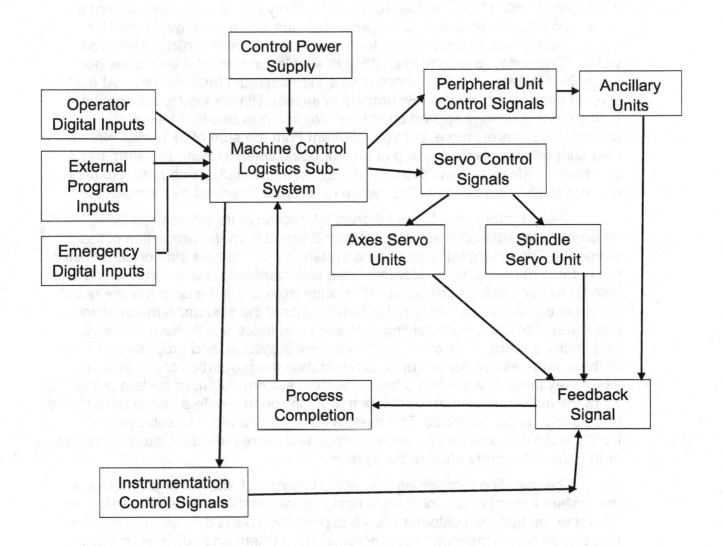

Figure 2.8 Typical control system architecture

2.5 Emergent Properties of the System

The system concept and the emergent behavior of the system have been around for a long time. Philip Anderson drove these concepts home in his article in *Science* entitled "More Is Different" (1972). A system's properties are not a linear combination of its elements' properties due to the synergy created by constituents of the system while interacting. In this seminal article, Anderson wrote: "The ability to reduce everything to simple fundamental laws does not imply the ability to start from those laws and reconstruct the universe... At each level of complexity entirely new properties appear. Psychology is not applied biology, nor is biology applied chemistry. We can now see that the whole becomes not merely more, but very different from the sum of its parts." Nature also supports his claim. A group of ants or bees, when working together, has a completely different output than when they act individually. A gathering of people expresses different behavior than when its individuals are all by themselves.

Every system, due to the inherent interaction of its elements, has emergent properties that are substantially different from the properties of the elements. The emergent output of the system is not a simple sum or difference of the outputs of its elements. Moreover, the emergent output of the system cannot also be traced back by just sum or difference rules since the output is the result of non-linear interaction among the constituents of the system. A machine tool has slides, spindles, work holding, a loading and unloading system, tailstock, bed, index system, electronics, and an electrical system, and properties of each of these sub-assemblies are quite different than the properties of the system when they interact with each other to create a system output or system property. All of the individual elements act together in unison to create a shape or form of a component being machined. The system inputs trickle down to sub-system inputs, each sub-system creates an output, and these individual inputs interact to create the emergent output of the system.

The question remains as to why the emergent behavior or properties of the system are important for system design considerations. This is the behavior or properties that the customer sees and pays for. This is the system behavior that should match customer requirements. The system and sub-systems are designed to create these properties. Hence, these properties are very important for acceptability in the market. These system properties need to be optimized, and sub-system properties should be selected so as to create these system properties. Such systemic treatment will force us to identify the nature of interaction among elements or sub-assemblies.

From the system equilibrium point of view, the system remains in equilibrium at any point of time, and accordingly, the system elements are also in equilibrium under these forces acting on them. If the emergent behavior of a system is to create a profile on a part (emergent behavior), we have to understand the motions required for each of the system elements as well.

To change the emergent properties of the system, we have to simulate the inter-element interactions using system dynamics principles. For example, if we want to enhance one of the system properties, we have to change the sub-system property that has the highest effect on the emergent behavior of the system. Depending on the type of inaccuracies during machining, we have to understand the reactionary inter-element forces that tend to create the diversion of system behavior.

2.6 System Decomposition for Machines

In order to analyze a system and the interaction of its constituents, we need to decompose or dissect the system functions or assemblies in a hierarchical fashion. From such decomposition diagrams, we can understand the boundaries of the system elements and how they interact with each other. The system decomposition also helps to identify interdependencies among the interacting elements. The nature of interaction among elements will determine whether the emergent properties of the system are desirable or undesirable. The decomposition also helps to identify the dependencies of the system on its elements. For a very complex and highly interacting machine system, it is almost mandatory to decompose the system by its assemblies or functions.

The decomposition is also called partitioning. I like to call it fencing, where a sub-system has one property and it communicates to a neighboring sub-system through the fence. The fence is the interface of the sub-systems. The perimeter and number of fences should be minimized to improve the efficiency of the design.

So, for standalone machines such as grinding machines of turning, shaping, and machining centers, decomposition by assemblies is almost a norm in the industry. I have also experienced similar treatment during the design and development of machine tools. In the case of highly complex machine cells, functional decomposition might be more appropriate since each machine in the cell or group of machines performs different functions to create a system output as a whole. In a manufacturing cell, we have to understand the system of the systems. A manufacturing cell could be thought of as a conglomeration of several machine systems working in unison to create the desired output. Either way, a decomposition of the system is necessary to understand its behavior in totality.

As the name implies, a functional decomposition follows the hierarchy of principal disciplines or functions of the system. The functional decomposition depicts what the product or system is supposed to do, not what it does. Functions are described as a flow of energy, interacting forces and moments, material and information signals. For machine tool design, decomposition should show the flow of energy or of interacting forces and moments.

The complete systemic function is divided into primary functional disciplines that are then sub-divided further into dependent functions for each principal function. A typical functional decomposition for a machine is displayed in Figure 2.6. This is a picture of all functional disciplines of the system and the functions of each sub-system. In this method, divisions depend on the functionality of each unit. This process of dividing functions into sub-functions is continued until the function of the smallest component is identified.

Rechtin has identified several guiding principles for generating the decomposition of any system. As a general rule, a function should not be split where rates of information exchange happen. If a function is highly dependent on other functions through sub-functions, or if a function of a system interacts with other functions, functions should be divided into non-interacting functions.

The functional discipline flow diagram could be the result of a system requirements matrix for the system. The system decomposition is very useful to understand task or requirement dependencies of each task/requirement on other tasks/requirements. This system requirement matrix (SRM) can be very helpful in a system requirement audit for the project during the project design review. It can show the manpower and other resources required during the project as well. The SRM can also help in sequencing the tasks to improve the efficiency of the project. The SRM is somewhat similar to the design structure matrix (DSM), but it is particularly related to functions, tasks, or requirements that have to be in alignment with customer requirements. The SRM could be also used as project deliverables to satisfy customer requirements.

In machine tool industries, the decomposition of the system, as per the sub-assemblies, is more commonly used. This method could be called structural design decomposition (SDD). Some authors have called it "assembly or design decomposition." SDD helps the design team engineers and designers concentrate on one particular sub-system. System engineering practices can help design personnel minimize functional or technical errors when designing interfaces, because they identify the nature of the interface required and the boundaries of the system desired. SDD also identifies the interfaces through the structural assemblies' interactions with other assemblies.

Interface design becomes very easy and without error or repetition. In system decomposition by assemblies of the machine, boundaries and interface connections are observed first. Guidelines for a quality interface specification state that the interface should be unambiguous, complete, and concise. System decomposition is also helpful in designing the architecture of a system. Minimization of interfaces is very crucial to keeping the system simple. To reduce the complexity of any system, the number of interfaces must be reduced. The system has to be designed around an interface. The assembly decomposition could also help to design a robust skeleton for the machine.

For example, take the method of connecting the spindle housing with the bed of the machine. Interface design calls for the attachment method, the shape and form of the attachment, the nature of the attachment method, etc. Eppinger calls these large assemblies "chunks." It is a common practice to bolt down the spindle assembly onto the bed of the machine. Bolting is preferred to others method of attachment, such as welding, due to the necessity of alignment during the construction of the machine and during servicing in case of accidents. The bolting can sometimes act as a weak joint between the machine bed casting and spindle assembly, and the head can move away from the tool during accidental impact, minimizing its effects and saving the bearings and spindle. The joint also needs to be structurally sound and robust to create a platform to transfer structural forces and moments during element interactions. A structurally sound system highly depends on the interface and boundary design.

Once the principal chunks have been identified, sub-assemblies of the principle chunk are identified. This top-down flow is continued until the smallest element or component is identified. The flow down is chunk by chunk, from larger to smaller chunk. The last step is to create a design structure matrix to show the interaction and dependencies among chunks. The complexity of such a method grows depending on the granularity of the system. It should be noted that this method gives importance to the boundaries and method of attachment more than the functionality of each chunk. The part-numbering system also uses the assembly decomposition method. A typical decomposition of this sort is displayed in Figure 2.7.

Some authors have also recommended system decomposition, depending on the service required by the system elements. From what I have seen or experienced in the industry, decomposition by service is hardly done as a separate initiative. The service engineers take the assembly decomposition diagrams and turn them into service manuals. The service manuals are written as per the spindle assembly, controls, electrical, work holding, work loading and unloading system, etc. I have not witnessed any machine system decomposition based on service during my tenure in the machine tool industry.

In reality, I have come across several types of decomposition methodologies during my tenure in the industry. Most of them were of a hybrid type. Some examples for turning machines and machines in general are given in Figures 2.6 through 2.9. Each one of these diagrams has a specific purpose during machine development. For example, the decomposition of machine functions used for stress and strain analysis is displayed in Figure 2.6. Such a diagram can be used to generate forces and moments for any sub-system for the purpose of further analysis. On the other hand, decomposition by motion analysis is also required to develop movements and the interaction among interacting units. The conventional functional decomposition of a turning machine is developed and shown in Figure 2.9. The main point is that there is not a single method established for the development of production machines. Users have to decide.

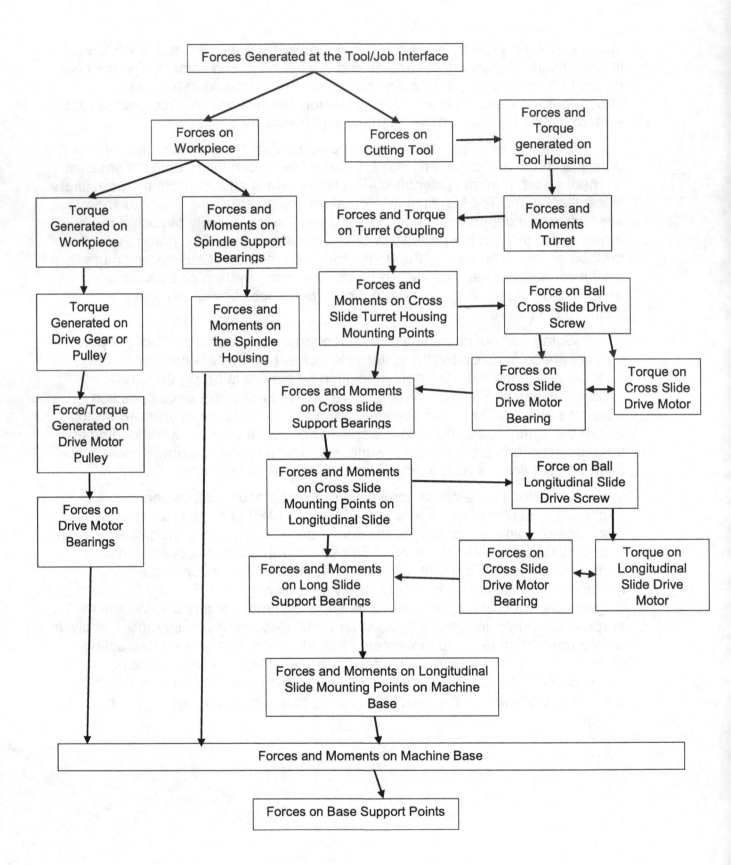

Figure 2.6 Decomposition by force analysis

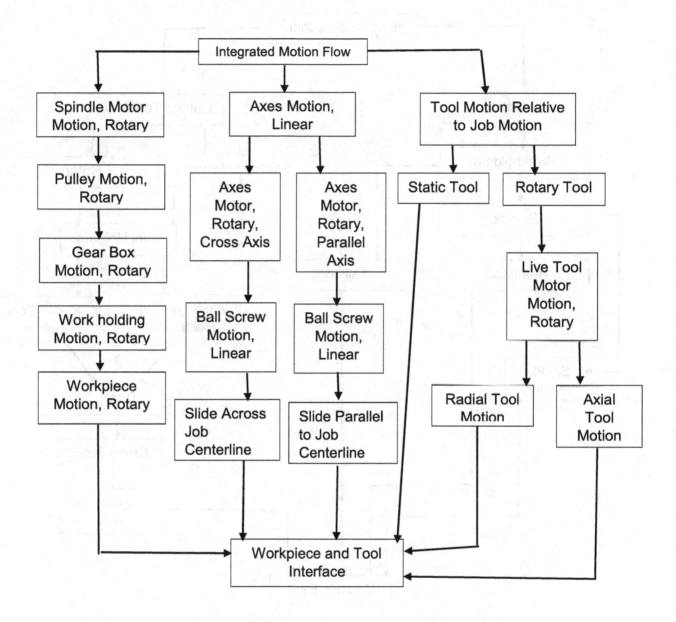

Figure 2.7 Decomposition by motion analysis

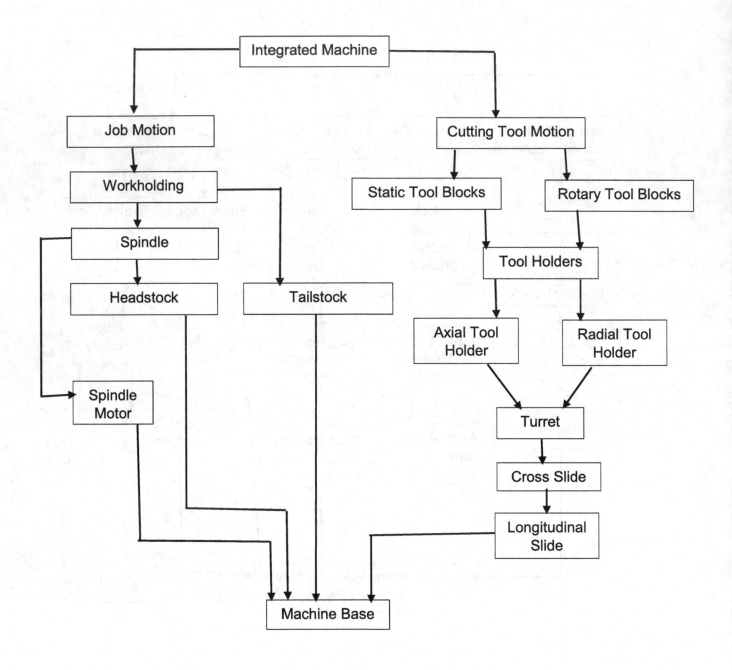

Figure 2.8 Decomposition by motion and function analysis

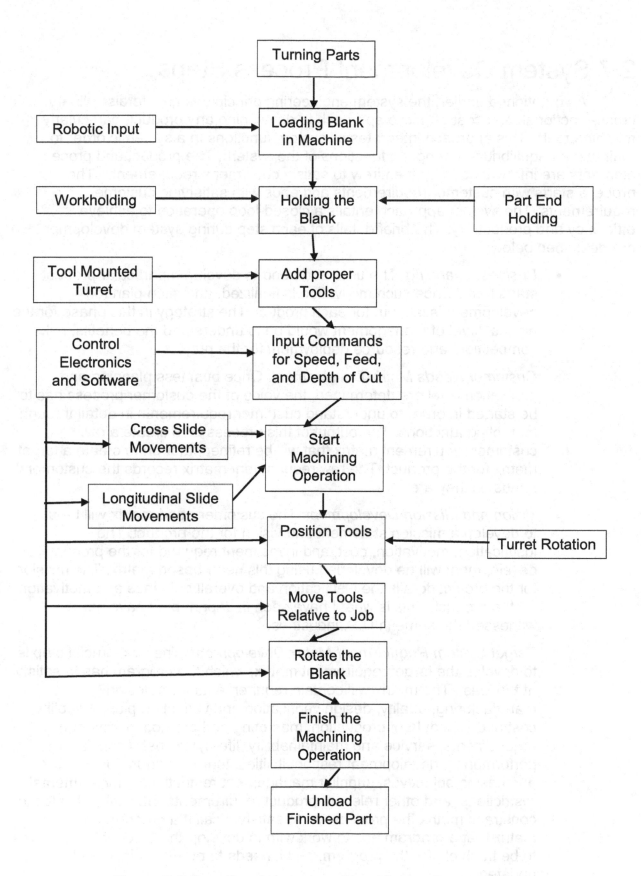

Figure 2.9 Functional decomposition of a turning machine

2.7 System Development Process Steps

As mentioned earlier, the system engineering principle is an interdisciplinary, cross-functional, and cost-effective approach to developing any product, particularly machine tools. This approach integrates all system functions in a systemic order to maintain an equilibrium among the functions of the system. The product and process elements are intertwined in their entirety to satisfy customer's requirements. The process starts with customer requirements and ends with satisfying customer requirements. The system approach entails a closed-loop operation to achieve efficiency and productivity. The brief details of each step during system development are described below:

- *Business Planning:* The urge and need to develop a particular product starts here. Once such motivation is finalized, an action plan for development is laid out for each product. The strategy in this phase for the highest level of management would be to understand the benefits, risks, competition, and resource deployment for the project.

- *Customer Needs Matrix Development:* Once business planning and motivation level are determined, the voice of the customer process has to be started in order to understand customer requirements in detail through marketing functions. The output of this process is to create a raw customer requirement matrix that will be refined later on to create a target matrix for the product. The raw requirement matrix records the customers' needs as they are.

- *Vision and Mission Development:* The customer need matrix will be used to develop a mission statement and vision for the product. The justification, motivation, cost, and investment required for the product development will be developed using this need-based matrix. The mission for the project details the justification and overall guidance and motivation for the product. This is what I have seen in Japan, but I have not witnessed the same in U.S. industries.

- *Target System Requirement Matrix Development:* The next crucial step is to develop the target requirement matrix, which the program has to satisfy at the least. The matrix will contain requirements from design, manufacturing, quality, design verification and validation plan, reliability, cost and schedule of production, marketing and promotion, revenue requirements, service and maintainability, life-cycle cost, financial performance, development responsibilities, team configurations, purchasing objectives, supplier management restrictions, environmental restrictions, and other relevant product requirements. Basically, this is the constraint matrix the product has to satisfy. I call it a program constraint matrix that a program has to work with to develop the product. This seems to be the bible for the program, and it needs to be maintained and updated.

- *System decomposition:* The target matrix represents the system requirements as a whole. Each requirement has to be a metric so that they can be evaluated down the line. When the target matrix is decomposed, the requirements for the system, control, electrical and electronics, mechanical sub-assemblies, loading and unloading, human interface, and safety considerations for each component are determined. The requirements are divided into functional, non-functional, performance, technical, manufacturing, and financial requirements for sub-systems and eventually flow down to the lowest component level.

  The product design has to satisfy these requirements when validated or verified through various form of tests. This decomposition process also helps to keep the traceability or source of the requirements for all sub-systems and components that they have to satisfy. This is an organized way of trickling down customer requirements into the requirements for the elements of the product. This organized process also helps to develop system architecture and subsequent development of components and their boundaries.

  Product architecture development gives rise to boundary or interface development for the sub-systems that act or function through this boundary interfaces. The boundaries for each interacting element must be compliant with each other for functionality. The strategy of this step is to minimize the boundaries of each element. For example, the interface between the spindle unit and machine bed needs to be developed based on the machine architecture.

- *Product Assembly and Integration:* Once all the components are designed and fabricated, they are assembled into sub-systems and eventually a system, which should be tested for verification against its requirements. These are the prototypes built for system testing. In some cases, sub-systems such as indexing, work holding devices, hydraulics, servo motors, spindle motor, conveyor, or fixtures can be separately tested as a sub-assembly against the requirements.

- *Requirement Validation and Verification:* The product system has to be qualified against the system performance defined earlier. The methodologies for testing the system are laid out in the design verification plan for the machine. At the end of this crucial step, verification compliance or deviation for the system is determined.

- *Management Product Control Functions:* Several program logistics and control points have to be developed to control functions such as design rationale, feasibility requirements, design and product configurations, and risk management and mitigation. The system engineering process is a very efficient and productive enabler for such control.

2.8 Manufacturing Process and Product Life Cycles for Machines

Product Life Cycles for Production Machines

There is a fundamental difference in demand pattern or production pattern between machine tools and consumer products. The principal reason for this seems to be the capital-intensive nature of machine tool production and the economic uncertainties in the market for machines. Another reason could be the fact that a machine can be maintained or updated internally for a very long time to give acceptable or compromised production output. So, machines are never discarded or obsoleted as soon as their main purpose is over. Machines can also be used for many other types of production even if it was purchased with a specific type of output in mind. The net effect is that the demand is in steps, and consequently, the production of machines is also in steps to match the requirements of the market.

It also has been found that when the economy turns out to be bad, the machine tool market is hit first since machine tool procurement is very capital intensive and it becomes very difficult to procure further financing. Hence, the tendency of machine tool user is to modify the machine and its setup instead of procuring a new one that might be more suitable for future requirements. So, the sales volume of machine tool production for most types of machine tools is somewhat "step up and step down" instead of the continuous production of consumable products. This particular observation is shown in Figure 2.10.

I witnessed such production behavior with several machine tool companies during my tenure in this industry. Even now, I do not see any other way. This particular production behavior is particularly applicable to general-purpose turning and machining centers. Some companies, such as Harding and Haas in the USA and Okuma and Mori Seiki in Japan, do show a tendency toward continuous production of machines overall, but several models show this step and down process. Another reason for such production is the uncertainty of market consumption due to economic fluctuation and the high risks involved in investment on the part of machine tool producers,

This very narrow approach for machine tool production is also due to the fact that the manufacturing base in the USA is evaporating very quickly and machine tool consumers overseas tens to support their machine tool producers. As a result, machine tool producers in the USA have taken a very cautious and restrictive approach toward producing machines on a continuous basis.

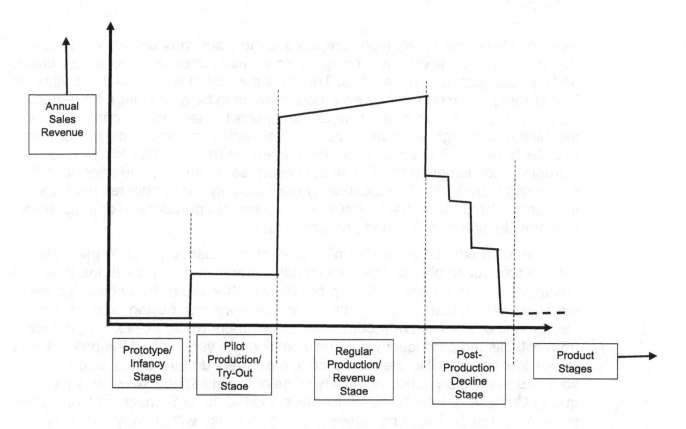

Figure 2.10 Machine product life-cycle stages

During the prototype production phase, the demand for the product is almost nil, and the prototype is used to explore design viability. Most of the time, these products are not for sale and are primarily used for exploring design feasibility and refinements further down the line. Due to the scarcity of financial resources at the start and design or quality concerns, one or two machines are produced. These prototypes are also used to develop the market for the machines. If the production is an extension of existing products, such as platform or legacy products, few more machines are taken up for production.

During this last phase, the primary strategy is to evaluate the design, test the performance, and use the prototypes for marketing. In order to avoid serious design, production, and performance flaws in the machine, system engineering methodologies should be used to address such issues before even ordering parts for building the prototypes. Nevertheless, prototypes are almost common among machine tool builders in the USA and overseas for assessing design, production, and quality feasibility.

Once the design, performance and quality studies look promising and customer feedback is very positive, a pilot production batch of a few more machines is taken up. These machines are normally purchased by either early adopters or established customers of the machine tool producers. The number of machines created during this pilot production phase depends on the initial assessment of customers' acceptance and feedback. The cost of these machines is normally very high compared to the actual production cost due to design changes, low volume, changes incorporated, undesignated supply chain, and risk associated with the sales. The profit generated is also on the lower side. Establishing a market for the new product seems to be a challenge for general-purpose machines due to stiff competition unless the technology content of the machine is very high, such as migration from analog controls to digital controls. In order to establish the product in the market, the machine has to have disruptive technology to provide higher speed, better quality, and special machining techniques. The technology used must increase chip removal rate, incorporate hard metal machining, combine several operations in one machine, and provide high flexibility and production rate.

If the product survives the pilot production phase, it gets into the high-volume production phase. The product gains ground in the market due to specific advantages over the competitors' products. Another aspect that I have learned over the years is that most customers are hesitant about buying products with new technology or from new companies. Customers would like to buy products from established producers when the investment is very high. If the product cost is very low compared to the market cost, customers do take risk, though with some apprehension about the quality of life of the machines. Now, very low-quality Chinese or Korean machines have flooded the U.S. machine tool market in the recent past. These machines are very low cost, without any quality or flexibility, and they are general-purpose machines most of the time. These machines are run of the mill, and, as a result, they are extremely inexpensive. The long-term rate of return from these machines is very low when compared with high-end machines due to their low quality and extremely short life.

To establish a market where machines can get to a very high volume or to gain ground in a competitive environment, price discounts, lease/buy options, and pay-as-you-produce schemes could be introduced. With such alternate-pricing methods, risk stays with the producers, but if the machines are of good quality, have a longer useful life, or are robust enough, this strategy pays off in the long run. At the present stage of global machine tool markets, U.S. machine tool companies have to have a very aggressive pricing strategy to establish the strength and quality of their machines.

During the production phase, the manufacturing process has to be uniform to ensure high quality and performance. The application of system engineering is almost a must for this phase since robust design supports a uniform manufacturing process. Manufacturing variations have to be reduced to a minimum, and that will ensure very low reworking. Standardization of the process and manufacturing flexibilities are very crucial at this stage.

This strategy will keep the after-sales cost very low and will increase profit. I think net profit is much more important than sales revenue for survival. Depending on customer satisfaction, I have seen a very slow increase of production rate during this phase, but I have never witnessed a significant growth of sales for general-purpose production machines, such as turning machines, machining centers, milling machines, and grinding machines.

The last significant phase is similar to that of other consumable products and could be called a stepwise decline. Actually, production and sale of most general-purpose machine tools will continue almost indefinitely. The stepwise downturn is due to the fact that demand for such machines in this phase slows down due to upgrades in technology, new entrants in the market, and the presence of more flexible or universal machines in the market. During this phase, companies use their products as cash cows and consume leftover inventories of parts and controls, expecting sales in the future as repeat orders.

It has been noticed that some machine tools' sales keep going up right from the start of production and continue in that fashion for a long time. For example, machines with computer numerical controls are still very good sellers even almost twenty-five years after introduction. In order to match the demand with production, each product has to be considered in its own context and nature of the industries it serves. The special-purpose machines for bearing industries belong to such a category. The demand for such machines remains almost constant over a long time. In a nutshell, the life-cycle curve for machines is not the same as those for highly consumable products and should be used with caution to predict demand.

Manufacturing Process Life Cycles for Production Machines

Some authors and researchers have divided the complete manufacturing process cycle into three main stages: early, middle, and mature. For machine tools, I would like to define the manufacturing phases slightly differently based on what I witnessed in actual industries. I would like to keep the production phases the same for both the product life cycle and manufacturing process cycle. The nature and trend of relative manufacturing cost during these phases are shown in Figure 2.11. Even if a particular type of machine does not show such behavior exactly, particularly for a well-established machine manufacturing cycle, this figure shows the manufacturing cost conceptually, and this strategic framework could be used to improve the cost during early phases of production.

The first phase of machine tool manufacturing could be called the "prototype build phase," or "infant stage," of machine tool production. Many unknowns are discovered in this phase. During this infancy stage, machines are built to prove the design and its performance. The manufacturing process followed is disjointed and similar to a job-shop environment.

Typically, during the infant stage of manufacturing, one or at best two machines are planned based on the maturity of the design and investment required to produce the prototypes. Variety of design changes are incorporated during these phases. This is the phase where the digital prototyping process could help immensely to keep the prototype building time and cost down to a minimum. Most of the design changes could be incorporated in the computer before buying the actual components. Digital prototyping is part and parcel of system engineering, and it should be adhered to in order to make the product more mature and error-free before prototype initiation.

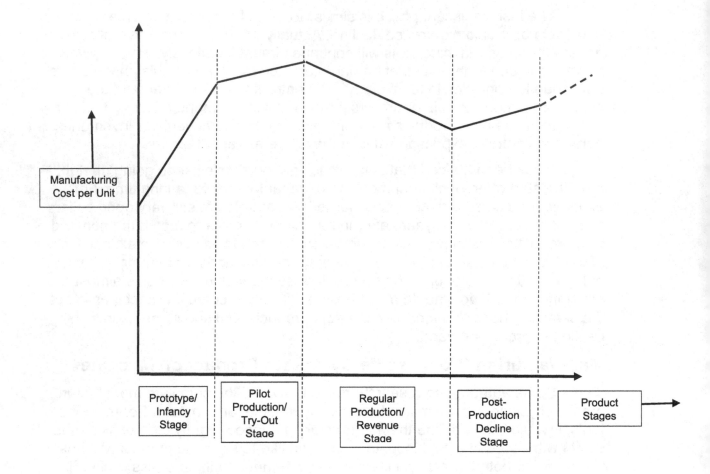

Figure 2.11 Machine process life-cycle stages

Due to design variations and the high procurement cost of special control and electronic components, the initial cost of prototype development is comparatively very high. Frequent design changes have also been found to be a primary cause of very high cost to start with. Due to more design flaws and immaturities, manufacturing cost tends to go up with time during the infant stage of prototype production. Prototype manufacturing somewhat follows job-shop characteristics.

In order to keep the cost to a minimum, the prototype development phase has to be eliminated totally by using digital prototyping or robust design principles. The design has to minimize the use of costly fixturing or other processes. The design has to incorporate methodologies for a flexible design or platform sharing. New technology has to be tried as a separate prototype effort to establish the technology, and then the bookshelf technology has to be used for the new design. The risk and uncertainties in using a new process or component have to be mitigated before incorporating them into the design.

System engineering methods should help this process, reducing the cost of prototype development. In line with the high cost of the job-shop environment, where volume is too low and risk of manufacturing is very high, dedicated suppliers should be used and design changes have to be questioned before implementation. During the pilot production phase, a similar trend occurs, but the rate of cost increment is much less since manufacturing changes are much lower and some vendor management is in place to reduce the procurement cost. The cost of new fixturing gets lower, the manufacturing process gets established, and the risk of unknowns decreases, too. Another reason could be the fact that technicians or workers learn more about the manufacturing process and design philosophies during this stage are rarer.

This stage of prototype production could be coined "limited batch production," and normally, the cost follows the trend of batch production. This is the last stage, where all the design and manufacturing changes are incorporated into the product and it gets ready for regular production, which could be "assembly line production" or "high volume production." The Warner and Swasey Company in Cleveland, Ohio, designed such an effective and highly productive assembly line to produce their high volume of CNC lathes in the 80s to cope with the high demand for such machines. In this phase, all the service, installation, and warranty issues must be ironed out. Service manuals should be finalized for final shipping.

Obviously, the production cost gets lower due to automation, use of fixtures, and improved manufacturing processes. Another reason for the lowering of cost is due to the familiarity of the design and manufacturing processes by the technicians. This could be termed a "reduction of cost due to experience effects." As the design and process mature, technicians' experience comes into play to find a better way of assembling the machines in a much more productive way. I also experienced the fact that technicians or assembly line folks get involved in the design and process refinements and suggest improvements as to how to reduce the cost of production. From what I have witnessed, such a trend is far more common in Japan than in the United States.

In the production stage of the process, most of the complicated alignment procedures are accomplished using fixtures or automated methods. Production and inspection processes get standardized. In such a stage, the flow of operation becomes much smoother and automated.

This is where the revenue and profit phases reside. Here the primary strategy is to get as many orders as possible and deliver the products with the highest possible accuracy and as quickly as possible. Errorless execution in this phase is very critical to the product's success for a very long time to come. This is also where the company should keep track of customer feedback on the product. Marketing and sales take the lead to promote the product among potential users and establish the product in the market niche that the company is aspiring for. This stage is very critical for the continuation of production over a longer period. In my opinion, this is a very critical stage, where company management can take the lead in promoting the product among potential users and gain ground on the competition.

The cost of production seems to again go higher after the production stage due to the low volume of orders flowing into the factory. In order to counter this effect, additional flexibilities must be induced in the design so the machine can be used for a variety of applications. An application engineer could take a lead here to identify a new use for the machine and demonstrate its capability to users to get more orders. During this post-production stage, keeping in touch with existing customers is very crucial. Quality and reliability feedback from customers also tend to be very critical for further success of the product. A time might come when a continuation of the product might not be a profitable proposition. For example, old Warner and Swasey turret lathes are not produced anymore even though the product was highly successful during or after the Second World War. CNC lathes replaced these turret lathes, providing greater flexibility, accuracy, and capabilities.

It would not be out of place to mention the fact that some special machinery production, called special-purpose machines, or SPMs, might not follow this progressive or evolutionary product and process life cycle. Companies that produce special custom machine tools have to follow manufacturing principles similar to low-volume batch production, where fixed costs have to be kept very low to keep the cost of the product at a minimum. Costly fixturing or an expensive automation process could be avoided to keep the fixed costs down.

The product and process life cycle explained above are normally applicable to general-purpose machines, such as CNC turning or machining centers, where products have a much higher chance of maturing into a high-volume product in the near future. For the success of such products, accurate design evolutions and an optimized process flow are very critical. Focus also has to be given to holding minimum inventories, which are required for product assemblies. Obviously, a decline in demand for the product will increase the production cost. Sometimes fixturing, inspection gages, and assembly line equipment need further investment to maintain production. Management has to decide the fate of these products and might use these products as "cash cows" as long as they are profitable.

Product and Process Matrix for Production Machines

Hayes and Wheelwright, in 1979, explained the link between a product and its optimized process to manufacture it. They did so using a product-process matrix, where the process structure required for the process lifecycle is plotted for various types of products, starting with a job shop process and ending in a continuous flow process. For production machine tools, this matrix needs to be modified since, in my opinion, there is nothing in machine tool production that uses a continuous flow system. On the other hand, it has been observed throughout the industry that a particular process should be used for higher productivity and efficiency for manufacturing. The process structure and process flow must be suitably matched. The appropriate matrix for machine tools is shown in Figure 2.12.

Type of Product → / Process Phase ↓	Low Volume, Low Standardization, Singular Design	Several Products with Low Volume for Each Product Line	High Volume, High Standardization General Products
Undefined Manufacturing Flow: Job Shop Environment	Special Purpose Machines or Flexible Manufacturing Cells		Not Suitable
Customized Line Flow for Definite Batch of Products: Batch Production Environment		Mid Volume Several Lines of Products under One Roof Manual Lathes	
Assembly Line Production: Semi-continuous Production Environment	Not Suitable		General Purpose Machines with Very Low Customization, Turning or Machining Centers

Figure 2.12 Machine product and process matrix

This particular matrix is based on three phases in the manufacturing processes for three types of machinery. The types of manufacturing processes or process structure are (1) process for low volume of specialized machinery, (2) process for several products with mid-volume, and (3) process for very high volume with limited categories of machines per line. The other part of the matrix is the product types: (1) those with highly customized parts, such as bearing turning machines, (2) manual or customized products with mid-volume, and (3) high-volume products that could be produced on an assembly line for higher productivity.

Applications for such a matrix in machine tool industries could be several. For example, a product has to be matched with the right process for gaining efficiency and higher productivity. A process could be selected for the machine right from the beginning of producing prototypes and in the pre-production phase. When the machine sales gain speed, the job shop type process could migrate to the process for high volume. It is always best to move along the diagonal. On the other hand, a process for high volume and standardized product will not be suitable for the single purpose of highly customized machines.

The process for highly special machinery with very low standardization is located in the upper-left-hand corner of the product-process matrix. Such machines are one of a kind, and other processes will cost a lot of money to incorporate. If any such machine becomes highly popular and takes the shape of the mid-volume product, the process could migrate to the line flow for batch type of production. For example, column-type machines were introduced for heavy chip removal by the Warner and Swasey Company in the 70s. To start with, the demand was somewhat sporadic and was produced in a very small lot of five machines per month. Due to its very high production capability and metal removal rate, it became very popular in the bearing and other related industries as a standard chucker machine. Eventually, the batch type of production for machines with different capacities and capabilities was set up. After some time in production, it was found out that some machines were selling very well in the market. The volume was extremely high. These models were then assembled in the line flow pattern to cope with the demand. Other models were produced in batches. In this example, process flow had to be changed accordingly.

Some companies, such as Haas, Okuma, Mori Seiki, and Harding, to name a few, produce several types of machines in the same facility. Due to the nature and flexibility of these machines, a batch flow could be justified. Sometimes very heavy, high-capacity machines can also fall under this category. For example, vertical turning lathes or high-capacity plano-milling machines can also use disconnected line flows for each category. The batch-type line flow production provides customization of the products without losing the ability to produce at a much higher rate. I believe this could be the most suitable type of production process for most of the companies who manufacture several types of machines at the same time to generate higher revenue.

The last category down the diagonal of the matrix includes the companies that produce a standard line of machinery or high-volume manual or CNC machines. For example, Warner and Swasey had to introduce line assembly for the production of their new CNC lathe in the 80s to cope with the demand. The manufacturing and quality control were highly standardized, and the design was highly matured. Customization was done off-line for very few machines. Controls and electronics were standardized. They were making about 70–75 machines a month. The machine was moved from station to station using a conveyor type of assembly. This was particularly done to keep the unit cost very low and produce the machines at a very high rate to keep up with the high demand for such machines. Such a manufacturing process has very low flexibility for customization, as per customer demands.

The main purpose of this matrix is to match products with the appropriate manufacturing process. Product has to be matched with process to achieve high productivity. It should also be mentioned here that the company has to understand and justify the reason for choosing a process that does not match the type of product structure as depicted by this product-process matrix. For example, a flexible machine system cannot and should not have the same process as high-volume machining centers. It would be an extremely inefficient use of company resources.

The primary objective of any company is to design a product that can be produced at the highest possible volume. System engineering principles applied during the product development stage would enable the company to achieve this success since system engineering treats the product as a complete system and helps the team to design the product as a whole instead of one sub-assembly at a time. The product-process matrix can also be used in another way. If a company has an established market for a line of machinery but is losing revenue by producing machines at a much lower rate than the demand, the subject company has to change the process structure but not the product structure.

For a very effective use of resources, the stage of the product has to match with the process. When developing a prototype to explore the market or the feasibility of the design, the process has to be similar to the job-shop environment. High-volume production is not required. As the product gains maturity from its embryonic stage to a matured stage, the process could be changed to suit the volume of production. This also reduces the risk or uncertainty of market acceptance for the product. To justify the installation of an assembly line for any product, a sustainable demand for it has to be established first. In such cases, the product comes first, and then manufacturing follows suit for optimum use of company resources. Like anything else in physics, an equilibrium between process and the nature of the product has to be established at different product life stages.

2.9 System Engineering Integration for Machines

As mentioned earlier, system engineering requires expertise in various disciplines, and it is extremely rare that you can find a small team of engineering professional to cover all them. A system engineer is trained to coordinate these activities and take advantage of the benefits rendered by system engineering. These professionals, who could be called "system integration engineers," support and participates in the design process from many perspectives of system engineering. These specialty engineers are essential for successful implementation and maintenance of system engineering methodologies in the company as a whole.

If the product is very complex, schedule requirements are too demanding, or resources are very limited, or there is too much risk associated with failure of the product, the system engineer works with other especially skilled engineers to keep the program objectives and timing in place. This sort of coordination and amalgamation of several engineers and other functional folks with various disciplines, such as design engineers, manufacturing engineers, quality engineers, service engineers, finance folks, human resource personnel, etc., is monumental and highly demanding. The system engineers need special training and education to accomplish such an objective in a very efficient way. These specialty engineers have an overall idea of all the disciplines and allow other experts to excel in their own domain. The net result of such an integration is to create a synergy in product design and development.

The system engineer defines the requirements for the product. They interact with other specialists to design and develop a product that is compliant with its requirements. They coordinate and help the design engineers to understand the requirements and create an environment where the product can be designed and manufactured, and they determine the trade-off analysis and risk associated with such decisions. The system engineer also performs several analyses to find ways to satisfy all the feasible requirements for the design. In the end, the system engineer has the responsibility to make sure that the product satisfies the requirements that represent the customer's voice. A design compliant with the requirements will ensure customer satisfaction. This is particularly true for companies manufacturing machine tools. In my opinion, the primary reason U.S. customers decided to buy machines from overseas machine tool companies is their dissatisfaction with indigenously produced machines that did not provide what they wanted at a cost that they could afford. That is the bottom line for why U.S. machine tool producers, except for a few successful ones, went out of business.

In order to understand the mammoth task that these system engineers can or should fulfill, we have to understand the interacting disciplines of a project, as described below.

- *Reliability of a Product:* The engineer has to allocate a metric for the reliability of the product right at its inception. The reliability engineer of the design team has to create a reliability model for the system and sub-systems of the product. The design team will design the product to satisfy the reliability requirements. Once this reliability is incorporated into the components by the design team, the system engineer will integrate these individual component values to identify the system's reliability.

- *Maintainability and Availability of the System:* The reliability engineer has the primary responsibility of allocating system-level repair time to repairable items of the system. The system engineer tracks design team performance in responding to these allocations. During the testing of the system, the system engineer looks for features that will reduce the maintainability of the system.

- *Testing the System against Environmental Noise Conditions:* The system has to be tested against environmental conditions that cannot be controlled, but it has to operate under such conditions as extreme heat or cold, humidity, external vibratory inputs, etc. The system engineer looks after the verification and validation process for the system to measure and compare the system performance against the requirements.

- *System Ergonomics:* The system has to be ergonomically viable and should be safe for operation. The system engineer, along with safety engineer, takes care of system features that will enhance the system safety even when it operates under extraneous conditions.

- *Human Engineering:* The system engineer ensures the fact that the system operates within human capabilities even when it is subjected to very critical and extreme conditions. The engineer also determines the ease of actions for operating and maintaining the system.

- *System Safety and Human Hazards:* The system engineer, along with the safety engineer, has the responsibility of identifying safety requirements based on customer requirements and advises the design team to incorporate design actions that will minimize or eliminate unsafe and hazard conditions during operation of the system.

- *Environmental Analysis:* In order to understand the impact of environments on the product, the system engineer, in conjunction with the design team, will characterize the environmental impact on the product and the system requirements to counter the environmental effects. The environmental effect of humidity, temperature, sympathetic vibration, biodegradation, chemicals, etc. can be analyzed. The integrator will make sure and coordinate efforts to mitigate environmental effects. The integrator has the responsibility of making the system environmentally friendly and compliable with the environmental regulations of the country.

- *Productivity and Life-Cycle Cost:* The integrator might take the responsibility to incorporate design, which would enhance the productivity of the system. Ease of loading and unloading, operation of the conveyor system, ease of maintenance, ease of accessing the system during operation and maintenance, attending the machine under repair, etc. will be analyzed, and the engineer will advise the design team accordingly to make the system compliant.

In addition, the system engineer can also assess the life-cycle cost of the machine, i.e., the total cost of the system from installation to salvaging, and will implement design functions to minimize the total cost of the system over its life. Costs like the recurring cost of the machine for maintaining, testing, and training, energy costs, logistics support cost, heat disposal cost, etc. should be taken into account during the design and development of the machine. This could be entered into the requirement matrix for the machine. The reduction of the life-cycle cost would also help the machine establish the product against its competition.

- *Supply Chain Management:* A procurement strategy for all parts, such as vendor selection, cost of procurement, procurement time and cost analysis, selection of the minimum number of parts, and off-the-shelf parts availability, can be analyzed and shared with the design and purchasing team for minimization of the cost of the machine. The integrator, along with the purchasing, design, and manufacturing team, can decide on make-buy-lease programs for the components or assemblies.

    Planning and development of the vendors and suppliers are very crucial for the success of the product. Management of quality among the supplied components is very much required for the quality of the system. Selection of controls, hydraulics, and electrical suppliers sometimes makes or breaks a product during manufacturing or use of the products in the customer's shop. I believe that one of the crucial steps Japanese manufacturers take is to develop suppliers according to the products' requirements.

- *Value Engineering:* During the manufacturing stage of the machines, the system engineer might look into the process elements and manufacturing cost and ensure the cost of production is within what's stipulated in the requirement matrix. The integrator will suggest ways to control the manufacturing cost to a desired minimum and, if necessary, suggest a process or design that will control the cost as desired. This is a joint effort among teams from several disciplines, and the system engineer could take the lead to get it done.

- *Quality Engineering Efforts:* One of the primary responsibilities of the system engineer is to bring quality ideas from the requirement matrix to the quality engineering team. Any deviation from the target quality desired by customers is not advisable. Producing a higher-quality product will cost more money than what customers can afford. On the other hand, a quality that is less than what customers are expecting will also reduce revenue because fewer products will be sold. So, management and sustenance of quality for a particular product is absolutely necessary for long-term sale of the machines. The system engineer will help the integrated team to deliver the product desired by the customer to the marketplace.

As can be imagined from the above discussion, the system engineer has a very important role to play during the design, manufacturing, installation, and service phases of the machine. Actually, the system engineer could use his expertise from birth to death of the machine within the company, and outside the company in the user's shop as well. The system engineer must interact continuously and concurrently with other team members to ensure the product maintains its integrity and performance throughout the life cycle of the machine. It is, in my opinion, a continued effort that must be maintained for the product to succeed in the market. Such efforts will continuously enhance the design, manufacturing, and service requirements for the product.

I have witnessed the most experienced and knowledgeable engineer of the design team take the role of a system integrator for a line of product, from design through manufacturing and service, including training requirements for the product. It is an excellent proposition for the product to survive as long as possible. Customers also feel that they are not left unattended when they need help.

2.10 Effects of Learning & Experience Curves

It has been observed for the last few decades that experienced technicians and engineers are moving away from machine tool industries due to low payment and lack of recognition for their efforts. Other jobs look more attractive and comparatively easier. Similar observations have come to my notice for other industries, such as textile, steel, and mining, to name a few. Personally, I have experienced the same situation in my own personal career. Machine building companies get sold several times until the assets are depleted to extinction. So, even our government did not think it necessary to keep the basic industries alive for the progress of the nation. The primary question is whether we have enough manpower to build accurate and precision machines or not. In my opinion, it takes much more time and effort to train people for a specific trade to excel again. In light of this concern, the effects of learning and experience curves are described below for the common knowledge of the readers.

Learning Curves: It is well known that Mr. Ford thought it was more efficient to bring the material to the person instead of bringing men to the material. To achieve higher productivity with the Model T, he introduced a conveyor-type assembly line. At each station, people were trained to perform specific jobs in a specified time to keep the assembly line moving forward. So, it is a fact that experience does enhance productivity. Although the machine tool production volume is way less than that of automobiles, the same principle of gaining efficiency and productivity through "learning or training on the job" applies. When this is used in any shop or the industry as a whole, the whole machine tool building community benefits. The production process becomes gradually more efficient. The same principle is practiced in Germany and Japan to a great extent.

Promoting an atmosphere of learning throughout the company can incorporate efficiency into its culture as the production continues to build up. In the machine tool industry, the prototype is a learning phase, and the pre-production phase brings in more maturity about building the machines in the production phase. Even in the production phase, I have seen technicians find more efficient ways to finish their jobs using automation and fixtures, loading and unloading heavy units, implementing an inspection process, packing the finished goods, providing installation and service, etc. This results in the reduction of the unit cost of production and optimization of production capacity for the facility.

On the other hand, people on the floor have to be allowed to express their opinion about how to design a product for ease of manufacturing and building. The new ideas from the shop floor need to be incorporated in the design before production starts. This will also increase technicians' motivation to build more efficiently down the line.

To promote ease of learning and gaining experience, management can also incorporate a few other ideas, as outlined below:

- Inculcate product design, eliminating setup completely
- Promote product design for ease of manufacturing and assembly
- Incorporate just-in-time inventory process and scheduling
- Plan and schedule well ahead of production
- Install equipment and assembly tools to reduce fatigue
- Empower line workers and shop engineers to improve the process
- Listen to the workers to improve the design

It has been observed in the machine tool industry and others that the number of hours required to complete a machine diminishes as the workers get more exposure to the design and manufacturing process. The assemblers get more efficient and productive as more working hours are spent on their jobs. This slowly turns out to be a profit enabler as the process is optimized. The marginal time required to produce another additional item declines due to this effect. Such a phenomenon can be observed, and the data can be plotted as a series of learning curves for each product. The nature of the data seems to be exponential. Linear regression can be used to fit the observed values for any particular machine model over time. It is also important to note that this improvement is not valid for an indefinite amount of time. After some time, the hour per unit required gets to a constant value and does not change with time or amount of production.

Learning curves have to be used with caution since this a concept that is product and process specific. This also points to the importance of work personnel understanding the design and learning about the process. This could be extended to engineering personnel, too. The team has to learn about the design and the process associated with it. It also has to have motivation without any personal agenda to optimize the process and eventually reduce the marginal hours required to produce another additional unit. Then and only then will this concept bear results that could be used to optimize production hours, which will lead to the reduction of production cost per unit.

Experience Curves: While learning curves can be considered a way of determining the marginal hours and, consequently, the labor costs associated with the machine, experience curves determine the net effect of cumulative experience gathered by the team during the initial phase and the production phase as the process matures. This seems to be more effective while a company is producing several lines of machines at the same time. This accumulated experience will obviously affect the cost of the machines and, subsequently, the profit of the company. As the name suggests, this effect is geared toward the experience gained by the team while the product is being manufactured.

Moreover, a new process and design have many unknowns that can be observed during the actual manufacturing, testing, and use of the machines. So, such curves might not be very effective for products that have been produced and marketed for a long time. For such products, the process and the cost might have reached equilibrium already. This concept has been recommended for newly designed products. If any company has been producing, say, CNC lathes for a long time and thinking about manufacturing machining centers, this principle could have more effect on the cost of the machine as the product moves toward maturity.

Due to the very low life cycle of any type of machine when compared to, say, that of electronic circuit boards, I have not seen much of an effect of this phenomenon in the machine tool industry. Nevertheless, it can be of some value to the producers when it takes a much longer time to design and manufacture highly complex machines. For example, this concept could be used to reproduce flexible manufacturing systems as a standard line of machinery for various applications having the same core manufacturing process.

Unlike learning curves, where time per unit manufactured is measured, experience curves are measured in terms of total cost per unit of production, which affects the cost of the machine as well. From what I have seen in the industry during the production of very complex machinery, senior technical personnel from other established lines of machines are given the responsibility of taking up the new machine during the initial stage of production until the product matures. Eventually, the matured process is used by less experienced people to carry on the product until the decline. This is a very efficient way of handling the initial manufacturing of a new line of machinery, as I have experienced in Japan.

The companies can use the effects of learning and experience curves in a very competitive market where the demand is very price sensitive. In order to cope with the competition or gain market share in a turbulent market, revenue depends on the cost of the machine. An all-around effort has to be made to keep the cost of production as low as possible. As a greater number of similar units are produced for a comparatively longer period of time, the machine tends to become a commodity item, such as a manual general-purpose shaping machine, lathe, or milling machine. In such cases, the application of experience and learning curves should be used to understand the ultimate cost of the machine to stay ahead of the competition, and the price of the machine has to be considered instead of the cost of the machine. When using learning curves, the cost has already been optimized due to a longer time of production. When the price is an issue, experience curves have to be considered to help face the stiff completion ahead.

Next, the question arises: how we design a manufacturing strategy using both learning and experience curve ideas? I think that for a newly designed product, we should use the learning curve idea since the design and process of manufacturing have not matured yet. The product goes through several cycles of refinement before maturity with respect to both design and manufacturing.

The use of system engineering principles should help reduce such effects of learning and experience, and system engineering applications should help the team integrate the product before it is even prototyped. The requirements and manufacturing process, quality control methodologies, and cost of components are thought of together for the system as a whole. Moving away from a compartmentalized process and toward an integrated approach should eliminate most of the recycling effects, which I have seen in automobile industries. The team evaluates a design from all perspectives, and product maturity is obtained during the system engineering process itself.

Moreover, for a revolutionary design or when using a disruptive technology, the product will go through a period of unforeseen and unsought changes during design, manufacturing, and introduction. Application of the concepts of experience and learning curves might be misleading for such cases. Even conclusions for new products should not be drawn from experiences from previously established products to avoid confusion and misleading conclusions.

Another note of caution for machine tool producers could be the notion that cost has to be reduced independent of other aspects of machine tool design and production. The machine selling cost has to be commensurate with design, manufacturing, quality, reliability, precision, flexibility, etc. The reduction of cost for the sake of making a profit, with considerations given to other competitive factors, is definitely a self-defeating proposition that must be avoided at all costs. In my opinion, cost and quality have to be combined to get customer satisfaction, which is the ultimate objective. Customers are first, and everything else is secondary.

2.11 Capacity Growth: A Strategic Factor?

I have never experienced a capacity issue in the machine tool industry, probably since the demand for machine tools seldom grows infinitely or is highly erratic and fluctuating. As mentioned earlier, machine tool demand is highly dependent upon the economic atmosphere and manufacturing environment of the country. The cost of production is also an issue. This is the U.S. scenario. On the other hand, some companies, like Okuma, Makino, Wasino, Akebono, HAAS, Mori Seiki, and some Korean machine tool companies, have sustainable growth for their businesses. In general, there has been hardly any growth in machine tool industries across Europe and in the US. For companies that have the opportunity to grow, capacity planning could be a strategic or competitive tool.

Customers also feel comfortable going with a company that takes risk and has a plan to grow in the future. A short and very brief discussion on how to plan for extended capacity to accommodate new lines of machinery is in order for common readers of this book.

Moreover, a planned reduction of capacity is somewhat easier than capacity growth planning. when capacity is reduced, the strategy could be to minimize loss of return on investment by managing it in a systemic and gradual manner. The following discussion pertains to the growth of machine tool industries in general.

By definition, the capacity of a plant is the amount of production that a facility can handle in a given planning time. To gain customer confidence, sustainable growth in the capacity of machine tool production is necessary. Moreover, for additional profit, growth could be a strategic tool for the company. Several competing factors have to be taken into account while planning for capacity growth, as outlined below:

- Nature of demand trend of manufacturing activities
- Demand trend of machines of interest in near-past and future time periods, i.e., planning horizon
- Total investment required from planning to completion
- Investment vs. time requirement
- Total time required from conception to completion
- Facility construction cost and operation ramp-up cost until production starts
- Legal and business environment of the country
- Sustainability of demand over the planning horizon
- Product and process matrix
- Technology intensiveness
- Supplier's chain-management policies
- Availability of human resources
- Competitors' approach to growth

As in other disciplines of business, capacity planning is mostly dependent on the revenue growth of the company. For enhancement and sustenance of profit for the company for some time to come, several questions have to be kept in mind before actual investment is made. It is very common among established machine tool companies throughout the world to use existing facilities for further production expansion until production, demand, and customer acceptance levels are established for the machine under consideration. This strategy reduces the fixed costs and amortizes the cost of equipment already available in the plant. This definitely reduces the cost of production, enabling the company to have a lower brake-even point until the product is established in the market. This is definitely a less expensive alternative than starting a brand-new facility, its support structure, and the development of production logistics.

The cons of such a strategic planning attitude are also numerous. For example, restricting growth by restricting expansion is a self-defeating policy for management. A new facility and its activities could help promote the product in the market and gain ground against the competition. Okuma started establishing businesses in the US in the 70s, and once the company gained ground there, it started created assembly and service facilities to gain confidence among its customers. It showed the seriousness of its intent to do business in this country. It was a very bold, strategic move that helped them establish their business in a foreign land. Steady growth is a motivating factor for employees as well. It helps the company retain its talent. Capacity expansion could also be a new dimension for the company to achieve success over time. Eventually, no growth is a precursor to death.

Once the business planning decision has been taken for further expansion of production capacity, the following questions, although not exhaustive, need to be answered:

- Is demand consistent during the planning horizon?
- What is the gap between total demand and total existing supply during the planning horizon?
- What is the total production capacity and rate of production required?
- What is the niche of the market that the product will serve?
- What is the total time required in between the start of the facility and getting the product out to the customer?
- Is technology ready for implementation?
- Is manpower available in the area?
- Is labor cost consistent with the cost model for the product?
- Are suppliers available for material supplies?
- What is the business environment of the locality?
- What is the political environment of the locality?
- Are any tax benefits or subsidies available?
- Is the area conducive to starting a new facility?
- Is the cost of living reasonable enough to sustain labor cost?
- What should be the size of the facility required?
- Does the new facility capacity support growth in demand in the future?

Basically, these questions pertain to when and how much capacity will be required. Overcapacity is a waste of resources, and under-capacity also restrains production and supply, constraining revenue and profit growth over time.

Another question that arises very often is whether the company should have the capacity to make its own parts inside the facility or buy from vendors. This is specifically for components that the design uses for sub-systems and integrated systems. For example, the Warner and Swasey Company had enough manufacturing facilities, machinery, and heat-treatment facilities, etc. available to make most of the parts required for constructing machines. Almost all the parts were produced in-house to keep the quality as desired and at a cost that the company could bear. It was also thought that the internal production of components added flexibility and capability for production. Another idea that I have always promoted and liked is to use company-produced machinery to make the components. I think this is a great selling point for the company. The customer can see the machines being used every day in a production environment. This instills customer confidence in buying the product.

Another aspect of adding capacity expansion is whether a company should use different facilities to produce different lines of products. The examples of such a strategy would be to produce lathes in one facility, machining centers in another facility, and grinding machines in another facility. Combining various products under one roof sometimes creates an undesirable attitude among the workers due to business conditions. The products that are generating more revenues might be where all the focus is concentrated on. This might create tension and an unfriendly atmosphere among workers of each line of production.

Another question comes to mind: whether companies should buy the components or make them in-house during prototype development. If the company is not practicing system engineering principles, the prototype goes through several iterative product design changes to refine it and achieve desired performances. This is simply what is called a "make-or-buy decision." This is very applicable and significant during prototype production because it can reduce prototype cost and also develop suppliers for final production. For components, the company has the option of buying them from outside sources or making them inside the facility if the machinery is available to produce them. A simple analysis for such a decision is considered below for the general knowledge of the readers.

K_1 = purchase cost from vendor per unit, independent of the quantity ordered from suppliers.

K_2 = internal production cost per unit of the same component.

Even if $K_2 < K_1$, the company has to invest in new machinery to make such a part. The additional investment is, say, K dollars, required to buy the new equipment. If a unit of assembly needs x number of components, the total cost of the firm to produce is $K + K_2 \cdot x$. The total cost of buying the units from outside the supplier is $K_1 \cdot x$. So, the number of units where these two costs are the same can be found by equating these two costs, i.e., $K + K_2 \cdot x = K_1 \cdot x$.

. This amount of production is called the break-even quantity. K is the fixed cost to be amortized over units of production. As x increases, the cost to buy and produce increase. There is a point of x where the cost to buy exceeds the cost of making the part in-house.

Hence, the break-even quantity (B/E quantity) = x_b = K / (K1 − K2), which means that if the required quantity is more than x_b, make the part in-house. Otherwise, buy it from a vendor. The relationship can be shown in graphical form, as in Figure 2.13. The graph shows tradeoff values for the investment.

The break-even quantity is useful for getting a quick estimate of the production amount beyond which the parts should be manufactured in-house. This approach assumes part requirements are static, which is hardly true over time. So, for the unanticipated demand of any machine in an emergency situation or to attend to a service problem, parts may have to be purchased from outside the company due to the dynamic nature of the requirement situation. In most of the cases I have seen, the most complicated parts tend to be manufactured in-house. In some instances, to keep the trade secret inside the company, complicated or precision parts and critical parts are manufactured in-house to have control over them. Some of these important issues were discussed in detail by Mann (1967–68).

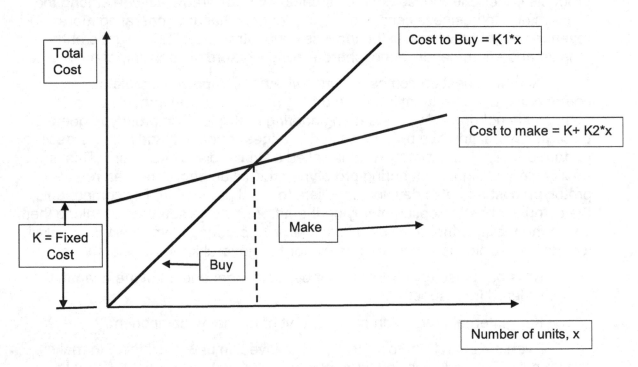

Figure 2.13 Break-even curves for prototype parts

The next topic of discussion with regard to the decision of adding another facility to increase production capacity is the effect of a new location for the new facility. This discussion is somewhat non-parametric in nature. The location of the new plant depends on many complex and related issues that are very difficult to quantify, such as the availability and cost of the real estate, the enormity and size of the plant, the types and number of products to be produced, the process to be used, the technology to be used, and so on. Other significant factors could be as follows:

- *Availability of Workers and Engineers:* This is more relevant when the technology and design for the machine are very complex and specific knowledge is needed for manufacturing. People are hesitant to move to a location where social interaction is at a minimum and living standards are very low. Both the numbers and quality of the workforce need to be addressed for the new location. Extreme weather conditions and environmental factors could be a deterrent for people.

- *Environmental Factors*: Areas with high levels of pollutants are not conducive to getting a required workforce for the facility. Similarly, for highly accurate machines, extreme temperature and humidity conditions make it difficult to maintain the quality of machines. Under such circumstances, maintaining uniform working conditions for the machine and the workforce could be quite expensive.

- *Umbrella Effects*: Putting a plant near the existing facility producing different machines is sometimes difficult to justify. The technical collaboration among engineers and workers becomes quite difficult when required. Use of common facilities could also be very difficult, resulting in a higher cost for the facility. Staying together within a boundary is more efficient for management and control of activities. For plants in different locations, a uniform umbrella effect is very difficult to achieve. Every plant has a tendency to develop a different work culture depending on the location.

- *International Locations*: In the recent past, many U.S.-based industries have opened up plants in Mexico, China, Korea, Europe, etc. because of the advantages of taxes, work culture, etc. Sometimes these factors seem to be attractive in the short run, but the operation becomes quite expensive when other costs such as control of design translation, custom charges, shipping charges, material movement, currency fluctuations, delivery uncertainties, the political environment of the country, etc. are considered. For many machine tool companies, international locations have turned out to be very costly propositions if the products are imported back to the market where they are to be sold. They eventually are closed down or sold to local owners.

- *Logistics Control*: Development of management of suppliers, transportation needs, and quick delivery are very time consuming and expensive. For example, adding manufacturing aids to the line of production is somewhat cheaper in Third World countries, but the useful life is also less, leading to a higher cost per unit produced. Even if it sounds very easy to control operations from remote locations using centralized computer controls, in reality, it becomes more difficult to achieve higher productivity and efficiency in an overall sense. Interactions become very difficult.

- *Customer Base*: It has also been observed that manufacturers would like to have new facilities where their customers are. This decision helps them to sell more machines and the service, installation, and warranty costs go down. Customers also prefer the idea since they get very quick service in case of any trouble with the machine. I think that instead of having a new facility in this area, a small sales and service center would be economically more viable. Many China- and Korea-based machine tool manufacturers have been very successful in following this route.

Several attempts have been made to optimize costs associated with location using mathematical models, but in the real world, they have turned out to of very limited value. This is mainly due to the fact that any such academic representations of real-world situations consist of assumptions and constraints that are difficult to quantify. At the end of the day, a new location, most of the time, is decided arbitrarily by the operation chief or the owner in a very haphazard way.

I think the more realistic and hybrid approach should be to perform a very simplistic and approximate analysis to get a preliminary direction regarding the location possibilities, and then final location could depend on the experience and choice of management personnel. Nevertheless, I believe that locations should be selected based on economic and social feasibilities, such as labor costs, political activity, customer base, vendor management, quality of life around the location, and cost and availability of support logistics.

Many companies tend to group manufacturing and design activities in the same geographical location, whereas administrative offices are placed at different locations. The decentralized control seems to be better for consumer products but not for capital-intensive products such as machine tools. For machine tools, I think, administration and operations should be in the same facility for synergy and effective control, whereas remote locations for sales, service, and warranty could be at the customer base. The primary focus for a new location should be on building a product that satisfies the customers. Any location must support this ultimate end.

In reality, the real-world situation differs from academic treatments of the problem of capacity expansion. Some of them are outlined below:

- *Product Life Cycle*: Every machine tool product, in reality, has a very finite life of existence. So, capacity or facility expansion has to consider the issue. Unless a future product is planned, the facility might not have use beyond its life cycle. Hence, facility life and product life have to be compatible and comparable to each other. Such a phenomenon has been noticed very often in the automobile, steel, and machine tool industries. Factors like labor costs, supply costs, etc. might have a direct effect on the type and size of the facility required for a product.

- *Variability of Demand:* Does the product have a cyclical demand pattern, or does it have an upward or downward trend in demand. For general-purpose machines such as CNC lathes or machining centers, due to their comparatively low cost and increased flexibility, the demand has a tendency to be flat or slowly rising over time, whereas demand for very special machines or flexible machining cells has a highly cyclical pattern. The gestation period for flexible machine cells is also very high due to its complex nature.

  Facility construction could also depend on this issue. Demand is also affected by new competitors from foreign countries, especially with general-purpose machines. In general, a high level of competition tends to slow the demand for any machine unless the product is improved and sustained for higher production or flexibility.

- *Disruptive Technology or Technology Development:* Developments in process or product technology does affect the decision for constructing new facilities for machine tools. New safety, ergonomics, and health requirements also affect construction. New building technology has a bearing on the capacity extension. When a process is altered substantially or a new product technology is introduced, the facility has to be modified to accommodate such changes.

  For example, many facilities had to be redesigned when control changed from NC (numerical control) to CNC (computer numerical control) for machines. Another example is when traditional casting was replaced by epoxy casting for ease of manufacturing and pollution control.

- *Nature or Type of Product:* Some machines such as grinding and fine boring machines need special facilities to maintain accuracy and precision. The facility should never be built around a place where a press plant is situated.

- *Governmental and Legal Issues:* Safety regulations, such as for noise, humidity, and sound, do affect the design and the construction of facilities. Environmental regulations also dictate building codes and construction over time.

- *Overhead Cost Reduction:* There are numerous examples of this issue, such as energy costs, air and water supply costs, cleanliness requirements, lighting conditions, natural lighting, etc. In order to keep the cost of the machine as low as possible, such minute details must be considered to be competitive.

- *Tax Subsidies:* Since tax payment does affect the net profit of the company at the end of the day, companies should build facilities in an area where local or state-level authorities provide tax incentives to locate in their areas. This adds to the expansion of the local economy. It is sometimes a major factor to be considered for facility locations. The Makino Company decided to build a brand-new facility in Cincinnati, Ohio, due to such considerations.

2.12 Summary

This chapter focused on system engineering principles and their effects on business as compared to the classical view of developing and manufacturing machines in a compartmentalized way. System engineering principles have been in use for a while in the aerospace and automobile industries, and they provide a proven process that could be very well adopted in machine tool design and production to gain productivity, efficiency, and competitiveness in the worldwide market. Machine tool design and manufacturing could be treated as an integrated system of activities to gain share in the marketplace.

System engineering principles should be applied to industries where time-based competition is very dominant. The time from concept to market has to be minimized to gain effectiveness in marketing. An integrated system engineering process will ensure machines are produced as and when required without mincing quality aspects. Perfect quality at an affordable cost is the ultimate objective of the integrated system engineering process, and it should be followed by machine tool designers, manufacturers, and management personnel.

In order to understand the process required for a particular type of machine tool, life cycles should be understood and applied for building new machines. These processes go through five stages of production: prototype, pre-production, production, post-production, and decline. The incremental production steps are specific to the machine. It is also necessary to understand which process for each of these production stages is best for minimizing investment and gaining productivity. Hayes and Wheelwright have been instrumental in the development of a product-process matrix. In a nutshell, product life cycle and process life cycles have to be matched to create synergy for a machine.

To understand the effect of workforce learning and experience on the reduction of marginal labor time required to produce machines over time, experience curves should be used. These curves follow exponential decay over time, which means that it takes less time to produce a unit than it took for the previous one. This is due to the experience gained while working on the machine and the adoption of improved manufacturing processes. These curves are very approximate indicators of the cost of production or time required to produce an additional unit, and they should be evaluated for each machine.

We discussed very briefly the expansion of the capacity to produce more machines, i.e., how to enhance the capacity of production. Capacity expansion is a very crucial strategic decision for the machine tool industry. Break-even quantity can be calculated for any machine and can be used to decide whether a component should be manufactured internally or offloaded to a vendor. The pros and cons for such decision were also discussed.

The last portion of this chapter discussed adding a new facility in a new location to enhance capacity. The issues associated with developing a new location for a product or a group of products were also covered. The issue of location with respect to machine production is very problematic and critical since the investment required is often hard to come by. I believe new locations are to be sought only when the product is well established in the market and does not need any refinement attention during manufacturing or management.

2.13 References and Bibliography

Buffa, E.S. and Sarin, R.K., 1991, *Modern Production and Operation Management*, eighth edition, John Wiley & Sons, NY.

Dertouzes, M.L., Lester, R.K., and Solow R.M., 1989, *Made in America: Regaining the Productive Edge*, MIT Press, Cambridge, MA.

Eckes, A.E., Stern, P., and Haggart, V.A., US International Trade Commission Commissioners, 1984, Report to United States International Trade Commission on Investigation No. 332-149 Under section 332 of Traffic Act of 1930, "Competitive Assessment of the U.S. Metal Working Machine Tool Industry."

Elm, J., 2013, "The Value of System Engineering," posted on May 20.

Freidenfelds, J., 1981 *Capacity Expansion: Analysis of Simple Models wh Applications*, Elsevier North Holland, NY.

Grady J. O., 1995, *System Engineering Planning and Enterprise Identity*, CRC Press, Ann Arbor, MI.

Hayes, R.H. and Wheelwright, S., 1979, "Link Manufacturing Process and Product Life Cycles," *HBR* 57(Jan–Feb), 133–140.

Hayes, R.H. and Wheelwright, S., 1984, *Restoring Our Competitive Edge: Competing through Manufacturing*, John Wiley & Sons, NY.

Kankuro, K. and Izawa, K., eds., 1965, "Proceedings IFAC Tokyo Symposium on System Engineering for Control System Design."

Maier, M.W. and Rechtin, E., 2000, *The Art of Systems Architecting*, CRC Press LLC, Boca Raton, FL.

Manne, A.S., ed., 1967, *Investments for Capacity Expansion: Size, Location, and Time Phasing*, MIT Press, Cambridge, MA.

MIL-STD-499A, 1974, "Military Standard, System Engineering Management" (May).

Nahmias, S., 2000, *Production and Operations Management*, fourth edition, McGraw-Hill Erwin Companies, NY.

Porter, M.E., 1990, *The Competitive Advantage of Nations*, The Free Press, NY.

Skinner, W., 1978, *Manufacturing in the Corporate Strategy*, John Wiley & Sons, NY.

Tooze, A., 2009, *Machine Tools and the International Transfer of Industrial Technology*, "The Global History of Machine Tools, Knowledge, Narratives and Fiction," 30-31, King's College, Cambridge.

Ulrich, T.K. and Eppinger, S.D., 2011, *Product Design and Development*, McGraw-Hill Companies, NY.

Womack, J.P., Jones, D.T., and Ross, D., 1990, The Machine That Changed the World, Harper Perennial, NY.

2.14 Review Questions

- Can we take up a machine development project as a system development task? If so, what are the pros and cons of this new approach? What are the risks associated with accepting this method?

- What are the salient features of system engineering? Can system engineering principles be applied to a sick industry? Is system engineering necessary to revive the machines tool industry in the USA? What are the pros and cons of such a revival? Would an integrated approach be more beneficial than a conventional compartmentalized process for the machine tool industry?

- In your opinion, what are the actions that a sick industry has to take to excel, grow, and stay ahead of the competition?

- Do we need a disruptive technology and process for the machine tool industry to use system engineering principles to improve the status of the industry as a whole?

C H A P T E R 3

3 Machine System Design & Development

3.1 Introduction

Product development is an art in itself, supported by engineering and scientific principles. The product development process is a disciplined approach to generating a product as per requirements. Any generic new product development process consists of a series of pre-meditated steps that help to convert inputs into the desired output. Accordingly, this process consists of sequential and concurrent steps that a company has to follow to be successful. These are scientific and well-proven steps. Nevertheless, the process also involves an artistic approach to developing the architecture of the product. For example, the Taj Mahal is an outcome of an artist's expression supported by scientific principles.

In general, the new product development process can be thought of as those activities that any company can follow to conceptualize, engineer, and market a product. These activities are intellectual by nature. Following this disciplined approach might or might not produce a product that satisfies all customer requirements, and the outcome might not be what the company or its customers wanted to generate. Even then, this series of process steps might be used to modify the new design to satisfy most customers. Such generic steps could also help the company optimize its resources to create an optimized product design. The common framework of product development processes could also be customized depending on the level of complexities and the nature of products being designed.

Following a well-proven, systemic product development process for generating a product can help companies achieve the following:

- **Product Quality and Uniformity:** Since the overall process consists of logically connected process steps with quality checkpoints in between, and since any product development process in the company is geared to go through this rigorous process, product quality and reliability will emerge.

- **Performance and Cost Optimization:** Since the primary objective of this process is to deliver the product as per customers' requirements at a cost customer can afford, this process will lead to a successful and desirable product. This process will ensure the design and development of a product that will deliver desired performance at the desired price.

- **Customer Satisfaction:** Since the product performance and cost are optimized to satisfy the end customer requirements, customer satisfaction is another dimension of this product development process. This generic process helps to transform the voice of the customers into the product requirements that the end product satisfies.

- **Coordination among Engaged Members:** For a very complex multi-national project, it is necessary to gain higher efficiency and productivity. Since the many people involved in the project could be in several countries, uniformity in the process can only help bring together all the involved engineers and management and lead to a successful end.

- **Management Control:** When a disciplined product development process is followed, it is much easier to track and control at any point in time. This could also help management control product development time and cost. Management can also identify the inefficiencies and project delays to take necessary steps to avert the overall delay of the project and minimize the cost.

- **Activity Planning and Coordination:** In order to control the interacting and cross-functional activities of the project, project progress could go through several milestones, satisfying the requirements at each. For each milestone, resources required to achieve deliverables could be controlled and managed effectively and in a much easier fashion.

- **Documentation:** For each element of the process, documents are kept for future reference and improvement. Since process should fit the product to be developed, the applicability of each step could be traced and improved if the process documentation is recorded and maintained properly.

Development Stage →

	Product Planning	Requirements Development	System-Level Design	Sub-System-Level Design	Component-Level Design	Testing and Design Update
Actions ↓	• Develop mission needs • Explore potential market niche • Obtain customer wants and needs • Review and benchmark products from market leaders • Develop product development strategy: legacy or new platform • Explore available and proven technology • Develop product architecture • Explore production and process capability • Identify manufacturing constraints • Explore make/buy strategies • Strategize supply chain management and development • Explore R&D requirements • Set financial and human resource constraints • Product cost/benefit analysis • Explore legal environments	• Gather raw data from customers' feedback • Interpret raw data in terms of needs and wants • Prioritize the needs into hierarchical fashion • Benchmark the requirements against competition • Establish financial implications of requirements • Develop target requirement list • Formalize the data on target requirements	• Develop system-level design requirements • Functional Decomposition into sub-system and component level requirements • Consider alternative designs • Select final design configuration • Finalize product architecture • Develop geometric layout, including options • Develop manufacturing flow process diagram • Develop final assembly process diagram • Define quality constraints • Identify manufacturing resource constraints • Identify human resource constraints • Evaluate service and warranty requirements • Re-evaluate cost constraints	• Create/update sub-system level requirements • Finalize sub-system architecture • Develop geometric layout, including options • Develop manufacturing flow process diagram • Develop final assembly process diagram • Identify manufacturing resource constraints • Define quality constraints • Identify human resource constraints • Re-evaluate cost constraints	• Develop complete specifications: materials, design, tolerances of parts • Identify parts for off-load and manufacturing inside • Define quality constraints • Process plan and tooling development • Develop supply chain vendors • Control part documents • Establish financial impact • Long lead Item management • Revaluate cost and budget • Assemble sub-systems	• Assemble system • Prepare prototype for testing • Reliability and accuracy testing • Performance testing • Durability testing • Validate and confirm requirements • Update design to satisfy requirement • Finalize part, sub-system, and system design/architecture • Re-evaluate cost and budget • Plan for production details • Finalize product introduction plan • Finalize product service and warranty plan

Fig. 3.1: Machine design development stages and actions for each stage

3.2 Machine Development Process

In a system-engineering-based process for the development of any machine, there is a series of action steps in each design stage. It starts with business planning and ends up with the final tested reliable product ready for introduction into the market. The system engineering principles, as laid out in a "system V" process, could be deployed to plan, design, develop, test, and market a new or legacy-based product in the market. It is also a fact that it is not absolutely essential to develop a technology-based product using system engineering principles. Nevertheless, most products could use this systemic process for higher productivity, less development time, and higher financial efficiency. Such s process becomes internal to the organization and could be customized depending on the experience of the company personnel and type of the product that the enterprise is trying to develop. In order to repeat such a process for products down the line, it is essential to use the systemic principle of product development. The process then could become an intellectual property of the company and can be used more than once for product development. Some modification might be required for this process to be applied to different types of products, such as turning centers, machining centers, or grinding machines. The overall process could remain the same, but some of the elements at different design stages could be changed or customized to suit the product.

The product development process, as shown in Fig. 3.1, has six design steps. The process starts with the business planning phase, which consists of marketing, technology development, and customer feedback. The output of this step is the mission statement of the product, a determination of its viability, and justification for such development. In brief, this step should answer whether it is worth it for the company to develop this product. The output of this process becomes the input for the next phase of requirement development. The end product of the development process is the machine that is more required by the customer than its competition.

The systemic product development process could be thought of as a filtration process to manufacture and market a product that customers would like to buy to reach their goals. Each milestone during the process becomes a filter to refine the concept into the desired product. The initial feasible concepts and architectures start with the mission for the product and customer requirements feedback. The filtration process is then applied to these alternatives to narrow the product alternatives into the best possible design with available financial and human resources. In order to keep the financial objectives within desirable limits, a financial evaluation has to be done at every stage of the process, and in the case of an out-of-control situation, a red flag must be raised to bring it under control. In extreme cases, the product has to be abandoned.

In other words, such a systemic process could be thought of as a means of optimization of company resources to manufacture and market a product. A company should not pursue all products simultaneously. The width of the product portfolio does not guarantee success or profitability. Heald very successfully and profitably produces an internal grinding machine without manufacturing other machine tool products. To maintain control of the time cost of development, a system process is necessary. The product development process could also be used as an information development and gathering process. Once such a process is developed, it can be reused for future products with minimum difficulty. In the long run, this can become a company or organization culture for developing products. In that case, this can be taken as a process tool to further develop products. If any product does not fit this process or framework, it can be abandoned to save money and time.

The six phases of the product development process are further explained in Table 3.1. The generic nature of each phase is described below:

Business Planning: This stage is also referred to as the "mission development phase" since the output is the mission statement or document for the product that will be followed by the design and development team further down the line. The product has to satisfy the mission of the product. This is basically a management plan for the product, and it depicts the approval of the product to be designed, developed, and marketed by company personnel to enhance the financial health of the company. It is a strategic filter. It is a sufficient and mandatory document to initiate any product development. A typical mission statement for machine tools is given in Fig. 3.2. The mission statement summarizes the overall strategic direction that must be followed by the product development team and other functional teams.

Requirements Development: In order to market a successful product, it is mandatory to understand customers' requirements. These inputs from the target customers are sometimes called the voice of the customer. The product has to turn these requirements into the product target specifications that the product has to satisfy when launched. This process is particularly true for market-pull products. In the case of technology-push products, such as the cell phone or iPad, customers learn about the product when launched. Since competition is comparatively very high for market-pull products, customers' requirements must be satisfied; otherwise, they will turn to one of many other options to buy products from. For machine tools and accessories, the design and performance of any product have to satisfy what customers want. In order to get feedback from the customers as to what they specifically want, market clinics, audience analysis, existing customer feedback surveys, distributor feedback surveys, etc. can be conducted. The output of this phase is a concise requirement or specification list for the product. I consider this document to be the bible for the product.

Mission Statement for a Multi-Purpose Turning Machine with Live Tooling	
Product Description	• Four-axes CNC hard turning universal machine with live tools, tailstock, steady rest, axes servo control, and on-line inspection capability
Key Business Goals	• Market leader for the CNC hard turning machine replacing grinding operation for hard material parts • New generic platform for similar future products • Capture 20% of the market segment for turning machine users • Networkable among machines and central remote library of parts location • Product Introduction latest by last quarter of 2021
Primary Market Segment	• First-time users not experienced in CNC machining technology • Lead users of automatic turning and grinding operation for high-volume parts • Mid-volume producers of turning parts for automobile companies
Secondary Market	• Batch production users • Group-cell users • Hard turning part producers
Assumptions and Constraints	• New product platform • Universal design for two-axis and four-axis capability • Digital servo control • Statistical part control on-line with 1.33 process capability • Automatic loading and unloading • Electricals suitable for the USA, Canada, and the UK • Servo and control to be supplied by GE-Fanuc • Costs less than $30,000 per machine without options
Stakeholders	• Vendors and suppliers • Manufacturing and quality operations • Service, warranty, and installation • Distributors and service providers • Investors • Bank

Fig. 3.2: Mission statement for a universal CNC lathe

System-Level Design: This is the most crucial step during product development. Basically, this step initiates the design, taking the target requirements into account. First of all, from the customer needs and wants list, a hierarchical and prioritized target list is generated. Engineers and management agree to the customers' wants and needs that the product has to satisfy. This could be termed a system-level requirements list or a system-level specification document for the product, and it should be in alignment with the mission statement of the product. The system-level specifications are decomposed to create sub-system-level specifications.

Next, a product architecture and/or system layout is completed to satisfy the system-level requirements. While developing the system architecture, several dimensions of product development have to be considered, such as quality, manufacturing, assembly, servicing, and supply chain considerations. The output of this design phase is the system-level specification list, sub-system-level specifications, system-level product architecture, and the system layout drawing of the product. The process flow diagram for this phase is summarized in Figure 3.3. At the end of this critical phase of product development, a design audit by the stakeholders is warranted since the subsequent phases are highly dependent on the outcome of this one.

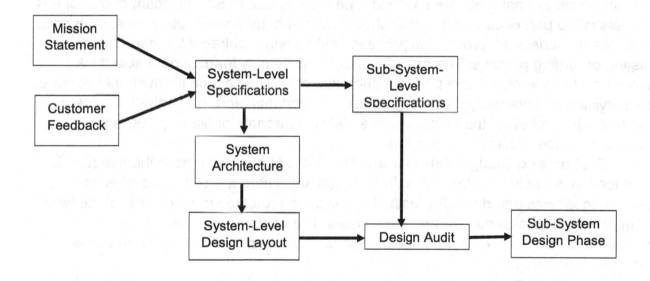

Figure 3.3: Process flow diagram for the system-level design phase

Sub-System Design: After the system design phase is done and a design audit is completed, the system architecture, along with sub-system requirements and system-level design layout, is available for sub-system design and development. For example, once the architecture for a four-axis CNC lathe is completed, the envelope for the index system cross-slide or tailstock could be taken up. Since the sub-system requirements for an index system are available once the functional decomposition from the system is completed, the boundary or interface for the index system could be taken up independently of other sub-systems and overall system design. It is a top-down or systemic approach for design. During this phase, architecture and sub-system design is taken up to satisfy the requirements of the sub-system

Detail or Component-Level Design: The component-level design starts with the requirement for the components in any sub-system assembly. At this stage, the complete specifications or requirements of the component of the sub-system are clearly known. The engineer has to assign the materials, geometry, and tolerances of all the unique components that the sub-system consists of. Also, parts can be used from previous designs as long as they satisfy the new requirements. In the case of new parts, the manufacturing and quality team assesses the capabilities required to produce the part. They also have to decide whether the part should be purchased from outside vendors. Since the cost constraints for the machine have to be satisfied, cost implications for unique or newly designed parts must be considered in the design stage itself. Since the part volume is not as much as it is for, say, automotive components, simpler component process sheet should be developed, including the quality constraints. The process plan should include machinery required, machining time, inspection requirements, the flow of materials, throughput time, etc. Once the process is established, a make or buy plan can also be established at this stage as long as the cost and quality constraints are satisfied. The primary and most significant output of this process is the part documents, which should consist of geometry, tolerances, materials, heat treatment desired, production process, and tooling required if the parts need casting or forging process. The manufacturing or assembly team should also think about how to assemble these parts for sub-assembly. The service team should consider the servicing and warranty issues with these components, and the analysis engineer has to analyze whether the components satisfy the strength or life requirements for the real-world usage profile of the machine.

Testing and Design Refinement: This is the last phase before the machine is produced in quantity for introduction into the market. There are two types of tests: performance tests and durability tests. The machine prototype must consist of the latest components, sub-assemblies, and assemblies. The components have to be manufactured as per the desired specifications and assembled as per the documented assembly procedures.

These assemblies and components can be called "production intent," and the machine can be called "design intent", which will be repeated again and again for selling. In the machine tool industry, very few prototypes are produced, one or at best two. This functional prototype has to be extensively tested for durability and performance evaluations. For example, the performance evaluations for a CNC lathe could be as follows:

- Position accuracy test
- Repeatability test
- Spindle acceleration and deceleration tests
- Cross-slide acceleration and deceleration tests
- Spindle and tailstock alignment test
- Spindle deflection test
- Slide friction torque test
- Acceleration and deceleration torque for slides
- Tool positioning accuracy test
- Index repeatability test
- Control heat generation test
- Thermal mapping test
- OD cutting Test
- Face taper test
- Axes taper test

The durability tests for the machine could be as follows:

- Indexing unit test
- Hydraulic unit test under running conditions
- Machine continuous cycling/durability test
- Electrical cabinet heat generation test

Actually, the nature and extent of such tests depend on the complexity of the machine. Moreover, these tests are customized for the type of machine, as will be discussed in the reliability test chapter of this book later on. In some cases, I have seen this design-intent machine put under real-life cutting conditions in the customer's shop, producing parts on a continuous basis. In this case, the prototype is extensively evaluated and tested by the customer, who gives feedback on the performance and reliability of the machine. In case the manufacturer has an in-house manufacturing facility, it could be a good idea to have the prototype manufacture the parts of the machine. Another important outcome of such tests is to find the non-conformance of the machine behavior regarding the required or desired customer requirements. Some assembly is designed or modified accordingly to change the output to satisfy customer requirements. Documentation during this phase is also very crucial for development.

Once the design changes are incorporated, finalized, and documented, production units are assembled during production Again, the system has to be evaluated as a whole, and system changes have to be in place before production runs are taken up. During prototype production and testing, training has to be provided to the assembly technicians, manufacturing personnel, and inspection personnel. All the stakeholders on the floor have to be trained, and their feedback about the machine, assembly, and inspection have to be documented and incorporated if warranted. Once the system and internal stakeholders have been tuned to near perfection, the production process has to start. It is also necessary at this stage to run an economic analysis on the final machine to determine its economic viability.

The machine development process described above is somewhat generic, and as such, it could be applied to any machine. Such a process is normally customized by the company depending on its manufacturing capability, intellectual capability, type of development required, and resource availability. This generic process is particularly suitable for the market-pull situation, where the manufacturer develops a product based on customer requirements and incorporating the best available technology at the time of development. For example, digital servo control was adopted by almost all manufacturers as soon as customers demanded such technology. In such cases, the company begins with the niche market opportunity and then adapts the available technology to satisfy the customer at an affordable cost. Machine development is a mixture of market pull and technology push. Due to the low volume of machine requirements in the marketplace, manufacturers always incorporate the best technology available to get ahead of the competition. The resultant product must satisfy the customer at an affordable cost. For example, the adaptation of digital technology has been very efficient in all types of machinery by all manufacturers to generate a higher rate of production.

The manufacturer should combine market requirements and the technology available to efficiently create products that satisfy customers irrespective of their type: Since the market is very small and specific for machine tools, competition is very intensive. This combination of market requirement and technology adaptation should become a survival tool for the machine tool industry. In some cases, the design and manufacturing processes might depend on a certain type of product. For technology-push products, more focus can be given to adaptation of available technology and less to process orientation. The competition is less for technology-push products, whereas for market-pull products, it is very intense and, in order to make a profit, the process might be more important and cost efficiency a must.

Some manufacturers would like to use a common platform for their products. Lathe manufacturers, for example, would like to use the same headstock for all the lathes they produce. Future products are built around the existing technology or design that they have developed. For example, spindle-bearing arrangements and headstock could be the same for all the variants of the lathes. Similarly, tailstock design could be the same while mounting or the boundary interface is changed to adapt the standard design to several configurations. For machining centers, the design and architecture of tool changers could be similar or almost the same. This type of assembly sharing reduces the necessary investment and the cost of the machine.

More platform sharing or part commonality is strongly recommended to enhance profit. In the case of platform-based products, technology remains the same, which turns out to be investment amortization across the products. In some sense, platform-based products are similar to technology-push products since all products share the same technology. For CNC lathes, live tool/indexing units could remain the same for high- or low-horsepower machines. Design and development time also get substantially reduced. Since the assembly has been used before and incorporates proven technology, reliability and serviceability of the unit also get better, and the customer has to store fewer parts for all of their similar machines. Products based on a common technology platform are quicker and simpler to develop, and this practice reduces the time from the drawing board to market to a great extent. Customers also get a benefit from such a practice since the cost of the product will be less than if it were designed starting from scratch. This is another predominant benefit of using system engineering principles for designing the system, since the design becomes independent of the designer's choices. Since requirements supported by the assembly remain the same, this assembly could be used anywhere across the product line as long as the requirements remain the same. For example, an indexing mechanism for static and live tooling could remain the same, and design tool heads could be the same for all turret lathes.

Examples of customized products could extend from lathes all the way up to flexible machining centers. General-purpose machines (GPM) normally share the same technology, but special purpose machines (SPM) are manufactured mostly per the customer's specific requirements. Such a design and development process should be highly structured and use desired design variables, materials, and dimensions. The manufacturer has to pay special attention to developing the customer's need matrix and eventually design target requirements. In such cases, platform-based assemblies could be used with minimum design changes. The other requirements could be satisfied when they are designed around the existing technology or design. It has been noticed that special machine builders have stayed away from developing general-purpose machines. Actually, SPM builders could use their developed technology to generalize their machines and promote them as GPMs. An SPM builder could migrate to a GPM builder much more easily and could be highly successful in a very short time. System engineering principles could become more important to SPM.

3.3 Machine Development Organization

In general, machine tool builders have very loose organizations, except for a few who are very large and make several products simultaneously. Nevertheless, a company has to have an efficient organizational structure, especially for product development. There are several types of organizations that can be built for product design and development. When survival is at stake, the organizational structure might not be a factor in earning more and surviving. In the long run, a proper organizational structure is necessary to be efficient and productive.

The product development organization has to have the personnel organized in such a way that their bonds get stronger while they are not restricted from thinking freely. The organizational structure should promote an environment conducive to creativity and productivity. The connection among product development people could be formal or informal and should clearly dictate reporting relationships, roles and responsibilities, and financial implications. The organization also should promote efficient resource allocation, performance evaluations, and budgetary regulations.

For a machine tool organization, individuals can be grouped together according to their expertise or function they perform. This type of organization is called a function-based organization. The other type is called a project-based organization, where the engineers and supporting staff are put together for specific projects. The functions of a machine tool development department can be electrical and electronic, mechanical, casting and forging, design and drafting, analysis, reliability, hydraulic, inspection, system integration, marketing and manufacturing, quality control, and purchasing. For example, SPM builders tend to have a project-based organization, and GPM builders have functional organizations. In a functional organization, experts with specific knowledge are called from the pool as the development team needs them, and the team consists of engineers from several functions. In a project-based organization, the team works together from start to finish of the project.

In a functional-based development team, each function will have similar professional people and budget of its own. For example, control engineers will consist of all the control engineers doing software and control functions of a machine at the same time. Similar functions for hydraulics, mechanical design, and manufacturing would be put together. This type of organization is not very suitable for machine tool builders due to budgetary restrictions. In the project organization, the development team would consist of engineers from several functions. Each group is responsible for finishing a separate project altogether. Each such group reports to the project manager, and the project managers report to a chief engineer who heads the engineering stream of the company. Start-up companies or SPM builders often go with such project-based organization.

A functional organization promotes the development of expertise and specialization. An engineer dealing with a specific type of control would only work with it. On the other hand, connection and interaction among different functions, such as the connection between control engineers and mechanical design engineers, could become weak, so the challenge for such an organization is how to bind or integrate these functional engineers together to create synergy. A system integrator could become the liaison among all these functional engineers to promote system engineering principles.

For project organization, resource allocation could be very efficient, and project efficiency and productivity tend to be very high. Since these engineers are focused on their projects and requirements of the projects only, they might build on developing expertise over the long run. They will only have experience doing a particular type of project. The knowledge horizon could be very narrow and limiting. Development of functional expertise among these engineers and sharing the knowledge across the project boundaries could be limiting factors.

Since both types of such organizations have pros and cons, as mentioned above, it is better for a machine tool company to have a hybrid type of organization where functional managers and project managers are cross-linked for machine tool development. In such cases, the company could build on the expertise required to develop either SPM or GPM as and when required. The schematic of such an organization is shown in Figure 3.4. Since available resources are very restrictive for a small or medium-size company, such a hybrid organization would be very efficient and productive as against the project based or function-based organization. Team individuals are compared to others. This is what I have witnessed in one of the Japanese machine tool builders. Such a team could be called a cross-functional project team, a product development team, or a hybrid PD Team.

The hybrid project development team promotes coordination and management of projects and simultaneously fosters specialization and skill development. System integration becomes much easier. In such cases, the organization tends to become larger, and investment also tends to be higher than in other types of organizations, though it also sometimes gets more difficult to balance resources among functions and projects. It is very difficult to choose a particular type of organizational structure for any company. It depends on the nature and size of the company, company strategy, the market the company serves, and the type of products made, to name a few.

Functional organizations are particularly suitable for GPM builders, whereas project organization is found to be suitable for SPM builders due to the nature of the business the company is dealing with. Hybrid, or matrix, organizations stand in between and tend to be highly suitable for system development. For hybrid organizations, integrating knowledge among professionals over a long period of time is an issue. Also, knowledge sharing across professionals working on different projects could be an issue in the long run.

There are several factors to be considered when selecting the type of organization. First, there is the nature of the machine the company is trying to build. If the machine is of very complex nature, such as a flexible machine system (FMS), where you need expertise from all the disciplines, a project-based organization could be more suitable. If the company is a start-up or a new venture company, a project-based development team could be more appropriate due to limited resources. If the company is dealing with SPM, GPM, and FMMS, a hybrid organization is definitely more suitable.

Proper selection of an organizational structure depends on company strategy and policies, company resources, product width and depth, intellectual resources available, market niche, longevity of the product, and other factors. Most companies in machine tool industries have a very loose organizational structure. If the company has a wide product range, such as Okuma or Mori Seiki in Japan, it is better to have a hybrid organizational structure. On the other hand, a specialized-product-based company, such as Heald Internal Grinding Company, should have a functionally oriented organizational structure. In the case of such a company, functional integration is not an important factor, and it should depend on developing skill depth in a specialized product.

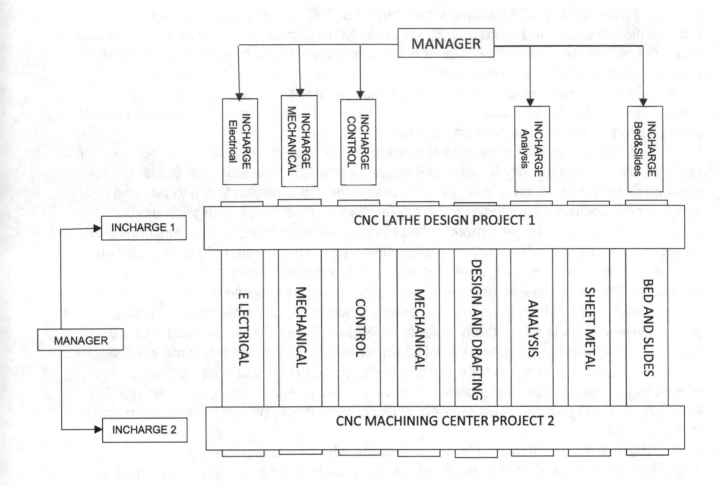

Figure 3.4: Typical hybrid organization for a machine tool company

Another factor is the market niche that the company is trying to serve. If the company is trying to serve a wide range of customers in the market, it is better to have a hybrid organization. In this case, functional expertise must be blended with resource optimization of the intellectual abilities of the company. The other predominant factor when selecting an organization is the availability of human resources and depth of knowledge of the engineers and technical personnel in the company. If the company cannot afford to have many engineering personnel for various projects concurrently, it is better to have a project-oriented organization. Sometimes the nature of nosiness dictates the type of organization. If the company is in the business of developing FMS and would like to focus on this discipline, a hybrid organization could be better if it addresses several projects in the same timeframe.

In order to reduce time to market or time from concept to market, the efficiency and productivity of each person on the product development team become very important. The organization has to be very thin and lean, delivering the product in the shortest possible time, to be at the leading edge of the market and to stay ahead of the competition. In order to enhance the speed of product development, a functional or project-based structure would be more effective. In such cases, coordination among engineers, transfer of knowledge and information, and sharing responsibilities would be more effective if the structure is functional in nature. Hence, a project-based structure would be more effective. The Warner and Swasey lathe division had a hybrid structure where several projects were being dealt with around the basic CNC lathe. All the projects, custom or conventional, were cross-linked with functional departments to deliver the product. This approach allowed the PD team to develop new and custom products within a very short time. The resource optimization was also obtained because the same expertise was being used by both custom and conventional PD teams.

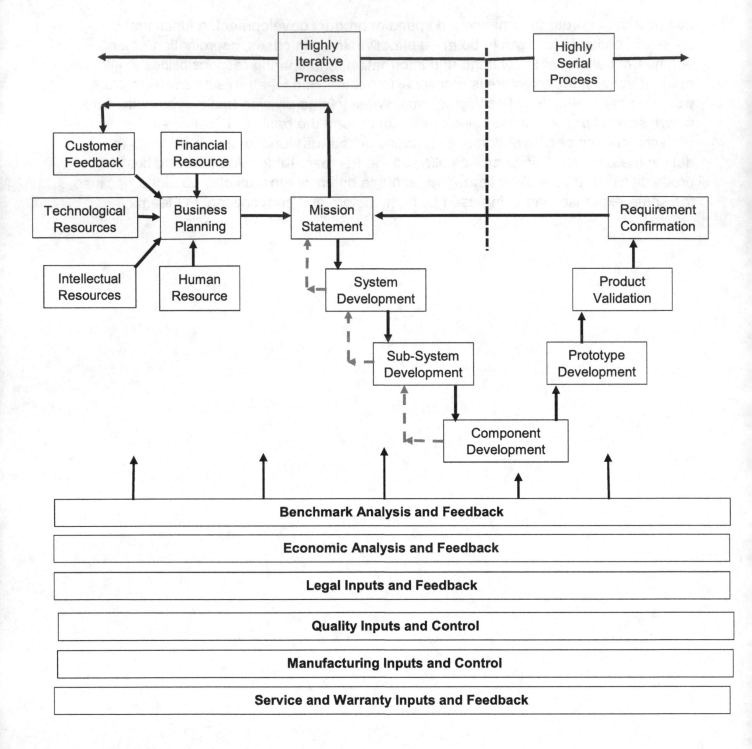

Figure 3.5: Systemic development process layout for machine development

3.4 Summary

For machine tool development, system engineering should be adopted throughout the organization to gain efficiency and productivity. To incorporate the system engineering principle, it is very necessary to think about the PD process and its organization at the same time. PD organization is based on the company strategy in general, whereas the PD process could be systemic. In general, in order to be system oriented, the organizational structure has to be hybrid in nature.

A system-based product development process consists of a series of steps adopted by the PD team to convert customer requirements into a product that customers can afford. This brings in customer satisfaction and is a vital tonic for company growth and financial viability.

A well-defined system-engineering-based PD process helps the company develop a product that ensures product reliability and quality within budgetary constraints. The system-based design also allows the PD team to be iterative in the beginning to include significant customer needs and wants and serial at the end to reduce time and resources required. The system engineering principles also support the continuous improvement process for the product.

In general, the machine tool design and development process consist of nine steps. This system development can be customized as necessary. The process starts with business planning and ends in requirement confirmation. In between, starting from the mission statement, which is the outcome of the business planning, there are seven more critical steps, such as target requirements planning, system development, sub-system development, component development, prototype development, performance validation, and requirements confirmation. At the end of every step, an audit process must be completed and formally documented for future reference. A financial look is also warranted to make sure the budget constraints are satisfied.

The system-engineering-based development process is more suitable for companies dealing with the design and development of several types of products concurrently. Other start-up companies can modify this generic process for their purposes. The effects of types of products on this development process have also been discussed. In the author's opinion, the system engineering development process is somewhat independent of the types of products, whether market pull or technology push. Most of the machine tool products are market-pull products anyway. For product derivatives or legacy-based products, the generic development process can be modified and documented for future reference.

The design and development process can only be followed to the advantage of the firm if and only if the company culture and environments are conducive to system engineering. Moreover, development team personnel must be capable and eager to follow such a process to complete the task. Company management also has to promote this process to thrive and gain market share in a very competitive and highly complicated business world. A proper organizational structure is necessary for a very smooth, efficient, and productive business development. The PD team members have to be linked among each other and to management to create a synergy for development. Roles, responsibilities, reporting relationships, and conflict resolution must be very clearly defined as part of the organizational structure. Each PD team member has to understand the financial constraint and deliverables at each stage of the development process. A proper organizational structure can ensure such mixing and matching in the company.

Rather than a functional organization or project organization, a hybrid organization is recommended for most of the machine tool industries. It has become a very necessary condition for survival to become a general-purpose builder instead of a special-purpose builder. I believe an SPM builder needs to have a hybrid organizational structure, where flexibility is the essence and the GP machine is just another project out of many such possibilities. In the case of start-up companies, either a functional based organization or project-based organization could be sufficient, but as the company grows in size and revenue, it should migrate to a hybrid organization to become more efficient and productive. This will also reduce concept-to-market time and enhance the profitability of the company. Management of personnel for a hybrid organization could be complicated and should also be managed properly and with care.

3.5 Reference and Bibliography

Abernathy, W.J., and Townsend, P.L., 1975, "Technology, Productivity and Process Change," *Technological Forecasting and Social Change* 7(4), 379–96.

Andreasen, M.M. and Hein, L., 1987, *Integrated Product Development*, Springer-Verlag, NY.

Blackburn, J.D., 1991, *Time Based Competition: The Next Battleground in American Manufacturing*, McGraw-Hill Erwin Companies, NY.

DeVenny, T.M., 1987, "Entry and Learning," *Management Science* 33(6), 706–24.

Drucker, P.F., 1991, "Japan: New Strategies for a New Reality," *Wall Street Journal*, October 2.

Eckes, A.E., Stern, P., and Haggart, V.A., US International Trade Commission Commissioners, 1984, Report to United States International Trade Commission on Investigation No. 332-149 Under section 332 of Traffic Act of 1930, "Competitive Assessment of the U.S. Metal Working Machine Tool Industry."

Galbraith, J.R., 1973, *Designing Complex Organizations*, Addison-Wesley, Reading, MA.

Galbraith, J.R., 1994, *Competing with Flexible Lateral Organizations*, Second Edition, Addison-Wesley, Reading, MA.

Hill, T.J., 1999, Manufacturing Strategy: Text and Cases, McGraw-Hill Erwin Companies, NY.

Hopp, W.J. and Spearman, M.L., 2000, *Factory Physics*, second edition, McGraw-Hill Erwin Companies, NY.

Kidder, T., 1981, The Soul of a New Machine, Avon Books, NY.

Krugman, P., 1994, *Peddling Prosperity: Economic Sense and Nonsense in the Age of Diminished Expectations*, W.W. Norton and Company, NY.

Nahmias, S., 2000, *Production and Operations Management*, fourth edition, McGraw-Hill Erwin Companies, NY.

Tooze, A., 2009, *Machine Tools and the International Transfer of Industrial Technology*, "The Global History of Machine Tools, Knowledge, Narratives and Fiction," 30-31, King's College, Cambridge.

Ulrich, T.K. and Eppinger, S.D., 2011, *Product Design and Development*, McGraw-Hill Companies, NY.

Wheelwright, S.C. and Clark, K.B., 1992, *Revolutionizing Product Development: Quantum Leaps in Speed, Efficiency, and Quality*, The Free Press, NY.

3.6 Review Questions

- Is it necessary to explore new technology for a machine tool to be profitable, or should the company adopt existing and proven technology for future products? Does technology adoption support future profitability?

- Is there a commonality between an automobile company and a machine tool development company? Why can an automobile company survive in a very competitive business while a machine tool company cannot? Is an automobile company a functional or hybrid organization?

- What would be the best suitable organizational structure for a highly skilled technical group of personnel working on a single large project? Can a hybrid organization support the personnel skill development necessary to retain highly skilled engineers?

- What would the system diagram for a flexible manufacturing machine tool builder be? Can it be the same as for the GPM or SPM machine tool builder? If not, why not?

C H A P T E R 4

4 Integrated Business Planning for Machines

4.1 Introduction

The ultimate objective of any business is to satisfy its customers by delivering a product that will help them achieve their goals. In short, a business has to satisfy its customers to be productive. The company must understand its customer base and their needs well enough to capture market share. The success of any product or the company hinges on doing the necessary homework before the design starts. The company must do market research, have good intuition, administer quick product development, and control the budget and time of development. The company and its management must start an integrated business plan as mentioned before. Out of this plan comes the mission statement of the product, as discussed before.

Once the mission statement is developed for the product, management must focus on a top-down system engineering development model that starts with the customer requirements, which then must be expanded into a system requirements document. System requirements as envisioned at the outset are called preliminary system specifications. The system architecture, product concepts, system simulation and dynamic modeling, operation analysis, environmental impact analysis, product benchmarking, trade-off analysis, and legal analysis are conducted to refine the preliminary specifications into the product target specification, which the product has to deliver at the end.

The product architecture and final concept must be driven by target specifications. Even if the product is the brainchild of a genius idea or innovative technology, a formal business plan and systemic approach are necessary for the overall success of the product in the market. Product development without integrated business planning might lead to a good product that is not profitable in the long run or does not satisfy customers in the desired market for a continued period. In many cases, a product without proper business planning often dies a premature death and is not sustainable in the future.

The business or product plan must be at the forefront of product development initiatives. In the area of machine tools, some products, such as a template copying lathe, do come out of innovative scientific principles or highly creative minds. In such cases, the product does show signs of excellence in the industry for a short amount of time. These products need to be nurtured by an integrated systemic business plan for further development. The system approach applied to these innovative product concepts can help re-engineer or modify the initial product to accept the new dynamic conditions in the market. In order to change the system for future use, new or modified system architecture with new functionalities and boundaries of operation might be necessary. The elements of previous concepts or architecture could be changed or replaced to achieve the new goals for the revised system. This system could be called a legacy-based product. Once this is carried out, the system engineering principles can also be applied to the revised architecture just as the process would be used for a brand-new system without any legacy. In such cases, instead of a completely top-down approach, the process can start in the middle and continue further, following a system "V" approach as described earlier in this book.

Even if the systemic integrated approach was not followed while developing the concepts in the initial stage or first-time go-round, system engineering practices might help the company to reconstruct or rethink the system according to new requirements of the customer. The point is that a systemic and integrated business plan is required for system development. Most of the time, machine tool development has escaped such an integrated approach, and as a result, few machine tool companies have become successful as developers in the long run. This process of linking customer requirements to the product and company abilities needs a disciplined approach whether the system is big and complex or small and simple.

Common questions are, when should the business planning be started, and how long it should continue? It has to start in the initial phase of the program and should continue until the program is completed, although the nature and intensity of the planning can change depending on the stage of the system engineering process. The type of planning required for collecting customer data will be quite different from the planning required to introduce the product. The type and nature of the business plan could also depend upon the type and complexity of the product being developed. For an innovative product with new technology, planning has to be more involved to reduce the uncertainties, whereas, for a legacy product, the role of planning can be less intensive. The planning can also depend on the complexities of the product and investment required at every stage of development. For a new product, market risk can be a determining factor for planning, and for an extended product, market penetration or gain in market share can be the primary reason for business planning. The main point is that an integrated business plan is strongly recommended to be successful as a business. In between end states of the product development, business planning can also be used as a tool for the continuous improvement of the product from a business standpoint.

Another important factor for management to have an effective integrated business program plan is to understand the strengths and weaknesses of the company itself. Management has to understand themselves, their capabilities, their established business practices, and their customers. In turn, management's responsibility is to convert the inner strengths into product capabilities to satisfy their customers. Moreover, they also have to find ways to improve their weaknesses by improving their business practices. This transformation of company knowledge into product capabilities and functionalities is what makes the company a great one. Customers look for the company to instill their strengths in their business. A proper business plan, in terms of the company's strengths and capabilities, will help the company to create an integrated product development environment. It is also important to keep in mind that any business plan is only as good as the capabilities and knowledge of the people administering the planning program.

For example, Sony's business and market planning for each of the electronic products that they have introduced into the U.S. market could be a remarkable example for machine tool companies. Sony's customers look for the name behind the product and not always its capabilities when they buy them. In other words, they are convinced that the product must be good if it is from Sony. That is the built-in customer confidence that is required for the continued success and profitability of any business. For machine tools, Okuma from Japan has also established itself in a similar fashion. Customers are convinced that Okuma products are good to go from the day the machine is installed in their facilities. In the recent past, George Fischer also had similar fame as a good copying lathe manufacturer. Many machine tool companies come and go, but the companies that follow a good business practice for their products will always win the race in the long run.

4.2 Systemic Business Plan Development

For a very large and complex machine tool development program, an orderly business development program similar to the customer requirements plan is necessary for business success. Similar to requirement planning, we can apply a flow diagram for proper management and execution of the planning part of the business. The top-down system development, called a system "V" approach, is very much accepted in the business community. However, the business planning flow diagram is not that well accepted yet among machine tool businesses. Every business has its process elements that have to put together to be effective. Sometimes customer requirements planning is confused with the business planning process for the product. Product planning, or customer requirement planning, is not business planning in a broader sense. Business planning needs further attention, as described below. A typical program plan is shown in Fig 4.1.

It has also been noticed that business and marketing managers often get too carried away with customer requirements planning, as they tend to become totally responsive to the customers' needs and wants. In the process, managers lose their identity as business managers, and eventually, they start working for the customers only instead of working for their own companies. A forward-looking company will always apply its business practices and processes to improve business parameters on a continuous basis.

Business planning is even more necessary when competition among machine tool builders is very fierce. Currently, competition is more between foreign builders and U.S. builders. Foreign builders have won the game since U.S. builders are not interested in competing on performance, quality, and cost of the machines. The builders from Japan and Korea invaded the U.S. market and basically took over and established themselves as precision machine tool builders. China also participated in this competitive process, but they only had a cost advantage without any accuracy or technology capabilities. Chinese builders were phased out of competition eventually. Such competition actually energized U.S. builders but they decided strategically not to compete. Instead, the machine tool user community decided it was cheaper to import. In order to regain the market, U.S. builders have to start from a clean slate, incorporating system engineering principles and a system-based management approach. They need a massive re-engineering approach to accommodate new or changed conditions.

A fresh look at the systems available is required. U.S. builders have to identify new functionalities and the technology required to achieve functional capabilities. For an existing machine, they can apply a similar procedure to identify system capability gaps and determine the elements of system architectures that must be changed or modified to achieve the new system. Whether a full system process is triggered depends on the system complexities and existing architectures and their interface boundaries. In any case, a systemic approach is necessary for building new machinery or modifying existing machinery. Moreover, whether the system users need such a pragmatic total approach for the development of machines is in order or not, machine builders must take a holistic system approach to regain what has been lost so far to foreign builders. Such a systemic approach will be a fairly complex integration process, linking customer requirements and the machine builder's capabilities, and it needs time and attitude to apply.

As laid out in the system planning diagram in Fig. 4.1, the company first must understand and evaluate its resources, capabilities, existing practices and policies, business attire, and business environment at the outset. The second step is to find how to link these business systems to customer requirements to achieve customer satisfaction. The company has to establish a way to transfer the systemic process to internal customers to achieve productivity and efficiency. Once such a process is established to energize the company's base elements, an ongoing effort of continuous improvements steps has to be in place to monitor and control each program development.

This process provides an integrated business environment that will lead to world-class capability in the product line. The business system integration applied to achieving customer satisfaction or delight is the ultimate goal of any business. In other words, business resources are optimized and geared toward achieving customer satisfaction, which promotes sustainable revenue generation. This is a perpetual self-generated business cycle that should be followed to start a business or modify or renew an existing system. This endless loop of the business process needs work and support from all the team members, external and internal, of the business to be successful in the long run.

Overall Program Plan: The overall program plan as laid out in Fig. 4.1 needs a little bit more discussion to clarify my intentions behind it. It is also true that such an overall program plan is required for a product development plan for efficiency and productivity. As for any other approach, universal acceptance of such a diagram has not been established throughout the industry. Such activities might be carried out in a company without laying them out in detail or documenting the procedure. It is also a fact that a progressive and forward-looking company will always follow a procedure similar to such a systemic process, although it might not be fully followed, as shown in the diagram. Some companies use this document for continuous improvements as a company strategy.

A top-down systemic and structured approach for programs will always ensure consistency, repeatability, and productivity. Similar to product development, the program plan has to have an architecture for consistency and execution. The company should also promote the use of such a holistic approach throughout the company for future use, refinement, and program improvement. In that sense, automobile companies are way more disciplined than machine tool companies. Adherence to a program plan approach will also allow company managers to be more functionally oriented and become experts in their own functions.

It has also been observed that the companies stick to their line of proficiencies and try to produce what they have done before. Most companies try to use their experience and learning curves in their future efforts. They would like to stay within their comfort zone. Companies will rarely try to explore other product lines using marketing efforts. Even in such cases, for new ventures or new products out of the alignment, the company is encouraged to follow a systemic program plan led by the functional managers who have developed the best methods over time due to repeated use of such methods. The company's best methods should be implemented for each program to get better.

The second line of defense for using such an overall program plan, even it is complicated and painstaking, is to understand that a disciplined approach is necessary before a product is developed to compete in the marketplace and satisfy customers. The system program development plan reduces the risk of program development, execution and introduction of the product into the market place. Following a preplanned approach will also allow management to understand the reasons for deviations from the expected and to set necessary steps to bring the program back within desired limits.

Following this overall plan will help management understand schedules, program steps to be followed, roles and responsibilities for the management team, time for the execution of any step, and interaction required for successful implementation of the plan to deliver a product. This step is required ahead of the product development plan. This is also called a grassroots approach toward product development efforts. The fundamental difference between piecemeal program planning and an integrated planning approach is that integrated program planning addresses all the functions of the company concurrently, which brings in customer satisfaction. The integrated business plan calls for an integrated master schedule and deliverables for all the programs of the company.

For a company handling program developments, such as a CNC lathe, manual lathe, copying lathe, or machining centers, each program will also have a program schedule for each product, which will stem from the integrated business management plan. Each individual program will follow the overall standardized program plan and timing with some program-specific modifications, but it will not violate the overall company program plan. In some cases, engineers and managers might like to develop a generic plan for their line of work without any considerations given to other plans running simultaneously. This is not an acceptable practice for the company as a whole.

Such a disjointed planning practice will only bring chaos and inefficiency for each program and should be avoided for all practical purposes.

Companies that build and market a single product but have several product lines under one general product line have to develop a generic business plan, treating each line manager as a machine tool developer. For example, a company dealing with grinding machines might have centerless grinders, surface grinders, universal grinders, and internal grinders as the product lines. Even if all the products do come under a common product type, in this case, grinders, the company can treat each manager of an individual product line as a functional head and develop an overall business plan for the company that they can follow as a common tool for planning their activities. For a company with one product, an overall business plan must be drawn to manage the resources very productively and efficiently for furthering profit for the company. A similar practice can be repeated for individual future projects to refine and optimize them for future benefits.

To summarize what has been discussed in previous sections, any product development effort should start with a properly documented business program plan. In such a process, it is fundamental to understand and establish the link between program requirements and customer requirements. Program design and development stems from business planning initiation. The program design, irrespective of nature and complexitles, should clearly define the system target specifications, work decomposition structure (WDS), statement of work (SOW), product planning and management (PMM), and product scheduling and milestones (PSM). The fundamental premise of this step is to deliver what the customer wants in the first place. The business planning, product planning, and product development steps are the in-between steps required to satisfy the customers' requirements. This is the simple but most important message that U.S. builders neglected to address.

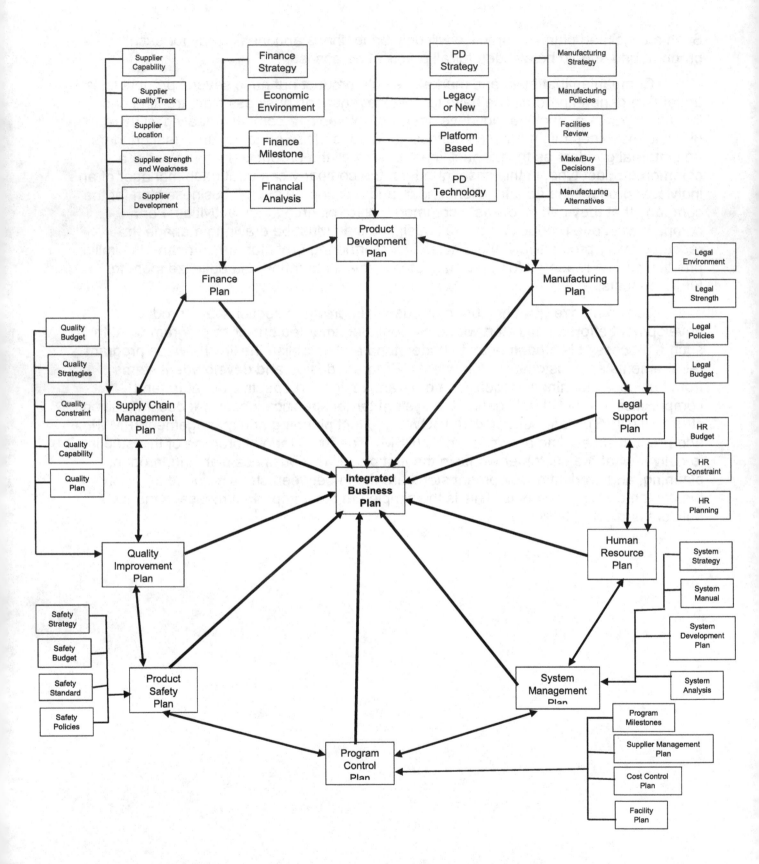

Fig 4.1: A typical program plan for machine tool management

4.3 Integrated System Management Review

This systemic product management process requires developing some fundamental documents for the program through which the product will be created eventually. The relationship between these documents is shown in Fig. 4.2 below. In order to reduce the time and resource requirements for a product development program, the following eight primary documents need to be prepared:

- Mission Statement Document (MSD)
- Customer Requirements Document (CRD)
- System Specification Document (SSD)
- Work Decomposition Document (WDSD)
- Statement of Work Document (SOWD)
- Integrated Company-Wide Product Master Plan Document (IPMPD)
- Integrated Product Master Schedule (IPMSD)
- Product-Specific Schedule (SPSD)

The mission statement document (MSD), the customer requirements document (CRD), and the system specification document have already explained. The other documents will be described briefly in this chapter. These documents have also been explained in detail in MIL-STD-499A. Some modifications have been done to fit the program requirements for machine tool development. In addition to the preparation for this document, some programs might need to use PERT/CPM, Gantt chart, and time optimization methods to optimize the program resources.

The mission statement document and customer requirement document are driven mostly by customer requirements and company policy requirements. The system specification document is generated from the MSD and CRD. A statement of work (SOW) normally should describe the work details that need to be accomplished for the machine/product. The system requirements give rise to possible system-level architecture, which will lead to the generation of a concept of operation and work decomposition document (WDD). Once the concept and its architecture are finalized, the data is fed into the company-wide master plan document (IPMPD). The IPMPD is then customized for generating the product master schedule (IPMSD) and product-specific schedule (SPSD), which are then handed over to program-specific team members to carry out the product development activities. The master documents give an idea of all the project details and schedule for the whole company. The IPMPD, IPMSD, and SPSD details are required to manage and control all the programs concurrently and effectively.

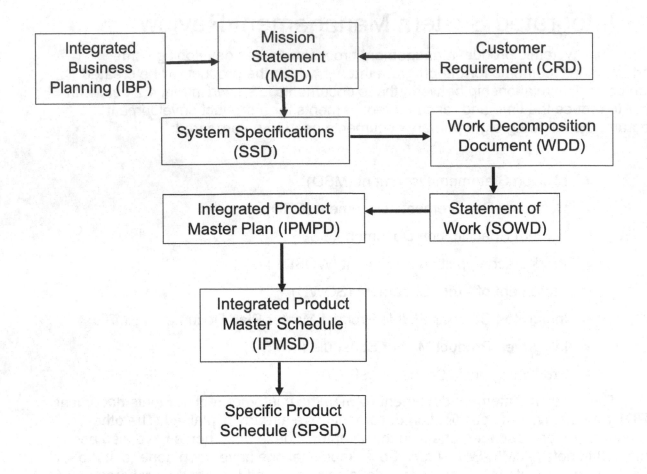

Fig 4.2: Product planning document recommendations

Work Decomposition Document Development (WDD): Basically, the complete work is decomposed into several manageable sections/chunks so that each team can handle each section and all the work can be integrated into the program. Each section must explain the task details, deliverables, and responsibilities in a hierarchical function. It is a flow diagram for the complete program work to be completed. Each section is divided into several other chunks that team members have to accomplish to finish the program as a whole. In other words, WDD outlines the total work for the program. It details the top-level deliverables, which are further decomposed into sub-chunk deliverables. A typical example of WDD is shown in Figure 4.3 for VNC lathe development.

The decomposition is carried out until the task can be managed and delivered by a single team member. Each functional manager can be responsible for a chunk and can divide the chunk into sub-chunks, and this division of work is continued until all the work is handled by the team members.

It is a top-down approach, where holistic tasks are divided into sub-tasks, and when all the tasks are completed and integrated together, the program work is completed in totality. Such a document links work decomposition structure and organization breakdown structure into a pictorial form.

In such documents, each task could have work percentages and budget allotted to finish the work within a standard period of man-hours. Once all the costs and time elements are integrated together, the resultant data shows the budgeted cost and time allotted for complete work. This structure could also be used to understand the variation or deviation of actuals from standard time and cost allotted for the project. This total cost could also be termed "burden center cost."

This WDD has numerous benefits if followed strictly for a complex machine development program. First of all, it defines the organizational structure, program work details, budget, and time allotted for each task. Eventually, budget for time and cost can be determined for each functional department. In case of overruns during the program executions, a quick estimate can be obtained at any point in time as to what and where it went wrong. Subsequently, the budget can be reallocated to fix the deviations. So, the WDD can also be used as a control document as the program is executed.

The work decomposition document can also be used to identify the program risk with respect to program deliverables, time, and cost. If the total cost of the project is insurmountable as laid out in the WDD, tasks can be revisited and redefined to determine the modified time and budget that the company can afford to spend. The total time allotted for the complete program can also be compared to the concept-to-market timing required to be competitive in the market.

This document can also be used as a common ground to establish a communication link for all the functional managers to execute the program within the stipulated budget and time. Moreover, functional managers can use this document to determine the human resource requirements for finishing the task. A productive organizational structure can be an outcome of this document if it is followed and interpreted properly.

Last but not least, the WDD can be used to identify the root cause of program deviation with respect to time or cost. Also, the WDD can pinpoint the region or elements where the program is lagging and the potential causes of such deviation. Corrective steps can then be taken to keep the overall project timing within the desired limit. It can also be used to identify the weakness of the system to deliver the program tasks and any teams that are not delivering what they should. So, this document can be used as a program progress map to identify the green and red areas of the project where it is lacking and needs reinforcements.

Creation of the design and development of the WDD is not a very straightforward path, and input from all the stakeholders of the company is required to have a fruitful document. A brainstorming session among all stakeholders of the program can be arranged to generate this document. Development of the WDD is also very iterative and always subject to arguments. The team approach for developing this document is more productive.

In order to develop a WDD, some general guidelines should be followed to reduce the time and enhance the effectiveness of the process. The general guidelines to create such a document could be as follows:

- Top-level function could be the product as final deliverable of the project
- Sub-functions or deliverables are assigned to each functional manager or team leader
- Each function is decomposed into sub-functions in a hierarchical fashion
- Time, work content, and budget for the functions, sub-functions, and elements are estimated and displayed in the document
- Cost allocation should not be arbitrary and must have legacy data to justify
- Time duration for each task should be as minimal as possible
- Each function should be independent of other parallel functions as much as possible. In case of interference, tasks could be divided into functions as appropriate, depending on functionalities
- Repetition or duplication of work should be avoided across all the functions and sub-functions
- If possible, each task must have a designated task bearer who will be responsible for the task
- Any sub-function requiring a time of two days or less should not be included in the WDD to keep it simple
- Each function level, L1, L2, L3, etc., should state the duration, budget of money and time, and person responsible for that function
- A systemic approach must be followed to create this document

In general, the WDD must cover all the significant functions of the program, and the content must consist of all the functionalities required for the program. The numerical coding system for each function and sub-function can be selected as desired but it must be consistent, as shown in Figure 4.3. The numbering system should be as simple as possible so that interacting team members can remember it easily. The

WDD is a management tool that can be used to control the program, and it should be updated as new functions are added or existing functions are deleted. It is a living document for the company and the program. Sometimes, it can be color-coded to denote the completion, delay, or holding status. The WDD can be viewed as a composite document representing all the system-, sub-system-, and component-level functions. It is also true that the WDD should contain only those details that are required to control the program, such as budget, responsibility, time allotted, and control levels.

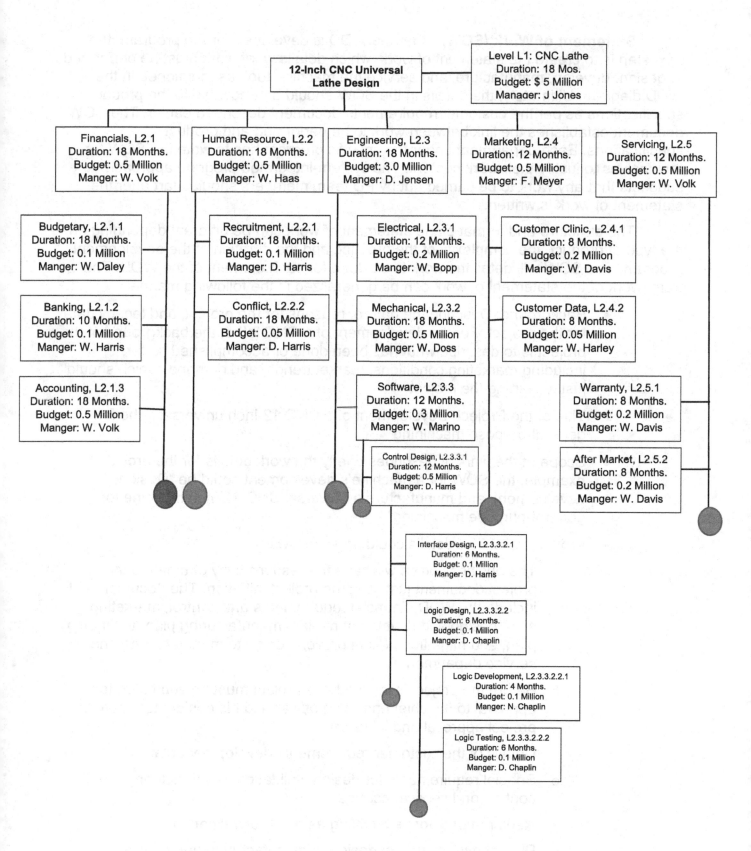

Fig 4.3: Work decomposition document (WDD) example

Statement of Work (SOW): After the WDD is developed for the program, the next step is to create the statement of work, which details what work must be performed to design, develop, manufacture, and service the machine tool, as mentioned in the WDD diagram. Ultimately, the details in the SOW should be traceable to the product specifications as per the customer requirements document described earlier. The SOW document establishes the link between work to be performed and product specifications. Each department under the specific program has to write the SOW for themselves to justify their work content. The system-integrated program approach suggests that any work to be carried out by PD team members should start after the statement of work is written.

The structure and format of the statement of work can be different depending on the type of product, the complexity of the product, and the duration of the development program. The SOW will detail the work to be done for each element of the WDD. The framework of the statement of work can be generalized in the following manner:

- Background: Describe the nature, reasons for the project, and legacy data for the project. For the development of a CNC lathe, the background of the project is to describe what has been done or accomplished so far, including marketing conditions, market trends, and demand, which should justify starting the project

- Title of the Project: Development of a CNC 12-inch universal lathe for general-purpose machining

- Scope of the work: It describes briefly the work details for the project. For example, the SOW for a machine's development could be "Design, development, and manufacture a universal CNC 12-inch machine for general-purpose machining"

- Work Details: The details could be as follows:
  - The business office will have the responsibility of creating the project document justifying the project initiation. The document will include the WDD, financial requirements and control, marketing plan, design and development plan, manufacturing plan, technology plan, and migration of final proven design to manufacturing and service departments
  - Before the project is initiated, the project must be submitted for approval to the chief operating officer and his designated team for project approval and initiation
  - Details of the customer requirements development plan
  - Special requirements for design, validation, manufacturing, quality control, and service facilities
  - Requirements for field testing as a part of validation
  - Plan of migration from design to manufacturing and quality
  - Progress report presentation schedule

- Schedule of the Program and Deliverables: First of all, the goal of the project must be defined. Then the deliverables of the project must be listed to satisfy the customer requirements or needs. It is also necessary to define and determine the stakeholders and the end-users of the project. Once the goal, deliverables, and stakeholders are determined, all the functions must be listed with time sequences in a Gantt chart for further execution. The Gantt chart shows the task against time for the whole duration of the project. The deliverables and their timings must be realistic. The timelines for each function must be scheduled in sequence using the project evaluation and review technique (PERT), if necessary. The PERT will determine the critical path, slack time for each activity, and time of completion for each activity required to finish the project. A typical example of the PERT process will be discussed in the next chapter

- Duration of the Project: From the work decomposition document (WDD), time of completion of the project for each function must be mentioned. The timelines for the functions can be obtained from previous similar tasks completed by the teams

- Facilities and Location: The SOW also should mention the facilities and their locations required to execute the program deliverables. This is needed to estimate the amount of support logistics to be planned for during the program and after the program

- Hierarchical Reporting System, or Chain of Command: The SOW should also list the reporting chain of command for all the functions. It also should describe the nature of the report, the content of the report, and reporting time and frequency

An example of a SOW for developing a typical machine development project is given below. Any company can also use this template if the company decides to use external contractors. It must be mentioned here that the use of external services for design and development and in-house manufacturing services has been very rare in the machine tool industry in general.

If a company does not have much strength in design and development but specializes in manufacturing activities, such a combination could be found more efficient and productive. In such cases, the SOW must be written by the contractor in consultation with the primary company to avoid confusion regarding deliverables, timing, and cost associated with the consulting work. The parent company must also explain very clearly and unambiguously the requirements for the program to the contractor as to their participation. This will allow the team to establish more traceability for customer requirements.

Prior to finalizing the external consulting company's participation in the program, a WDD with program overall functions, budget, work contents, desired schedule, and responsibilities must be in place in an almost a rigid framework, since any change in these parameters might affect interaction and create confusion down the line.

The following SOW example is written for a consulting company that has been asked to submit a quotation for design work for a program.

A Typical Example of a SOW for Machine Tool Consulting Companies

Statement of Work

Parent Company Details

Contracting Company Details

Program Details: CNClathe12

Program Manager: XXXXX

Executive Abstract: The statement of work covers the design and development of a CNC 12-inch chuck universal lathe for general machining for the parent company, XXXXXXX. This is a draft write-up for initial discussion on the project, and a final SOW will be written when requested by the parent company in due course. The program is being initiated to design and develop a CNC lathe as described later and has a total estimated cost of USD XXXXX and an estimate of time of completion within 18 months from the start of the project. For details in price, delivery, and schedule, consult Section V of this report as mentioned in the Table of Contents on page XX.

Table of Contents

1. Scope and Disclaimer:

The consulting work with regard to design, development, and analysis services will consist of specific deliverables described in this report. The work detailed herein will represent the full scope of this proposal. Further work on this project might point toward modifications of the deliverables and schedule originally agreed to. The nature and extent of modifications required to fulfill the requirements might need further negotiations and agreement.

2. Introduction and Work Summary

The contractor, XXXX, will help the parent company, XXXXX, with design, development, and analysis and might extend to developing the first prototype in-house for the product XXXX. The deliverables will include only the main development, design, and analysis of the product final concept. The optional work, if any, will have to be considered separately from this work. The present work only consists of the main product, and it will not include any variants of this main product.

The principal component of this work will be to conceptualize and finalize the design to satisfy the requirements provided by the parent company. This will also include, development of four possible concepts, trade-off analysis, and architecture developments and will analyze the concepts to determine the best possible concept satisfying the requirements. This will also include design and detailing the components that will be used to manufacture and develop the prototype.

The overall goal of this contract work is to create a working prototype in the parent company's facility, ####, within 18 months from the day the contract is awarded and payment is made as per the agreed-upon schedule. The quotation does not include any work on software at this point in time, but this might be considered afterward.

The milestones and deliverables at each milestone will be controlled and certified before the next phase is taken up. There will be three such milestones: preliminary design review (PDR), interim design audit (IDA), and critical design review (CDR). The prototype building cost will be provided by the parent company after the final acceptance of the design.

3. Client-Supplied Specifications and Materials:

3.1 Requirements Specification Details: Parent company has to provide the original and subsequent changes, if any.

3.2 Material: Only for building the prototype once approved by the parent company.

4. Proposal for the Work: The work details will be primarily guided by the WDD. The delivery at each milestone is described below:

4.1 Services to be Provided:

- Milestone 1: Requirements analysis, architecture development, concept selection
    - Benchmark analysis with competition
    - Legal requirements analysis
    - System requirements
    - Sub-system and component requirements

- o Three concept designs, minimum
- o Trade-off analysis
- o Concept selection
- o Comparative analysis
- o Final concept approval and discussion
- Milestone 2: Preliminary design review
  - o Concept selection presentation
  - o System and sub-system design
  - o System layout
  - o Detail layout for system and sub-system
  - o Preliminary design analysis
- Milestone 3: Detail design review
  - o Final optimized design
  - o Mechanical and electrical system details
  - o Tolerance analysis
  - o Manufacturing details
  - o Assembly concepts
- Milestone 4: Critical design review
  - o In-depth design analysis and review
  - o Concept and design approval
  - o Cost analysis
  - o Quality and reliability analysis
  - o Cost containment
  - o Project timing details
  - o Marketing plan for introduction
- Milestone 5: Initial build and prototyping details
  - o Prototype build plan
  - o Prototype inspection plan
  - o Prototyping manufacturing plan
  - o Prototype testing plan
  - o Prototype introduction plan

- Milestone 6: Production build and manufacturing plan
  - Jig and fixture plan
  - Assembly plan
  - Break-even analysis
  - Supplier development
  - Gages and instrumentation
  - Training and documentation
  - Servicing product plan

4.2 Reports: All the documentation of design, layout, analysis, and testing must be submitted for approval. Each design phase documentation must be submitted for company approval. The prototype build plan must be approved before the first build.

4.3 Revisions and Updates: After the contract is awarded with down payment, any change in requirements, WDD, and SOW must be approved by the contractor before execution and further reconsiderations. Any associated change in cost must also be discussed and approved by both companies.

4.4 Meeting and Discussion Schedule: Contractor must meet company representatives on a regular basis at least twice a month.

5. **Deliverables, Pricing, and Schedule:** The quoted values for each phase are based on the requirements given in Appendix A

   - **Phase 1:** Deliverable: Report; Price: ####.##; Schedule: X-X after receiving order

   - **Phase 2:** Detailed Design: Deliverable: Report; Price: ####.##; Schedule: X-X after receiving order

   - **Phase 3:** Prototype Build Activities: Deliverable: Report and prototype; Price: ####.##; Schedule: X-X after receiving order

   - **Phase 4:** Production Build Activities: Deliverable: Report; Price: ####.##; Schedule: X-X after receiving order

6. **Optional Design Items:**

   - **Option 1:** Servo index instead of mechanical indexing: Cost: ####. ##; Time for delivery: Three months after order

   - **Option 2:** Hydraulic index instead of mechanical indexing: Cost: ####. ##; Time for delivery: Three months after order

7. **Payment Schedule, Terms, and Conditions:**

7.1   **Payment Schedule:** Initial payment:10% with contract award; Milestone 1: 35%; Milestone 3: 35%; 20% after CDR report is submitted

7.2   **Payment Terms:** Net 15 days

8. **Warranty Services and Redesign Efforts:**

8.1 **Warranty Services:** All the contract work and consulting services will comply in all respects with the requirements specifications and written guidelines provided by the parent company. This warranty is valid only for 90 days after the final document is submitted for approval. This warranty does not cover any loss of report and other documents by the personnel of the parent company. The contract company has no responsibility and liability after the contract period is over.

8.2 **Return of Product:** In case the submitted report, design work, or analysis does not conform to the requirements or guidelines provided by the parent company, the contractor will replace the product or services free of cost to the parent company. If not satisfied, the contractor will return the money within 30 days after the dispute is settled.

9. **Client Responsibilities:** Client agrees to inform the contractor about any changes that might affect specifications of the designs or any deviations from the original SOW and shall work with the contractor's personnel for product design, validation, and verifications. Any product questions must be answered within a reasonable time so that the contractor's efforts are not affected by unnecessary delays. If any option is desired by the parent company, add that in the purchase order with notification to the contractor.

10. **Terms and Conditions:** If the SOW is found acceptable and agreed upon, work will start with immediate effect. If the contract is taken off, a penalty of 25% of the purchase value has to be paid within 30 days after withdrawal of the contract.

11. **Concluding Remarks:** The contractor will work with the parent company in a confidential manner and will work with dignity and honesty with the parent company in all possible ways.

The statement of work is very much generic by nature, and this format could be modified to suit the nature and complexities of the project and the company. In general, sometimes any functional department can take the help of an outside consulting company. In that case, the subject functional department can write their own statement of work specific to their requirements. Nevertheless, the elements of the SOW can be the same, and the end output should relate to the requirements of the product. The purpose of the SOW remains the same.

As mentioned before, a system engineering department can take charge of the integrated system management activities, including coordinating the SOW and other integration activities for the program as a whole. Since functions and deliverables of each department can be substantially different, an integrating department such as a system engineering department might make sense to integrate the whole program in addition to making sure that the end products satisfy the customer requirements. Under the banner of the system engineering department, functional deliverables of manufacturing, quality, design, reliability, testing, operational and simulation analysis, and finance can be integrated together to deliver what the customer wants. This will definitely create a synergy in the process. To manage such an enormous coordinating and managing task, the responsible system engineering manager can put together a task vs. department matrix to keep track of coordinating activities, as shown in Fig. 4.5.

Phase	Task Details	End Delivery Date	Functional Department
Requirements Review (!)	• Customer Clinic, 1.1 • Customer Selection, 1.2	##-##-####	Marketing, Primary
Preliminary Design Review (2)	• System Review, 2.1 • Sub-System Review, 2.2 • Product Architecture, 2.3	##-##-####	Design (Primary), Marketing, Manufacturing
Critical Design Review (3)	• Design Detailing, 3.1 • Component Detailing, 3.2	##-##-####	Design (Primary), Manufacturing, Quality
Design Audit (4)	• Manufacturing Audit, 4.1 • Cost Analysis, 4.2	##-##-####	Quality Audit, Design, Finance, Manufacturing, Management

Fig 4.5: Task distribution and control matrix

Integrated Product Master Schedule (IPMSD): By this point in time, we have gotten familiar with the mission statement document (MSD), customer requirements document (CRD), system specification document (SSD), work decomposition document (WDD), and statement of work (SOW). The next step for the integrated program approach is to put together the integrated companywide product master plan and product master schedule. It has to be kept in mind that these are program control documents and they are primarily required to deliver customer wants using minimum resources and efforts. In general, it is important to know each major and minor task and the responsibility of each task.

Some companies do put together a master plan and schedule for all their product development efforts when many such efforts are taken up simultaneously. For one product at a time, a master plan is not that important, but a company wish list for all the future products must be put together. All products are not created equal and do not need a detailed plan. For further explanation of integrated master plans for products and schedules, you can consult MIL-STD-499A. This U.S. military standard assumes every program follows system engineering principles. The IPMPD and IPMSD are basically expansions of the WDSD and SOWD, giving details on tasks, roles and responsibilities, timing, cost, etc.

Such critical documents are also necessary to reduce the concept-to-market timing, and this is necessary since other competitors are also working on similar products at the same time. To gain a competitive advantage over competitors, product introduction must be made as early as possible to capture the desired niche of the market. The marketing department should have a good idea of the timing for competitors in the market. This date defines a point in time when the program must complete the program in all the aspects and introduce the product into the market for the desired customers. Once such date is known, the system integrators, in consultation with the marketing function, can put together a schedule for the product, and adjustment of the original WDSD, SOW, and schedule might be required. So, this customer activity, if known, acts as program control feedback, making it a closed-loop program control. Such a dynamic change in the production schedule is often necessary to be competitive and to stay ahead of the competition.

Hence, the modified customer requirements, work schedules, and responsible departments might affect the product design and architecture in extreme cases. To enhance the management potential for these integrated tasks, the integrated system management and control documents require the planner to refocus on the activities and their timing to steer the program to success. This is a very critical responsibility of the system integrator.

All of these, when linked up together for any project, will create a system of documents, manual or computerized, that can be used to manage any project in a systemic way, which provides implementation guidance to the team members doing the functional tasks and management folks health-monitoring tools to control the program. Integrated management planning relates or links customer requirements to company activities to deliver a suitable product to satisfy intended customers at a cost that they can afford. This link establishes the work responsibilities of each team member. Such an integrated program approach might also dictate the organizational structure of a company.

The company that is able to integrate all of this information into a program system document will also have a distinct advantage in concurrent engineering practices, which needs a very effective communication link. The series of timely updated program documents eliminates guessing by team members as the program matures. Direct use of planning database content will facilitate system functioning, giving rise to a productive and efficient product. It will also speed up planning and information access processes to a large extent. Eventually, it is possible to change the document, if necessary, in accordance with an effective continuously monitored process, which can also be used to trace and monitor the program at any point in time. Such a document enhances the program traceability process, ensures completeness, and avoids omissions, duplications, and redundancies.

4.4 Summary

- The product architecture and final concept must be driven by the target specifications. Even if the product is the brainchild of a genius idea or innovative technology, a formal business plan and systemic approach are necessary for the overall success of the product in the market. Product development without integrated business planning might lead to a good product that is not profitable in the long run or does not satisfy customers in the desired market for a continued period. In many cases, a product without proper business planning often dies a premature death and is not sustainable in the future.

- Some companies do put together a master plan and schedule for all their product development efforts when many such efforts are taken up simultaneously. For one product at a time, a master plan is not that important, but a company wish list for all future products must be put together. Do recognize the fact that not all products are created equal, and a detailed plan for each is not always needed.

- Business planning is even more required when the competition among machine tool builders is very fierce. Currently, competition is more between foreign builders and U.S. builders. The foreign builders have won the game since U.S. builders are not interested in competing on performance, quality, and cost of the machines. The builders from Japan and Korea started invading the U.S. market and basically took over and established themselves as precision machine tool builders. China also participated in this competitive process, but they only had a cost advantage without any accuracy or technology capabilities. Chinese builders were phased out of the competition eventually.

- In general, the WDD document must cover all the significant functions of the program. The content must consist of all the functionalities required. The numerical coding system for each function and sub-function could be selected as desired but must be consistent, as shown in Figure 4.3. The numbering system should be as simple as possible so that interacting team members can remember them easily. The WDD is a management tool that can be used to control the program, and it should be updated as new functions are added or existing ones are deleted.

4.5 References and Bibliographies

Chapman, W.L., Bahill, T.A., Wymore, W.A., 1992, *Engineering Modeling and Design*, CRC Press, Boca Raton, FL.

Eckes, A.E., Stern, P., and Haggart, V.A., US International Trade Commission Commissioners, 1984, Report to United States International Trade Commission on Investigation No. 332-149 Under section 332 of Traffic Act of 1930, "Competitive Assessment of the U.S. Metal Working Machine Tool Industry."

Grady, J.O., 1995, *System Engineering Planning and Enterprise Identity*, CRC Press, Boca Raton, FL.

Grady, J.O., 1994, *System Integration*, CRC Press, Boca Raton, FL.

Hill, T.J., 1999, *Manufacturing Strategy: Text and Cases*, McGraw-Hill Erwin Companies, NY.

Kidder, T., 1981, *The Soul of a New Machine*, Avon Books, NY.

Tooze, A., 2009, *Machine Tools and the International Transfer of Industrial Technology*, "The Global History of Machine Tools, Knowledge, Narratives and Fiction," 30-31, King's College, Cambridge.

4.6 Review Questions

- What are the steps to be taken for integrated management approaches to a legacy-based product?

- What are the economic advantages and disadvantages of following an integrated systemic approach? Is it cost effective to follow this approach?

- What are the advantages and disadvantages of the systemic approach in light of continuous improvement of systems? Can we use this approach to improve the business planning process?

- For a company with multiple products at multiple locations, is this approach very effective for control of concurrently running projects? If not, why not? Is it cost effective for a company developing a single product at one time?

C H A P T E R 5

5 Customer Requirements

5.1 Introduction

I remember very clearly the day when the Warner and Swasey Company decided to start the design and development of a new series of CNC lathes for the future of the company. I had to ask myself one question: is it necessary for the market? Eventually, the company spent most of its remaining financial resources designing and developing this product. Its design was based on what our chief engineer wanted. and the specifications were summed up when the product design and development was almost over. No wonder the subject company with a background in manufacturing machine tools for more than 70 years is bankrupt today, while all the Japanese machine tool companies are thriving. Knowing what I know now, it was basically management's fault that they did not know what to design and who would be their future customers. I am sure there are many such examples in every sphere of business, where products are developed assuming customers will like them when launched. Obviously, this is not what we want machine tool builders to do if we want them to stay in business for a long time to come.

Hence, such examples justify collecting data from the customers about what they want and what will satisfy them. In order to be successful, customer needs and wants must be evaluated and responded to. The present chapter highlights some futuristic steps as to how to collect the data and incorporate it into the product to be designed and developed. This particular step needs the help of sales and marketing department personnel and associates before the product design thoughts are in place. Moreover, customer requirement data is somewhat difficult to obtain since customer answers from the feedback are often not very clear or specific. The team needs to translate the feedback into a set of concrete sets of requirements that can be designed into the product.

There are several reasons why this step is very crucial for a successful product:

- Make sure that the product is a true representation of what customers want
- Make sure customer feedback is clearly and well understood
- Make sure a wide set of customers in the desired market are considered for the feedback sessions

- Make sure that the product specifications are true representations of what customers need in the future

- Make sure feedback is analyzed properly, taking into account negative feedback about current products

- Make sure the customers become true stakeholders in the process of designing and developing the product

- Make sure customers are treated as designers of the product who will buy, use, and be happy with it

- Make sure every member of the PD team well understands the customer requirements in every possible fashion

- Make sure the feedback and process are well documented for future and present reference

The primary purpose of this step is to link the customer with the product. This is another way of allowing the customers to become stakeholders in and designers of the product. Unless the PD team members get a chance to build this link or create this bond between the product and the customers, mistakes will be made in designing a product that the customers would like to have. Anybody can design a product in their own way, but the question is whether the company can make money or profit selling the product to its customers. A successful product is what makes a customer happy and satisfied. The PD team members have to treat this set of documents as a bible for their product, and the end product must satisfy these requirements; otherwise, the product has failed to deliver what customers want.

As mentioned in the product development process chapter, customer requirements are a key element of system engineering product development. The subsequent process steps, such as system requirement evaluation, sub-system requirement development, and component requirement selection, depend heavily on customer requirements. Even the concept development and system design, or product architecture development, depends on the evaluation of customer requirements. The concept development of a system starts with its requirements, which come from the customers.

Moreover, even if the development of a product's target specifications stems from customer requirements, the needs and wants for a system are different than the target specifications of the system. Customer needs represent what a typical user would like to see or have in a product. The engineer's job is to design and develop a product that satisfies the requirement. There are many such design possibilities that will satisfy these requirements, and sometimes the product can do more than what the customer wants. It is the responsibility of the OD team to deliver the product that matches the attributes of the product with the target specifications, which come out of the customer requirements. The PD team has to select a concept and final design that can satisfy the requirements and make money for the company when launched. The product has to be optimized in many possible dimensions, and customer requirement is one. The needs and wants are different from product designs. The target specifications and products have a one-to-one relationship.

For example, a customer would like to have hard-turning capability in a CNC lathe. Once the nature of hard-turning material and machining process is known, the PD team has to provide enough horsepower and speed so the machine can turn a hard metal and obtain the desired surface. So, the responsible engineer has to understand what the customer exactly meant by "hard-turning" capability, which the customer might not know technically enough to be specific. The PD team has to explore further and design the machine to specifications that will allow the customers to have that capability. Concepts and architecture will depend on the specifications that the concept will satisfy. A successful product also has to satisfy financial, manufacturing, quality, and reliability constraints in order to be profitable for the company.

5.2 Customer Requirements Development Process

The identification of customer requirements is a very rigorous process and needs proper attention and understanding to get fruitful results. There can be up to seven steps for this process, although every company has a different way of handling it. As mentioned before, the system engineering process is very much dependent on the outcome of this process. The seven steps can be as follows:

- Step 1: Select the market for the product

- Step 2: Select a wide spectrum of the customers in the desired market

- Step 3: Select the survey process or methodology

- Step 4: Gather, synthesize, and interpret the raw data collected from the field in terms of customer wants and needs through brainstorming with PD team members

- Step 5: Organize and prioritize the customer requirement data

- Step 6: Summarize and document the results for further processing

Let us take an example of a CNC machining center to explain the important points of this process. We will simplify the data collection process for the machining center since it is a very complex product to start with. Nevertheless, this particular process can be massaged use with other machine tool design and development. It is also necessary to remember the fact the system engineering process is top down, starting with customer feedback after management has decided to go for a new product. Sometimes it is also necessary to identify the missing data from the survey. In case the data collection is incomplete and needs further input, it is better to collect the data again instead of avoiding the same. The collection of data should start after the business mission for the product has been finalized, as shown in Fig 5.1 below.

Mission Statement

Descriptions	Primary Mission	Additional Comments
Product	A four-axis computer numerically controlled machining center for general-purpose machining of prismatic parts	-Product architecture should support future legacy products
Business Objectives	-Product to be launched in 2020 -Return on Investment to be 25% minimum at launch -10% market share in one year after launch, with market growth of 5% minimum per year for five years -Market leader in five years in the machining sector market -Introduction cost to be 5% less than the market leader in machining center	-Special market promotion for new products to be addressed before launching the product -Budget must be within stipulated limit
Market to be Addressed	-General-purpose machine Shops with low and mid volume	- Old machinery replacement
Customer Focus	-Customer satisfaction to be higher than 90% after one year of launching	- 3 mo. MIS, 6 mo. MIS, and 1 yr. MIS reports to be controlled
Primary Assumptions	-In-house CNC control with digital servo control -Machine horsepower to be 30 HP and 30,000 RPM -Four-axis machining with ATC of 40-tool capacity -Hydraulic and air assist	- Comparable to outside commercial control provision -40 HP is an option - Six-axis machine is a variant - Power consumption to be minimum
Principal Stakeholders	-Direct buyers -Dealers -PD team, quality and manufacturing -Sales and service -Bank and venture capitalists	-External and internal customers

Fig 5.1: Mission statement for a four-axis general-purpose machining center

The mission statement for the machining center, shown in Fig.4.1, depicts the overall scheme for the product with specific constraints and stakeholders. The mission statement is a layout of the company's objectives, goal, and plan for the product. The mission statement can be compared with the mast of a ship. You need it for overall direction for the product. In my mind, any business must have a vision and mission for the future product. The visionary view for the company about the product is laid out in the mission statement. The focus is on the customer. In order to get feedback from the customer that the company is trying to address, the selection of customers must be undertaken in a very systemic fashion. The market segments for the product must also be considered.

Moreover, the competition profile must be prepared among the present builders serving the market, and this will be used to benchmark the company product against the market leader. Requirement planning starts with the data from the customer, called the voice of the customer (VOC) in management or marketing terms. After the mission statement is formed, a system engineering process starts with VOC, as shown in Fig. 5.2. This is a two-dimensional way of defining the generic product development process. The upper diagonal process steps are iterative (double-ended arrow), and the lower diagonal process steps are serial in nature (single-ended arrow).

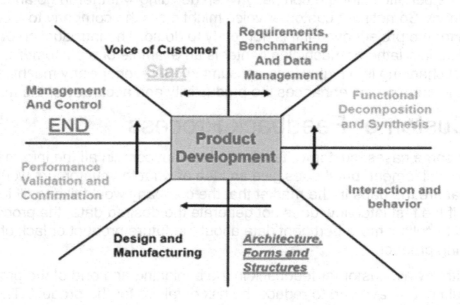

Generic Product Development Guidelines

Fig 5.2: Voice of the customer in PD generic process

The machining center is a very well-established product and has been in use for a decade or more in its present form. Control technology and capabilities, however, have changed substantially in recent years. In order to be successful as a new entrant in the market, the PD process structure must be strictly followed. Disciplined systemic process development is the key to the success of introducing such products into the desired marketplace with a minimal gestation period. Customer satisfaction is a given outcome of the PD process since consumers must express the highest level of satisfaction when using the product. This initial structured and formal step of collecting customer requirement data is a necessary and sufficient condition for a successful product.

For an entirely new product incorporating new technology, this structured process must be followed to reduce risk and enhance the probability of success. The venture product needs more customer data to fully understand the requirement. For a technology push product or new technology products such as hybrid cars, a structured process is even more required to explore uncharted paths. For an existing or legacy product, customer data will reduce the time for customer acceptance in a competitive market, and this will reduce the time to get the money invested \. Hence, the collection of customer requirement data is a mandatory step for any type of product. The risk when not collecting such data before product development is initiated is very high, and the success of any type of product could be in jeopardy.

It is also true that customer or market clinics, or interaction with customers, will not create the desired data all the time and in all cases. Even if customers are not very specific in expressing their views about a new product, general feedback will suggest a positive and negative view of any product. A smart assessment of the customer voice can only be beneficial for the company when deciding whether to go ahead with product development. Sometimes customer voice might steer the company to a completely new direction for the project even if it is not ready to do so. The introduction of a live tooling mechanism in a lathe or machining center is an example of a customer-driven product. Since tool changing is a highly time-consuming operation for any machine tool, automating the process enhances the productivity and accuracy of the machining.

5.3 Customer Feedback Process

In some cases, customer feedback does not contain all the information for product development, but it does give an idea of what the customers would be happy to buy. What are the gaps in the market that the company would like to fill to make money? If the first interview does not generate the desired data, the process can be repeated to collect more pertinent data about the future product or lack of capabilities in the existing product.

Either way, customer feedback in the beginning and end of the process is a must and should not be avoided to reduce the risk of failure for the product. The necessary steps are as follows:

Step 1: Select the market for the product

Market selection for the product is a very difficult task. If the market for the product is rising, it is somewhat easier than if the market is saturated with similar products. For machine tools, the market is matured and demand is somewhat constant or slowly rising at best. For a new start-up company, the selection is even more strategic and critical. Let us discuss a few points on how to select a market for a new product in a saturated market, which is typical for machine tools.

First, the marketing team has to find out how many existing customers are there and what the potential number of buyers in the near future will be. The global market for machining centers must be found out first. Then market segmentation data also has to be determined. The number of sales for machining centers and their area of sales concentration must be determined. Now the company has to decide whether there is an opportunity to sell the particular machine that it is trying to develop. Next, the company has to decide whether the sale will be direct to the customers or through the established dealers in the area. This data can also be collected by going to the machine shops in different regions and asking for it.

Second, can these potential customers pay for your products? This will help you to find the price cap for the product and its margin of profit. Normally, in a saturated market, the profit margin is relatively very small. For a new entrant into the market, the margin is even smaller, and the period to make some reasonable profit is also longer. The company has to decide whether it can sustain itself through such adversity for a continued period. To gain ground in a saturated market, the product has to be of very high quality, enriched with many options, and the price has to be equal or smaller than comparable machines from established or larger builders.

Third, is there a wide gap between what customers are using and what they are looking for? In other words, can the new product satisfy the customers' requirement at a cost that the buyer can afford? Are they looking for a change? Are they comfortable with using somebody else's product since that change might force them to alter their present set-ups, tooling, operators, etc.? The team has to ask whether the new product can solve customers' needs. Here, the question of capability and price has to be critically reviewed. Are these buyers favorable to purchasing machines from a new company?

All of these are relevant questions that marketing must answer before a decision is made to go ahead with the design and development of another similar market-pull product. Even if getting answers for all of these questions is difficult and time-consuming, the answers must be found out to make a correct decision for going ahead with the project or not. The company has to have a product that solves the customer's needs, which is very critical in making a decision.

Fourth, how many existing builders are supplying a similar type of product in the desired market? Who are the competitors? What are their strengths and weaknesses? If the market is already oversaturated with long-established builders, there is a very small chance that users will go for the new product from a new company. It is somewhat easier for the established company to introduce a new product if and only if the customer's requirement is satisfied.

Last but not least is the important question: do you have enough resources and resilience to sustain and support yourself through adverse conditions for a very long time? Even if you have selected the desired segment for your product, identified the customer segment in the desired market, and assessed your competition's strengths, you have to still answer the question of whether your product can survive the enormous difficulties and adverse conditions. Here, the gestation period, machine value proposition, performance strengths, and cost proposition are to be blended to justify the development of another product. If the product is a technology-push product or one that gives significantly higher productivity and capabilities than existing machinery, the chances are much better for the company to be successful. For example, when digital control was introduced into the market, any machine with such capabilities could be sold at a much higher price than what they are now. Market saturation and competitive strength are very critical points to consider before another product is developed using technology similar to what's already out there.

Step 2: Select a wide spectrum of customers in the desired market

Once the target market has been selected, the next assignment for the marketing team is to select the customers in the desired market in such a way that true feedback is received for the product. There are many possibilities, as outlined below:

- **Meet existing customers:** For the company that is already in business, the first easy way to get feedback is to select present customers. Their feedback about the present product could be positive or negative. A patient hearing is always advisable to convert unhappy customers into repeat customers. If a customer has a product from a competitor, their feedback is even more important for the new product design. If you listen to such customers, you can win back their confidence and develop their interest in the new product.

- **Consult websites for potential customers:** Sometimes a company's website will contain the information about the products that they are using and the good and bad points about their existing machines and tools. The website might also contain information about their business standing and potential purchasing plan for fixed assets or equipment. The website also often contains a company's balance sheet, which might shed some light on their financial strengths and weaknesses.

- **Telemarketing, direct mailing, and video presentation:** Newsletters, video clippings, and TV advertising could be other avenues to get feedback from potential customers. Free product and service incentives can also be used to get customer feedback. Sometimes potential users can be encouraged to create a video as to how to use a product to satisfy their needs. This is another way of getting help from users in designing the right product.

- **Focus Groups:** In this method, a group of users is gathered, and a conversation is conducted by marketing personnel to initiate feedback discussion and collect data, either through a questionnaire or verbally. Sometimes this is also called a "market clinic." The session is led by a moderator, who represents the PD team. Sometimes participants are paid a financial incentive to be in the market clinic. The process can also be deemed part of market research for the product. A prepared market clinic session is better than a random session of a general nature. The following questions might be presented to participants:

  - Are they are using such products?
  - What is the frequency and nature of use?
  - What is the primary intent of using this particular machine tool?
  - Can they display how to use the machine properly?
  - What are the difficulties they face when using this product?
  - Is there a productivity or efficiency issue in using the product?
  - What do they like about the product?

- Is accuracy enhancement required?
- What do they dislike about this product?
- When do they plan to replace the machine in the future?
- Who is the decision maker for purchasing the product?
- What improvements would they like to see in the product?
- Is the present product reliable enough?

Several interviews and interview modes might be needed to get most of the customer feedback. It is generally recommended that, for a very complex product like machine tools, a minimum of 12–15 interviews be used to get most of the information from customers. The design of the questionnaire is another important dimension of the interview process.

Interviews should be conducted sequentially among the same customers to get most of the data, and the process can be terminated when most of the crucial data is obtained. The customer consideration segment of the interview is also another very important factor for successful data collection. The interview process should be repeated across the segment in the market if required. This will also give a consistency of data across the segments.

The data collection process can start from the existing users or early adaptors in the desired target market. The existing users are very helpful in detecting the future needs of similar products, and they sometimes help the PD team design the product in their own way. This feedback is particularly helpful for the PD team since these users know the gap between what they have at present and what they would like to have in the future. In many cases, as I have experienced, that often customers already have a feasible solution for the problem that they have encountered. These are very useful for the PD team to take advantage of.

The system-based design, when combined with accurate customer feedback, will result in a successful product that will stay ahead of the competition if it embraces some new additive features that are lacking in the competition. The data must be collected from lead users, early adopters, actual users, dealers, service centers, and purchasers. As proposed by Professor Eppinger in his book *Product Design and Development* (2011), a customer selection matrix can be used for such purposes. The problem is that the selection or identification of these people in a given strategic market segment can be a very difficult task.

I would think a random selection of actual customers in the desired market segment would be also very effective in getting feedback. The selection of the proper customers to get feedback from is almost more of an art than a science. Some teams are very efficient in conducting such interviews. Moreover, a face-to-face interview is preferred to the indirect approach, such as mailing out a newsletter.

It is also important not to include customers who are not willing to give feedback. In the feedback process, it is generally advisable to follow these guidelines when selecting customers for proper feedback:

- Customer must be willing to give feedback voluntarily

- Customer must be selected at random
- Feedback must be taken over a long period and in a systemic fashion
- No bribing of the customers for the feedback: No incentives provided
- Clearly telling the customer the intention behind the interview
- The number of customers must be statistically significant to draw a productive conclusion from the feedback
- The questionnaire must be to the point, simple enough, and not time-consuming to answer
- Customers' answers must not be challenged
- Stay focused on the questions that provide the best possible answer
- Do not discuss how the machine will be designed or when
- Avoid words like "must," "should," "would," and "how"
- Be positive even if you get very negative/unexpected answers on any question
- Watch for customer's normal reaction to any question

Step 3: Select the survey process or methodology

In my opinion, a survey is the most effective way of collecting raw data from customers about the product. It is used almost exclusively for many household products, and I think it would be also very productive and efficient for machine tool applications. The survey is very easy to set up, and the data can be analyzed very quickly. There are two types of surveys in general: the short survey and the long survey. Another key point about the survey is to keep in mind that too much dependency on the survey answers can be misleading, too. They have to be taken in the proper context. Surveys are just starting elements of the PD process. In general, surveys should have the following common-sense approach:

- The survey questions must be properly designed, and then it can be sent to the customers, sometimes before the actual face-to-face meeting. This allows the customers to think about the answers beforehand

- The survey must be short and sweet. Ask a few pertinent questions specific to the intended product. About 15 to 20 questions should be good enough for the feedback

- Ask questions that the customers are knowledgeable enough to answer and that are to the point

- Design the questions in such a way that the answers are not ambiguous

- Questions should not be debatable or political

- Ask the questions for which you are looking for answers

The primary assumption here is the fact that the audience is familiar with a similar product or have used such products before. The target audience must be familiar with some machines of the same nature as that of the product to be designed. For example, users must be familiar with manual milling machine operations or the machining process in general. If the process is absolutely new or has a brand-new technology, then it could be termed a technology-push product, for which such detailed customer feedback might not be necessary. In the interview process for an existing type of product, the focus could be on both its use and the type of product.

For a brand-new product concept, the focus might be only on the use for such a product and not on the product itself. Since machine tools have been in use for a very long time in some form or the other, focus on both the use and type of product would be more applicable. For example, if you are updating the same machine with more user-friendly high-end control software, survey questions could be on finding out only about the use of existing control software.

The next question to address is how to capture the interaction process between the marketing team and the customers. There are several proven methods to keep records of the clinic sessions for subsequent use and audience analysis. Some of the most commonly used methods are described briefly:

Handwritten notes: This is the most common practice. The team might designate a member to note down the proceedings, and he or she might be the scribe for the team. The scribe records the interaction without being biased or judgmental against any comments or questions. Sometimes it might be impossible to record everything that the customers say. Other times, the scribe might misunderstand the answers. To avoid any misunderstandings and to help clarify answers or questions, a team session discussing the outcome of this process should be conducted.

Audio and Video Recordings: Unauthorized audio and video recordings might be illegal sometimes, so permission for them must be granted by the customer. Nevertheless, this is the most effective ways of recording a market clinic or focus group session, and the recordings can be exactly interpreted afterward. This might also be the costliest process, however. Such recordings can be also used for remote viewing by people who are not present during the sessions. This allows the team to view the interactions or customer answers and expressions multiple times to capture what the customers really meant. Identification of some hidden requirements, which are not communicated by words or expressions, is possible by viewing the video.

Session Still Pictures: This is another way to keep records of the clinic session. Since these are static recording of the session, customers' expressions can be captured. This can also be a document for future reference or repeat sessions, helping to decide whether to include some customers or avoid them in the future.

Step 4: Gather, Synthesize, and Interpret the Raw Data

The success of a systemic product development process depends on the quality of data collected from the customers and the results are called "customer requirements." Collection, synthesis, and interpretation of the data to create a target list of system specifications is a very important step of the development process, and it should be followed religiously. The output of the market clinic sessions is the raw data, which should be analyzed critically to get the underlying meaning of what the customer meant. A data template can be designed to capture customer statements and the team's interpretation of them. One such template is shown in Fig. 5.3.

This table must be filled up right after the interaction session is over. It is also necessary for team members to get together to interpret customer feedback. The first column of the table could have many possible elements other than what is mentioned here. The customer statements must be critically reviewed by the team members and documented in the second column of the table. The last column shows the outcomes of the customers' responses. These outcomes are just possibilities at this stage and do not constitute any final action.

Clinic Session Nr. _____ Date and Time: _____ Interviewer: _____

Category of Questions	Customer's Answers	Team Interpretation of Customer Wants	Comments
General use of the machine	Would like to have higher feed rate during cutting	Higher feed rate capability	Required for all the axes engaged during machining
	Need more HP for machining harder metals	Higher spindle power required	Would affect spindle and belt design
	Sometimes I use the machining center for round parts	Circular interpolation capability required	Control function capability
	Machine should be productive and efficient	High speed with high level of automation	Loading and unloading fixture with ATC: minimum manual operation
Strength of the existing machine in use	Like the user-friendly software for control	Software use must be very simple and interacting	Higher-level software required
	Like the flexibility of the machine	Machine must be capable of prismatic and circular parts	Machining software update required
	Servicing is very simple and easy	Components must be easy to replace	Simple and easily available parts to be used
Weakness of the existing machine in use	Not accurate enough for most jobs	Higher milling accuracy desired	Automatic inspection devices required
	Vibrates while cutting at higher speed	Machine rigidity to be increased	Spindle and axes stiffness to be enhanced
	Coolant leakage is a major problem	Sheet metal design is not adequate	Sliding guards must be improved
Potential future capabilities of such a machine	Hard milling capability is a must	Higher stiffness for the machine with high speed	Ball screw and spindle bearings
	Conveyor interface is a must	Machine capable of accepting automatic conveyor	Machine bed has to be redesigned
	Loading/unloading capability during cutting	360-degree rotating fixture outside cutting envelope	New loading/unloading fixture, servo controlled
	120-tool capacity for the ATC	Servo-controlled 120-tool capacity ATC	New 120-tool ATC to be designed

Fig 5.3: Typical raw data recording and possible interpretation analysis

Each customer response must be critically analyzed to understand the customer's real needs. It is quite possible that any customer statement could be interpreted by the team members in a quite different fashion, so it is necessary that this raw data be reviewed by more than one team member to draw a definite conclusion about the customer's wants and needs. This will also allow the team members to unearth the hidden needs of the customers. The guidelines for writing the actual needs of the customers could be as follows:

- Most customers express what they would like to see in the future product, i.e., they seek a solution to problems that are not resolved by the machine that they are currently using. The responsibility of the PD team members is to understand the real issue and incorporate design features or architecture into the future machine that will solve the problem. So, need interpretation should give the possible technical solution for the problem.

- No specific assumptions should be made on the raw data. It should be synthesized as it is noted during the session. Needs that come out of the raw data should represent the customer response.

- Negative customer responses to any question should be turned around as a positive aspect of the future machine. For example, the customer responds, "The machine does not have to be so heavy to cut the parts." This negative statement could be used to reduce the weight of the machine and save material without losing stiffness and rigidity.

- Needs must be translated to define the features or options for the machine. A need for higher accuracy could be translated into an increase in stiffness for the machine. A need for automatic loading and unloading could result in the design of a rotating fixture that will help to load and unload the part while the machine is making another part.

Frankly speaking, there are really no fixed guidelines as to how to write the clinic session or survey questions. It is an art, and a very tricky job, too. Even then, the following points should be followed:

- Only ask product-related questions

- Do not ask how to design the product

- Be very specific about the question

- Frame the questions to get mostly positive answers

- Questions should elicit answers that lead to specifications for the product

- Avoid ambiguous questions

Step 5: Organize and Prioritize Customer Requirements Data

This is the most crucial step in this process since the product specifications will directly reflect what the customer needs. Since the database could be huge depending on the number of customers interviewed or the number of repeats of the interviews, a priority list of requirements is necessary for further use of the data. Needs can be organized and segregated into primary, secondary, and tertiary. For a simpler product, such as a legacy-based of facelift product, primary needs could be sufficient. For a complex machine such a CNC lathe or machining center, all the sets must be prepared to get a total picture of customer requirements. The data can be then customized for a specific machine since some requirements might be general requirements for all machines. The organization of data for a machining center is shown in Fig. 5.4 below. The list is not exhaustive, though.

Needs identified	Primary Need	Secondary Need	Hidden Need	Priority
Machine has plenty of power to do milling on hard metal	Machine must have high horsepower	Machine must have high horsepower at a higher speed	Machining of tough and hard metals	1
Machine must have user-friendly control	Control software must be menu driven	Software must have pull-down capability	Control screen must have a large display area for easy viewing	1
Machine must have automatic loading/unloading capability	Loading and unloading fixture must be accessible while machine is in cutting mode	Loading and unloading fixture must be out of cutting enclosure of the machine for easy accessibility	Machine must be highly productive, with reduced cycle time	1
Machine must be highly productive and efficient	Machine idle time must be reduced	Time for critical functions must be reduced	Tool-changing time and rapid speed must be reduced for all axes	1
Higher milling accuracy desired	Enhance accuracy of radius profiling	Machining rigidity must be increased	Automatic inspection methods to be implemented	1
Sheet metal design is not adequate	Redesign of machine enclosure to stop cooling escaping cutting envelope	Enclosures must be of higher rigidity	Higher sealing capability of enclosure without sacrificing styling	2
Circular interpolation capability required	Software should include circular interpretation module	Curve interpolation extended to include circular profiling	Machining of prismatic and round part capability	3
Components must be easy to replace	High-service items must be easily available for replacement	Local and easy availability for service parts	Parts should be indigenous and not imported	3

Fig 5.4: Needs priority list for customer requirements

A disciplined and systemic approach to generating the priority list should be followed when the data is too much for a quick resolution. It is also true that the process is somewhat inorganic, and data management and interpretation could be an issue. This process should also be carried out by as many team members as feasible to reduce bias. A brainstorming session among team members is also recommended to understand and interpret customer answers. Some recommended steps are as follows:

- Eliminate duplicate or repetitive questions and answers throughout the data
- Combine similar questions into one category, and answers for each question should be grouped together
- Treat each need statement and corresponding answers separately, one need at a time
- Once completed, review and revisit each need and answer
- If more than one team is available, rotate the question and answers among the teams and create a consensus for the interpretation.

This overall process could be complicated and time-consuming when the Q&A sessions are for a brand-new product with a lot of technical requirements. If the data is from different segments of the market, things could get complicated in a hurry. So, In order to get statistically significant data from the customer, consider some of the following dimensions:

- Different segmentation of the market
- Different products
- Different customers in each segment

For further details about the market and the product, the survey should also focus on the following areas:

- Present and future potential requirements
- Urgency of needs and wants
- Optional features
- Basic features
- Cost issues
- Types of controls

In order to make trade-off decisions during the development of the specifications for the new product, the priority of each requirement is required. Moreover, trade-off decisions also have to be made when the prioritized list of requirements is benchmarked against the specification of similar products in the market segment. The prioritized list after benchmarking also provides the gap in the data collected. If the answers do not contain significant data for the product, repeating the survey might be necessary. There are certain internal factors that can be traded off during specifications development, such as quality, cost, and legal and resource requirements. The survey should focus on competitive external factors that are essential for the new product. Hence, the survey should focus on factors that will make the customer happy and satisfied. The survey questions should also give the customer scope for comparison feedback, such as a scale of importance on the performance of the product.

In the case of large data from the survey, a statistical comparison can be established by calculating the mean and standard deviation of the responses. If the data has a central tendency or is normally distributed, the mean value of the responses and the range of the value of such response will help the team identify a desired value and range of the value for the performance. Depending on the statistical analysis, data could be arranged in order of importance, and a priority scale could be established. In order to determine the hidden or unexpressed needs of the product that customers expect to have, survey questions should focus on futuristic or unique approaches for the use for the product.

Step 6: Summarize and Document the Results for Further Processing

The final step in this phase is to summarize the results and document the process. As mentioned before, this step is more of an art than a science, although some procedural steps can be followed in order to reduce cost and time for this phase. Since the whole system engineering process stems from customer requirements, this stage is very crucial to identifying market segment, conduct survey or market clinics, analyze and prioritize the data, and document the same for further processing.

This is also necessary for the team to prioritize the needs and critically review the requirements before jumping into the next step of the systemic process. The team must also be ready to retake the data in case of any substantial doubt on its outcome. If a gap or non-correlation is found in the data, it must be verified again using different input conditions for the survey. Since this survey step is very time and cost intensive, the outcome must be consistent, statistically correlated, and justifiable.

Sometimes a peer review is done to make sure the conclusions drawn from this step is consistent with the legacy knowledge of the company. the more complex the future product is, the more necessary it is to analyze the data before further action is taken. Before this process is finalized, the PD team should ask themselves the following questions:

- Have we considered all the potential customers in the desired segment of the market?
- Have we captured the true intentions of the customer about the use of the future product?
- Are the answers consistent enough to carry out a statistical analysis of the data?
- Have we asked appropriate and unambiguous questions in the survey?
- Have we gotten answers that reflect true customer requirements?
- Should we revisit the customer to get more data?
- Should we repeat the survey to fill gaps in the data obtained?
- Have we documented the process enough?
- Is this most cost-effective way of capturing the required data?
- Can we be more efficient when conducting the same process in the future?

- What did we do wrong in this process?

- Was the process too long and did it alienate the customers?

- Did we complete a peer-review process?

- Is the outcome consistent with the knowledge base of the company?

- Did we consider all potential team members?

- Did we need this step at all to save money and time?

- Did we already know the outcome before we conducted the survey?

- Did we expect the answers that the survey resulted in?

- Was it worth the time and money that were spent during this process?

- How do we use the output of this step more effectively for system development?

5.4 Summary

Customer needs determination is an integral part of and the first step in the systemic engineering product development process. The resulting prioritized list document will migrate to the generic product specifications, which will be refined to create target specifications that will be followed by the PD team for the future product. The PD systemic process starts with the requirements prioritization list.

- The process of capturing customer needs and wants consists of the following steps:
    - Customer must be voluntarily willing to give feedback
    - Customer must be selected at random
    - Feedback must be taken over a long period and in a systemic fashion
    - No bribing of the customers for their feedback: no incentives provided
    - The customer must clearly be told the intention of the interview
    - Number of customers must be statistically significant to draw a productive conclusion from their feedback
    - Questionnaire must be to the point, simple enough, and quickly answered
- This process creates an essential link between what customers want and what the PD team should incorporate into the new product. PD team management will work together to ensure each of the specifications are met to obtain customer satisfaction, which is the ultimate goal of product development. The creation and accurate understanding of the requirement list is a preamble to the product development process using system engineering principles.
- Early adopters in the market segment and existing users of company products are a good source for generating raw customer data. The PD team must embrace both positive and negative responses from such customers for future products. Responses from such customers will contain the necessary data that could be converted into the specifications for the new product.

Sometimes it is also necessary to get feedback from a segment of the market that uses competitors' products. The response from such users will reflect on what is lacking in the company's present products.

- It is very necessary and important to capture and understand hidden needs from the customer responses. The survey questions have to be designed properly to get such answers more clearly. The hidden requirements are more important than the expressed needs since hidden needs, if incorporated into future products, will result in essentially strong features. Sometimes the hidden requirements can be taken care of by introducing optional features of the product at an extra cost, which enhances the revenue.
- Survey questions should focus on what should be designed into the product and not on how to design the product. Customers do sometimes help to design certain features of the machine due to their current use of the product. These answers should not be neglected, and such customers should be taken into confidence by the PD team to design the future product. Nevertheless, the "what" type of questions leads to generating several architectural concepts of the product. The "what" is what customers want, and the "how" is the responsibility of the PD team to deliver for a successful product.
- The primary benefit of this particular step is to ensure that the product is what customers want in the future. Incorporating these customer expectations into the product will guarantee customer satisfaction at the end. A successful product ensures customer satisfaction at a cost that they can afford. Moreover, there should be a strong consensus among all the PD team members that these requirements must be incorporated into the product specifications unless there is a strong, valid, and justified reason for not doing so. Developing a customer-requirement-based product will also result in the overall satisfaction of all the PD team members and management and forge a very strong connecting link between PD team members and other parts of the organization. Delivering a product that satisfies the customers also generates revenue in a repetitive manner, which becomes an added benefit for the whole organization and for the sustainability of the business in the future.

5.5 References and Bibliography

Abernathy, W.J., and Townsend, P.L., 1975, "Technology, Productivity and Process Change," *Technological Forecasting and Social Change* 7(4), 379–96.

Burchill, G., et al., 1977, "Concept Engineering," Center for Quality of Management, Cambridge, MA, Document No. ML0080.

Blackburn, J.D., 1991, *Time Based Competition: The Next Battleground in American Manufacturing*, McGraw-Hill Erwin Companies, NY.

Drucker, P.F., 1991, "Japan: New Strategies for a New Reality," *Wall Street Journal*, October 2.

Devinney, T.M., 1987, "Entry and Learning," *Management Science* 33(6), 706–24.

Eckes, A.E., Stern, P., and Haggart, V.A., US International Trade Commission Commissioners, 1984, Report to United States International Trade Commission on Investigation No. 332-149 Under section 332 of Traffic Act of 1930, "Competitive Assessment of the U.S. Metal Working Machine Tool Industry."

Griffin, A., and Hauser, J.R., 1993, "The Voice of the Customer," Marketing Science 12(1), 1–27.

Hopp, W.J., and Spearman, M.L., 2000, *Factory Physics*, second edition, McGraw-Hill Erwin Companies, NY.

Hill, T.J., 1991, *Manufacturing Strategy: Text and Cases*, McGraw-Hill Erwin Companies, NY.

Kidder, T., 1981, *The Soul of a New Machine*, Avon Books, NY.

Kinnear, T.C., and Taylor, J.R., 1995, *Marketing Research: An Applied Approach*, fifth edition, McGraw-Hill, NY.

Krugman, P., 1994, *Peddling Prosperity: Economic Sense and Nonsense in the Age of Diminished Expectations*, W.W. Norton and Company, NY.

Nahmias, S., 2000, *Production and Operations Management*, fourth edition, McGraw-Hill Erwin Companies, NY.

Payne, S.L., 1980, *The Art of Asking Questions*, Princeton University Press, Princeton, NJ.

Ulrich, T.K., and Eppinger, S.D., 2011, *Product Design and Development*, McGraw-Hill Companies, NY.

Urban, G.L., and Hauser, J.R., 1993, *Design and Marketing of New Products*, third edition, Prentice Hall, NJ.

Tooze, A., 2009, *Machine Tools and the International Transfer of Industrial Technology*, "The Global History of Machine Tools, Knowledge, Narratives and Fiction," 30-31, King's College, Cambridge.

5.6 Review Questions

- Do the customer requirements lead to concept development for any type of product, market pull, or technology push? Is this process only applicable to a commercial product? Is it only necessary for a mature product in a mature market? Can this step lead to creating an innovative product or services, such as a cell phone?

- Who should lead this process, PD functional engineers or the marketing and sales team? What should the optimum number of team members be for carrying out this process? What should the time and budget constraint for this step be? Is this process required at all for a legacy-based product? How do you resolve a conflict issue when interpreting the customer response and putting it in the product specification?

C H A P T E R 6

6 Product Planning for Machines

6.1 Introduction

For any company to be competitive and successful, it has to have a long-term business and product strategy. It also has to have a strategy for transforming a technology into a viable product that satisfies customers' wants and needs. For a rapidly changing and highly volatile market, any business has to have a long-term and effective product plan. The challenge is to plan for platform-based products that have a common design philosophy, manufacturing abilities, and quality strategies.

The product planning process comes into being after the business planning has taken place and before any design and development project for machines is started. Both the business and product planning have to be in place before substantial company resources have been utilized. Product planning is an umbrella process wherein all the company's products are considered in totality and then future products are thought of to complement existing products and technology. The product planning process considers the portfolio of projects that the company would like to develop in the future. It must be a long-term vision for the company. The product planning strategy must be in alignment with the business and customer strategies for the company. It should address the following questions:

- Are future product development projects in alignment with present products?
- What should the product mix for the company be to satisfy the customer base?
- Is the time ripe for taking up another project?
- What is the optimal product mix for the company?
- Is the company ready to venture into a new product market?
- Is the new product an extension of existing products?
- Is the new initiative going to be a platform-based product or a derivative or facelift of the existing product?
- Are the products in the product portfolio in sync with each other?
- What are the timing and sequence of product introduction in the market?
- Does the new product address a new market sector?

- Is there a real need for the new product in the market?
- Are customers hungry for such a new product?
- Are there any capacity constraints for future products?

As mentioned before, each new product must support the overall company vision and mission. The new product must be developed along the lines of the company mission, and the product mission must be developed in conformance with the overall objective of the company. Before a product development team is formally put in place, the team has to have a product mission for future guidance. The product mission statement must address the following questions:

- What are the cost and return-on-investment targets for the product?
- What is the budget, timeframe, and milestones for the product?
- What are the quality strategies for the product?
- What is the manufacturing, service, and warranty strategies for the project?
- What are the critical constraints for the product with respect to finance, manpower, concept-to-market timing, and intellectual ability of the company?
- What are the constraints in the application of the technology for the future product?
- Does the product use bookshelf technology or explore a new technology?
- What is the nature and requirement of customers the product has to comply with?
- Is the product innovative or a "me too" product?
- What is the nature and extent of the competition for the product?
- Are there any government regulations?
- Are there any tax incentives for the product?
- Are there any environmental issues?
- What are the safety and ergonomic concerns for the product?
- Are there any international considerations and constraints?
- Are there any supply chain management issues for the product?

Another aspect of product planning is to consider the importance of operation strategy for the company and how it is connected to the special products under consideration. One current important question of high relevance is whether the product will be manufactured offshore or indigenously. The importance of manufacturing activities must be considered beforehand since it might affect design considerations for the product. If the machine depends on high-quality castings or forgings, it might appear that manufacturing overseas might be a better idea. But when the other aspects such as controlling the quality, transportation logistics and cost issues, the effect of economic order quantity, quick response to order delivery, workforce issues, etc. are considered, the decision might change. For low-volume production and high-quality products, offshore firming might be an issue that should be avoided. When planning for the product, four elements must be considered simultaneously: product and process life cycle, supply chain management, consistency in quality, and customer base. In order to be cost-effective, just-in-time philosophy considerations must also be considered for the product. Consistent evaluation criteria should be established to determine the effectiveness and productiveness of the strategy accepted for the product.

Another consideration of primary importance is the notion of the customer base for the product. Recently Japan has been very successful in establishing the idea that it produces the best machine tools even if this is not always true. As Michael Porter suggests, considerations such as demand-supply conditions, supporting industries, and firm strategy must be combined effectively to create a synergistic condition for the company. In my opinion, machine tool quality and features come before cost considerations. The importance of following lean six-sigma principles must also be considered for supply chain and in-house production facilities.

For a new and innovative product, the product and process life cycle must be considered, as mentioned before. New products go through four stages: start-up, rapid growth, saturation, and decline. The process that is suitable for all of these cycles must be considered. Depending on the volume and nature of production, a product-process matrix must be established for the specific product under consideration. For machine tools in particular, learning and experience curves are very useful in modeling the effect of labor hours or of production costs, as the experience of the existing workforce is considered. Finding a skilled engineering and supporting workforce is always an issue for capacity expansion.

More often than not, during a recession or decline in demand phase, the workforce gets affected more than management. This has affected machine tool industries in the United States, as I have witnessed in the recent past. Consideration of technical skill and the maintenance of the workforce is very much a factor for capacity expansion for existing or new products. Break-even curves might be considered for capacity expansion, but they do not consider the effect of skills available. So, break-even curves for volume consideration and experience and learning curves must be considered together for new products that might create financial and technical pressure on the existing system. The problem is definitely complex, but careful management of all these issues as a system consideration would benefit the company in the long run.

An overall product plan for the foreseeable future must be considered beforehand. There are several types of products that should be considered, such as innovative products, new platform products, facelift products, extension products, or special customized products. For example, the product formed by the addition of live tool capability to an existing CNC static tool lathe could be deemed "derivative." Increasing the spindle horsepower or axes speed of a product, or adding a safer sliding door, creates what could be called a facelift, or improved, product. An innovative product takes a very fundamental approach or explores completely new dimensions of the machining process. Hard-turning machines, which eliminate the need for grinding machines, could be considered an innovative product. Adding a CNC machining center to an existing CNC lathe manufacturer could be called adding a new platform product.

Although there is a fine line of separation among such types of products, planning for all must be considered in unison so that there is no strategic conflict among them. A futuristic timeline of such products must also be done for resource considerations and the timing of the introduction of the product.

6.2 Product Planning Process

The project portfolio vs. timeline diagram, as shown in Figure 6.1, connects the time of introduction of the product with the type of products to be produced. This is a long-term product plan for the company. The vertical axis represents the type of product, and the horizontal axis depicts the time of introduction and duration of the project.

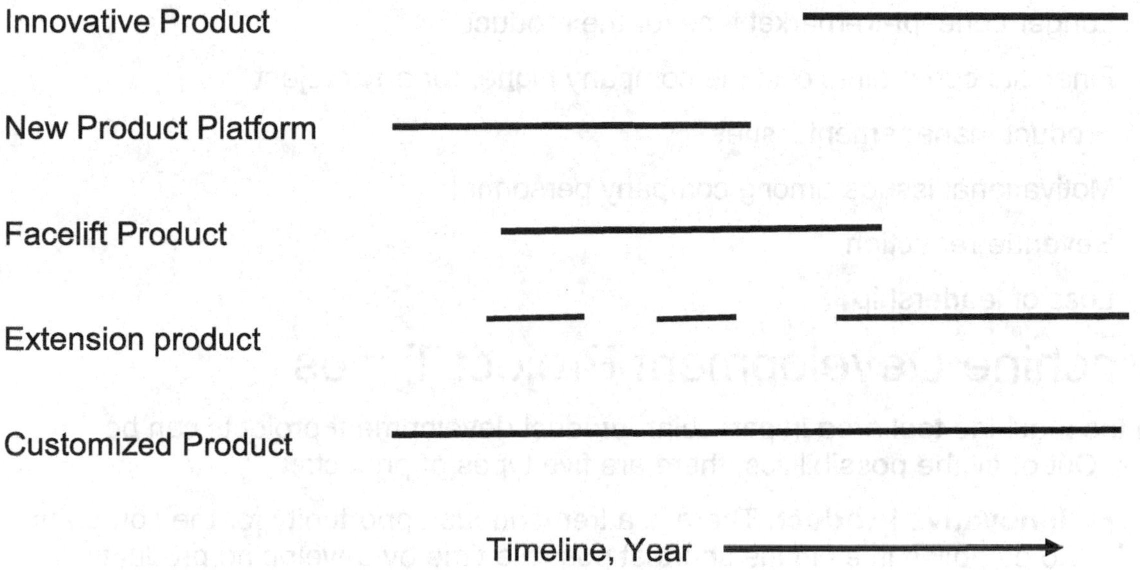

Innovative Product

New Product Platform

Facelift Product

Extension product

Customized Product

Timeline, Year ⟶

Fig 6.1 Product portfolio vs. time of introduction

The product planning process starts primarily with the customer requirements identified by the marketing or sales department. For new and innovative products, the process could start from any R&D or product development team. In light of resources available and intellectual capability of the company, all the projects, as shown in Figure 6.1, must be mapped together to identify opportunities for developing any product. With machine tools, only a limited number of projects can be run simultaneously, especially due to financial and manpower constraints. In such a competing project portfolio environment, management must be very careful when selecting a project or projects to continue. Barring emergency projects to support customer complaints in the field, long-term projects can be considered on the basis of return on investment. Financials for each project can be assumed or determined form legacy data and then used for its financial evaluation.

Such product plan mapping must be updated on a regular basis to understand the implications of any new project being taken up in the future. The continuation or delay in any project can depend on technology updates, competition, and customer requirements. In general, the product plan for a machine tool company should focus on revenue goals, intellectual capability, resource constraints, and target market constraints. In my opinion, a competitive environment, especially one caused by a foreign influx of machines, is a very dominant key to the success of a machine tool project, even if financial constraints are not an issue for the company. If the product plan portfolio mapping process is not considered, the company might face several uphill battles, as outlined below:

- Delays in product introduction into the target market

- Loss of customers' confidence in the company

- Longer concept-to-market time for the product

- Financial constraints cost the company higher for any project

- Product management issues

- Motivational issues among company personnel

- Revenue reduction

- Loss of leadership

6.3 Machine Development Project Types

In the machine tool area in particular, product development projects can be numerous. Out of all the possibilities, there are five types of projects:

- **Innovative Product:** There is a tremendous opportunity for the company to establish itself in the shortest possible time by developing products using new technology or new engineering concepts that satisfy customer needs that competitive products do not. Going from analog servo equipment to digital servo controls, fueled by Fanuc, was a major improvement made by Japanese machine tool companies. U.S. machine tool companies lacked the competition and lost market opportunities. The machining center, with its very high-speed spindle, higher axes speed, and other customer-friendly capabilities, could be considered an innovative product. It was a quantum leap from the standard milling machine manufactured by Cincinnati and others in the USA.

- **Customized Product:** These products could also be termed "special purpose machinery," or SPM. The company has to be extremely careful before taking up these types of projects. Due to very high customer expectation about machine capabilities and machine accuracies demanded, such projects turn out to be a nightmare in the long run. The customer expectations and demand changes very frequently, and the company falls into the trap and are unable to satisfy the customer. In the end, the company loses money in the process and gains a bad name as well, even if a sincere and hard attempt was made to deliver what the customer wanted. I would suggest that companies start with a standard line of machines and customize the standard to support customer requirements. I have witnessed many successes along this line of thinking. Again, the ROI method can also be used to determine the financial viability of the project.

- **New Product Platform:** These products are based on existing products and technology. Most of the time, such products can be modified to accept new technology to satisfy customer requirements. This is a very common approach for most machine tool projects. The development also leads to families of products using a similar technology or design philosophy. For example, the same technology and design philosophy can be very effectively used to add series products, such as CNC lathes with six-inch chuck capacity, eight-inch chuck capacity, 10-inch chuck capacity, or twelve-inch chuck capacity. This method could be used to some extent, and beyond a certain capacity, the methodology should be changed. For example, to design a CNC lathe with 18-inch chuck capacity based on the design philosophy of a lathe with six-inch chuck capacity could be hard to do. Moreover, the new product family should address the new customer market, new competition, and capacity constraints and should share the common parts as much as possible to reduce the cost.

- **Facelift Products:** These are existing products with added features, as per recent customer requirements, or that are in alignment with recently introduced products, for example, adding a steady rest for an existing machine or adding a live tool to a stationery CNC lathe. Adding a copy turning attachment to a bench lathe could also fall into this category.

- **Extension Products:** These products are added to an existing line to satisfy customer demand or address new demand in the market. For example, adding a four-inch CNC lathe to an existing line six-inch and eight-Inch chuck lathe.

6.4 Product Planning Process

The product planning steps are outlined below. This process starts with identifying market opportunities and ends with the specific product development. As shown in Figure 6.2, the process starts with identifying customer requirements.

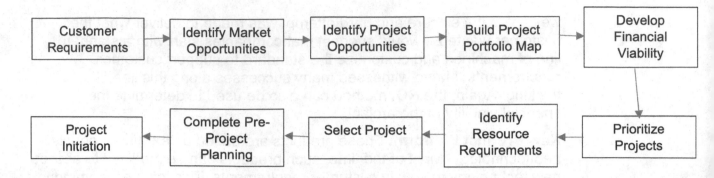

Figure 6.2: Product planning process

The project planning process starts with customer requirements or voice of the customer. The marketing function also identifies the market volume for the product and the approximate price for the product at the time of introduction into the market. For machine tools, getting into new markets might be difficult. One common problem is the type of tooling and work holding fixtures required for the production. Once a set of machinery is purchased from the manufacturer, changing the tooling for another manufacture is cost-prohibitive and time-consuming for most users. Hence, they tend to stick to one machine tool manufacturer. That's why it is sometimes better to design machines for common tooling to get wider acceptance and gain market share.

Once the requirements are known, management can put together all the potential projects and opportunities. Since all the projects cannot be tackled simultaneously, projects that are within the intellectual capability of the company and have an acceptable rate of return for the product life are selected. This is accomplished during the "project financial viability" phase of the planning process. Once the projects are prioritized and resources identified, a project is identified for further consideration. As mentioned before, a mission statement for the product is developed in discussion with the product development and marketing team. Once the project is established with a feasible and realistic mission statement, the project is taken up for further development by the product development team.

The planning process described above seems to be way more simplistic than what it is in reality. The reason is that the fact selection process requires various other inputs for finalization and feasibility. The process is highly iterative and complex. This gets even more complex when a new product with new technology is planned for development. The risk and uncertainty associated with the outcome of the project must be considered. Many times, a milling machine manufacturer might like to move horizontally to enter into, say, the turning or grinding business. So, instead of vertical integration, the company might select horizontal integration. In such cases, my suggestion would be to buy an established company in the lathe business and get into the established market. Instead of developing a line of machines that are completely new to the company, it is better to buy a company that has established itself in the desired market segment. Another reason for iteration is the selection of the projects competing for available resources. Unless resource allocation, i.e., time, personnel, and money, is managed very efficiently and productively, most projects might get affected to the point that some of them cannot be even completed.

Moreover, the project planning process might be updated frequently depending on the market conditions and competitive situations. If the upcoming economy is not conducive to new product entry or if the target market sector presents unwanted difficulties for new products, it is better to delay the project introduction. The other side of the argument is that new product entry is advised at a time when the economy is turbulent and the market is unstable to some extent. This will allow the market to get used to the new product and its features during this unsettling time, and as soon as the market turns around, customers will buy the new product. I have seen both of these approaches used very effectively by indigenous and foreign manufacturers. Nevertheless, a close watch must be kept on what is happening in the target market, and project maps should be updated accordingly. This is a very common practice in the automobile sector, and many started projects do not see the light of the day at the end. It is also a fact that project planning cost is way less than actual product development, manufacture, and marketing. Hence, it is always preferable to stop a project during the planning stage instead of carrying it to manufacturing and marketing stage. Such adjustments must be made for the company to be more profitable.

In order to develop a product plan, the following suggested steps could be followed:

- Collect customer data
- Identify market opportunities
- Build a project portfolio
- Create a project map with timeline
- Develop financial feasibility for each potential project
- Select financially and intellectually possible projects
- Develop project mission statement
- Plan for downstream product development

6.5 Product Planning Details

The product planning process starts with identifying opportunities in the market. The primary question to ask is: "What product should we develop?" The answer should come from the intended customers of the target market. Let the customer design and conceptualize the product the way they want. In addition, product ideas can come from various other resources, such as marketing and sales personnel, the engineering team, the quality and reliability team, the research and development team, existing customers, etc. Nevertheless, I have found out that sincere or existing customers are the best product initiator. Many products fail because the customer did not want it. Another effective way to understand the general trend of market requirements is to conduct a benchmarking study in the product sector.

Once customer feedback is obtained and assimilated, the products should be evaluated and prioritized for further downstream execution. The project portfolios should be prioritized based on resource requirements, intellectual capabilities, available expertise, capacity available, etc. Once the projects are prioritized, resource allocation and the planning horizon have to be planned accordingly for further downstream development. Once the resource requirements, i.e., finance, people, and capability, are established, the product development process has to be initiated.

For innovative products, the company has to take the risk of starting a development process. They have to identify a gap either in technology or with any product in the target market. The initiation of a brand-new product could begin by taking the following steps:

- Understand the product gap in the existing product stream

- Understand customers' needs and wants

- Understand bookshelf-ready technology available for the particular product

- Understand how to mitigate the risk of such an innovative product

- Understand how the new product will help customers make more money

- Understand existing competitive products and their shortcomings

- Understand the efforts, resources, and intellectual capabilities required to use available technology in the new product

The opportunity basket will have many possible products that can be developed in the future. Some of these could be outside the reach of the company's capability. For example, a manual lathe builder could decide to stay away from building high-speed CNC machining centers even if there is tremendous market potential for them. For any company, several projects might be running simultaneously and competing for available resources. Hence, a careful evaluation has to be done before another project is taken on without finishing the existing projects that are currently being pursued.

In order to evaluate and prioritize the basket of opportunities, several avenues could be followed, such as competitive strategies, cost strategies, brand differentiation, technology to be deployed, and product platform strategies. For example, China and Taiwan have established themselves as providers of low-cost machine tools for U.S. industries. For a well-established company in the market, product differentiation might work out better. In order to re-establish machine tool builders in the USA, innovation is the only tool left at present. The companies have to provide innovative products or unique features.

The other possible strategic dimensions could be quality and reliability, speed and power, features and flexibility, capabilities and ease of operation, etc. In my opinion, the quality and reliability of U.S. products are sub-standard to Japanese and European products. The quality has to be understood and incorporated in the design itself. Moreover, the customer defines the quality, and not anybody else in the chain. Speed and power could be another brand feature or specialty. Enhanced speed and power reduce the machining time without sacrificing any quality of machining. This also means more chip removal per unit time, and that relates to making more money per unit time of operation. Similarly, flexibility offers a range of capabilities when the machining tasks change over time. For example, a turning machine can turn, mill, and grind parts without part transfer. The new product must have the ability to respond to new demand out of the machine without any further investment. In most cases, attending to all of these dimensions will raise the cost to the point of pricing the product out of the market, so the company has to decide which specific dimensions the new product will excel in to separate itself from other products, i.e., the company has to take a stand on positioning itself against the competition.

Other classical views of product strategy could be gestation time for development and introduction, product manufacturing aspects, and product characteristics. First of all, product development time is very crucial for new products. Total time from concept to market has to be reduced to a minimum. The time dimension also affects decisions on finance and manpower requirements. Time is also very crucial since the demand and supply conditions might change substantially during the product development time. Moreover, management must make a decision about timing for all projects. Time consideration is also very important when the supply chain has to be developed for new technology suppliers. The next consideration is manufacturing issues for the product.

As suggested by Skinner in 1974, the product has to focus on several issues, such as process technology, market demand, product volume, quality level, and manufacturing task. All these factors must be thought of in unison for the new product. If the product needs a new facility, such considerations become even more crucial for the success of the product. Management also has to consider professionalism in the plant, changes required for the new product, and manufacturing tasks. Next is the consideration of cost, quality, and customer satisfaction. These aspects are required to develop and maintain a satisfied and loyal customer base. Any product goes through a loop starting with the customer and ending with the customer as well. A very clear company strategy has to be established about all of these issues to provide guidance to other functional streams of the company, such as marketing and service, finance, design, quality, and manufacturing.

Recently machine tool products have become very technology intensive. The latest machine tools have digital controls, servo-controlled axes and spindle power, remote sensing devices, maintenance software, statistical quality-control software, thermal sensors, tool wear sensing devices, etc. As described earlier, technology S-curves are a conceptual tool to help decide on the technology to be used, keeping the future in mind. The technology curve displays the life concept of any technology. S-curves for any technology display the start-up, rapid growth, and maturity stages of any technology with respect to time. Eventually, the technology could become obsolete.

6.6 Product Platform Considerations

Another recent trend is to design products based on existing lines. The platform concept for the design allows the company to share assets across all the product lines. For example, a similar spindle design or tool indexing unit could be used for several products. In general, assets refer to assemblies and sub-assemblies. Using any platform, you could have a variety of similar products, called derivative products. In this way, design cycle time can be reduced substantially, and technology transfer becomes much easier. In derivative products, customization to suit specific customer requirements becomes much easier. Once a platform has been created, derivatives come off the platform at reduced cost and time.

Designing a platform depends on technology available and research development efforts inside the company. Along the same line of thinking, any company could develop a technology roadmap for the future. A technology roadmap shows the availability of ready-to-use technology for a product being developed. For a critical part of the system, such as control, servos, etc., a technology roadmap can be very useful, as the product and technology are eventually synchronized with respect to time. This diagram will show how the critical systems are aligned with the technology available at any point in time. Product-Technology road-mapping could be a strategic tool for the company for product development. This is also very useful when technology is rapidly changing for any function of the machine tool, such as high speed and reliability management.

One of the most stringent tasks for senior management is to balance the product development project portfolio properly. Company managers have to consider the financial outcome of any project, the risk to be undertaken, the risk of failure in the marketplace, and the risk of negative feedback from loyal customers. As mentioned before, Wheelwright and Clark proposed a product-process change matrix to balance the project portfolio. The authors suggest tracking down the changes required for the product and its process of development. Professor Eppinger and Ulrich also explained this method. The process change diagram is reproduced in Figure 6.3 with some modifications.

The matrix elements have been slightly modified to suit the nature of typical machine tool development projects. This figure depicts the relationship between product and process changes required to suit the types of products. For example, a glass grinding machine is substantially and technologically different than a regular grinding machine. The glass grinding machine will need an innovative process change to manufacture. Another example could be the development of a machining center as opposed to developing a mechanical milling machine. Similarly, the design of a 12-inch chuck lathe could be an extension of a previously developed 6-inch lathe. In such cases, the existing process could be renovated to suit the 12-inch machine.

For some projects, no process change is required to solve customer-related issues. The company might decide to spend a few dollars to keep the old products alive, and such products do not need any process change. These products could be called cash-cow products, which generate revenue without any further investment. Such a perspective can be useful to identify project-process mismatch, and a company might decide not to pursue a project for which a breakthrough process is required. Some companies would like to be a close follower of the competition and might decide not to go for any breakthrough product or process change. In some cases, a company might like to come out of its comfort zone and venture into breakthrough projects or processes to stay competitive in the marketplace.

Such diagrams are only directional in nature, and there is no hard and fast rule for pursuing any fixed direction. In my opinion, a company should pursue projects in line with their experience, intellectual ability, and personnel motivation. It is definitely a complex process. For the U.S. machine tool industry, it will be very difficult to survive further unless some disruptive type of product design is taken up to establish leadership again. Hence, the company's choice to be a close follower or a leader in the industry will dictate the types of projects that it might pursue. To be a leader, a company might have to take projects that need a breakthrough in technology or process. When a product reaches maturity over time, the company might have to embrace a completely different technology, called a disruptive technology, to gain leadership in the market.

Product leadership emerges through product innovation. When the industry is saturated with competition or products, technology or product innovation is the only way to establish market leadership. When the product is innovative, revenue generation can also be substantially different due to the fact that customer acceptance is more and customers will pay more to have a desirable product. For innovative products, a low-cost strategy might not be valid until competition follows suit. Given the fact that disruptive technology implementation is riskier than incorporating traditional technology into the product, it is worthy to pursue that risky path when revenue growth becomes alarmingly low. So, firms that would like to have new technology development projects to stay ahead of competition must take the additional risk as well. Moreover, it is also a fact that all the products with brand-new technology might not see the light of the day more often than not, and it also has to be understood that the planning of technology development is quite different than product planning. Technology development must be done ahead of product planning If the product uses that technology.

	Innovative Process	Renovated Process	Minor Change in Process	No Change in Process
Innovative Product	Breakthrough Project			
Platform Product		Platform Projects		
Derivative Product			Platform-Based Projects	
No Change in Product				Cash Cow Projects

Product Design Change (vertical axis, upward arrow)

Manufacturing Process Change →

Figure 6.3: Product-process change matrix for machine tool projects

Getting back to product planning, most of the time, a machine tool company cannot afford to invest in all projects at the same time due to finance or manpower limitations. Too many projects will compete for available human or financial resources.

Hence, it is necessary for the company to be realistic when managing its project portfolio. Unless such activities are planned properly, project completion, quality, and cost could be at stake. Composite, or aggregate, planning could be done to manage the project portfolio within available or budgeted resources. In the aggregate planning process, the requirements of financial or human resources to complete a project must be assessed with reasonable accuracy. Aggregate planning might help the company not be in dangerous or risky situations when resources are overcommitted to projects. Such critical consideration is particularly applicable for machine tool builders since resources are very limited. Hence, the company must decide which projects to pursue in the future during the product planning phase. A realistic project schedule must be considered instead of being in dreamland.

6.7 Product Concept-to-Market Timing

Product introduction timing is also a very critical decision a company might have to make. Traditionally, companies introduce new products during machine tool shows around the world. I am not sure that a machine tool show is the right platform to display new products, considering the fact that the cost of attending such shows have gone up exponentially in recent times. In-house shows for loyal customers could be a better way to introduce new products. Nevertheless, it has been observed that the early introduction of new products is a better strategy. On the other hand, launching a half-completed, poor-quality product early might be a very bad strategy. A robust application-ready technology must be used for the product to ensure product quality and reliability are maintained. Product introduction timing might also depend on the economic conditions of the market. Introducing a product too quickly might frustrate loyal customers who would like to keep up with the technology but do not have enough financial strength at the time of introduction. On the other hand, an introduction that is too late might persuade customers to go and buy competitors' products with the latest technology. I also strongly believe that product introduction timing is not an important factor when a product ensures lower cost and higher quality and reliability, i.e., the product must have the best technology embedded in the product to get the best quality, features, and reliability at the best possible cost. In other words, for machine tools, quality and cost are more important than time of introduction.

6.8 Project Constraints and Limitations

Before any project is finalized, intellectual capabilities, manufacturing capabilities, warranty and service possibilities, and legal and environmental conditions must be considered. Normally, these factors determine the viability and suitability of the projects. More often than not, these constraints limit the possibility of downstream development for profitable projects. For example, if a product has assemblies that can be serviced only by factory-trained technicians, marketing these products in faraway places or overseas could limit the product's development. If a CNC control is used for a product and it cannot be serviced, or if spare parts cannot be obtained to service the product, it might not be a good candidate for development. In such cases, time and cost of repair are very high and cannot be sustained over the long run.

For extremely complex products, such as machining centers or glass grinding machines, the design of manufacturing systems or a special process might cost much more than the cost of design and development. Castings of beds or columns could be very difficult to procure within budgeted resources. So, supply chain considerations must be addressed before any such project is taken up. Getting the castings in small quantities could be cost-prohibitive to the point that the project might have to be discontinued. The manufacturing facilities for critical parts such as spindles, ways, etc. must be considered before the component design is finalized.

In order to consider the manufacturing aspects of any product, the following questions should be asked:

- Are the parts to be manufactured in-house or out-of-house facilities?

- What is the criticality or risk of part manufacturing?

- What is the design of the supply chain for raw materials, manufacturing, and heat treatment?

- When should the key suppliers be involved in the design?

- Does the product use brand-new technology?

- Is the existing facility good enough or capable of producing parts of the desired quality?

- Does the manufacturing system have quality personnel to inspect the parts?

- Does the manufacturing system have inspection equipment to inspect the parts?

Another important point to consider is the provision for servicing and repairing the machine within the shortest possible time. For machine tools, servicing and providing spare parts are very critical to keeping downtime to a minimum. The company has to have a strategy of servicing the machines. Service and warranty cost strategy might be a significant factor for designing the number and type of parts. A simple assembly with a minimal number of parts is very easy to maintain. A design for assembly and manufacturing must be given during the design phase. A strategy for warranty cost and serviceability must be considered before the design is released. Assemblies must be designed for replacement in the field. Modules or sub-assemblies must be designed in such a fashion that they can be replaced, aligned, and tested with minimal difficulty. The design should satisfy the mean-time-to-repair criterion for any module.

The next point to consider for any machine tool project is the economic and legal environment of the country. Also, environmental considerations must be acknowledged to produce parts that satisfy requirements or maintain sustainability. For example, a machine must be designed to accept biodegradable coolant, and parts must be designed to have materials that are environmentally friendly. All components could be rem-manufacturable or recyclable or both. Recently another consideration has come to the surface: energy management and efficiency. The machine must be designed to be energy efficient. The heat management of any machine tool will also enhance the accuracy of the machine overall. Heat distortion, heat dissipation, and heat management must be considered in the design of the product. In short, the machine must be energy efficient to reduce power consumption during machining.

6.9 Summary

Product planning is a very critical job that every manager must carry out before any product is taken up for development. The management team has to consider the planning phase as a basic strategic tool that needs to be deployed for a successful venture. The process is also very complex and multi-disciplined. In general, the team has to satisfy the following questions:

- Is the product planning in alignment with the overall competitive strategy of the company?
- Is the product planning in alignment with customers' requirements?
- Does the product have a specific well-defined market target?
- Is the product planning in alignment with the business strategy of the company?
- Does the product satisfy the company's financial goals?
- Can the company live without taking up this product in the near future?
- Does the product have competitive advantages over competitive products?
- Does the product satisfy company resource availability?
- Does the product strain the resource position of the company?
- Can the product be produced in collaboration with other companies?
- Does the company have the intellectual capability to design, manufacture, and service the customer?
- Did the team consider financing alternatives for the project?
- Is it an exciting product that will shake the customer base?
- Is the technology ready and proven for use in the product?
- Did the team consider risks associated with the product's introduction?
- Are the company's personnel excited about the new product?
- Has the mission statement of the product been completed, reviewed, and agreed upon by most of the stakeholders?
- Has the planning process been documented for future use and improvements?
- Does the development team have enough knowledge and capability to design this product?

- Is an innovative product better than a derivative product?

As mentioned in the previous chapter, the planning phase ends with the generation of a mission statement for the product. An audit of the whole process must be carried out to make sure that the process did not leave any stone unturned. This must be an iterative process, creating a product that the internal team will love to design and that customers will be proud to own. Time and resources must be spent to go through this process since the cost of downstream processes will be way more. Hence, every step and every decision must be questioned to get the best answer. The next step of the process is to deliver the mission statement to the product design and development team for further design exploration. Any flaw or shortcoming in the planning stage will amplify eventually during future steps, costing the company more than what the project should.

This chapter explains the product planning process, which could differ from company to company depending on the nature of the products or machines the company would like to deal with. In general, the generic product planning process is somewhat a systemic one that could be applied to various machine tool applications. Nevertheless, the planning process is an essential and critical step for the company. The process must be open for review, criticism, iteration, and re-iteration to generate a flawless plan for the product. The plan must be devised concurrently with existing projects, and a total product planning map must be generated for the company for many years to come. The product plan could be a roadmap to the company's success in the market. The product planning process for any product must identify limitations, constraints, risks, and ability of the company to produce or develop another product.

The product planning process must consider the following:

- It is a periodic process, and the roadmap must be continually updated as necessary depending on market conditions and the company's financial position.

- The planning process must consider the product portfolio of the company in totality.

- The product planning process involves several systemic steps, which include: customer and target identification, project evaluation and suitability, resource allocation, mission statement generation, and product portfolio mapping of the company.

- Separate considerations must be given for different types of products: innovative products, derivative products, facelift products, and platform products.

- Consideration must be given to whether to buy or make the product. To establish the company's image, the company must design, manufacture, and deliver innovative products and should not become a dealer for other companies.

- Potential product development projects should be evaluated against the financial strength of the company, the intellectual capability of the company, the historical background or legacy of the company, the technological background of the company, and the existing product platform, among several other things.

- A balanced product portfolio must be developed wherein the company should have innovative products, platform-based products, derivative products, facelift products, and cash-cow products. The primary objective is to generate the best possible revenue and profit for the company.

- The profit generated must be optimized under financial and manpower constraints.

- Aggregate planning programs can be used to develop a project portfolio for the company, where all projects compete for company resources

- The mission statement for the product must include and satisfy financial targets, the competitive landscape, cost and quality targets, manufacturing targets, stakeholders, and features of the product.

6.10 References and Bibliography

Abernathy, W.J., and Townsend, P.L., 1975, "Technology, Productivity and Process Change," *Technological Forecasting and Social Change* 7(4), 379–96.

Cooper, R.G., Edged, S.J., and Kleinschmidt, E.J., 1998, *Portfolio Management for New Products*, Perseus Books, Reading, MA.

Devinney, T.M., 1987, "Entry and Learning," *Management Science* 33(6), 706–24.

Foster, R.N., 1986, *Innovation: The Attacker's Advantage*, Summit Books, NY.

Griffin, A., and Hauser, J.R., 1993, "The Voice of the Customer," *Marketing Science*, 12(1), 1–27.

Groeneveld, P., 1987, "Road mapping Integrates Business and Technology," *Research-Technology Management*, 40(5), 13–19.

Hopp, W.J., and Spearman, M.L., 2000, *Factory Physics*, second edition, McGraw-Hill Erwin Companies, NY.

Kinnear, T.C, and Taylor, J.R, 1995, *Marketing Research: An Applied Approach*, fifth edition, McGraw-Hill, NY.

McGrath, M.E., *Product Strategy for High-Technology Companies*, McGraw-Hill, NY, 1995.

Meyer, M.H., and Lenhard, A.P., 1997, *The Power of Product Platforms*, Free Press, NY.

Nahmias, S., 2000, *Production and Operations Management*, Fourth Edition, McGraw-Hill Erwin Companies, NY.

Porter, M.E., 1985, *Competitive Advantage: Creating and Sustaining Superior Performance*, Free Press, NY.

Ulrich, T.K., and Eppinger, S.D., 2011, *Product Design and Development*, McGraw-Hill Companies, NY.

Urban, G.L., and Hauser, J.R., 1993, *Design and Marketing of New Products*, third edition, Prentice Hall, NJ.

Wheelwright, S.C., and Clark, K.B., 1992, "Creating Plans to Focus Product Development," *HBR* 70(2), 70–82.

6.11 Review Questions

- Explain the technology S-curve for the digital control technology of machine tool applications. How can a company use the technology curve to develop a machine tool product in the future? What is the effect of the technology curve on the financial aspects of the company developing a brand-new machine using a technology that is in its adolescence?

- What is the effect of engineering manpower availability on an R&D-oriented company that would like to develop a high-technology product in the shortest possible time? How can the company adjust the engineering manpower requirements for the project? Is it absolutely essential that the company have highly trained engineers to develop a platform-based product?

- Create a product-technology roadmap for a grinding machine or machining center. Should a company stay in its own area of expertise or venture into a new line product to increase its revenue? Does the availability of technology affect the development of a platform-based product?

- Compare a product line between a Japanese machine tool company and a U.S. machine tool company? Do you recommend developing machines in the area of expertise? Is the development of a brand-new innovative product critical for the success of a machine tool company?

C H A P T E R 7

7 Specification Development for Machines

7.1 Introduction

Once the product mission statement has been developed and the project has been handed over to the product design and development team for downstream work, the first task that the PD team has to complete is to determine the target specifications for the product. This normally means developing the system target specifications the product has to shoot for to satisfy customer needs. Let us say the PD team wants to develop a CNC lathe and the budget and timing have already been approved by management. This particular product is a new product for the company and its PD team. The company has been very successful in designing mechanical turning lathes, and it has decided to go for a new computer-controlled turning machine. The firm has taken a stand to widen the product portfolio using the latest analog control and servo systems with the highest speed and power possible in this particular segment. The product should create a high value for the customer who wants higher production for parts. The mission statement also suggests that the machine has to combine turning and milling operations in a single machine to add more value for the customer. The company basically is migrating from making mechanical lathes to a highly productive and automated lathe business.

At this point in time for the project, customer requirements have already been documented for the specific machine. The team has reviewed the benchmarked products and talked to lead and legacy users about their experiences with the existing machines. The team has been able to identify gaps between features and what is required for the future product. These could be technology gaps, feature gaps, capability gaps, quality gaps, reliability gaps, and cost issues. These assessments must be done before design and development activities are taken up. The team talked to the dealers and supply chain elements and vendors to identify their potential to imbibe new technology or new processes required for the new product. In a nutshell, the team has a pretty good idea about what the team has to deliver to satisfy customer needs. The PD team also has a product mission statement document as a guideline of the new product. The team now has to plan the downstream functions.

The downstream challenges could be numerous, and some of them could be outlined as follows:

- The first and foremost challenge for the team is to find ways of converting customer needs into a viable product

- The team has to define precisely what a successful machine is, i.e., the criteria for assessing a successful product in the marketplace

- How to design a product that will satisfy customer requirements offering quality and capabilities at an affordable cost?

- What is the new technology the product must have to be competitive or stay ahead of the competition?

- How to gain market share and enhance revenue to stay in business?

- What are the priorities for product features that the PD team has to satisfy?

- What is important: cost, quality, capability?

The present chapter deals with the generation of product target specifications that the team will strictly follow and deliver. The customer needs for a CNC lathe are given and prioritized in Table 7.1. Most of these specifications are very typical for any CNC lathe development. Nevertheless, specifications for the machine must be developed beforehand to avoid confusion down the line. This table represents very clearly the customer wish list in comparison with the available product in the market. It is important to note that customers very seldom give feedback as to how to design the product or incorporate these specifications into it. Sometimes, for some specific fixturing or loading/unloading devices, customers can help the design team develop the assemblies as per the customer requirements. It may not be a bad idea to let the customer design the product to some extent. Customers might feel a part of the product design team in this way. Another aspect while translating customer requirements into the specifications of a product is the interpretation of the customers' comments.

Proper interpretation and explanation might be necessary to understand customer requirements. The requirement interpretation should be specific and objective in nature. Another discussion point is to lay down the specifications that could be precisely evaluated and tested. Requirements must be measurable quantities; otherwise, they are not specifications. Basically, specifications provide an unambiguous definition of the product's properties. Hence, the specifications must represent the exact features that the product will have when designed and produced. Sometimes customer requirements are ambiguous. For example, the customer says, "The machine must have sufficient power and speed," but he or she does not specify any specific values for the parameters. The team has to come up with the values to make the product more competitive in the marketplace. In some companies, product specifications are also called "technical specifications" or "engineering specifications." The label is not important, but the content is.

Item Number	Assembly	Customer Needs	Priority	Remark
1	Machine bed	Lightweight	5	Cost reduction
2	Machine bed	High stiffness	5	Value addition
3	Machine bed	High sampling	4	Needed for hard turning and tool life
4	Machine bed	3-point support	3	Competition has 3-point support
5	Machine bed	Thermal distortion and dissipation	2	Required to enhance accuracy
6	Indexing unit	Indexing time	5	Less than 1.5 sec. for one station indexing
7	Indexing unit	Number of tool stations	5	Minimum 8 stations
8	Indexing unit	Servo control	3	Hydraulic alternative
9	Slides	Axes speed	5	Minimum 30 m/min
10	Slides	Linear bearing	5	For high speed
11	Slides	Ground ball screws	5	For accuracy
12	Spindle	Speed	5	6000 RPM or better
12	Spindle	Bearing arrangement	5	Fixed-floating
13	Spindle	Power transfer	5	Belt-driven
14	Spindle motor	Servo-controlled	5	20 HP or better
15	Spindle motor	Mounting	5	Foot-mounted/adjustable for belt tensioning
16	Control	Digital	5	Servo-controlled spindle and axes drives
17	Hydraulic power supply	Integrated unit	5	5 HP or less for indexing

Table 7.1: Product-specific customer requirements matrix with priority

As mentioned earlier, if a requirement does not have a metric or measurement method, it is not a requirement. For example, "index time should be very fast" is not a requirement, but "index time should be less than 1.8 seconds" is definitely a requirement since index timing could be precisely measured and validated against the requirement. Moreover, the value also must have a unit of measurement, such as second for time, horsepower for power, kilograms for mass, etc.

The product specifications must be established before the actual design of the product starts. For some items, fixing the specifications before the PD stage might be difficult. For example, the weight of a machine can never be assessed unless a prototype is built and the weight is physically measured. For such items, there could be two requirements: target specifications and actual specifications. Another example could be the amount of drag force for the slides. Unless the slide is built, the drag force cannot be measured. Target specifications for the items tend to be a wish list in the beginning of the project. Target specifications must be reasonable and realistic to achieve. In extreme cases, these values might have to be adjusted depending on the outcome of the test of the product. In any case, a value will be established before the product is introduced into the market. Target specifications and actual specifications might help the PD team to compare several possible designs, and the design with the highest number of satisfied specifications might stand out and be taken up for further design and development.

Another aspect of fixing a specification, one that I have witnessed several times, is the cost containment of the product. Normally, using a brand-new off-the-self technology might price the product out of the market. This Is a serious issue to deal with. So, there is an underlying risk when satisfying a target specification if it is not reasonable. Another point is that the product must have actual producible specifications for it to be marketable. The target specifications normally try to put the product ahead of the competition. So, a target specification must not be selected at random just to look better. For example, using a technology that has never been tried before has a tremendous risk associated with it. The technology might not be proper or mature enough for the product at the time of introduction. Instead of spending lots of time and money to integrate the technology into the new product, a proven technology could be used before introduction and eventually upgraded with new technology after the product is successful in the marketplace.

Another valuable point during specification finalization is the trade-off. A company might have a legacy of producing very high-speed spindles that are not available from other competitors. In that case, other specifications might be traded off to keep this particular feature in the new product. In any case, specifications must be revisited several times during product design and development. In general, if a technology or specification does not enhance quality or reliability, or if it increases the cost substantially, it should be avoided for all practical purposes. The general process is to set up the target specification at the outset. The prototype is tested to prove the measure or specification. Once proven, the target specification becomes the final specification if the cost is not prohibitive.

Any specification must have a cost advantage added to it. If the product is not cost effective, it should be avoided even if it adds some apparent value. So, two phases of product specifications might be a better approach instead of fixing arbitrary specifications that a customer cannot afford. At the end of the day, the PD team delivers the product with agreed-upon specifications that customers would like to have. The PD team has to satisfy specifications, resource requirements, and return on investment for the business. The specification or requirement matrix becomes the bible for the PD team until the product is developed and marketed successfully.

7.2 Target Specifications Development

The target specifications should normally be established after customer feedback is obtained and before concept generation is completed. In any case, target specifications must also be specific and designable. They should have metrics similar to final specifications. It has been found that more aggressive target specifications make the product better than the competition, but the specifications must be realistic and obtainable.

The process to determine target specifications should start with customer needs. Also, collect the specifications for the similar established products available in the market, including cost. Select the specifications that satisfy customer needs. Compare these values and establish their range for each specification. Also, decide on their priority. A list of metrics for the turning machine is displayed in Table 7.1. In some target specifications, the range might not be necessary, and the mean or most-accepted value is inserted to reduce the confusion. In this target specification table, the most important element is the relationship between customer needs and specifications or requirements, which should be measurable. The promise is the fact that the satisfaction of these metrics will lead to customer satisfaction with the product at the end (see Fig. 7.2 also).

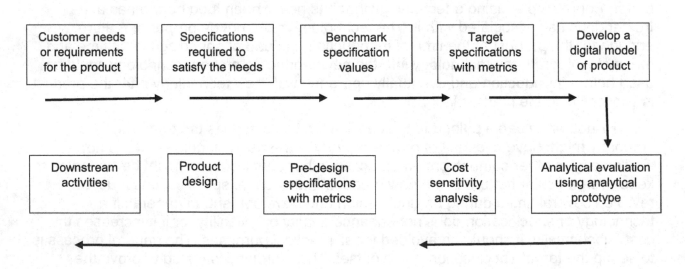

Fig. 7.2: Product development process block diagram

The most difficult metric to attach a value to is the time required for servicing the machine. Most of the time, it is denoted by mean time between failure i.e. MTBF. The servicing efficiency could also be denoted by mean time to repair, or MTTR. For servicing, metrics for MTBF and MTTR must be added. A typical value for MTBF is around 2,000 hours or less. The product design has to ensure that servicing could be done with minimal effort and within the shortest possible time. Servicing inefficiency is one of the primary factors that make customers very irritated and unhappy. This could be one of the reasons why customers slowly switched to Japanese machines from American-made machines.

The availability of spare parts for servicing is another important factor. The budget must include the cost associated with the service and warranty of the products. The design has to be robust enough to reduce the time between failures, or it should be geared up for no failure at all within a reasonable expected time of use. If machine failure is very frequent and at random, customers get very unhappy even if the machine's productivity is very high. They look at the cost of the machine over its life. The unexpected production downtime makes the machine highly unprofitable. The design must be very robust to increase the time between failures, and it is the key element for customer satisfaction.

Sometimes needs can conflict with each other. For example, in order to have high bed damping, the machine bed has to be comparatively heavier to absorb vibration due to cutting conditions. To satisfy this need, the weight of the machine will have to go up, and the cost will go up proportionately, too. In such cases, a happy medium depending on the previous experience must be sought for. On the other hand, increasing bed weight might increase bed deflection. So, weight, stiffness, damping, and associated cost are interacting parameters that need to be optimized by the design team. Nevertheless, each specification must have a metric for validation.

The metrics should include all the customer requirements as much as possible. If the list of requirements becomes too lengthy, it should be prioritized, and important ones should be dealt with first. In general, the product should satisfy customers and make the company more competitive and profitable in the market. A single requirement might need several specifications. Moreover, one specification might satisfy more than one need simultaneously. For example, machine weight is a dependent variable depending on the shape, materials used, and cross-sectional properties desired. The importance rating of any specification depends on the importance rating of the requirement.

For the must-have requirements, the importance rating and priority should be very high and must be included in the product design. The composite ratings of the requirements and their specifications must be considered in consultation with all the team members or stakeholders of the product. On the other hand, the material used for the bed could be an independent metric. In general, performance metrics are the parameters that a machine is designed and evaluated by. They are a guideline for the design team.

The designers should have the freedom to select a material or a design form or shape to satisfy the weight requirements. Satisfying some of the requirements is not good enough; all machine requirements must be totally satisfied for highest customer satisfaction. Moreover, requirements and desired performance metrics must be realistic.

The bottom line is the cost of a machine as delivered to the customer. The performance of the product must be evaluated for conformance to the specification and its metrics. In other words, each requirement must be related to specification and its metrics, and the machine must be evaluated against all these specification metrics for confirmation. Any deviation after product validation must be noted, and the reason for the deviation must be determined. These final specifications might be different from the target specifications. The machine might have to be redesigned to include deviations.

Frequently, some customer needs cannot be converted to metrics or design parameters that could be evaluated or designed in the product. For example, the fact that owning a precision machine is a matter of pride cannot be easily translated into machine specifications for the product. In such cases, the design has to have uniqueness, or machine features have to be something special that the owners can be proud of. So, ownership pride can be indirectly satisfied using the contents and design of the machine. Most of the time, the evaluation of subjective requirements is left to customers. Feedback can be obtained for further design refinements in the future. A panel of judges can be established to evaluate the machine against subjective requirements.

Requirements and specifications with metrics must be in alignment with other competitive products. In other words, the product must be competitive, but that does not mean that it has to be in line with low-end products. The product must be, in my opinion, competitive with respect to cost, features, quality, and reliability. The product and its usefulness are decided by users. Nevertheless, an evaluation of the product by an independent audit team before introduction to the market must be made to gain confidence and suitability in the desired marketplace. The audit team can start the evaluation using the mission statement, the specification matrix, and performance measures. The design characteristics and performance criteria could be established beforehand for comparison with other similar products in the market.

In order to show relationships between a requirement and associated metrics, a house of quality can be developed, as shown in the "Product Quality" chapter of this book. This house of quality is also known as quality function deployment, or QFD. It provides a roadmap for the company to translate customer requirements into specification requirements. QFD also translates customer requirements into technical requirements that the company must meet to satisfy the customer.

In order to complete the specifications and performance metrics, in general, some guidelines should be followed, as described below:

- **Performance metric must be specific to the requirement and unambiguous:** Each customer requirement will correspond to a specification, which must have a metric value. In case a customer requirement needs more than one performance metric, it must be split into several definable segments with a performance metric for each one. A metric is a measure of performance for the product against a specific requirement. Specifications can be also dependent on each other. For example, the weight of a machine is dependent on the design and material selected. In such cases, the weight should be a dependent metric, and designers should select the material and design to satisfy it. The performance metric for the weight is a specific value that the product's end performance is measured against. In other words, each specification must have a metric value for validation purposes.

- **Performance metrics must be realistic and reproducible:** The list of specification and performance metrics should not specify how to obtain them, but it should pinpoint the requirements in clear terms. For example, the index time for a turret is a metric to be satisfied, and the design should conform to this requirement at the end.

- **Performance metrics can be used for comparison with benchmarked values:** This is a very important aspect of the target specification for the product to satisfy a specific market. This makes the design a practical, productive, desirable, and feasible product that can compete against other products.

- **Performance metrics must be obtainable within reasonable cost:** It should be mentioned that the cost consideration for each metric must be well thought out before implementation. The cost containment of each metric is a very important consideration. The amalgamation of cost and design is the single most important step in product development. If the product is not cost-competitive, it cannot be sold in a highly cost-competitive market.

- **Metrics must be measurable and objective**: Some customer requirements are subjective by nature, and translating that type of requirement into a quantifiable number is often very difficult in practice. Subjective metrics, such as look, customer pride, etc., could be evaluated by a panel of experts for further explanation. Metrics must contain the industry benchmark criterion so that they can be compared against the competitors' values. Metrics must have specified units as well so that buyers can compare the product's value against all the competition.

In order to understand the dependency of each performance metric against the need, a need vs. performance metric matrix (See Fig 7.1) can be developed. This combined matrix will show the relationship between each performance metric required to customer requirements. This also displays whether a performance metric is related to several needs or not. For example, damping is dependent on the weight of the bed, material selected, and design of a machine structure. If the matrix is diagonal, the need and its performance metric are singly related and such a situation is the best to start with.

The priority of performance metrics also has to be established. The priority rating of the performance metrics will depend on the importance of the customer needs that they satisfy. If the importance of satisfaction of any specific requirement is relatively more important than others, the priority setting of the performance metric must display that as well. For a one-to-one relationship between a need and its performance metric, the priority of the metric will depend primarily on the priority of the need. In any case, needs must be prioritized, and the performance metric must be realistic and measurable using a predetermined method of evaluation.

The next step is to collect benchmark data for the product, as shown in Table 7.3. For some companies, the benchmark is its own product. It is extremely important to understand competitors' products in relation to the technical requirements of the product. This link determines the commercial viability and success of the product in the marketplace. Target specifications must be compared with those of competitors' products in order to position the new product against existing ones.

Users in the machine tool industry often show very strong loyalty to some manufacturers because of their good experiences with their product. In order to break that bond, competitors' products and their features must be understood. For each requirement, benchmark data is collected and recorded for comparison as shown in Table 7.3. This is also a significant portion of the quality matrix.

This is a simple idea, but it shows a lot of information for the product. Gathering such data for the competitive product is time-consuming and sometimes very difficult as well. The process of collecting such data might involve several departments, such as marketing, sales, service, and purchasing. Even if this process of generating data for competitive products is difficult, it is also very important and necessary for the new product to be successful. It is also sometimes necessary to buy the most competitive and well-known successful product and dissect it to get the metrics. This is a very good step for understanding the product. The published data might not be sufficient for the machine, and the data needs verification to be sure. Moreover, the benchmarking chart is very simple to construct once the data is available.

Another survey could be conducted to understand why some users prefer some manufacturers. The preferences could be also shown in the form of a table. This chart sometimes shows the customers' preferences and perceptions for the products. It is also true that one product cannot satisfy all the requirements.

Another table could be constructed where the preference of companies satisfying each need can be shown. The methods and tools for measuring customer satisfaction were analyzed by Urban and Hauser in 1993. One such table is shown in Table 7.4. One product cannot have all the best possible metrics even if it is from a very reputed manufacturer. Moreover, a new product should also not be designed to have all the best features embedded in it, because the cost would be prohibitive. The idea is to identify what customers would like the most and what the new product must have to be competitive in the marketplace at a cost that customers can afford.

The next step is to set the best and the second-best value of each requirement by comparing competitive products and customer preferences. Hence, all the available data should be combined together to come up with a synthesized target specification value for each requirement for the new product. It is also sometimes useful to have the second-best value or range of values for each metric. The best value is the best possible value for each metric. The second-best value could be the minimum that each metric must have to satisfy the customer. These are required to make the new product commercially viable and cost-competitive. These target values will be eventually used for concept generation and selection.

Once the concept has been selected for further redesign, some of these values could be adjusted or compromised to make the product cost-competitive. Quality is never compromised, though. The values of each metrics could be expressed in many possible ways at the outset. For example, spindle power could have a 20 HP minimum, or at least 20 HP. The idea is to identify the minimum power for the machine, so in this example, the machine will not be suitable if it has less power than 20 HP. Another way is to have the highest possible end of the metric. For example, stiffness for a bed could be, say, 150,000 lbs./inch. Anything more than that would make the machine very heavy, and material would be wasted. For a normal cutting operation, a higher value than this would be unnecessary. This would only add to the cost of the machine.

Another way of expressing the target value would be to specify the range for the metric. The station-to-station index time could be between 1.6 and 1.8 seconds. Any value less than the lower value would make the machine non-competitive, and any value more than the higher value would add cost without adding to the commercial value of the machine. Sometimes it is necessary to have an exact value of a metric. For example, the number of turret faces has to be 12. No further deviation from this metric would be allowed for the new product. Such values impose additional constraints in the design of the product and should be avoided. For machine tools, such exact values of a metric are very commonly used. Sometimes a discrete value can also be used. For example, the same machine can have 8, 10, or 12 turret faces.

Sometimes the metric values are dependent on each other. For example, the higher the weight of the machine bed, the higher the damping value. More weight sometimes lowers the stiffness if not designed properly. Such details could be avoided in the beginning for the target values, but they can be tuned to exact values once the product is prototyped and tested. At that point, the values will be called final specifications for the product. The finalized target specifications for the machine can be summarized in a table, as shown in Table 7.5. The team creates the target values for each requirement. One must keep in mind the cost effects of each requirement.

Keeping in mind competitors' values for each requirement and product capabilities, the product mission statement, the product cost, and the company's ability to reproduce these values, the team should prepare the target value list. Each requirement has a range of values to be satisfied. These will make the product commercially competitive. Sometimes it is also necessary to keep the immediate future trend of each metric and use the highest possible value. For example, the higher the axes speed, the greater the productivity of the machine. So, the design should accept the highest possible value of the axes speed. If possible, the value could be set slightly higher than what it is possible now or available in the market.

It should also be remembered that these values are only targets for each requirement and not the final specifications. Final specifications can only be made once the product is designed, the prototype is built, and the metrics are validated. The team has to have these values in mind while the concept is generated and a specific concept is selected for further detailed design. These values could also be used to select a particular concept out of many possible concepts.

Performance Metrics ⟶

Needs ↓	Weight less than 10000 lbs.	Stiffness higher than 2e6 lbs./inch	Damping 3 times higher than steel	Non-deformable support	Indexing time less than 2 seconds
Lightweight	X	X			
High Stiffness	X	X			
High Damping	X		X		
Three-Point Support				X	
Indexing Time					X

Figure 7.3: Needs vs. performance metrics matrix

In order to create the best possible target values for the requirements for the specific product to be developed, the team can accept the best possible value, the marginally accepted value, or values in between. There is no norm for such a selection. For lathes, it has been found that the accuracy of machining depends highly on the damping and stiffness of the machine at the cutting point. So, the new machine must be designed to have the highest possible values for these requirements. On the other hand, conveyor capacity or the color of the machine is not a customer priority. So, a minimum value for these requirements can be selected depending on past experience or to keep the costs at a minimum. There is no ideal target value for any metric. The composite values of all the requirements must be selected in unison to develop the best possible product with a cost that is competitive.

It is better to have an acceptable range of target values, a minimum and a maximum. The product must be commercially viable and competitive in the market to be successful. Anybody can design a product, but designing a cost-competitive and reliable quality product is always a challenge. Machines with the highest possible target values do perform better, but very few buyers can afford them due to their cost. This is especially true when the market is very competitive and too many builders are offering similar products. The highest quality possible at an affordable cost should be the norm for designing any new product, and the quality and cost should be specified very clearly before the machine is designed.

The process of selecting the target specification matrix is a highly iterative process. The target value selection should also be a very thoughtful process. At every step of this process, before the target values are selected, team members must reflect on the values of each metric and make sure it is feasible to produce them. The team should select aggressive values for the metric, which will make the product most productive and cost-competitive. It is also possible that all the best possible values cannot be baked into one single product. The product design can be done in phases. The introductory phase of a product can have most of the design features, and then a facelift product can pick up the rest, depending on feedback from customers about the introductory product.

When the product needs to be too versatile in nature to be competitive, it is better to proceed progressively, and multiple products instead of a single one is a better choice. Sometimes I have also seen that in spite of all the attention given to it, a product might miss a crucial specification. Hence, iterating the selection process is very helpful for filling all the holes before the product design is taken up. At the end of the process, the specification list must satisfy most of the customer wants at a cost that most of the customers can afford.

The proposed target specifications for the product are shown in Table 7.5. It shows the range of the metrics for each requirement. Once these specifications are available, the team builds a digital prototype to confirm the specifications theoretically and conducts finite element analysis, reliability analysis, strength analysis, buckling analysis, manufacturing related analysis, etc. Once the specifications are confirmed, an intermediate target specification list is created, and the team starts conceptualizing the design to satisfy these metrics.

Metric Number	Need Number Reference	Performance metrics description	Priority	Values with Units
1	1,3	Machine bed weight and damping	1	Less than 10,000 lbs. and 3 times more damped than steel
2	2	Machine bed Stiffness	1	Higher than 60,000 pounds per inch in vertical direction
3	4	Bed supports	3	Equal distance from the CG of the bed
4	5	Bed thermal dissipation rate	2	1000 BTU per degree F
5	6	Indexing time, one station	1	1.8 seconds or less
6	7	Total turret tooling stations	1	8 or 12 Stations
7	8	Servo index drive	1	5 HP servo motor at 2000 RPM
8	8	Servo control	2	Fanuc control
9	9,10,11	Axes speed	1	Minimum 30 m/min
10	12	Spindle system	1	6000 RPM maximum speed
11	13	Spindle power	1	20 HP minimum
12	14	Spindle speed	1	6000 RPM or better
13	15	Spindle control	1	Fanuc
14	16	Machine control	1	Fanuc 0T
17	Hydraulic power supply	Integrated unit	5	5 HP or less for indexing

Table 7.4: Product-specific customer requirements matrix with priority

Metric Number	Need Number Reference	Performance Metric	Priority	Product A, Company A	Product B, Company B	Product C, Company C
1	1,3	Machine bed weight and damping	1	12,000 lbs.	8000 lbs.	11000 lbs.
2	2	Machine bed stiffness	1	85000 lbs./inch	65000 lbs./inch	95000 lbs./inch
3	4	Bed supports	3	Triangular, 3-point support	Triangular, 3-point support	Triangular, 3-point support
4	5	Bed thermal dissipation rate	2	1500 BTU per degree F	2500 BTU per degree F	1200 BTU per degree F
5	6	Indexing time, one station	1	1.6 seconds	2.0 seconds	1.6 seconds
6	7	Total turret tooling stations	1	12 stations	12 stations	12 stations
7	8	Servo index drive	1	5 HP servo motor at 2000 RPM	5 HP servo motor at 2000 RPM	3 HP motor hydraulic at 1400 RPM
8	8	Servo control	2	Fanuc control	Own control	Fanuc control
9	9,10,11	Axes speed	1	30 m/min	32 m/min	28 m/min
10	12	Spindle system	1	6000 RPM maximum speed	6000 RPM maximum speed	8000 RPM maximum speed
11	13	Spindle power	1	20 HP	20 HP	20 HP
14	16	Machine control	1	Fanuc 0T	Own control	Fanuc 0T

Table 7.5: Benchmarking chart for competitive products

Metric Number	Need Number Reference	Metric	Priority	Product A, Company A	Product B, Company B	Product C, Company C
1	1,3	Machine bed weight and damping	1	Most preferred	OK	OK
2	2	Machine bed stiffness	1	Not satisfied	Best	OK
3	4	Bed supports	3	OK	OK	OK
4	5	Bed thermal dissipation rate	2	Best	Worst	Medium
5	6	Indexing time, one station	1	Best	Worst	Slowest
6	7	Total turret tooling stations	1	Highest	Highest	Highest
7	8	Servo index drive	1	Best in market	OK	Best in market
8	8	Servo control	2	Best	OK	Best
9	9,10,11	Axes speed	1	Medium	Best	Worst
10	12	Spindle system	1	OK	OK	Best
11	13	Spindle power	1	OK	OK	OK
14	16	Machine control	1	Best in market	OK	Best in market

Table 7.6: Benchmarking chart for satisfaction of customer requirements

Metric Number	Need Number Reference	Metric	Priority	Minimum value	Maximum value	Comments
1	1,3	Machine bed weight and damping	1	10,000 lbs.	12000 lbs.	Higher the better
2	2	Machine bed Stiffness	1	90000 lbs./inch	95,000 lbs./inch	Higher the better
3	4	Bed supports	3	3-point support	3-point support	4-point support not required
4	5	Bed thermal dissipation rate	2	1500 BTU per degree F	2500 BTU per degree F	Higher the better
5	6	Indexing time, one station	1	1.6 seconds	2.0 seconds	Lower the better
6	7	Total turret tooling stations	1	10 stations	12 stations	Higher the better
7	8	Servo index drive	1	5 HP Servo Drive	5 HP servo drive	No hydraulic
8	8	Servo control	2	Fanuc control	Fanuc control	Should accept other control
9	9,10,11	Axes speed	1	30 m/min	32 m/min	Higher the better
10	12	Spindle system	1	6000 RPM maximum Speed	8000 RPM maximum Speed	Higher the better
11	13	Spindle power	1	15 HP	20 HP	Higher the better
14	16	Machine Control	1	Fanuc 0T	Fanuc 0T	Standard Control

Table 7.7: Target specifications for the product

7.3 Pre-Design Specifications Development

After the initial specifications and metrics are developed, another iteration of these specifications needs to be made before concept generation. In this step, the specifications are revisited for a reality check and feasibility of manufacturing. Each specification is refined to a value instead of a range. Another critical step is to allow trade-off among the specifications and requirements. The technical specifications could be divided into must and wish categories. In the must category, a refinement of metrics is necessary to reduce cost and manufacturability. In the wish category, specifications could be eliminated if their cost is prohibitive.

For example, a machine tool bed could be fabricated or cast. If fabricated, the bed material is steel, and the stiffness is very high. If cast, stiffness is less, but damping is comparatively higher than the steel structure. In order to enhance the damping of the fabricated bed, a concrete filling is used at a very low cost. For improved cutting properties, a steel structure with concrete might be a better proposition. Other advantages would be lower cost and time for production. Molds are not necessary for fabricated structures. For a cast bed, a higher volume of production is absolutely necessary to reduce the cost of the machine per unit of production. These conflict with each other. If the production volume is high, a cast bed is preferred because the cost is lower than that of a fabricated one. So, cost is the ultimate criterion for the selection of specifications in the wish category. A trade-off analysis of each specification or performance metric must be made to reduce the cost and enhance manufacturability and technical feasibility.

In order to complete this critical step, a digitally built system is necessary. The approximate concept of the system is designed to satisfy the requirements. The concepts could be borrowed from existing systems. Computer-aided analysis is applied to evaluate the digitally built prototype against each requirement. The primary purpose of the digital prototype is to predict the values of the specification metrics after analytical evaluation. Instead of physically building the prototype, the analytical prototype serves for such approximate evaluations of each metric. For example, an approximate bed structure could be built analytically to evaluate weight, stiffness, ease of manufacturing, location of the center of gravity, location of support points to minimize bed deflection due to its own weight, and other imposed systems.

Similarly, an approximate spindle configuration can be built showing type and number of bearings, distance between bearings, location of the bearings with respect to cutting point, etc. Then the spindle sub-system can be analyzed for force analysis, life analysis, and deflection analysis. Once these values are theoretically obtained, they can be compared with the requirement specification metrics for theoretical confirmation. If any adjustment or changes are required to satisfy the metric, it is much easier to change the structure at a minimal cost. Such analysis also can be done for the index system or tailstock sub-systems. The system-build provides envelope and interface designs for each sub-system. The digital prototype also allows designers to develop an interaction diagram among all sub-systems. Due to the primary nature of such a build, adjustments and refinements are very easy to achieve at minimal cost and time.

Because of the availability of system software and hardware, such builds have become almost mandatory for developing a machine with reduced cost and time. The approximate digital prototype can help to create a dynamic model of the system to develop resonant frequencies, modal analysis, damping characteristics, and impact analysis. The digital prototype also can be used to evaluate the bending and torsional stiffness, stress, strain, and strength properties of the approximate structure. For critical components subjected to dynamic loading during its use, a fatigue analysis can be performed to satisfy the life requirement. In short, both static and dynamic performance of the system could be obtained using this digital prototype even if the system and sub-systems are approximate at this stage of development.

The advantage of a digital prototype is to allow the team to predict the system and component specification metrics analytically without physically building the system or component. For each system or sub-system, an input-output diagram can be built. The system can be analyzed against the inputs to predict the outputs, which can then be compared with the desired specification metric for confirmation. Sometimes it is easier to build sub-systems first, and then a system configuration is arranged using the analytically proven sub-systems. Since this step is to refine the specification metrics to determine feasibility and trade-off requirements, the sequence of building is not that important as long as the relationships between system and sub-systems are maintained. It is also true that each sub-system need not be built for confirmation and that only critical ones need to be built for a quick check.

During my career, I always built an approximate structure and mechanism for the spindle structure, bed structure, index mechanism, and tailstock. Sometimes cross-slide configurations were also analyzed to predict acceleration and deceleration capability, servo motor power and torque requirement, frictional drag, etc. In general, this is a very effective and powerful method of analytically proving the specification requirement, and it often helps to build confidence in the design. In my opinion, such an analysis is a must for a successful product. In case an analytical model cannot be built for evaluation, a physical build has to be completed at the end of the design or during it to conform to the specifications. The analytical prototype also helps to determine the interference aspects of the system and interaction among elements of the system. Once an interference-free system has been developed, the sub-system and system envelope, boundaries, and interface can be approximately fixed for further design activities.

The author has used the analytical model below to determine the following aspects of any concept design:

For spindle configuration, the following metrics were determined before the actual design:

- Optimum bearing spacing to reduce the deflection at cutting point
- Force analysis of each bearing using duty cycle forces
- Optimum bearing lives and resonant frequencies
- Optimum bearing configurations
- Bearing preload required to sustain axial force

- Number of bearings required to sustain external loading

For slide systems, the following properties were determined:

- Servo system requirement for load and time duty cycle

- Friction drags

- Slide acceleration and deceleration capabilities

- Slide bearing requirements

- Lubrication requirements for the slide bearings

- Optimum slide bearing locations

- Ball screw force and life analysis

For tailstock configuration, the following could be determined theoretically:

- Optimum bearing spacing to reduce the deflection at cutting point

- Force analysis of each bearing using duty cycle forces

- Optimum bearing lives

- Optimum bearing configurations

- Bearing preload required to sustain axial force

- Number of bearings required to sustain external loading

For the bed structure, the following aspects were determined:

- Optimum three-point support locations

- Location of center of gravity

- Weight optimization

- Mode shape and resonant frequencies

- Deflection due to gravity

- Stiffness and damping

Hence, at the end of this step, the confirmation of the specification metrics of the requirements have been theoretically accomplished. Moreover, if any metric is not possible to achieve, the value has been readjusted for the specific requirement. Every attempt has to be made to achieve the metrics for the most-wanted requirements of the customers. The target specification list has been redesigned and can be called the "pre-design specifications," or the "must and wish requirements." If some of the wish list requirements cannot be satisfied, they can be traded off for requirements that can be achieved more easily. This is the list of specification metrics that the design team will start designing concepts against.

The next step is to evaluate the cost of the approximate system. This should include the cost of manufacturing and assembly of the system, plus the expected profit. The cost should also include some part of the design cost for the system. The objective of this step is to ensure that the cost of the approximate system does not exceed the target cost, which is a specification metric to start with. The target cost of the product should not exceed the market value of the product. In case the cost exceeds the present market value of the product, the system or sub-system has to be redesigned in order to contain it. Since the design at this stage of the process is not final, the estimated cost must be somewhat lower than the final product cost. The effect of inflation and cost of capital also have to be kept in mind.

To be more accurate and realistic in estimating the cost, cost rationing of components or assembly might be necessary. To contain the cost, the design, materials, manufacturing process, or volume of the production might have to be adjusted accordingly. A cost sensitivity analysis might also be necessary to determine the effect of different materials, designs, or methods of manufacturing. For example, a machine bed might have to be cast instead of fabricated because it would be cheaper. A different manufacturing process might have to be selected, such as casting instead of forging, turning instead of grinding, heat treatment instead of no heat treatment, etc.

Since the design is not concrete yet, the cost associated with the present design will be very approximate. The range of cost for each component could be maintained for future record and adjustment. Several designs could be proposed for a component, sub-assembly, or sub-system to find the effect of the design on the cost of the system. The system must be designed to satisfy the requirements, which include the cost of the system, too. The cost is the ultimate index of success of any product for the company. A successful design must satisfy requirements, performance, and cost. For carry-over design, such analysis becomes somewhat easier.

In case of a new innovative design, a bill of material (BOM) for the assemblies can be created to track the cost of each assembly. This BOM can be updated as the design process develops, and the cost of each assembly should be tracked and controlled on a regular basis. In the past, cost was not a critical element since it was a seller's market and competition was not that common for machine tools. Now it is completely a buyer's market, and cost sensitivity is very high in the machine tool market.

Since the design is in its infancy, the BOM cannot be complete and perfect. Nevertheless, a cost evaluation must be carried out at every stage of the design to make sure it is cost effective. The other way of approaching this issue is to get an estimated selling cost of the product from the sales and marketing department. Then the cost of each major assembly can be rationed or budgeted accordingly and tracked as the design matures. The cost matrix can help the design team be innovative and find cost-effective alternatives. Due to intense global competition, cost has become a predominant index of the design. The cost-effective design that does not affect quality and reliability is a productive design; otherwise, it is just a design but not a productive and successful one. A finalized specification matrix is shown in Table 7.8

Metric Number	Need Number Reference	Metric	Priority	Minimum value
1	1,3	Machine bed weight and damping	1	12,000 lbs. and 0.8 lbs./inch
2	2	Machine Bed Stiffness	1	95000 lbs./inch
3	4	Bed supports	3	3-point support
4	5	Bed Thermal Dissipation Rate	2	2000 BTU per degree F
5	6	Indexing Time, One station	1	1.4 seconds
6	7	Total turret tooling stations	1	12 stations
7	8	Servo index drive	1	5 HP servo drive
8	8	Servo control	2	Fanuc control
9	9,10,11	Axes speed	1	32 m/min
10	12	Spindle system	1	8000 RPM maximum speed
11	13	Spindle power	1	15 HP
14	16	Machine control	1	Fanuc 0T
15	17	Cost of machine	1	$20,000.00

Table 7.8: Finalized target specifications for the product

Once the target specifications are iterated using the analytical or digital model, a final target pre-design specification matrix can be finalized, and the design team has to follow it as the design process is carried out. Final target specifications must be compiled by the team members. The design team has to agree to the feasibility of these specifications. The marketing and sales team has to agree to the market price of the end product at a given volume. The manufacturing team has to deliver the volume at a predetermined manufacturing cost, and hence, they might have to manage the manufacturing process to keep cost and quality under control. The synergy among team members is very much required to make an efficient design that will satisfy customers. In other words, the team will deliver the desired product at a very competitive price and the best quality possibility to the desired market nitch for the product. This will ensure that the company generates revenue for future growth and development.

Once the target specification and its metrics are finalized, another comparison should be made with respect to the existing products in the market. The design specifications and cost can be compared to the existing market leader's specifications and cost before proceeding to the design. It is also a good idea to take some loyal customers and dealers in confidence, share these values with them, and get their feedback for the specifications. Such competitive analysis might be more necessary when the company would like to enter an established market crowded by many global manufacturers. This also gives the company a better chance to position the product in the desired marketplace before actual production starts. This is an idea that automobile companies frequently follow for their future products, using market clinic events. Such comparative feedback might also pinpoint the weaknesses and strengths of the specifications and give the team one more chance to refine the specifications before baking them into the product design. I have experienced such final adjustments for several products during product development.

In some high-volume products, such as automobiles, a conjoint analysis is also carried out to align the specifications with the customers' requirements. For machine tools, this analysis is not warranted and might not be productive enough since the market is very narrow and thin. It is also true that getting full and honest feedback from customers is very difficult since customers are biased against their previous purchase patterns. So, asking a small base of customers repeatedly for their feedback might not be a productive tool, since surveys cost a lot.

Instead, sharing the finalized targeted specifications, cost, and time of entry can be shared with loyal customers and dealers in a localized and customized setting. Such a process is also very economical and productive. Customers, in this case, feel that they are part of the design and development process, and more often than not, they become the first or lead users of the new product since they helped design it along the way. This could be called a "customer-driven design," and I often witnessed such a process carried out in Japan. In such cases, the vendors also participated, and they also helped the design team design the machine in a way to suit their manufacturing process. In this way, vendors, customers, and company become a productive team together. Japanese companies' success as quality machine tool manufacturers and suppliers depends very much on such cooperation and coordination.

Hence, it becomes a concerted effort to make the product very successful. Such feedback can only help the design team be innovative and productive when designing the new product. This also helps to incorporate specifications into a product that will maximize the market share and enhance revenue. The next challenge is to develop sub-system and component specifications from the system target specifications.

7.4 Sub-System and Component Specifications Development

As described earlier, once the system specifications are developed, the sub-system and component specifications need to be detailed before the design of the sub-systems is started. This is not always a very straightforward task, and it imposes challenges on the design team members. For example, the system-level specifications for the machine bed are already developed. This sub-system already has weight, stiffness, damping, and support specifications, as shown in Table 7.6. The bed has to be divided into sub-systems such as support corners, motor support sub-system, way support sub-system, etc. The requirements for each sub-system must be developed without violating the bed system specifications.

For example, all the support structures, when put together, must not weigh more than the specified weight of the bed. It should also not reduce the stiffness or resonant frequency of the system. The boundary and interface between the way block and bed must also be developed. Once such sub-system specifications are developed, the way system design team can follow these specifications for designing the way blocks for the machine. This is in accordance with the left side of the V system diagram described earlier. The development and flow down of the system specifications into sub-system specifications can follow the same process as the overall system development.

This process is also highly iterative and more detailed. Hence, each specification for the sub-system must have a metric against each specification. Similarly, for the servo systems, motor torque, max revolution, acceleration, and deceleration capabilities, duty cycle capabilities must be specified with metrics and followed. The specification for the control should include the dimensions, power requirement, cooling requirements, heat generation and dissipation, component reliability and life requirement, type of electrical and electronics components, and they must be developed and specified in detail for its further design.

In addition to the challenge of developing sub-system specifications that are consistent with the system specifications, another challenge in the development of sub-system specifications is the interaction between them. For example, specifications of the way block must be consistent with the specifications of the slide sub-system. The length and width of the way surface, the cross-section of the ways, the mounting of the ways, and the location of the ways with respect to the cutting point must conform to the specifications of the slide sub-system.

Once the specifications of all the sub-systems are completed, a cross-check has to be made to ensure consistency among the specifications of all the sub-systems and the system. A trade-off might also be necessary for the sub-system specifications. Again, the cost of each sub-assembly must be considered and kept on track before the design is initiated. Hence, the cost of each sub-system must also be specified for the individual design team in addition to the performance metrics of the sub-systems. Once the sub-system target specification development is completed, the specifications will flow down to develop the component specification, which should include, weight, manufacturing method, quality control aspects, materials, stiffness, fatigue life, strength, etc.

The component specifications are the metrics that the components will be designed against. Once the component specifications are completed, a consistency check must be made again. Component performance must be understood before the final component specification is developed. For example, the selection of material for any component might affect the performance of the component, such as fatigue life, vibration aspects, the weight of the component, the machining of the component, etc. At this point, the design and detailing of the components should start. Another critical performance criterion is the damping of a structure, which depends heavily on the component design and material selection. A cast component could have more damping but less stiffness and fatigue life. Sometimes a theoretical model can be put together and analyzed to understand the component behaviors.

Such interacting and conflicting aspects of properties must be understood in detail before designing the component. The individual design team member has all the information required for the design. In order to maintain and satisfy the specifications, the team member has to be very innovative and productive. One more important aspect of designing the components must be kept in mind: the design team has to understand how the elements, when put together, will interact with each other to give rise to the properties for the sub-system.

Such interaction is not always very clear, and theoretical analysis must be made before the component design is finalized. A component is the smallest element of a sub-system. The elements behave quite differently when put together with other elements. This is sometimes quite a challenge for the design team members. At this point, the components are completed and then assembled together to give rise to each sub-system. Now we are on the right-hand side of the "V". Integration is the key process. It is highly serial in nature, and things are solidified and cannot be changed that easily. Once the design is complete, a cost check for the sub-assembly's performance criteria and cost must be made to be consistent.

7.5 Concluding Remarks on the Process

The ultimate test of the success of this rigorous systemic process is a cost-effective, quality, and reliable product that is comparatively easy to design. Since the specifications of each element of the system are pre-defined, it is comparatively easier to design them and, eventually, the whole machine.

The team has to answer an important question to determine if the product is a success or not. Does it satisfy customer requirements at an affordable cost? Is the product competitive enough to generate revenue for the company? Does the product show innovativeness and productivity? Does the product have some unique characteristics that will attract new customers? If some of these important questions remain unanswered after this process is completed, the team has to redo it to answer these questions in more complete form. The team has to revisit the specifications to bring some more added features and performances to the product by enhancing the specifications. So, an iterative approach has to be taken before moving to the next step of design and development.

In case any specification has been avoided or a new specification has been added to the product specification, what is the risk taken for the product's success in the marketplace? Will the added feature or specification delay the project by a considerable amount? Is the added specification really necessary for the product to be successful? Is the risk avoidable? The risk, probability of occurrence, and uncertainty must be considered for each action at this stage of the game. What are the steps that the team has to take to mitigate the risk and increase confidence in case some of the assumptions made do not come true down the line? Can the risks be avoided to some extent? Such risk analysis for the product must be made before the actual design is taken up.

Most of the time, a machine tool's design looks like others already present in the market. If the design look of the machine and other performance features of the machine are substantially different from others, success could be difficult. So, the team has to critically review the preliminary concept in comparison with the other benchmarked products or standards for the industry. Can the concept be sold to the customers for the price it demands? What are the innovative features of the design that make it a different product? It is also possible that the product has too many features, making it way more costly or complex to use than other products in the market. Can the concept be simpler to reduce the cost?

If the product is highly innovative or far superior in performance to competitive products in the market, should the company pursue another market? The product has to conform to market demand. If the product brings a new focus or disruptive technology to the market, can it create a new market? Sometimes a highly progressive design that is way ahead of its time does not sell well because customers do not understand the nature of the product or the advantages it brings. In that case, the product design and concept have to be toned down to reduce cost. Is the product push type or pull type? The team has to understand the marketing aspects of the design before its introduction.

Another aspect of product design and development is to understand the aggressiveness of the design of the product. It is a common tendency to design products with all the best performance features possible. The cost and time to manufacture and design become secondary priorities. In doing so, the company takes an enormous risk. In order to satisfy all the features or specifications of the product, the quality control or manufacturing process has to be substantially different from what the company is used to. This might give rise to additional fixed costs for the product, which reduces profitability. The product might generate substantial revenue, but the net profit could be zero or very small. Hence, such a step is not productive and should be avoided. The design team has to deliver what the customer wants.

7.6 Summary

This section is an application of system engineering principles depicted earlier in this book. The principle is applied to generate the target specification, which will eventually help the design team design a product in a systemic way to satisfy customers. This is the starting point of any design activity. The product is designed based on these target specifications. The developed specification metrics should be verifiable, either analytically or through physical testing. The specifications must reflect customer requirements and be in alignment with the specifications of competitive products in the target market. Also, the developed specifications will make the product manufacturable and economically viable in the marketplace. The specifications must have metrics for cost, quality, reliability, time to manufacture, and performance.

The preliminary target specification is developed once the customer requirements are known. These specifications are then refined or revised using an analytical or digital model, which also could be called a system layout. A very crude and approximate analysis is completed using the digital concept of the product to confirm the initial target specifications. These pre-design specifications are used for the detailed system design. The cost to build and time to manufacture also become specifications for the product. A bill of material is also created out of the digital concept sketch. A cost sensitivity model for the system is also developed at this stage. The cost runs are modified along the way when firmed up or more matured designs are made during the design and detail phase of the process. Every sub-system has a budgeted cost and specifications attached to it. These finalized target specifications ultimately become the design constraints to be satisfied by the design.

The specifications represent the performance features of the product when designed and built, although a confirmation prototype has to be built to confirm the specifications through physical testing at the end. Accordingly, few of the specifications might fail in the end, and some of them might exceed the specifications. If the system fails the "must satisfy" category of customer requirements, a redesign effort might be necessary. If some of the nonessential specifications are not met, a critical look at the specification is in order and could be eliminated from the final specification list of the product. Such adjustments are almost a norm for all the machine tool products that I have come across during my professional career.

The first step of this process is to select the target specifications dependent on customer requirements. Each specification must have a metric and a unit for verification purposes. The list of specifications is compared with benchmarked products. A range of values or metrics for each specification is developed. It depicts the least and best values of the metrics for each requirement. Feasibility and reality also have to be considered for each specification so that these values can be produced and validated down the line.

The next step is to develop a digital concept sketch of the product with as much detail as possible. This model is used for refinement of the initially chosen specification values. Theoretical models and simulations are conducted to confirm these values. If not possible, the metrics are adjusted accordingly. Such an evaluation makes the metrics more realistic. Moreover, an approximate bill of material for each assembly is created and an approximate cost model is prepared. This BOM is updated as the design matures. The theoretical analysis also helps the team conduct a trade-off analysis among various specifications. This is a refinement step applied to initially chosen target specifications, or pre-design specifications, which the design team has to follow. So far, the refined specification list is based on the system. Once the system specifications are selected and finalized, they flow down the left side of system "V", and the specifications for the sub-system are generated. Once the specifications for the sub-system are finalized, the component-level specification metrics are generated. At the end of this step, the performance requirements for all the elements, sub-system, and system are completed. This process is highly iterative, and all the specifications must be generated in a consistent and logical way.

Since this process is highly iterative in nature, a systemic way has to be developed for the whole process. Tables, graphs, spreadsheets, analyses tools, brainstorming sessions, etc. need to be developed to keep track of the progression of this process. Benchmarking tools and conjoint analysis can also be used. A consensus among team members and management has to be developed to instill confidence in the specifications. The decision has to be unanimous for all the specifications. In order to be realistic in developing the system specifications, the team has to accept the help of customers, dealers, and vendors wherever possible.

The knowledge of the target market and intended customers is a must. The cost of the product at the desired level comes from discussion with the customers. Several design alternatives might have to be developed to contain the total cost of the product. The list of specifications and metrics represent the combined thoughts of the design team, manufacturing team, sales team, marketing team, quality team, and management. This is a very critical step for any successful design.

7.7 References and Bibliography

Aaker, D.A., Kumar, V., and Day, G.S., 1997, *Marketing Research*, sixth edition, John Wiley and Sons, NY.

Cooper, R.G., Edgett, S.J., and. Kleinschmidt, E.J., 1998, *Portfolio Management for New Products*, Perseus Books, Reading, MA.

Cooper, R., and Slagmulder, R., 1999, "Develop Profitable New Products with Target Costing," *Sloan Management Review* 40(4) summer 1999, pp21-35

Griffin, A., and Hauser, J.R., 1993, "The Voice of the Customer," *Marketing Science*, 12(1), 1–27.

Groeneveld, P., 1987, "Road Mapping Integrates Business and Technology," *Research-Technology Management* 40(5), 13–19.

Hatley, D.J., and Pirbhai, I.A., 1998, *Strategies for Real-Time System Specification*, Dorset House, NY.

Kinnear, T.C., and Taylor, J.R., 1995, *Marketing Research: An Applied Approach*, fifth edition, McGraw-Hill, NY.

McGrath, M.E., 1995, *Product Strategy for High-Technology Companies*, McGraw-Hill, NY.

Meyer, M.H., and Lenhard, A.P., 1997, *The Power of Product Platforms*, Free Press, NY.

Porter, M.E., Competitive Advantage: Creating and Sustaining Superior Performance, Free Press, NY, 1985.

Ulrich, T.K., and Eppinger, S.D., 2011, Product Design and Development, McGraw-Hill Companies.

Urban, G L., and Hauser, J R., 1993, *Design and Marketing of New Products*, third edition, Prentice Hall, NJ.

Wheelwright, S.C., and Clark, K.B.,1992, "Creating Plans to Focus Product Development," *HBR* 70(2), 70–82.

7.8 Review Questions

- Develop a system specification for a vertical machining center. Then develop a sub-system and component specification table for it. Which is more difficult to do, and which is more critical for product development activities? Also, can you develop a specification metrics for an index system for a CNC 12-tool turret index mechanism? Is it worth following a system engineering concept for the development of the specifications? What are the advantages and disadvantages of this process?

- For innovative machine tool development, how can you get customer and market feedback? What have you got to do to create a target market for an innovative product? What are the risks of developing a machine tool that has never been done before? Can you use the same system engineering principle to develop an innovative machine tool?

- If you are part of a marketing team, how can you convince the design team to develop a machine tool with features that are very difficult to contain within the cost constraints? Can you use trade-off analysis to combine features and specifications without exceeding the budgeted cost? What is more important in your mind: cost of a product or features of a product at any cost? Which one would be more successful in a highly competitive market?

- How do you prioritize the performance criteria for a product? How do you determine which are more important than others? In case of conflicts among the team members, how do you manage and resolve them? When do you apply brainstorming in a team environment? Is it a good idea to design a machine that has all the best possible specifications but does not satisfy the cost criteria? What is more important: technology adoption with or without cost considerations?

- While developing the system or sub-system specifications, should you consider the capability and intellectual ability of the team members? What are the new steps you must take while integrating a new technology that needs a new manufacturing system for implementation? How does this step affect the cost of the product? Is new technology always necessary to design and develop a new and successful machine tool for an existing market?

C H A P T E R 8

8 Concept Development of Machines

8.1 Introduction

At this stage of the product development process, the team has finalized a target specification for the product that satisfies most or all of the customer requirements. The specification also has been compared with those of the benchmarked product to get more confidence in the target specification list and its metrics. The next step is to develop several system concepts that will satisfy the specifications. The target market and the cost of the product are also very well defined. The design team takes the lead in developing several possible concepts. It is also necessary to develop and conceive several system concepts that will satisfy the requirements.

During the finalization of the target specifications, a very crude digital or analytical model of the system was completed. The team can start from that sketch or develop new alternatives of that sketch. The primary function of the design team is to develop alternative designs, and it has to develop a method for selecting a design that best satisfies the target specifications. The alternative considers new technology available for such a design. The team has to select proven and bookshelf technology for use in a very cost-effective way.

Alternative solutions must simultaneously consider technology, cost, and manufacturing methods. The concept must also be developed to consider in-house or out-of-house manufacturing methods. A comparative list of all the possibilities must be made to ascertain the suitability of possible designs regarding the final specification list. Similar to the analytical methods employed during finalization of specifications, these analytical methods and simulation could also be used for this step to see which model satisfies the metrics the most. For the system "V" concept, we are still on the left side of the "V," and this step is also highly iterative.

A systemic method of developing concepts must be followed for time and cost efficiency. Any successful design must embrace cost, manufacturing methods, specification and its metrics, quality, reliability targets, and time to manufacture. All these qualities must be satisfied and agreed upon by all development team members for further processing beyond concept generation. The selected concept will be used for detailed design in the next step.

8.2 Concept Generation Process

A concept is the form of an engineer's idea of how the system should look like. The concept might consider new technology or a new architecture that satisfies customer requirements and subsequent specifications. Normally, a concept displays machine assemblies and structures in a logical fashion. A concept generally includes a specific shape, form, and architecture combined together. The variation of concepts depends on how all the sub-assemblies are arranged together to have an optimized system. During concept generation, qualification of the concept, good or bad, is not necessary as long as it includes all the sub-assemblies and it supports the overall specifications. Several concepts are required to compare them technically, and eventually only one approved concept is put forward for further processing. The most important thing is to create as many feasible ideas as possible.

Concept generation is comparatively inexpensive and less time-consuming if done in a digital environment, which is very common nowadays. Since the end product depends heavily on one of these concepts, a good amount of time should be spent on thinking it over again and again. In Japan, a considerable amount of time is spent on creating concepts of various shapes and forms, almost 20 to 25% of the total project time. Both system and sub-system specifications must be kept in mind while generating these sketches. This is a very creative process, as I witnessed during my professional career.

As outlined earlier, concept generation starts with the customer requirements and specification list and ends with a viable and feasible sketch. Several of these sketches are made by different engineers independently, and in the end, one or some combination of these ideas is accepted for further development. In general, three to five sketches or concepts are generated in the beginning by engineers. Once the sketch shows some promise of being feasible, the selection process starts. Most of the time, the final sketch is a product or combination of these several concepts, and in rare cases, one sketch or architecture stands out very strongly against others, and the concept is accepted as the final sketch or design.

The primary reason for generating several concepts is to make sure that all ideas are explored in designing the product. Since one sketch or design is selected out of many, it creates enough confidence among engineers that it is the best possible concept. The final form or design does represent the intellectual ability of the engineers, so they should be extremely creative during this phase. This is the most creative phase of the design and development process.

This initial selective and iterative process also eliminates the chance of finding another possible concept down the line that is far superior to what was considered in the beginning. Sometimes the selected concept becomes a subject of discussion with vendors, technology suppliers, distributors, dealers, etc. to get their feedback. If carried out systematically and methodically, the resultant concept can ensure the best possible performance. Such a concept does not show any design details of the sub-assemblies.

A critical series of questions should always be debated and answered before a structured product concept generation activity using system engineering principles is considered. Some of these questions could be as follows:

- Is this structured generation process required for machine tool development?

- Why do we need such a process since any process costs time and money for the company?

- Does a structured concept generation process help with product development, since there are many instances where such a process was not followed and the company still thrived for a long time?

- Does it depend on the size of the company or of the design team?

- Is it only necessary for a complex product?

- Does it save money and time for the company?

- Is it a constraint for engineers to be creative?

I believe that irrespective of the type or complexity of the product or size of the company, a systemic structured concept generation process should be used to avoid future confusion among team members and to ensure the success of the product. The concept generation step is the most critical step for this success. If alternative concept designs are not considered, concept optimization is not possible. One concept might lack generalization and might not answer the total solution for the intended product. One or two concepts might not be able to include all the requirements, but a combination of these concepts might be a better solution to satisfy them all. Instead of depending on one or two team members for design solutions, it is better to allow all the engineers to be creative and to generate as many concepts as possible and then select the best using a structured elimination process.

Since the concept is generated by getting information from several sources such as marketing, benchmarking, user community, dealers, etc. the structured concept generation process reduces the risk of designing a product that will not satisfy the target market. The alternative concept designs give an opportunity to the team members to evaluate each concept against the requirements and select the best possible concept out of all those available. This prevents the team from accepting a design that only partially satisfies the requirements.

Such an integrated approach also motivates all the team members and encourages them to participate in this selection process, enhancing team coherence and eventual success. This step also allows the team members to get familiar with the concept that will finally be taken up for further design details. This helps the team to finish the development project much earlier than anticipated. The team feels much more encouraged and energized to get involved with the project, and such synergy will bring ultimate product success. So, this concept generation process allows the team members to be very cohesive and focused on the design, which results in a successful product for the company.

Before the steps for developing concepts are explained, two more fundamental concepts must be discussed. These two concepts are system architecture development and functional decomposition of the system. The next two sections will be dedicated to explaining these concepts.

8.3 System Architecture Development

By definition, an architecture of a product or system is the configuration or arrangement of functional sub-systems in which these modules, elements, sub-systems, or sub-assemblies are combined or configured together in physical block diagrams that will interact with each other to create an intended output of the system or product. The configuration must be arranged in such a way that the system elements or sub-assemblies, when interacting with each other, satisfy the design requirements or specifications, system constraints, and external or internal environment. The assemblies must interact with each other in such a way that the system output satisfies the desired output requirements. The specifications of the system can also be called the "constraints" of the system. The engineers follow this configuration or architecture to satisfy the system requirements or specifications. In the machine tool world, the architecture of a machine is also called the "system layout," "system diagram," or "system concept layout." The architecture of a product or a concept layout that shows the architecture of the product has the following underlying features:

- A functional relationship exists among functional blocks
- It is not a physical structure of the system
- It is a concept layout of the system showing the relationship among different sub-assemblies
- It displays the organization of a system with boundaries and interfaces

Architecture development is an emergent output of the system's functional analysis and decomposition, which is discussed in the next section. Architecture development focuses on the primary question: what does the system have to accomplish? It does not focus on how the architecture will accomplish that. Functional decomposition basically leads to boundary and interface analysis and development. The same architecture could have different emergent outputs when subjected to different environments. A lathe could act as a machining center when subjected to a different environment. Any architecture should start with system functions or requirements and end up with a concept structure or form.

A system concept layout is required to determine the suitability of the types of configurations, e.g., modular or integral. A modular assembly is added to the main structure, as opposed to the integral assembly, which becomes part and parcel of the main structure. A modular structure is very often preferred for machine tools for maintenance purposes. Once designed, a modular sub-system can also be used for a series of products.

The system layout should show boundaries, interfaces, and other peripheral sub-systems. In case the team is planning to design a platform of products down the line, it is necessary to have a modular structure with a common interface to carry the design across the platform. The concept layout also helps the team to identify manufacturing concerns, if any, promotes the design for manufacturing and component integration, and helps eliminate redundancy of design.

The concept layout also shows the physical relationship among various modules in a conceptual form and displays the approximate interface locations with respect to each other. The modular structures support simultaneous or concurrent engineering in a sense that, with a given interface and well-defined boundary, teams can work on several assemblies at the same time. This helps increase efficiency and promotes team efficiency.

Another important aspect of the concept layout is to determine whether the structure should be integral, modular, or hybrid by design. A purely granular structure or architecture is one where the machine concept layout shows the modules completely separated and each module performs a specific requirement all by itself in a stand-alone situation. Clearly defined interactions or links exist between modules or sub-assemblies, and design activities can be performed independently.

Before architecture development, the following questions must be answered:

- Product complexity and variety
- Cost of the product
- Project duration and time of introduction into the market
- Manufacturability and its methods
- Management and control aspects
- Product innovation aspects

A generic product architecture development is also shown in Fig. 8.1. Functional decomposition should be carried out earlier than architecture development, and design activities should start after architecture development is finalized and an architecture is selected for design consideration. On the other hand, the concept could be designed in such a way that more than one module or sub-assembly react with each other to create a function or requirement. Each module can be multi-functional and react with others to generate one or several functional requirements. The module interactions are not separable. They act together. The interface or boundaries between modules are somewhat transparent or unidentifiable, and sometimes they are non-existent. Such architectures are more common for machine design. In the real world, machine tool designs are hybrid by nature, as will be explained later on when we discuss an example.

The generation of concept layout or product architecture development activity is completed on the left side of the "V," as shown in Fig. 8.3 below. In general, once the system requirements are developed, concept generation, including machine architecture activity, starts. As shown in this diagram, the development of the concept needs both requirements and business inputs such as legal inputs, financial inputs, etc.

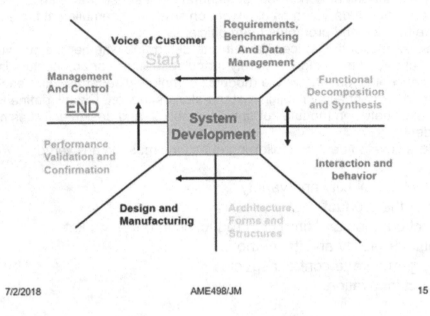

Fig. 8.1: Product architecture development guideline

The system architecture development process starts with the verified requirements or finalized specification list with metrics. These requirements are put into action by developing a system concept or product architecture. The architecture development is also highly iterative. The architecture must be functional and should lead the design efforts. A flow diagram showing the architecture development process is shown in Fig. 8.4. The diagram depicts a verification of the system concept, or architecture, with regard to functional requirements and design feasibilities. That is the reason why several concepts are developed and then one is selected for further downstream processing.

SE Concept Application using System "V"

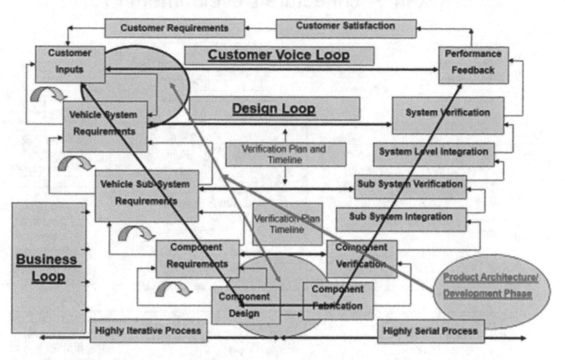

Fig. 8.3: Product architecture development and system "V"

For a legacy product, this process might not be as important, but for new development, this process should be followed for efficiency and productivity. Moreover, once a concept is selected, confirmation of requirements, manufacturing feasibility, and financial requirements must be confirmed for further processing. The final selected concept could be called the "verified system architecture." Another salient feature of this step is to create a physical form of the design that the individual design team will pursue further. Any successful product architecture defines the functional requirements and performance simultaneously. It has to satisfy customer requirements as much as possible within the stipulated budgetary constraints, and it also has to satisfy functional performance requirements. The resultant physical form or architecture also has to satisfy functional requirements. Physical forms must represent the sub-assemblies in required perspectives. The product architecture is dictated by the product design philosophy of the company. This phase is the first of product design.

System Architecture Development Process

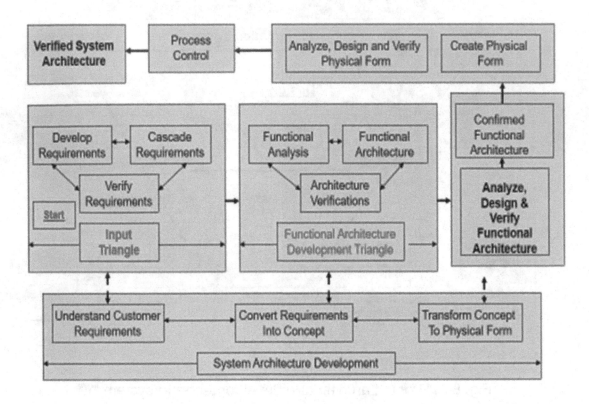

Fig. 8.4: System architecture development process

System design guidelines for machine tools can vary for different companies. During my professional career, I found nothing that was commonly used. The system functional decomposition, as explained in the next section, has to be completed first irrespective of the nature of the structure or product. Another important aspect of any structure is to how to divide the assembly into logical sub-assemblies or components. In other words, how to maintain modularity in the system architecture will depend on several factors:

- How innovative is the design?
- What technology is being used?
- What is the level of component standardization across the product platform?
- What is the extent, nature, and diversity of product performance required?
- What is manufacturing technology to be employed?
- What is the company's design and innovation legacy?

The product architecture also depends on the following aspects:

- Marketing philosophy followed by the company
- Production and manufacturing management
- Technology management
- Supply chain management
- Intellectual capability of the company

Another aspect of the product is the granularity of the architecture, which allows a change in one module without major alterations in other interactive modules of a structure. If the product or process change is imminent, the granularity of a structure should be high. The next consideration is the product variety or platform considerations, wherein the same modules are used in various legacy products to reduce cost. This is also similar to product standardization, which allows the same components to be used for various products to reduce cost. The granularity also depends on the nature and extent of performance desired, as well as maintainability aspects or after-sale service considerations. The performance factors can be speed, efficiency, life, accuracy, etc.

Let's take an example of developing a system architecture for a CNC lathe, as shown in Figures 8.5A through C. The process of developing the architecture is to start with a concept interaction sketch, as shown in Fig. 8.6. This sketch shows the relationship among several modules in a very crude form. Normally, the team starts with the basic elements of the system. Once that is done, the concept sketches of the system are completed. As shown in Fig. 8.5 A through C, the different modules are arranged in such a fashion that they interact and create the conditions to satisfy the requirements. For example, a CNC lathe has several modules that function as a metal removal device. It has an indexing unit for positioning the tools with respect to the spindle unit. It needs to have a spindle unit to hold the job and rotate it with respect to the tool mounted on the turret unit. The spindle unit must have a work holding unit attached to it for holding the job in position, a device called the "chuck," which is mounted on the rotating spindle. The turret unit is mounted on a cross slide, which moves up and down across the job.

The cross slide is also mounted on a longitudinal slide for movement along the length of the job. Both of these, the cross slide and longitudinal slide, are driven by the servo motors with the help of ball screws. For long and slender jobs, the system needs to have a tailstock to support the job at the center of the spindle while removing metal from the job. The tailstock unit is mounted on a tailstock base so that it can be positioned with respect to the job's length. All of these independent modules are mounted on a machine base, which is supported on a three-point support system. Hence, the finalized concept sketch or machine architecture displays all the modules positioned with respect to each other in such a way that it becomes suitable for metal removal from a part.

The architecture, or concept, layout does represent the interaction in such a way that will satisfy the requirements. The architecture shows a physical form at the end. It also shows the boundaries of each module and their relative positions. Once the interacting elements are taken care of, the outer periphery of this assembly is conceived. The concept of the surrounding sheet metal shrouds, control cabinets, cooling, etc. are also shown in the complete system concept, as displayed in Fig. 8.5C, which is the ultimate form of this machine system that the design team will take up for further detailing. Once developed and agreed upon, the architecture remains stationary and does not change. In order to optimize such a structure, it is necessary to have several of these concepts or layouts generated.

In some complicated cases such as the development of a lathe suitable for flexible manufacturing systems, the author has witnessed several forms of the interaction diagrams completed before the concept layout is initiated, as displayed in Figs. 8.7 and 8.8. The sequence is not as important. The concept of the system does represent the following aspects, as described below. As mentioned earlier, the system must have context, perform functions, and possess forms or structures. Each of these system properties can also be mentioned for the CNC lathe.

- System Context or Background:
  - External environment: Temperature, humidity, vibration
  - System inputs: energy, material, control Inputs
- Behavior or Functions:
  - System functions: It removes metal
  - System specifications: Requirements with metrics that the system must satisfy
  - System response: Store the input file and make the spindle and tools active for material removal using control functions
  - System transitions: Accept the input file through control and activate the servo motors and spindle to interact to create a job profile
  - System output: A finished job as intended
- System Structure or Physical Forms:
  - Internal elements: Assemblies such as indexing unit, spindle unit, cross slide, longitudinal slide, etc.
  - External elements: Controls, shrouds, cooling units, conveyors, power supply, transformers, coolant tank
  - Module interface and boundaries: The spindle units are mounted on the machine base using bolts, for example
  - System architecture: It shows the interacting modules together and the mode of interactions.

The system concept, or architecture, should also take inputs from sources other than just technical requirements and the specification list. As shown in Fig. 8.9, the system architecture should satisfy technical requirements, business logistics, and customer needs.

The architecture creates a synergy among all these functions. For example, a CNC lathe or machining center must satisfy the technical specifications, which comes out of the customer requirements, and in addition, it also has to satisfy functional aspects, such as legalities, market environments, and budgetary decisions. The point is that the development of system architecture is a very critical function of the whole product development activity and should be very well thought out. The architecture development is a combination of all these activities efficiently put together. In my experience, it should be mentioned that most machine tool builders do not follow all of these critical steps while developing a machine. As a result, the end product most of the time is not what the customer wants. In such a situation, it is very difficult to trace back its path and make changes. If the critical steps of system engineering are not followed to some degree, the result will be a very costly mistake: the process becomes irreversible at some point, and changing direction costs the company time and money. In some cases, this means abandoning the product development process altogether.

In conclusion, it can be said that a product system should have a context, functional behavior, and physical form, or structure. The architecture for a machine tool should show the concept of the system and the distribution of functional requirements for the physical elements of the system. The primary characteristic of machine tool architecture is the degree of granularity or modularity of its sub-systems or building blocks. The product architecture should be established before the actual design process is implemented. Architectural concepts and product design can also be concurrent. Last but not least, a successful architecture for any product starts with the system functionalities or requirements and ends with physical forms. To conclude this discussion of developing system architecture, it needs to be mentioned that cost is an issue that the design team should also keep in mind, as I experienced while working in Japan. The functional analysis for the structure must be performed at every stage of the development process or every time a substantial design change is made. The functional analysis stems from customer requirements, which contain a cost requirement, too. The satisfaction of each requirement has a cost associated with it.

Some architecture is much cheaper than others. A financial analysis of any structure must be made to understand the financial impact of any structure. In some instances, budgetary decisions dictate a design or help the design team to create a structure that satisfies all the financial constraints. It is critical to fulfill the technical aspects of the architecture, but only if the budgetary constraints are satisfied.

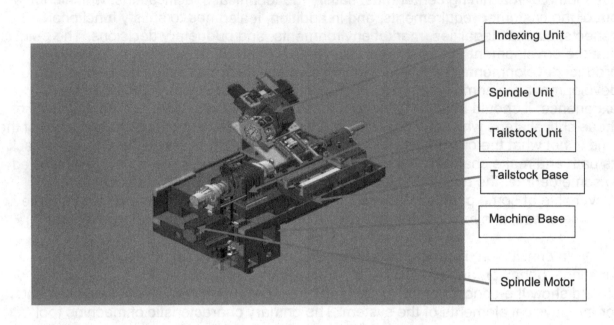

Indexing Unit

Spindle Unit

Tailstock Unit

Tailstock Base

Machine Base

Spindle Motor

Fig. 8.5 A: Finalized machine tool system architecture for CNC lathe

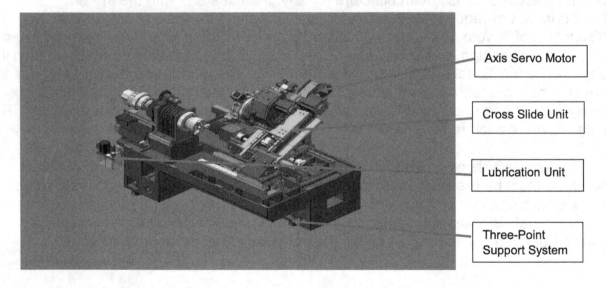

Axis Servo Motor

Cross Slide Unit

Lubrication Unit

Three-Point Support System

Fig. 8.5 B: Finalized machine tool system architecture for CNC lathe

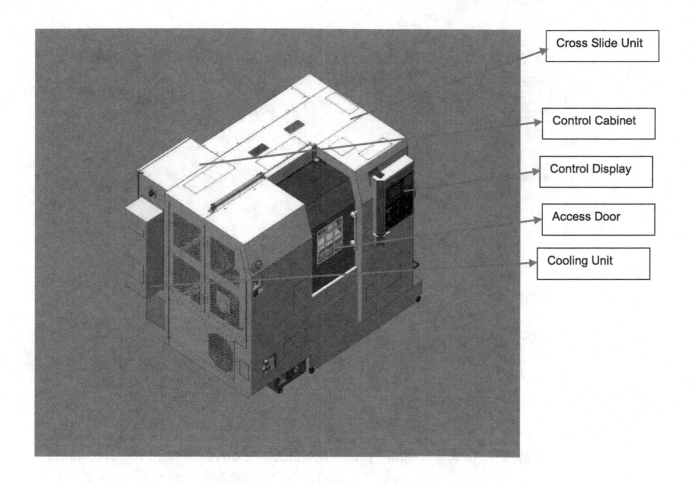

	Cross Slide Unit
	Control Cabinet
	Control Display
	Access Door
	Cooling Unit

Fig. 8.5 C: Machine tool system architecture for CNC lathe

System Architecture of a Machine Tool-Example

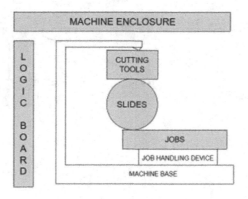

Fig. 8.6: Machine tool system architecture for CNC lathe

Product Architecture of a Machine Tool-Example

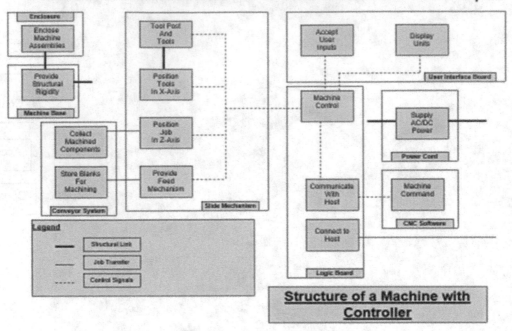

Fig. 8.7: Machine tool system architecture development for CNC lathe

Product Architecture, Interaction Diagram

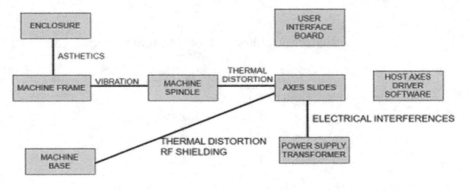

Fig. 8.8: Machine tool system architecture development for CNC lathe

8.4 Functional Decomposition of a System

Functional decomposition is a dissection process to separate a complex system into smaller, manageable modules for better understanding of the whole system. For example, the budget of a company is divided into several burden or cost centers, such as human resources, manufacturing, design, accounting, etc., for better management and control of finances. Product system functions are broken down into the functions of each sub-system, such as the spindle unit, cross slide, tailstock, and machine base. The sub-functions of each sub-assembly are identified in order to make the whole system work. Functional decomposition creates or identifies functions for each module separately. The sub-system requirements are obtained from the system requirements. Once the functions are sub-divided, each module can be optimized or designed to satisfy its requirements.

System decomposition starts with a proper understanding of the environment into which a system has to function, as shown in Fig. 8.9. The functional decomposition of a machinery system should include three aspects. It should address customer requirements, business requirements, and design requirements, and then it is complete.

The functional decomposition process starts with customer requirements, migrates to system requirements, then moves to subsystem requirements, and ends up with the required architecture, as shown in Fig. 8.10. The design variants come out of several architectures possible for the machine. In reality, there is not one architecture that satisfies all business and customer requirements. Functional decomposition also helps generate several possible architectures satisfying some or most of the requirements of a business.

After the architectures are developed and the functional decomposition is done, a preliminary analysis should be performed to ensure that the structures or concepts are in alignment with the requirements. The development of architectures and functional decomposition can be carried out concurrently, but it is preferable that functional decomposition be completed before architecture development is taken up. Architecture development starts informally during the strategy development phase, from brainstorm sessions, crude sketches, or a migration from the initial concept sketches or existing products. It can also happen before the design phase is started, depending on the mature technology used or product legacy requirements. For a brand-new product or a highly innovative product, the architecture becomes the focal point of the design and development process.

In general, the product architecture should be established before the design phase is started, and the design of the product must be dependent on the agreed-upon product architecture. In some cases, product architecture development and the design phase can be combined for synergy, depending on the design team's intellectual capabilities. There is no hard and fast rule about this, but a systematic and consistent process must be followed for the product to be successful in the marketplace.

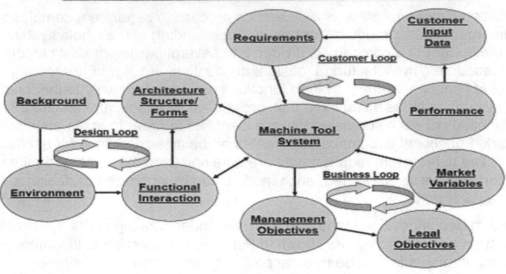

Fig. 8.9: Machine tool system functions

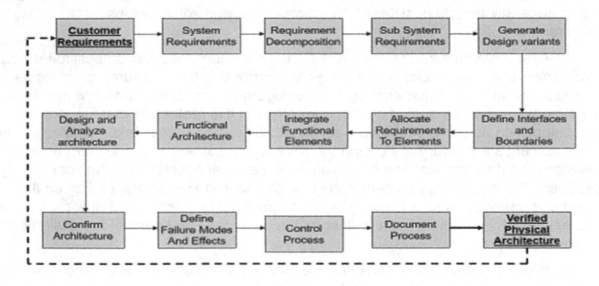

Fig. 8.10: System functional analysis

8.5 Concept Generation Guidelines

Once the product architecture development and functional decomposition process are well understood, the concept generation process becomes easier to deal with. The specification and identification of sub-system requirements become somewhat obvious, too. These initial steps precede the development of sub-system module design and detailing. First of all, the team has to clearly understand the objectives of the project along with its requirements. The team also has to understand the external and internal background and environment of the project. In addition, the team has to understand the business conditions, competition, and market the product is going to serve.

All of these are defined in the mission statement, discussed earlier in this book. The mission statement, customer requirements, and finalized target specifications are the initial documents that will lead the product architecture, or concept, efforts. So, the team members must revisit these documents before the task of concept generation or concept layout development. For example, let's say the task at hand is to develop a computer-controlled, fully automatic vertical machining center. Some of the initial assumptions for, say, a machining center could be as follows:

- The machine will be controlled by external software
- The machine will need part documents before machining
- The machine will have an automatic tool feed system
- The multi-point tool rotates, and the job is held fixed in relation to the rotating tool
- The operator interfaces the machine movement and part loading/unloading system manually or through the controllers
- The machine will have two axes of motion in addition to tool rotation
- The machine must have an automatic coolant and lubrication system

Based on these general requirements for the machine, the team has collaborated, benchmarked the existing products in the target market and identified the finalized target specification list with metrics that will satisfy customer requirements:

- Both axes should have a minimum speed of 30 meters/min
- The maximum spindle speed should 15,000 rev/min
- The machine controller will be purchased from an external source, FANUC
- A rotating fixture for holding the job and loading/unloading is required
- The machine weight should not exceed 15,000 pounds
- The tool magazine will have the capacity to hold 30 tools minimum
- The machine base support system could have three points or more, as needed
- The machine must be environmentally friendly

The next step is to functionally decompose the system requirements into smaller sub-system requirements. In other words, system requirements/functions are divided into sub-functions that each module has to work against. The functional decomposition is required to simplify the complex system requirements as much as possible into sub-system requirements. For the machining center, the design can be thought of as a collection following design functions:

- Tool Magazine Design

- Longitudinal/Horizontal Slide Design

- Machine Base Design

- Job Fixture Design

- Machine Shroud Design

- Control Interface Design

- Spindle Unit Design

- Tool Transfer Mechanism Design

Each individual module has to have a clearly pre-defined function. The individual module's requirements or specification must be consistent with the whole machine's requirements. They have to work in unison to give rise to the machine system function. Once the sub-system specifications and its interface with the system for each module are developed, the sub-system design can be independently taken up. The functional decomposition should also define energy management, coolant and lubrication management, control interface management, machine movement management, etc.

A general question comes up, and that is how far refinements or divisions of system functions should be carried out. No such norms are established yet. In general, the team has to feel confident about the fact that a module can be very well designed using the specifications and functions of the sub-system. The ultimate objective of functional decomposition is to define the functions required by the sub-systems. So, the extent of decomposition varies from task to task depending on the nature and complexity of the project, the team's ability to develop the machine, the time and cost of development, etc.

Functional decomposition does not fix the design, nor does it depict how to design the sub-system at this point in time. It just creates a list of functional constraints for the sub-system in alignment with the system functions. In most machine tool development projects, the technology to be used is somewhat specified at this stage, but how the available technology is implemented into a specific product is a matter to be decided by the design team. For example, epoxy resin technology instead of casting or welding is a technology used to enhance the damping of the system. How this proven technology can be incorporated into a system will depend on the team members. The team also has to decide whether such technology is required or not since it might impose a financial constraint on the cost of the system. As mentioned earlier, the system concept layout or system architecture layout only shows what is required but does not pinpoint the solution for each requirement.

At this stage of the game, the system can be thought of as a black box that transforms inputs into the desired output. This transformation box displays the system input and output relationship. This process does not show how the inputs are transformed into outputs but shows the start and end of the process. For example, some of the inputs to the machining center project black box could be as follows:

- Requirements
- Energy inputs
- Benchmark data
- Legal requirements
- Legacy requirements
- External and internal environment
- Cost requirements
- Specifications
- Control signal
- Team resources

The output of this transformation box can be as follows:

- Desired system design or product
- Noise, byproducts, or undesirable outputs that cannot be controlled
- System performance
- System variability
- System quality and reliability

The black box could be also called a "transformation box," and it converts the input into desired output and noise. For a whole machine, the inputs could be material/job and part control documents, and the output a machined part with some noise such as heat, sound, etc. There is no fixed way of designing this input-output process. Similarly, there is no particular solution for a functional diagram or functional decomposition of a machine. Also, a complete decomposition might not be possible for a product. The extent of decomposition could also depend on the complexity of the product. For example, the functional decomposition for a lathe might not match that for the grinding machine. Also, the nature and details of a functional decomposition analysis will depend on the team members. It might be a good idea to start with the functional decomposition of an existing product and then migrate to the decomposition of the new product under design. Nevertheless, a sincere effort to decompose the functions into sub-functions is recommended for any complex machinery, and this process helps design the modules and system together.

The primary purpose of functional decomposition is to identify the requirements for the critical modules of a system. From a set of requirements of a system, it is sometimes very difficult to separate the sub-system requirements unless a functional decomposition is completed, especially for a very complex product like a machine tool. While focusing on the sub-system functions, it is also important to identify the boundaries and interfaces of each module. Once the requirements for the system and sub-system are developed, the team can develop the priority of the sub-systems. In between, the concept generation step has to be completed to understand the boundary and interface possibilities of each sub-system or module. For a machine tool, the progression could be to develop sub-systems such as the spindle unit, indexing unit, cross slide, longitudinal slide, and tailstock in this sequence. The machine base is developed last. Once the system is put together, the sheet metal and peripheral component development are taken up.

In order to develop the architecture, or concept, layout of a machine system, the next step is to look for existing structures or systems in the market. They could be systems from inside or outside of the company. To get out-of-the-box ideas for a system, one can also consult the patents for similar system or sub-systems. The external search could start even at the outset of the project. The idea is to get examples of what is out there in the market. In some cases, customers get used to a particular system and its functions. In order to get customer appreciation for a product, it is sometimes better to follow the trend, unless another system is way more developed and productive and the system becomes a trend-setting product.

Currently, all machining centers and lathes look alike, and Japan set the trend for designing such architectures, even though U.S. manufacturers started designing these products before Japanese industries. Another reason for looking at the existing product is to understand the trend of technology used in such products. In the last decade, control technology has developed to a very high level, and this trend was set by Fanuc, Japan. The technology of combining turning and milling operations in a lathe is another trendsetter. In order to eliminate some simple grinding operations, machine and tooling systems have integrated the technology of "hard turning" for lathes. In general, lathes can now perform turning, milling, and grinding operations for very hard metals.

The external or internal search for better design will allow the designers to be very creative and productive. This search also makes sure that a design does not become a copy of an existing patent, which is a legal violation. The nature of the search could depend on the team members. Sometimes, the search for designs for different types of products could prove to be beneficial as well. This is basically an information-finding step to understand existing products and design trends. Such information could also be collected from customers, early adopters, machine shops, consultants, a patent search, trade shows, literature surveys, benchmarking competitive products, dealers, and vendors. In a nutshell, the idea is to collect information as much as possible before starting to develop a new concept. Another reason for collecting information from several sources is that some users or customers might have already adopted innovative solutions to a problem. Consequently, the concept generation process becomes very efficient and productive.

I have experienced very innovative solutions adopted by customers or lead users in the field for fixturing and part handling. In order to make the process more productive and less time-consuming, users normally come up with a customized way of handling different parts in their shops. So, to develop a loading/unloading sub-system of a machining center, it is better to start with such users and their ideas. The machine tool trade is very much a home-grown trade, and I do not believe it is a good idea to use consultants at this phase of product development if internal resources are available. The involvement of an outside design house could be more productive in designing sub-systems, but not the initial system concepts.

The patent search also has some severe limitations. Normally, the patented design is protected against copying. So, it is better to understand the limitation or constraints before starting any architectural concept. The design team members could consult the US Patent and Trademark Office for patents in their areas of interest. The design team can also get information from online databases for any particular patent and consult published literature on machine tools. The problem with the latter is that intricate designs are not published unless they are patented before publication. Journals might publish general trends in technology or system efficiency, system productivity, system engineering principles, etc. In any case, consulting trade journals, in general, is not a bad practice. There are many handbooks available in the market for consultation, such as *Mark's Standard Handbook of Mechanical Engineering*, *Machine Design Handbook and Mechanism Handbooks*, *Thomas Register of American Manufacturers*, etc. All these methods are part of benchmarking competitive or existing products.

The concept generation phase is very much dependent on the intellectual abilities of the design team members. Some members are more innovative in their approach towards a design. It is extremely difficult to ignore the fact that human capability is the prime source of innovativeness. Hence, conducting an internal search for ideas about a concept is the best method for conceptualizing a product. In order to be highly creative, the team members can create a productive environment wherein every engineer has the chance to promote their ideas without creating any conflict among team members.

In order to maintain a highly creative and friendly environment for designs, the team members must cease to be judgmental. They have to stop criticizing other team members and refrain from voicing damaging comments or judgments during this creative phase. Critical ideas can come from anywhere and at any time. All ideas are good ideas whether adaptable or not. Instead of criticism, encouraging suggestions to improve upon an idea is a better approach during this phase of development. As said earlier, the team has to generate a lot of concepts, and some of these ideas are combined to generate a better concept.

Moreover, one idea often leads to another, better idea. So, many concepts are better than a few concepts. The team has to explore the possibility of all the feasible solutions for a product. Hence, the team has to create an environment where team members feel very positive and comfortable in advancing their concepts and ideas without any hesitation. All suggestions are to be welcome and appreciated.

Another effective way of building a cohesive and productive team is to have an open channel of communication among team members. Regular team meetings and brainstorming sessions have to be conducted to discuss the challenges and design concepts. The team environment must not be compartmentalized at this creative phase of design and development. So, design freedom must also be maintained among team members, and fear of designing has to be totally abolished. Once a particular concept is developed by a team member, it could be discussed among members in a meeting to help him or her to make the concept better or to enhance the design to overcome the shortcomings of the concept.

For example, I have seen almost ten to twelve different ideas or concepts being developed for a machining center design project in Japan. A general meeting was arranged to discuss the concepts once a month in the beginning phase of design. As the design matured, meetings once or twice a week were arranged to finalize any concept. Every such meeting had a positive and definite outcome, promoting ideas or enhancing ideas to a better stage. More experienced team members always took the lead in making the design better instead of criticizing it. It is important to note that group meetings are far more effective than individual meetings for creating an effective concept.

In order to create an innovative architecture, everyone must be very open in his or her thinking and possess a high level of background information. You have to know what you are creating, why you are creating it, and what is out there already. I found one method to be very effective and productive. The team leader selects a competitive architecture that he or she thinks is suitable for a particular project. Then this concept becomes open to all for further enhancements and improvements. The team members become much more focused and get very excited about creating something of their own that will be appreciated by other members. So, allow members to be creative, and they will become creative. In order to promote the environment for creation, the leader could detail the competitive markets and products, help team member be open in the design space, or create activities that excite the team members and encourage them to act and react together. The team leader could set up the goals and objectives of the project during the group sessions.

Another important aspect of team interaction that I have witnessed in Japan is team discipline. The process of developing the design becomes very systematic and group-oriented. The leader might discuss several feasible or possible solutions to any problem or design and challenge the team members to be more active in reducing, say, cost, weight, etc. Leadership is sometimes very critical to the success of this phase. Proper guidance might help the team members to be creative. The leader has to excite the base to create something very innovative.

There is also not any realistic way of finding a right or perfect answer to developing any architecture. It is definitely a trial-and-error process. It is a highly iterative process, too. The team has to identify its own approach to the optimized solution of the problem. Sometimes in this process, the team might invest more time for an unimportant design that customers do not care for. For example, the design and selection of a conveyor is not an important event as long some means of carrying the chips out of the machine and holding coolant is provided. The team has to focus on critical modules first.

It is also possible that the specification list and its metrics might have to be revised due to problems developing a concise architecture, or concept. So, refinement of the functional decomposition or sub-system functions might have to be adjusted for feasibility. Sometimes the sub-system design might also have to be changed. Anyway, the idea is that concept generation is a difficult step and previous steps might have to be adjusted to suit the concept. For machine tool design projects, some considerations for energy sources are always given, and deviation from norms is very expensive. For example, the energy source is always fixed since components accepting the electrical, pneumatic, or hydraulic energy are fixed by the design of the components using the energy. In general, electrical energy is costlier than hydraulic energy. To reduce cost, hydraulic energy should be used as much as possible, although the design of a hydraulic unit supplying power is also costly.

Once many possible concepts are generated by the team members, the concepts are synthesized or combined for optimization, and best the possible architecture satisfying the system specifications is selected by the team. The team has to manage this selection process depending on company-based policies. There are no fixed guidelines. A comparative list of all the architectures and their advantages and disadvantages are noted. Each architecture can then be compared to each other.

At the end of the process, the team has to satisfy itself that the best possible solution has been picked. The team has to consider the following before this chapter is closed:

- The best possible solution has been selected out of all possible concepts
- All the design constraints have been satisfied
- All the requirements have been satisfied or adjusted to suit the concept
- Functional decomposition is as complete as possible
- Patent rights are not violated, and other designs are not copied directly
- All the business, financial, and legal considerations have been met
- The accepted concept has buy-in from all team members without any hesitation
- The concept generation process is time- and cost-effective
- Proper attention has been given to future platform considerations and standardization
- The team has become more cohesive and productive
- Last but not least, the concept generation process has been documented for future use

8.6 Summary

A product's architecture, or concept, is the designer's way of representing its specifications and customer requirements. The architecture also represents the technology adaptation for the future design. The suitability of the design for customers' use largely depends on the product concept layouts. A smart and efficient architecture always has a very high probability of being accepted by the customers. The architecture, or concept, does satisfy the financial constraints.

The index of success of concept generation and the architecture development process is the conformance of the concept with the requirements of the design. The ultimate test of the architecture is the performance, cost, and quality of the end product. For machine tools, accuracy, quality, and reliability are the three prime indices against which any product is measured. The enhancements of these measures are extremely necessary for a successful product. Most of the time, customers will accept some unimportant deficiencies if the machine performs beyond expectations.

The concept generation or the product architecture development process begins with customer requirements and specification metrics. The architecture accepts these constraints and delivers the concept that will satisfy these constraints. For machine tool applications, three to five concept variants should be good enough for further selection activities. The important point to remember is the fact that these variants are not always fundamentally different from each other, but as long as all of these concepts capture most of the feasible solution space, it should be good enough to move to the next phase of design.

The concept generation phase should be systematic and well thought out, as said earlier. Most of the project time should be spent here, and a critical review of all the designs must be made before a selection is made unanimously by the team. The problem, constraints, and design requirements must be understood correctly and in detail by each team member. The team members should individually look for a possible design solution, consulting patents, benchmarked products, internal legacy designs, and other dependable sources. This should enhance the knowledge of the team members as to what is out there, and the challenge is to overcome the limitations of such existing designs and make the present design better.

The process followed must be very systemic and procedural without hindering the productivity and innovativeness of the team members. An open and creative environment must exist at this stage. The process must be documented for future reference. The team leader must allow the engineers to be very creative and productive while also tracking the timeline of the project. A disciplined and predefined approach must be followed. The process should not so rigid and structured that it hinders the creative capabilities of the team member. The team leader has to find a fine balance between fostering a creative environment and producing project deliverables. It is important to do so. An open-ended approach at this stage will allow the team members to explore the full design space and come out with feasible solutions that will encompass all possibilities. The experienced designers should take the lead, and the others should follow them, and team coherence should be maintained by the team leader, as I witnessed it being done in Japan.

This concept generation phase is highly iterative by nature. In some cases, designers move back and forth between requirements and feasibility or reality. In extreme cases, requirements or metrics might have to be modified to make the concepts more manufacturable or performance-oriented and cost-effective. Whether the product is a brand-new product or a legacy product, the iterative nature of this step must be maintained to optimize the concept design. The system engineering principles must be followed as much as possible. The functional decomposition of the system requirements into sub-system requirements is an extremely important step for efficient design of the sub-systems, which, when put together, will give rise to desirable system performance. The noise of the design must be kept to a minimum.

I experienced the following phenomenon several times in my professional life in the machine tool industry: The team is led and dictated by the most experienced designer even if the leader is mediocre. The ultimate result is the creation of an average and mediocre product. The inexperienced engineers and designers are not always given proper opportunities to show their talents. This tendency and practice must be stopped since, most of the time, experience hinders creativity. Experience is definitely necessary for the optimization of a design but not for creating an out-of-the-box design.

8.7 References and Bibliography

Abernathy, W.J., and Townsend, P.L., 1975, "Technology, Productivity and Process Change," *Technological Forecasting and Social Change* 7(4), 379–96.

Cooper, R.G., Edgett, S.J., and Kleinschmidt, E.J., 1998, *Portfolio Management for New Products*, Perseus Books, Reading, MA.

Devinney, T.M., "Entry and Learning," *Management Science* 33(6), 706–24.

Foster, R.N., 1986, *Innovation: The Attacker's Advantage*, Summit Books, NY.

Griffin, A., and Hauser, J.R., 1993, *"The Voice of the Customer,"* Marketing Science 12(1), 1–27.

Groeneveld, P., 1987, "Road Mapping Integrates Business and Technology," *Research-Technology Management*, 40(5), 13–19.

Kinnear, T.C, and Taylor, J.R., 1995, *Marketing Research: An Applied Approach*, fifth edition, McGraw-Hill, NY.

McKim, R.H., 1980, *Experiences in Visual Thinking*, Brooks-Cole Publishing, Monterey, CA.

McGrath, M.E., 1995, *Product Strategy for High-Technology Companies*, McGraw-Hill, NY.

McGrath, J.E., 1984, *Groups: Interaction and Performance*, Prentice Hall, NJ.

Meyer, M.H., and Lehnerd, A.P., The Power of Product Platforms, Free Press, NY.

Paul, G., and Beitz, W., 1996, *Engineering Design*, Springer-Verlag, NY.

Porter, M.E., 1985, *Competitive Advantage: Creating and Sustaining Superior Performance*, Free Press, NY.

Ulrich, T.K., and Eppinger, S.D., 2011, *Product Design and Development*, McGraw-Hill Companies, NY.

Urban, G.L., and Hauser, J.R, 1993, *Design and Marketing of New Products*, third edition, Prentice Hall, NJ.

Hippel, V.E., 1988, *The Sources of Innovation*, Oxford University Press, NY.

Wheelwright, S.C., and Clark, K.B., 1992, "Creating Plans to Focus Product Development," *HBR*, 70(2), 70–82.

8.8 Review Questions

- How to do you develop a search procedure for designing an index system for a CNC lathe? How do you develop both an internal and external search plan for an index system with live tools? Is it necessary to conduct such searches? How does it help concept development?

- Is it necessary to develop several concepts for one design? How does this help the design process? Is there a limitation to developing several architectures for the same design? How do you maintain the conformance of a concept with the requirements? What leads the concept generation or architecture development process, requirement satisfaction or innovativeness?

- Develop a functional decomposition process for a fully automatic computer-controlled grinding machine for automobile part machining. What are the constraints of the design? What are the sub-system requirements for the grinder? How does the functional decomposition help the concept, or architecture, development phase? What are the limitations of such a process? Does it help or hinder creativity?

- Do the software tools help engineers to be creative and productive at the same time? Do they make it difficult for an engineer to be creative? Do they force the engineer to think in a restrictive way? What are the advantages of using software or a computer environment during concept generation phase?

- Should team members accept the help of lead customers when creating a concept? What are the benefits and limitations of getting input from the customers?

- Is the concept generation phase limited to a highly complex and costly product? Can it be applied to a simple product? Does this step help to promote creativity among team members? Is there any reason why machine tool development should have this particular phase? Is it necessary at all, since it costs time and money to carry out this phase?

C H A P T E R 9

9 Concept Confirmation

9.1 Introduction

At this point in the design and development process, the team has the specification metrics and sub-module specifications confirmed and more than one concept generated for further consideration. In order to move further, the team has to select one of the concepts. For example, a sub-module of the CNC lathe system, the live tool indexing unit, will be explored to explain this process of selection and confirmation. The specifications for this module have been generated in alignment with the system requirements.

As explained earlier, the process is still on the left hand of the system "V" and one step ahead of the design and detailing of the sub-system. The adjustment of specifications is still possible and can be done in case it is found that specifications for the sub-system are unrealistic or too costly to support or manufacture. Such iterations are required to reduce the cost of the system or enhance the manufacturability of the unit depending on the resource requirements. Nevertheless, customer requirements must be satisfied, and the product must be competitive among the benchmarked products in the target market.

Let us reiterate the requirements for the indexing unit again for further processing. The indexing unit of a CNC-automated and programmable lathe consists of static tools and live tools. The tools remove chips from the part to give it the intended shape. The live indexing unit has two prime servo drivers: one is for rotating the tools, and the other one is for indexing the turret. This indexing is required to position the tools with respect to the job as required to perform turning, milling, drilling, etc.

The operation of the turret starts with a command from the central computer interface unit, which is controlled by a program or an operator. When commanded, the turret rotates, positioning the tools with respect to the job for an intended operation. The advantage of such a complex unit is that it combines the turning and milling machine together. The turret is mounted on the cross slide, which moves either axially or radially with respect to the job. The units are completely programmable and also manual.

The requirements for such an indexing unit are as follows:

- Indexing unit has to have 12 stations
- Indexing unit must have static and rotary tooling
- Indexing time for one station is 1.5 seconds

- Indexing time for six stations is three seconds
- Unit bolted down on the cross slide of the machine
- Unit is servo-driven and computer-controlled
- Tool positioning accuracy is less than 1 angular second
- Ease of maintenance
- Mean time between failure is 4,000 hours
- Unit weight has to be less than 200 lbs.
- Easily programmable for tool selection
- Minimum life requirement is 10 years

The concept for the indexing unit is shown in Fig. 9.1. Three possible concepts for this unit were generated, and Fig 9.1 shows the design that was selected. Each concept satisfied most of the requirements, but this one had the highest promise or probability of satisfying almost all of the requirements mentioned above. The team has to have a selection process established to single out the best possible concept, or architecture, from more than one possibility. Such a selection process is sometimes very difficult and creates intellectual conflict among team members. Hence, it has to be procedural and systematic so that each team member has high confidence in the selection process.

The team has the responsibility to select one concept, or architecture, knowing very well that the design is very fluid at this stage of the game and could change substantially. The design is conceptual and not ready for detailing yet. The selection process should be very fair to each team member, without any prejudice or bias. This is difficult since it might make a team member very unhappy if his or her idea is not accepted for further processing.

In order to get a consensus on the selected design, a logical process is necessary. The selection process must be agreed upon by the whole team. In order to make the process very effective and productive, a method has to be identified to eliminate the weaker designs and select the best possible one. The weaker aspects of any design also have to be identified and discussed among team members.

Last but not least, the selection or concept confirmation process must be documented for future references. In short, a logical process of confirmation or selection has to be developed. For a machine tool business, such processes are almost non-existent or not used very formally and vary widely from company to company. Such a wide variance does promote inefficiency and leads to a high-cost design. The objective of any process is to standardize the design process so that it can be used repetitively in future design initiatives irrespective of the nature of the design.

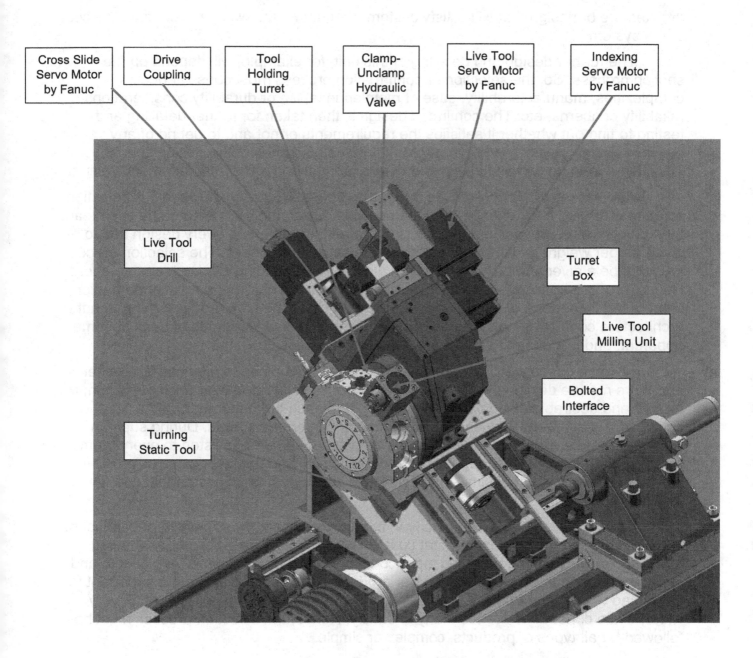

Fig 9.1: Live tool turret for a CNC lathe

(Courtesy of Target Machinery Pvt. Limited, Mysore, India)

It could be also said that the concept, or architecture, selection process is an integral part of the design and development process. The primary purpose of this selection or confirmation process is to satisfy customer requirements without exceeding the budgeted cost of the system. The containment of the cost is at the focal point here. The advantage of generating several possible concepts at the outset is to make sure that all the feasible solutions leading to customer requirements have been explored. The concept confirmation process identifies the strengths and weaknesses of any possible concepts and tries to eliminate the weaknesses and select the best possible architecture or design that will satisfy customer requirements without exceeding the cost of the system.

The further design and detailing of this unit, for example, will depend on the section process. So, the selection or confirmation process must consider design complexities, manufacturability, ease of maintenance, life or durability considerations, reliability concerns, etc. The confirmed design is then taken for further detailing and testing to find out whether it satisfies the requirements or not and to get rid of any possible design noise. The negative or unintended outcomes of the design should be identified during testing and eliminated before production for manufacturing and selling.

Another reason for subjecting the design and development through this selection process is to give a fair chance to all the possible design thoughts created by each team member. Since each design approach is fundamentally different, every design has to be given proper weight and importance during the selection process. The selection process must not be a diverging process, creating a difference in feeling among team members. Hence, process considerations must be highly iterative in nature, as mentioned earlier. Most of the time, none of the designs pop out, but when they are processed through this architecture confirmation process, the advantages and disadvantages of each become apparent, and it is easier to select one for further processing.

In some cases, several of these concepts are combined to give rise to a better and more mature design. Once these initial designs are iterated over and over again, a mature and desirable design becomes apparent to each team member, and a consensus is achieved very easily and without any internal conflicts. During such an iterative process, the team leader has to take a subdued role and listen to every team member and their design thoughts to come to a logical and unbiased conclusion. This is a digestive process that assimilates all possible concepts into an acceptable one for further processing.

For a simple stand-alone machine, this process is simply neglected or bypassed. In my opinion, when the design is not complex, there is a greater possibility of being more innovative and of generating more concepts. When the design is very simple and every company can come up with similar products, innovativeness is more important to get ahead of the competition. Complexity sometimes restricts the ability to think very widely since constraints are too many to satisfy. Hence, the selection process must be followed for all types of products, complex or simple.

9.2 Concept Selection Process

A concept, or architecture, is the form of an engineer's idea of what the system looks like to satisfy the customer requirements. The concept might consider a new technology or architecture that satisfies the requirements. It might include legacy designs or the current trend of designs. A typical process flow chart is shown in Fig. 9.2. Most of the time, the selection process is implicit in many companies. Even then, an underlying process is followed but not documented properly. In general, most machine tool companies do not have enough resources available to document every design process step. This is a mistake that should be corrected.

The methods of concept selection also vary from company to company. Some companies use external consulting companies to help with design and development, and the consulting company selects the best possible architecture, or concept. Sometimes the company shares the information with their lead users and gets their opinion about the best possible design based on their requirements. In most of the cases, though, the product leader selects the best possible design based on his background, experience, and training. In Japan, I observed that the best design was selected by the leader and then discussed with vendors, manufacturers, and trusted users. In other countries, the concept is selected internally using the intuition and consensus of the designers. The product leader takes a very strong role, selecting the best possible design.

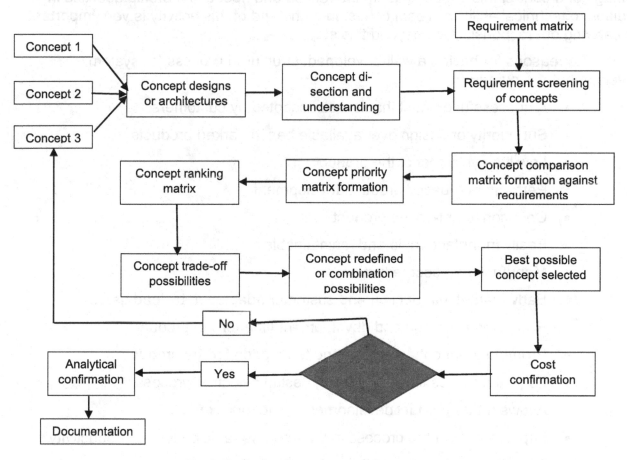

Fig 9.2: Selection process flow chart for machine tools

Irrespective of the methods used, explicit or implicit, a selection process must be followed and documented. In some companies, this process is internal to the design team and guided by the team leader, and the best concept is selected by mutual discussions and considerations suitable for the company. It is very important for this process to be highly iterative and discussion-oriented. During such discussions, pros and cons for each concept, or architecture, must be documented. The strengths and weakness of each design should be noted. Manufacturing considerations must be given for each design, keeping in mind the manufacturing resources available within the company, if available. The team or evaluating team, which could be independent of the design team, should have a mutually agreed upon selection criterion before the designs are subjected to such a process.

Product success is the ultimate index of any design and development effort. Product success means to design a quality, performance-based product that can be manufactured easily and introduced in the desired market with a cost that is competitive or even less than that of similar products. The selection process is definitely very complicated and critical to product success. It is also true that the selection process sometimes creates misunderstanding, mental agony, or tension among team members since some concepts are discarded during this process. The team leader has to manage and control this aspect of team interaction and needs leadership qualities in addition to technical abilities. Team cohesion at the end of this activity is very important for carrying out responsibilities beyond this step.

The reasons for having a well-developed, structured process for system development are numerous:

- A successful product that is well accepted by customers
- Superiority of design over available benchmarked products
- Low life-cycle cost of the product
- Performance-based product development
- Cost-constraint-based product
- Easily manufacturable and maintainable
- Manufacturing cost reduction
- Early market introduction and customer adaptation by lead users
- Reduction of design and development time for the product
- Promote team cohesiveness and team pride for the product
- Logical and less time-consuming design selection process
- Allows future product development using this example
- Improvement of the process in the future for efficiency and productivity
- Consistent revenue growth for the company

From the above list, it is obvious that a well-developed and structured selection process has numerous advantages. Hence, it is necessary to follow such a process in a disciplined fashion. It could be somewhat less structured, but a definitive process must be adhered to. As shown in Fig. 9.2, the total process consists of the following major steps:

- Identification of positives and negatives of each concept: comparative study
- Comparison of concepts against the requirements: requirement study

- Trade-off and combination of concepts, if possible: consolidation study

- Selection of dominant design, concept, or architecture: design selection

- Satisfaction of cost and performance constraints: re-confirmation study

- Documentation: recap and reflection

The first and foremost step of this selection process is to hold discussions among team members about the pros and cons of each design, which could be called the *comparative study* of the designs. Once design analysis is completed, a requirement study of each design is carried out. This step is called the *requirement study*, in which each design is benchmarked against the customer requirements, which are weighted or prioritized based on customer reviews. Each concept is ranked against these priorities. This will also allow a determination of the superiority of each design over the others for each requirement. If a dominant design is very obvious, the team might like to pursue that design or concept. In case this is not possible, a *consolidation study* is taken up and the good aspects of each concept are combined or merged with those of other concepts to see if the combined concept is better than the originals. After the consolidation study step is completed, the best feasible design is selected.

The next step is to ensure that the best design supports the budgeted cost and performance requirements for the product. In addition to cost analysis, a theoretical study is carried out to ensure that the selected design satisfies performance requirements. This step, the *re-confirmation study,* might need a readjustment to satisfy cost constraint. If the cost containment is not possible, new concept generation has to be initiated, starting the process all over. Otherwise, the best concept is put forward for further processing. The last step is to document the process for future reference.

In general, the process starts with the design concept possibilities, and the designs go through several checks and balances to determine suitability based on customer requirements. Obviously, the purpose is to select the best possible design that can be manufactured within a cost constraint specified by the market. In most cases, this stringent process will filter all the designs and will give rise to the selection of the best possible design. As mentioned earlier, the product has to satisfy at least two aspects: cost and performance. As long as any such process helps the team achieve these two indices, it is suitable for the team and the company.

Let us analyze the index sub-system concept design, which has three initial concepts:

- Index unit with complete hydraulic drive: Concept 1 (see Fig.9.3)

- Index unit with complete servo-hydraulic drive: Concept 2 (see Fig.9.4)

- Index unit with combined servo-electric drive: Concept 3 (see Fig.9.5)

Concept 1: Turret rotation and positioning are performed using a hydraulic drive, i.e., a hydraulic motor and mechanical cam attached to the main index shaft. The turret moves axially in and out during clamping and unclamping of the turret coupling. The turret position is possible using a combined four-cam combination. The clamping and unclamping operation of the turret is done using a gear train and face cam.

There is no electric drive attached to this unit for indexing. The live tool drive is separated from the indexing drive and connected to the tool rotator by a belt drive. The live tools are separately driven by servo-controlled motors. The sequence of operations for indexing is as follows:

- Operator commands indexing through control box

- The turret is unclamped using hydraulic piston

- Turret rotates to a desired indexing position

- Turret is clamped and positioned with two-piece coupling

- Index station confirmation with axial cams (four required)

The operation of this design starts with the command from the operator. The index station is selected by the operator. Once this input is received, the hydraulic unit directional valve is activated to supply hydraulic power to the piston to move left, disengaging the two-piece coupling. Once unclamped, the index hydraulic motor is activated for turret rotation to the desired position.

The turret positioning is held using the four mechanical cams fixed at the end of the turret shaft. Once the turret station is confirmed, a hydraulic valve is activated to move the piston axial to a clamp position. Hence, the indexing cycle is completed. The whole sequence operation is computer-controlled. The positioning cannot be done manually for this design.

The motor for rotating the live tools is also motor. Positioning is not possible. When the motor stops, the live tool stops. This is a very serious disadvantage of this design since threading operations need positioning accuracy; otherwise, the tap will break very frequently. For other tools, a hydraulic motor is OK. Also, in some operations, hydraulic motor revolution and power could be limiting, and the customer might not like the slow rotation capability.

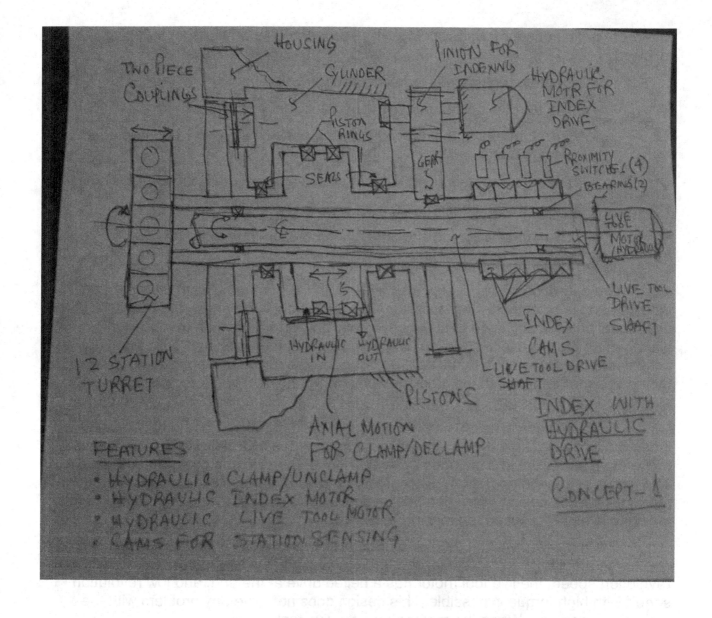

Fig 9.3: Index unit architecture, Concept 1

(Courtesy: Warner and Swasey Company, Cleveland, Ohio)

Concept 2: Turret positioning is designed using a servo-controlled hydraulic drive, two-step gear train for RPM reduction, and an encoder attached to the end of a hydraulic drive. The positioning cams are replaced with electric encoders. The clamping and unclamping operation of the turret is still done using a hydraulic drive and a two-piece coupling. The live tool drive is separated from the indexing drive and connected to the tool rotator using a belt drive. The live tools are separately driven by a hydraulic motor, and there is an encoder for positioning. The sequence of operations for indexing is as follows:

- Operator commands indexing through control box
- The turret is unclamped using hydraulic drive
- Turret rotates to the desired indexing position using the encoder and electric motor
- Turret is clamped using a hydraulic drive, spring, and a two-piece coupling
- Index station confirmation from encoder
- Live tool positioning and speed confirmation from encoder
- Machine ready for operation

The operation of this design also starts with the command from the operator. The index station is selected by the operator. Once this input is received, the hydraulic unit directional valve is activated to supply hydraulic power to the piston to move left, disengaging the two-piece coupling. Once unclamped, the index hydraulic motor is activated for turret rotation to the desired position. The turret face positioning is sensed by the encoder attached to the index drive hydraulic motor. The face cams are eliminated in this design.

Turret position is held using the encoders fixed at the end of the hydraulic drive index motor shaft. Once the turret station is confirmed, a hydraulic valve is activated to move the piston axial to a clamped position. Hence, the indexing cycle is completed. The whole sequence operation is computer-controlled. The positioning cannot be done manually for this design. Here, the positioning is more accurate, and the chance of mis-indexing is highly reduced, enhancing reliability.

The motor for rotating the live tools is also hydraulic. The positioning is possible due to the electronic encoder attached to the end of the live tool motor. To reduce the revolution speed, the live tool motor has a belted drive at the end, and low revolution speed with high torque is possible. This design does not have any problem with the threading operation. When the motor stops, the live tool stops.

The synchronization of the tool positioning and indexing station is controlled by the computer. Both manual and automatic operation are feasible for this design. Most of the deficiencies of Concept 2 are eliminated here.

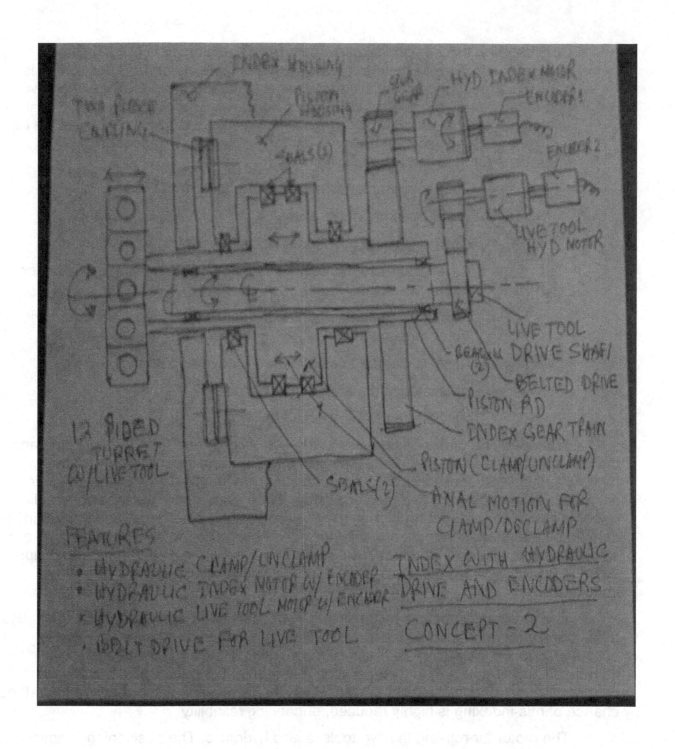

Fig 9.4: Index unit architecture, Concept 2

(Courtesy: Warner and Swasey Company, Cleveland, Ohio)

Concept 3: In this design, turret clamping and unclamping is done by the hydraulic unit. Turret rotation is done with an electric motor, and positioning is sensed by an encoder. The live tool drive is separated from the indexing drive and connected to the tool rotator using a belt drive. The clamping and unclamping operation of the turret is still done using the hydraulic drive. The live tools are separately driven by servo-controlled motors. The sequence of operations for indexing is as follows:

- Operator commands indexing through control box

- Turret does not move axially

- The three-piece coupling is unclamped using a hydraulic drive

- Turret rotates to the desired indexing position using the encoder and servo-electric motor

- Live tool is connected to the live tool drive using a servo-electric motor and encoder

- Turret is clamped using hydraulic drive and spring

- Index station confirmed from encoder

- Machine ready for operation

The operation of this design also starts with a command from the operator, and the index station is also selected by the operator. Once this input is received, the hydraulic unit directional valve is activated to supply hydraulic power to the piston to move left, disengaging the two-piece coupling. Once unclamped, the index hydraulic motor is activated for turret rotation to the desired position. The turret face positioning is sensed by the encoder attached to the index drive hydraulic motor. The face cams are eliminated in this design, and the turret does not move axially. The three-piece coupling disengages from the fixed portion of the coupling and engages with the turret-fixed coupling.

The turret positioning is held using the encoders fixed at the end of the hydraulic drive index motor shaft. Once the turret station is confirmed, the hydraulic valve is activated to move the piston axial to a clamp position. Hence the indexing cycle is completed. The whole sequence operation is computer controlled. The positioning cannot be done manually for this design. Here, the positioning is more accurate and the chance of mis-indexing is highly reduced, enhancing reliability.

The motor for rotating the live tools is also hydraulic. The positioning is possible due to the electronic encoder attached to the end of the live tool motor. To reduce the revolution speed, the live tool motor has a belted drive at the end, and low revolution with high torque is possible. This design does not have any problem with the threading operation. When the motor stops, the live tool stops.

The synchronization of the tool positioning and indexing station is controlled by the computer. Both manual and automatic operation are feasible for this design. Most of the deficiencies of Concept 3 are eliminated here.

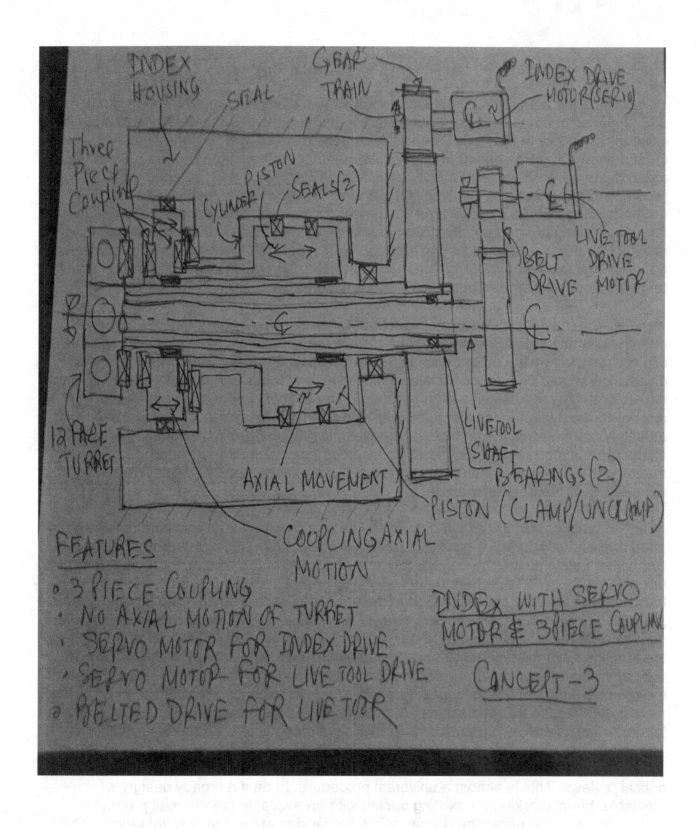

Fig 9.5: Index unit architecture, Concept 1

(Courtesy: Target Machinery, Mysore, India)

9.3 Concept Comparative Analysis

Once the concept design features have been discussed among group members and analyzed in detail, and once the salient features have been understood, a table is created, as shown in Fig. 9.6. This particular table displays the comparative analysis of each design's features based on the requirements for the design. The requirements are noted in the first column, and then whether each design satisfied the criterion is recorded. The success of this step is dependent on group participation and understanding without bias towards any design or concept. The discussion should be fair to every design, and a comparison has to be made, keeping the focus only on the satisfaction of requirements for the product. In this table, all the concepts are compared against each requirement for the product at the end. The process of comparison should be unanimous and agreed upon by each team member.

Since machine tool design concepts are very few by nature, all are kept alive until the ultimate selection is made. The discussion and critical review of the concepts are normally led by the most experienced team leader in most cases. Moreover, the discussion should be held on an intellectual level, and focus should not be given to final details or to how the design will be produced. The concepts should be compared relative to the design ideas just to differentiate one from the other. The analysis is definitely comparative by nature. While developing the comparison table at this stage, all the more important criteria should be included first, such as performance criteria and manufacturing criteria, without which the product would be useless against the competition in the desired market.

In most cases, the existing or legacy design is taken as a base concept since the good and bad aspects of the present design are pretty much known at this stage, say, Concept 1 in this case. The deficiencies and strengths of the design must be known and understood by each team member as well. The other concepts could be rated against the criteria of the new concepts. Sometimes the deficiencies against the benchmarked products are also considered at this phase. The idea is to hold on to the efficiency of the design, and the deficiencies are to be overcome in the new designs. That is the primary idea unless a completely new innovative design is being considered. It might happen very often that every member might not completely agree to the relative values for all the requirements. The team leader normally resolves such issues and moves on to the next stage. In case legacy designs are not available, benchmarked products or market leader products can also be used for further development.

For machine tools, performance and cost are the criteria always considered for critical reviews. This is almost a universal procedure. Even if a legacy design is available, benchmarking an existing design with an available commercial product is always done first to understand what points are lacking are in the present design. The idea is to compare the designs and learn from what has been successful in the market.

Another fact that has to be kept in mind is that the team is looking for overall ideas to satisfy almost all the requirements using available human and manufacturing resources in the company. It is not looking for details of the designs. Too much critical discussion and review sometimes kills innovative ideas since most team members cannot appreciate groundbreaking future ideas about a design. Nevertheless, a somewhat general review of the designs could be made without much effort and problem. Knowing for a fact that casting is cheaper than forging, if one design calls for forging and another calls for casting, obviously, casting has an edge over forging. That is the level of analysis at this stage. The idea is to eliminate some concepts out of many possible concepts. In this case, it appears that Concept 3 seems to be the best out of three possibilities.

The next step is to come up with the relative weights of each criterion using brainstorming sessions among team members. The team weighs each requirement against others to identify what is more important for the customers. The team leader can take a predominant role, conducting such sessions in a very constructive way so that there is no bias. A second table is created, as shown in Figure 9.7. The relative weights for each concept are recorded after discussion among team members.

Again, all the requirements are recorded in the first column, and the relative weight for each requirement is recorded. Then, for each concept, the weighted value for each requirement is calculated by multiplying the weight factor. The idea is to compare and rate each criterion for all the concepts against a requirement. It is better to take and finalize one criterion before the next one is taken up down the list. Another way could be to finish the table concept by concept. Now each score for one concept is multiplied by the relative weights to come up with a scoring number. This procedure is continued for all the concepts and all the requirements. Next, for each concept, the points are added to come up with the cumulative value for each concept. For example, the concept scores are 0.5, 1.8, and 2.8 (rounded values) for Concept 1, Concept 2, and Concept 3 respectively. These values just show that Concept 3 is more suitable than Concept 1. The absolute numbers are of less significance than their relative values. Moreover, these cumulative comparative points for each concept represent what the team members thought about the priorities of each concept.

It is important to mention here that the determination of relative weights and scoring for each concept against any requirement is a somewhat tedious job. Many a time, disagreements among team members do arise, and the team leader has the responsibility to resolve the issue without alienating any member. The process is somewhat approximate, but it is very effective most of the time. There could be some incorrect values for relative weights or scorings, but the net result of rating the concepts against each other remains pretty much the same. During such discussions, the reference, legacy, or existing design should be kept the same for comparison. The comparison process becomes much easier and more convincing when compared to a known design and its performance. The team consensus is very important during this approximate process.

As said before, the cumulative points for each concept only denote the fact that one concept is better than the other. In this case, Concept 3 scores the highest, meaning that the team thought it was far better than the legacy design (Concept 1) and somewhat better than Concept 2. Since Concept 2 and Concept 3 are better than the existing design, some good design aspects of the two will be combined to come up with another concept. This is what happens most of the time. In any design, one or two bad aspects always remain in the design, and they should be eliminated before moving on. Sometimes such bad points might have degraded the value of the overall design.

Hence, before a design is totally discarded, this bad point should be eliminated if possible. Then the ranking process must be repeated to make sure that final selection is still consistent. One example is the manufacturing issue. The team might think one concept is very difficult to manufacture, which might not be true in the end. Hence, a thorough discussion of this issue with the manufacturing engineers must be held before discarding the concept altogether. I like the idea of taking the best two concepts at the end, finding the common design aspects of both, and then creating another concept, keeping the good aspects and then eliminating the bad elements of each design in the combined concept. This purification, or iteration, step often brings the best results. The new concept, Concept 4 in this case, should be added to the ranking tables (Figs. 9.6 and 9.7), and the selection process repeated to make sure that the combined design concept is, in fact, the best possible. For example, it has to be ensured that the combined or cumulative rating point for Concept 4 should be the highest of all. The best concept design is shown in Fig. 9.8 for this index unit, which was taken up for design and detailing.

Before the team starts the concept comparison process, a baseline concept is selected. The baseline concept could be the existing design that the team is trying to improve to suit market demands. Another point is the clarification of the requirements, which cannot be ambiguous. The requirements must be very well defined and understood in the clearest terms by all team members. The list of requirements could be arranged chronologically or in a hierarchical fashion depending on the complexities and number of requirements to be considered for comparison. In most cases for machine tool application, performance measures come first, and then quality and reliability requirements come up in order of priority. The next issue is manufacturing. This order needs to be changed.

I believe that the design has to be made in such a way that manufacturing and quality issues get similar attention. The objective is to reduce the cost without compromising quality. The cost is of prime importance to be competitive in the market. That is how Japan beat the U.S. builders: their machine quality was given by design, and the cost of their products was comparatively lower than those from the US. Quality should never be compromised. The concepts really ensure quality products when manufactured to suit the design.

The team also has to create a list of priorities for the requirement and a weight value is attached to each requirement. The weight value reflects the customers' choice of requirements. The highest weight is given to the requirement that most of the customers want. If any criterion is mostly demanded by the customers, that feature is given the highest weight. These should be called "musts," and some of the features that customers do not care for can be given a very low weight. In addition, some companies have some aspects of their product that help the company to establish their brand. It could be cost, quality, design flexibility, performance, durability, reliability, etc. If the company has priorities, they are also given a higher weight. In any case, a scale has to be defined and agreed upon by all the team members well before the comparative matrix is formed. The new concept is always compared against the baseline design, with which the team members already have experience. For some aspects, the existing design might already be already, and those should be maintained all throughout the design concepts.

In some cases, the baseline design is not available. If so, any established product in the market can be used as a baseline design. The idea is to identify the deficiencies of this product, and the new design should improve upon these deficiencies or weakness.es In some cases, products from different manufacturers are also considered for comparison. One product could be taken up for performance, another one for durability and reliability, etc. The idea is to capture the product leaders and incorporate into the future product aspects that will enhance its performance in the marketplace.

Once the process is completed, the team has to make sure the best features of all the concepts are combined, if required. The team should focus on the best concepts but should not forget the best features of discarded designs. This could be called "concept optimization." For example, the best design concept was to accept the three-piece coupling, servo motors with attached encoders, hydraulic power for clamp/unclamp, separate live tool motor with belt drive, optimized gear train for indexing, and automatic coolant off and on for tooling stations. The unit seems to have much better features than what is available commercially. The team has to now recap the process and get a deep meaning and understanding of the unit's functions and limitations. In case the new comparative values do not show a significant and dominant design, re-evaluation of the weight values and scoring for each concept must be revisited to be realistic.

Now the selected concept has to go through a cost-evaluation step to ensure that the design satisfies the cost constraint. The cost of the unit primarily depends on the manufacturing methods, cost of purchased components, and cost of assembly, i.e., the total fixed and variable cost for the unit. The cost evaluation step is a very critical step, but it is one that teams very often neglect. The approximate cost of manufacturing must be considered.

I made a similar mistake, accepting a complex design without cost consideration. As a result, you get a product that is not competitive at all in the market. At that point, either you have to redesign the unit for cost reduction or discontinue the project, which is the ultimate step. The team members should feel very comfortable and confident about the selection process and its outcome. A consensus among the team members is almost a necessity before the design is taken for detailing. This agreement and consensus will help the team commit wholeheartedly to the design. Such an agreement is essential for the product's success.

While developing these tables, the team should look for the improvements of new concepts over the existing or base concept (reference concept), the changes in the new concepts that make the product better or improved. The team has to consider whether these new concepts are innovative or creative. The team has to measure the productivity enhancements that these new concepts bring to the customers. Once the best concept is selected, the team can play the "what if" game to confirm their selection. These results could be evaluated alongside a cost analysis as well. The cost analysis will optimize the design, minimizing cost, and the design has to be modified accordingly.

Since the concept, or architecture, is in its embryonic stage, the risk associated with any design must also be considered. If the new concepts depend on brand-new, unproven technology, the team has to find out the risk associated with the application of this technology in the near future. The risk and probability aspects are very important considerations that sometimes make or break a design. For example, switching the machine tool control from mechanical to numerical carries a lot risk with respect to durability and reliability since the technology was not matured at the point of introduction. Normally, a bookshelf or proven technology can be better for immediate use to mitigate the risk. If the new concepts tend to be risky, it is sometimes better to be a follower than a leader. Another consideration is the probability of success for this design, technology, or manufacturing method. Use of technology or design methods just for the sake of using them is not a good idea in my opinion. The risk of using such technology or manufacturing methods must be very seriously considered before implementation of any design. Sometimes it is strategically all right to be a follower in the machine tool industry.

Once the final selection is made, the team takes up the concept for further designing and detailing of the components. Once the design has gone through the cost gate and risk analysis, a theoretical analysis is conducted on its critical components to make sure performance requirements have been met. The methods of such analysis can be numerous, such as finite element analysis, reliability life analysis, and durability analysis using a real-world usage profile, etc. Once the analysis shows the performance requirements have been met with an adequate confidence level, the concept is ready for further detailing and manufacturing method development to build a physical prototype. The assembly considerations must also be discussed at this stage, and documentation has to be completed for this process.

As said earlier, this step is at the bottom of the system "V" diagram. Once this step is passed, the design sops being iterative, and the next steps are very much serial in nature. Changing the design after this step becomes very costly and demoralizing to team members. Again, the system and the sub-systems are finalized, and any substantial change thereafter is almost impossible since design flow up the left arm of the "V" is very difficult, to the point that the program might be stalled altogether. Hence, the team must be absolutely satisfied that the selected design concept is what the team, the company, and the market want for sure. The team has to make sure that the selected concept has the highest potential of satisfying customer requirements without exceeding the cost expectations.

One of the critical considerations in the manufacturing arena is to answer the following question: can the selected design be manufactured with high volume and repeatability? Producibility concerns must be analyzed, and various manufacturing methods must be considered to reduce the cost and time of manufacturing. These are the reality checks for the design, and eventually these checks and balances bring the product success. Sometimes it is also a good idea to share the design with trusted external members, such as manufacturing vendors, for their opinion. Expert opinions can also be sought before advancing to the next step of designing and detailing the concept. In case of any consistency or concern about the design, the team must resolve the issue with absolute confidence.

The team has to ask itself whether the selection process has helped them to select the best possible concept. The team should answer the following:

- Is there any better way of dealing with this process?
- Does it need any change for future use?
- Is there any disagreement left unresolved?
- Is the process very complicated, to the point where it increases the timeline?
- Does it support team development?
- Is the process very time-consuming and costly?
- Does the process enhance team cohesiveness and team building?

Requirements	Concept 1	Concept 2	Concept 3	Remarks
12 tooling Stations	YES	YES	YES	Must have
Static and rotary tooling	NO/Restricted	YES	YES	Must have
Indexing time for one station less than 1.5 second	NO	YES	YES	Preferred
Indexing time for six stations less than 3.0 second	NO	YES	YES	Preferred
Bolted interface	YES	YES	YES	Preferred
Computer-controlled	YES	YES	YES	Must have
Tool position accuracy less than 1 angular second	NO	YES	YES	Must have
Manufacturing resources available	YES	NO	NO	Need new machines
Mean time between failure less 4000 hours	NO	YES	YES	Hydraulic drive has more failures
Unit weight less than 200 lbs.	NO	NO	NO	Metric to be revised
Easy programming for tool stations	YES	YES	YES	Control function
Minimum durability life 10 years	NO	YES	YES	Hydraulic unit fails very often
Cumulative points	NO- 7 YES- 5	NO- 2 YES- 10	NO- 2 YES- 10	Relative values
Relative ranking	Base	Better than base	Best concept	Move on to next stage

Fig 9.6: Comparison of concept designs against requirements

Requirements	Relative Weightage	Con 1	Points for Con 1	Con 2	Points for Con 2	Con 3	Points for Con 2	Remarks
12 tooling stations	0.13	1	0.13	1	.13	1	0.13	
Static and rotary tooling	0.12	0.3	0.036	1	0.12	1	0.12	
Indexing time for one station less than 1.5 second	0.12	0.3	0.036	0.3	0.36	0.8	0.96	Con 3 best
Indexing time for six stations less than 3.0 second	0.12	0.3	0.036	0.3	0.36	0.8	0.96	Con 3 best
Bolted interface	0.01	1	0.01	1	0.01	1	0.01	
Servo-driven and computer-controlled	0.1	0	0	.1	0.1	1	0.1	
Tool position accuracy less than 1 angular second	0.15	0.6	0.09	0.7	0.105	0.9	0.135	Con 3 best
Manufacturing issues	0.05	0.5	0.025	0.3	0.015	0.4	0.02	
Mean time between failure less 4000 hours	0.05	0.6	0.03	0.75	0.59	0.85	0.04	
Unit weight less than 200 lbs.	0.02	0.4	0.008	0.8	0.016	0.8	0.16	Con 3 best
Easy programming for tool stations	0.03	0.7	0.021	1	0.03	1	0.03	
Minimum durability life 10 years	0.1	0.6	0.06	0.8	0.08	0.8	0.08	
Cumulative points	1.0 or 100%	6.3 out of 12	**0.5**	8.1 out of 12	**1.8**	10.4 out of 12	**2.8**	Con 3 best

Fig 9.7: Relative weighted comparison of concept designs against requirements

Note: 1 means 100% compliance with design requirements, and 0 means zero compliance (comparative scaling)

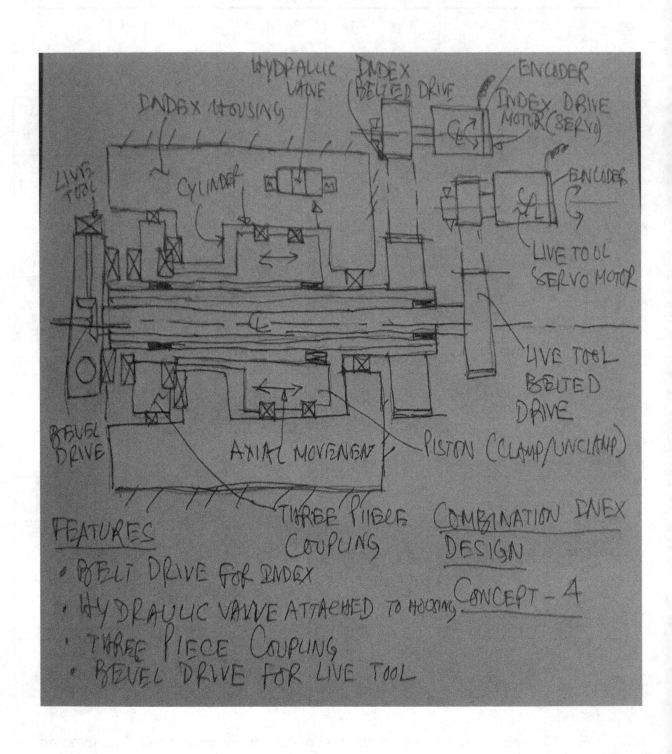

Fig 9.8: Combined architectural concept for the index drive sub-assembly

9.4 Recap on Selection Process

The selection discussed so far might be a debatable topic to some folks since the steps are not scientifically proven. It is not objective but subjective in nature. Let us discuss such issues:

- Functional decomposition is not an established formal process by definition. The success of such step primarily depends on how detailed the team wants it to be. The team members have to have a proper understanding of the system requirements without designing the system, and this is very difficult to do. One way of accomplishing this is to develop a functional decomposition of an existing system and evaluate the system properties based on the new requirements for the future product. The sub-system interactions and boundaries must be properly understood as well. In a nutshell, the decomposition process is somewhat abstract, and sometimes it is very difficult and almost impossible to decompose a system's functional requirements into sub-system functional requirements.

  Nevertheless, such a decomposition process might help the team to define the architectural boundaries of the system elements and how the sub-systems are supposed to work together for a successful product. The penalty for not doing such a functional decomposition could be very heavy and should be avoided for all practical purposes. Moreover, for a very complex electromechanical or software system, functional decomposition is almost a must for simplifying design activities. For example, the weight of a product depends quite simply on the weight of the sub-systems. Unless the weight restrictions or constraints of each sub-system or element is defined properly, it will be extremely difficult to satisfy the weight requirement for the product in the end.

  Decomposition also helps team members understand the cost of each sub-system and its effects on the whole system. Each sub-system must be designed with cost in mind. The cost distribution is rationed at the system level and then compared when the design is completed but before manufacturing starts. So, there are multiple advantages to taking pains to develop a system decomposition at the outset.

  For properly developing a complex system, a functional decomposition, even if it is not accurate in the beginning, must be completed and then reiterated at the end for future reference. The functional decomposition process breaks up a highly complex system into simpler sub-systems that can be independently designed and regrouped together. That is the primary purpose of functional decomposition: to make the sub-systems independent of the system when designing for the first time.

- The next point of discussion is to keep the criterion selection points as simple and objective as possible. The selection criteria must be simple to define and understand. They must be unambiguous and realistic. Comparison of the concepts with these selection points must be made very carefully and in a very systematic way. There should be no bias in play here. In order to be very effective, the team should have a minimum set of requirements selected for this process. Only very important ones should be selected and compared against each other. It is much easier to handle a reduced set of requirements that do not contain unnecessary and unimportant features.

  For example, minimization of tool vibration is very important to get a good finished surface. So, vibration amplitude at any speed is very important, but not the mode shape of the structure. The stiffness and damping of the structure are determining factors, but the tool shape is not that important. So, vibration amplitude is definitely a criterion, but not the tool shape, which should be kept off the comparison list. The subjectivity of the criterion list must be minimized as much as possible, and objectivity should be considered as much as possible.

- There is another pertinent issue for selection or comparative analysis. This process should be used to further the design and development of the architecture, or concept. Each concept should not be compartmentalized and treated in a block fashion. The good features could be extracted from each concept and ambulated into another concept to create a synergy so that the emergent behavior of the resultant system is the best of all. This is particularly true for mechanisms in the sub-system. Ease of manufacturing must be maintained throughout all concepts. Common features of all the designs must also be kept in the emergent design. Such an improvement process is necessary for product optimization and coordination in the manufacturing process.

- It must be emphasized that the objective at the end of product design is to deliver quality performance at a cost that customers can afford. The design activities decide the cost and quality of the product. Quality must be built in the design. Cost and quality of the end product is not an entity separate from design. Quality and reliability are the emergent properties of a well-coordinated design. The design or concept should encompass the cost or price constraint for a product. It has to be remembered that the cost of production is not a customer requirement but that the price is. The profit is more important than the revenue of the company. The revenue could be generated without any profit for the company. So, cost is the internal factor that could be controlled by design and manufacturing whereas price depends on the market conditions, competition, and customer acceptance. So, design dictates the profit generated from a product.

So, the profit is not the end function of a design development project, but it is the ultimate index of success of any product. The cost of a product is the ultimate constraint and the single most important determinant for marketing the product to generate revenue and profit. Profit is more significant and important than the revenue of a product for the company.

- Another issue is understanding how to use this process to improve a concept for the system and sub-systems. The team could use this process to generate a more developed and improved system by integrating the sub-systems. This process also gives the team members a chance to refine the concepts even at the element level. If a concept is not accepted due to the inefficient sub-system, this process could be used to improve that system. The idea is to separate the sub-system designs using the functional decomposition and segregation scheme so that the sub-system designs can be independently taken up. All the sub-system designers have to do is satisfy the requirements for the sub-systems, including the design of the interfaces. The functional decomposition might also help the engineers to identify the common elements across the designs. This reduces the cost of the system.
- The experience that the team members can gather from this process could also be used down the line across all the design and development steps. The comparative scheme could be applied for cost sensitivity analysis as well. For example, the same procedure could be used for material selection of the components. This process can also be used for selecting the best possible manufacturing method for a component and to compare the assembly process for the system and sub-systems. Hence, such a process can be used in a wide number of areas during the design and development process. A similar process could be used to select optional items such as conveyor systems, coolant systems, tooling systems, jigs and fixture selection, and others.
- Like anything else, exaggeration of such a process is also not advisable since this could lead to a waste of time and resources. Too much refinement could lead to a very complicated process that would very difficult to manage. Finer details could be left for the next design steps.

9.5 Summary

A product architecture, or concept, is the designer's way of representing the specifications and customer requirements. The selection process is somewhat subjective in nature but could be made more objective by selecting proper requirements and comparing each concept against them in a systematic manner, as described in this chapter. The inputs for this process are the design concepts and requirements. The output of this process is the identification of the best possible concept that most satisfies the requirements.

The selection process depends on weighing the features of the designs against these requirements. Customer satisfaction is the only objective in addition to satisfying the cost criterion. The end output of this process is the relatively best possible concept, architecture, or design that will be taken up for further design and development according to the system "V" guidelines.

Team involvement and understanding are very important for this process to be successful. The team should be able to evaluate the designs using a structured process outlined in this chapter. The team has to initially compare each design against the requirements. The team weighs the requirements using a relative scale. Then each concept is weighed against these relative weights, and combined scoring is determined for each concept. Based on these cumulative points, the design with the highest scoring point is selected for further reprocessing. Basically, this process narrows down all the possible concepts into one or two possible designs that show the highest promise of satisfying the requirements criterion and the cost criterion.

The process uses a rating scale that is determined by the team member through discussions and the brainstorming process. The whole process consists of several steps in sequence: selection of critical requirements for the designs, determination of the relative weights for each requirement, determination of relative weights of each design, determination of cumulative rating points for each design, ranking the concepts as per these cumulative points, and selection of one or two concepts that have the highest ranking.

The next step is to dissect the selected design and identify if the best features of these designs could be combined to generate a better concept. The discarded concepts are shelved for future reference. Once the selection process is over, the best design concept is also evaluated against the cost requirement to make sure the design satisfies the cost constraint. In some applications, cost constraints can be used as a requirement, and the selection is based on the relative ranking process.

This process is also recommended for use in other design processes where the team has to select the best out of the whole range of possibilities. The concept selection process needs a proper understanding and consensus among team members for it to be successful. The team leader has to monitor the progress of the selection process and has the responsibility to resolve any conflict or misunderstanding among team members. The team's decision is final in this selection process.

9.6 References and Bibliography

Abernathy, W.J., and Townsend, P.L., 1975, "Technology, Productivity and Process Change," *Technological Forecasting and Social Change* 7(4), 379–96.

Alger, J.R., and Hays, C.V., 1964, *Creative Synthesis in Design*, Prentice Hall, NJ.

Cooper, R.G., Edgett, S.J., and Kleinschmidt, E.J., 1998, *Portfolio Management for New Products*, Perseus Books, Reading, MA.

Devinney, T.M., 1987, "Entry and Learning," *Management Science* 33(6), 706–24

Griffin, A., and Hauser, J.R., "The Voice of the Customer," *Marketing Science* 12(1), 1–27.

Groeneveld, P., 1987, "Road Mapping Integrates Business and Technology," Research-Technology Management 40(5), 13–19.

Keeney, R.L., and Howard, R., 1993, *Decisions with Multiple Objectives Preferences and Value Trade-Offs*, Cambridge University Press, NY.

Kinnear, T.C., and Taylor, J.R., 1995, *Marketing Research: An Applied Approach*, fifth edition, McGraw-Hill, NY.

McGrath, J.E., 1984, *Groups: Interaction and Performance*, Prentice Hall, NJ.

Alger, J.R., and Hays, C.V., 1995, "Estimating Errors in Concept Selection," ASME Design Engineering Technical Conference 83, 396 –415.

Paul, G., and Beitz, W., 1996, *Engineering Design: A Systematic Approach*, Springer-Verlag, London.

Porter, M.E., 1985, *Competitive Advantage: Creating and Sustaining Superior Performance*, Free Press, NY.

Pugh, S., 1990, *Total Design*, Addison-Wesley, Reading, MA.

Souder, W.E., 1980, *Management Decision Methods for Managers of Engineering and Research*, Van Nostrand Reinhold Company, NY.

Ulrich, T.K., and Eppinger, S.D., 2011, *Product Design and Development*, McGraw-Hill Companies, NY.

Urban, G.L., and Hauer, J.R., 1993, *Design and Marketing of New Products*, third edition, Prentice Hall, NJ.

Wheelwright, S.C., and Clark, K.B., 1992, "Creating Plans to Focus Product Development," *HBR* 70(2), 70–82.

9.7 Review Questions

- Why is the selection of baseline design or existing design important for the success of the concept selection process? What would you do for a start-up company that did not have any baseline design? Can some of the features of the base concept be carried through the new designs?

- Is it necessary to develop several concepts for one design? How does it help the design process? Is there a limitation to developing several architectures of the same design? How do you maintain the conformance of a concept with the requirements? What leads the concept generation or architecture development process, requirement satisfaction or innovativeness?

- Can the selection process be applied to concepts of different types of machinery, such as stunning, milling, and grinding? Is it necessary to have similar designs for comparison?

- Develop a selection process for a CNC grinding machine that uses a brand-new control technology that has not yet been proven in the market. What are the risk factors that should be included in the scoring table? Also, develop a risk-probability graph for the different aspects of this new technology that might affect the grinding machine.

- Is it possible to select more than one concept for further design and detailing? Can two different versions of the same product be designed out of two generated concepts? How do you differentiate these two products while marketing?

- Should team members accept the help of lead customers when creating a concept? What are the benefits and limitations of getting input from the customers? Should the team share the concept with the vendors for their feedback? If so, what are the advantages and disadvantages of sharing the designs with external agents?

- Is the concept generation phase limited to a highly complex and costly product? Can it be applied to a simple product? Does this step help to promote creativity among team members? Is there any reason why a machine tool development effort should have this particular phase of architecture development? Is it necessary at all, since it costs time and money to carry out this phase?

- Can a product selected using this comparative scheme fail in the marketplace? If the product fails, what are the steps to be taken for re-introduction of the product? How do you find the reasons for such failure? How do you handle risk mitigation process for a concept selection process?

C H A P T E R 10

10 System Design, Analysis and Detailing

10.1 Introduction

At this stage of design, the exploratory concept and the system requirements are very well defined and developed. A concept has been selected for design and detailing. This activity is at the bottom of the system "V" diagram and at the critical juncture between iterative procedures and non-iterative procedure. From this point on, the team is very much confident that the selected concept will satisfy the customer requirements and the final product will be viable commercially. The system and sub-system design requirements also have been finalized. So far, the focus has been to satisfy customer requirements, and the next focus is on design, manufacturing, and detailing of sub-systems so that the system can be produced at a desired rate of production. See Fig. 10.1.

That puts the team into the right arm of the system "V," which consists of sequential steps. The team understands the fact that from now on, going backward is a difficult step and such a reversal would cost lots of money and might also result in discontinuation of the project. The team is in a creative mode to mature the concept design into a manufacturable system which will combine quality and cost and generate revenue and substantial profit.

The system design and verification flow diagram is displayed in Fig. 10.2. During the design and prototype build phase, there are two loops to be considered. First of all, the design loop consists of designing the components and then integrating them into sub-assemblies, and then the emergent design goes through the performance manufacturing and quality checks to ensure the parts can be manufactured within the constraints of quality and manufacturing methods. The performance checks are normally done analytically at this stage since the physical build is not available for physical testing.

The second loop can be called a "verification loop," where the physical prototype is available for testing performance and durability. Testing details will be discussed in the next chapter. Once the physical prototype is available, performance requirements have to be validated and confirmed for the product.

During the verification, or design, loop, if the requirements are not satisfied, the design has to be redone to satisfy them. The physical verification can be slightly delayed until the analytical methods conform to the requirements.

Once the design loop steps are completed, the physical build can be continued, and the verification loop completed. In order to save time and money, these two loop functions are carried out simultaneously following concurrent engineering methods. These two loops allow the design to be reiterated for further refinement and confirmation before the design is released for production activities. It is also not a bad idea to reconfirm the cost of the design once these loop functions are completed.

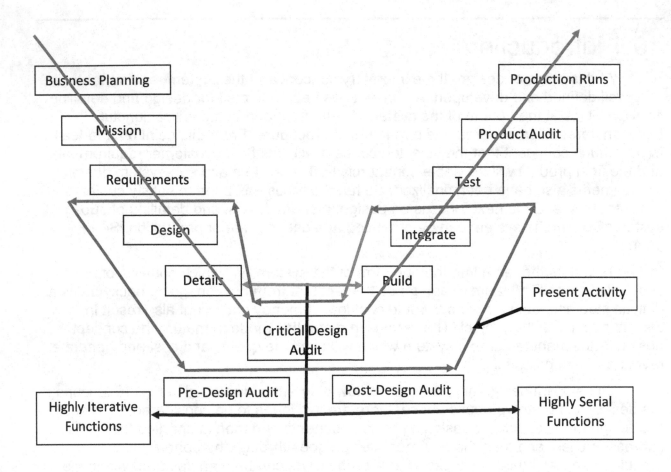

Figure 10.1 System V Life Cycle model system engineering for machines

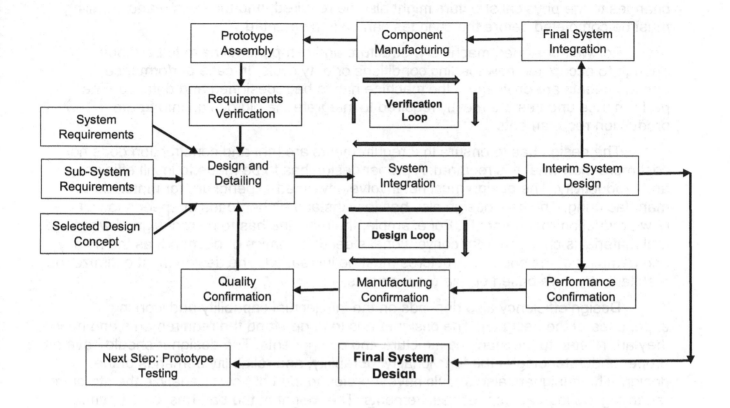

Fig. 10.2: System design and verification flow diagram

10.2 Machine System Design Considerations

Machine system and module design is an engineering art that converts concepts into physical engineering drawings that can be manufactured and serviced. The art is expressed in terms of detailed engineering drawings. It is documentation of the concept. As mentioned earlier, the concept could be a brand-new and innovative concept or based on legacy designs that are already in use. Irrespective of the nature of the concepts, whether new or an improvement over existing machinery, design and detailing must be completed before manufacturing.

Sometimes existing designs are modified to accept recent trends or challenges, such as adding tool changer to a machining center or adding a very high-speed spindle unit instead of a conventional spindle. To enhance existing machinery to meet new requirements, the concepts and design must be completed. For an existing machine, substantial changes might occur when higher horsepower or speed ability is added to the existing machine. To reduce the time for indexing, a new index unit might have to be designed altogether.

So, when design parameters or requirements are changed for an existing design, changes to the physical structure might also be required, and the design and detailing must be completed before the changes can be implemented.

For well-designed machinery, structure and components have to be robust enough to accept the new loading conditions or duty cycle. In case performance enhancements are envisaged, the machine has to be redesigned and detailed. The performance updates are mostly required to integrate the old design into future production requirements.

The design has to ensure that requirements are met and construction does not cost more than what it is required. A proper design has to be made to instill efficiency and productivity. The design must not involve unwanted expenditure for materials or manufacturing. The new design also has to withstand higher loads or speeds to suit the new production requirements. For example, the machine has to be redesigned when the tool material is changed from conventional steel to ceramics or diamond, as the rigidity and damping of the support structures must be increased. The design must optimize the resultant structure based on the requirements.

Design efficiency also depends on the intellectual capability and working experience of the designer. The designer has to understand the requirements and how they are related to the machine structure and components. The designer should have a proper understanding of the functionality, durability, and reliability demands of the design. The designers also should have a profound idea of how to analyze the structure regarding the load and speed requirements. The weight of the part has to be optimized without violating durability or reliability requirements.

The design should enhance reliability and durability by reducing stress and strain under external or internal loading situations. The designer also must have a clear idea about the material, manufacturing, and quality control methods for component inspections. The designer has to understand the effects of the shape or form of a part on the manufacturing or assembly of parts, and the parts have to be designed for ease of maintenance, servicing, manufacturing, and assembly. Finally, the machine has to be designed to reduce the life=cycle cost of the machine. To incorporate so many conflicting requirements into an optimized design, the designer must think about four engineering topics: mechanisms, strength, manufacturing, and ease of service. In a nutshell, the design of a machine is all about optimizing the structure for durability, reliability, quality, and cost. Once these requirements have been satisfied by the design, an experimental evaluation is required to validate the theoretical findings. The steps for designing a machine or mechanism are:

- Start with the concept already selected
- Create the base design and its performance values
- Understand the critical requirements for the design
- Analyze the enhancements required over base design
- Select critical functional elements first and then secondary elements
- Select available materials and component design variants

- Analyze new components based on requirements and baseline values
- Consider manufacturing issues and methods of the design
- Revise the design if necessary
- Finish the design layouts
- Finish the component drawings
- Documentation

Although the process seems to be a detailed one, some of these steps might not be required for all the designs. It depends on the complexity, cost, and nature of innovation for any particular design. Some of these steps can also be merged with others as necessary. For theoretical analysis of the design, several tools could be deployed, such as finite element analysis (FEA), computer-aided design (CAD), computer-aided manufacturing (CAM), reliability analysis, design and quality audit (DQA), etc.

For comparison of the new emergent performances for the new designs, baseline values must be obtained. The baseline values can be from the existing design or industry trends. There has to be a well-defined metric for each requirement for the sub-system. Subjective requirements can be set aside for such analysis. Without a definitive value, the design performances will be of no value during the comparison.

10.3 System Analysis Procedures

Components are to be analyzed using external and internal loading conditions. The forces can be externally applied or internally generated, and they can be enumerated as outlined below (also see Figs. 10.1 and 10.2):

- Duty cycle or real-world usage cycle (RWUC): load vs. time
- Gravity forces
- Thermal profile forces
- Frictional forces
- Inertia forces due to acceleration or deceleration conditions
- Dynamic forces due to motion: centrifugal force and rate of change of momentum
- Forces due to interference, shrinkage, and expansion
- Residual forces due to manufacturing practices deployed
- Forces due to dynamic loading: sympathetic vibration or earthquake
- Viscous forces
- Impact forces

The first and foremost considerations for strength and life analysis are the assumptions used for such analysis. The assumptions must be as realistic as possible.

If the assumptions made before the analysis are very complex and unrealistic, the analysis results will be obviously wrong. On the other hand, if the assumptions are very simple to start with, the results can also be misleading. Hence, consideration of assumptions is very important. If the design engineer is unsure about the working conditions of the unit or does not have enough information about the component to evaluate it, he or she must make simple assumptions first and then repeat the analysis to include more realistic assumptions.

In the end, the results must be confirmed under similar assumptions and input conditions. Sometimes legacy data is also helpful. It is also true that if too many unrealistic assumptions are made, the result will be far from real or actual results. If the legacy data or analysis methods are available, they should be followed. If any assumptions were made in previous occasions, the present analysis should also assume similar conditions. Moreover, the assumptions made should lead to a safer solution. The experience of the analyst is definitely a great help in determining the assumptions made.

The accuracy of the analysis depends on the determination of forces and the methods used. Primarily, the prediction of the analysis and eventual correlation with the test results depend on the type of analysis and the force/duty cycle used. The duty cycle for a machine tool determines the forces used vs. time. From these types of graphs, one can determine the maximum amplitude and number of uses for any component. Complex loads can be converted to simple forces and moments. The forces and moments can also be subdivided into static and dynamic components. The cyclical nature of the forces and moments must also be considered for components subjected to dynamic loading. The maximum amplitude of forces and the nature of the application of forces must also be considered. The results for the force analysis should be documented well before use. The equilibrium conditions for the assembly must also be checked against all the forces and moments applied to the assembly or component. The conditions of use should include the following:

- Machine size and type
- Rail spacing
- Rail mounting
- Mounting angle: horizontal, slant, or vertical
- Length of travel
- Speed of travel
- Usage profile
- Duty cycle
- Life requirement
- Accuracy and rigidity
- Environmental conditions

Let us take as an example the selection of linear bearings for machine tool application. In order to select a particular style of linear bearing for the slide system, the flow diagram shown Figure 10.3 could be used.

The selection of the type of bearings might include models such as SR, HSR, HRW, etc. from THK Bearings Incorporated. The selection of ball screw might include the type of ball screw, end bearings, motor maximum revolution, type of drive, etc. The life estimation should include nominal life requirements, etc. The analysis also should consider the number of bearings, number of rails, and spacing among them.

The accuracy and rigidity of the bearing system dictate the system deflection under load and directly reflect the accuracy of movements of the slides. The accuracy also depends on the bearing and mounting surface conditions. The rigidity or stiffness depends on the bearing accuracy, bearing preload, mounting method, and mounting surface rigidity. The lubrication requirements will depend upon the type of lubricants, method of lubrication, bearing and rail material selected, corrosion resistance of the bearing materials, and bearing protection method selected. The frictional drag force depends on the type of bearing selected. The axial force on the balls crew depends on the motor torque required and the frictional forces induced during movement.

In order to determine the life of the linear bearing in the slide system, forces on it and the amount of linear travel must be known under a duty cycle from a real-world usage profile. The flow diagram for such a life calculation is shown in Figure 10.4. The flow diagram shows how to calculate the life of a linear bearing step by step for the slide system. First of all, the nature, type of loads, usage conditions, environment, and frequency of load applications must be determined. Then the applied loads, radial and transverse, must be found out. From this data, the equivalent mean load on the bearing must be calculated. Using the life calculation formula, the worst-case life of the bearing must be calculated or estimated. The load calculations should include the following:

- Maximum magnitude of load and its direction
- Inclination or line of application of forces
- Bearing arrangements
- Duty cycle
- Stroke of linear movements
- Mean velocity of movements
- Required life
- The worst loading on the bearing

The applied load should include gravity load, thrust forces, reactionary forces, transverse forces, acceleration forces, etc. The mean value of loads on the bearing is used to calculate its life. The free body diagram for the component or assembly must be sketched out to understand the flow of forces.

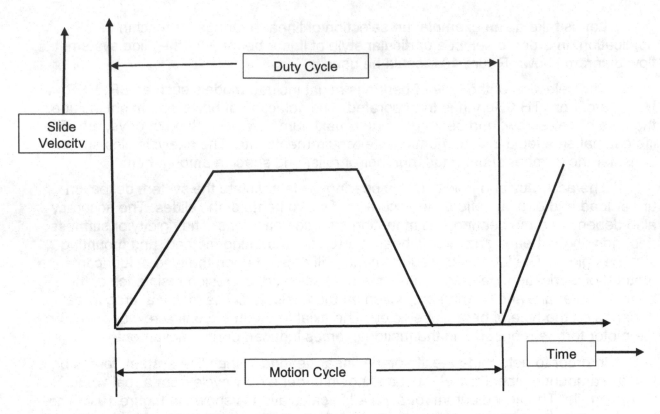

Figure 10.1: Slide velocity duty cycle for linear bearing life calculations
(Source: THK America Incorporated Catalog)

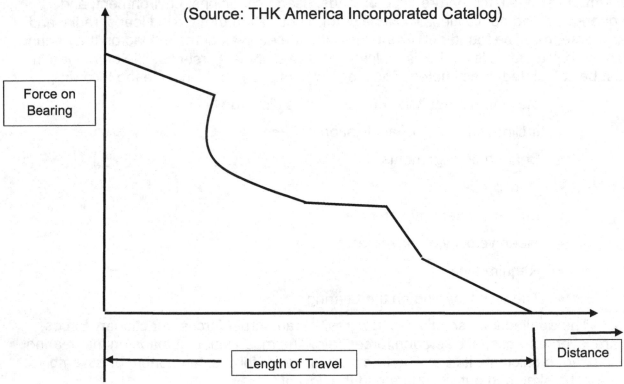

Figure 10.2: Bearing forces for linear bearing life calculations
(Source: THK America Incorporated Catalog)

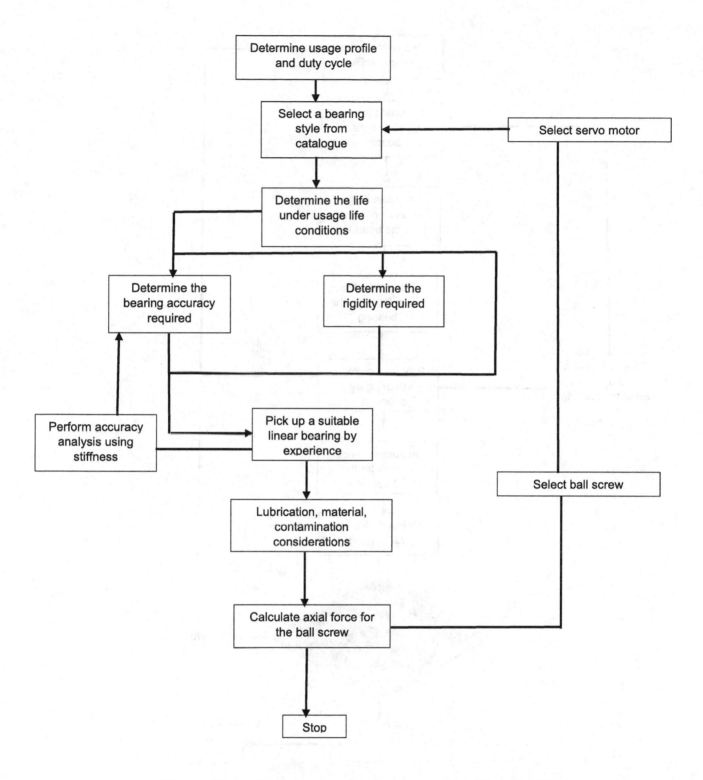

Figure 10.3: Flow diagram for linear bearing life calculations
(Source: THK America Incorporated)

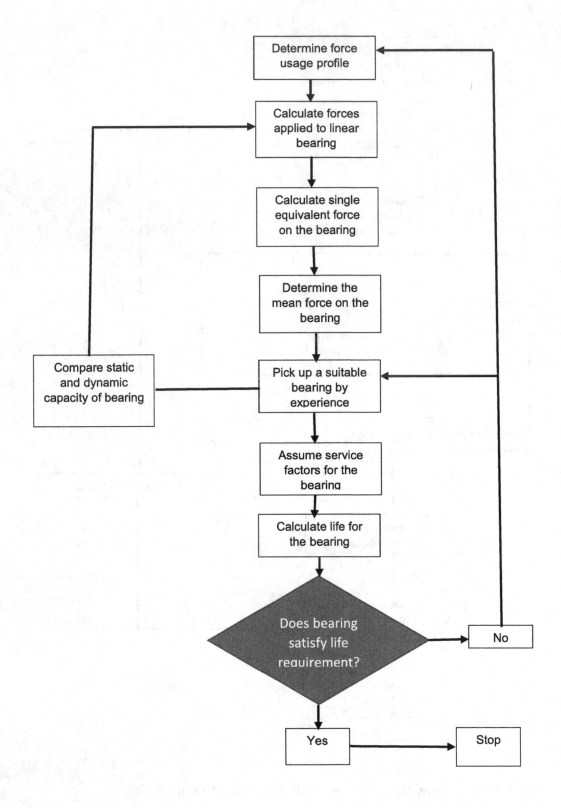

Figure 10.4: Flow diagram for linear bearing life calculations

(Source: THK America Incorporated Catalogue)

10.4 Inputs for System Analysis

The analysts have to select four important inputs before the analysis is started: allowable stress or strain for the material, factor of safety for the analysis, materials for the component, and selected design details. Once the design is selected, the free body diagram of the design representing the flow of forces and moments through the system must be understood and sketched out for further application.

The design has to be dissected and prepared for analysis. The details of the representation of the design components will depend on the types of analysis being performed. The components can be represented by a series of elements for performing finite element analysis. Some designs can be represented by, say, a four-bar mechanism for dynamic motion analysis. In most cases, components can be analyzed using uniform and standardized stress analysis. The components have to be represented accordingly and differently for strength, wear, resonant frequency, motion analysis, accuracy, resonant frequency, and deflection analysis.

The next point of discussion is the selection of material for each component. The selection depends upon material availability, manufacturing capabilities of the company, strength requirements, and budgeted cost. The material selection also depends on the shape of the part and nature of forces applied to the component. Weight, stiffness, or rigidity requirements also dictate material selection. Other factors such as corrosion resistance, damping factors, castability or formability, and machinability of the material, should be considered while selecting the material. The material selection also dictates the life cycle of the part. For example, the machine bed must be very heavy, highly damped, machinable, cost-effective, and formable for critical shapes and bends. That is why the machine bed is made out of cast iron. In case higher rigidity is required for hard turning, etc., the bed can also be a welded steel structure. In order to enhance damping, concrete is used inside the welded structure.

The single most important factor to determine the material of a component could be the cost. If two materials are suitable for the component, the material that provides reduced cost and higher machinability is selected. The manufacturing issues such as manufacturing capabilities, manufacturing methods, material handling capacity, labor cost for machining, etc. must be considered. If the manufacturing cost issues for a material outweighs other advantages, another suitable material is selected. Hence, the material selection is an optimized process involving several conflicting objectives such as weight vs. stiffness, machinability vs. rigidity, castability vs. formability, etc. The experience of the design engineer and legacy data are very crucial for this step of the process.

The selection of material also depends on the strength required for designing the part. The strength should be the ultimate strength of the material and the yield point of the material. If the part is subjected to lots of impacts or cyclical loading during its use, the material should be selected to conform to those requirements. For such cases, forging material could be better than castable material. The selection of material also could depend on load variation, maximum peak load, and stress concentration due to sectional changes after casting and forming.

The stress analysis should be able to identify stress concentration and stress under cyclical or static loading. The part should be designed to satisfy stress requirements while optimizing the weight of the component. Sometimes vibration requirements can dictate the material selected. In such cases, the stiffness of a component is very critical to the success of its use so that it does not resonate when subjected to cyclical loading at a higher frequency. The analysis should identify the strengths and weaknesses of various possible materials for a part.

10.5 System Analysis Sequence and Details

In the beginning, when the shape or form of a component or assembly is not firmed up, an initial assessment of stress, strain, or deflection can still be made using an approximate shape or form from legacy data. Such analysis will guide the designer at the outset. The purpose of such a preliminary evaluation is to determine the approximate dimensions of all the components. This will help determine necessary stiffness, strength, and load-bearing capability, which are very important for going further to finalize the shape or form of the part.

It is also true that the part shape or form depends on the working environment, operating conditions, and how the part under analysis is connected to other adjacent components in an assembly. More often than not, the dimensions of a component need to be adjusted due to manufacturing concerns and to avoid interference issues with other components. The reduction of weight and total cost of a component also depend very much on its shape or form. In some cases, the component is designed as per the service requirements and then adjusted to satisfy manufacturing concerns and operating conditions. This is an iterative procedure that should be followed to optimize the part dimensions, shape, and form of the component.

In order to optimize the weight of an assembly, the components need to be optimized first, and then all the components are put together for the assembly. When assembled, the cost and requirements should be rechecked again to satisfy the assembly requirements. This is a reverse process to be followed for analysis, i.e., a bottom-up procedure to be followed for optimization considerations. If the stress, strain, and durability are the only concerns for a component, its dimensions should be adjusted to satisfy the requirements first. It should also to be kept in mind that increasing the size or complexity of the part form will increase the cost of manufacturing beyond the normal cost of the component. Design optimization should reduce the complexity of a component and enhance its durability and strength.

The inputs to the analysis, such as material properties and stress concentration factors, do affect the outcome to a great extent. The numerical values should not be selected at random, and they must be justified before using them in the analysis. The designer has to have a thorough understanding of the operating conditions of the assembly, load-bearing capabilities required, manufacturing methods to be used, and assembly operations. If the parts are used for static conditions, a static analysis should be performed.

If the component has to work under extreme conditions, resonance frequency and fatigue analysis must be carried out for the component. For example, when the parts are to be used in a corrosive environment, the proper material has to be selected to handle that condition. Material with reduced friction and lubricating conditions must also be considered to enhance the wear resistance of the component. Areas of contact must be increased to reduce the pressure per unit area under the loading conditions. The load variations can also be minimized to reduce wear and tear. The idea is to understand the critical use of the part and then perform an analysis to predict performance based on critical requirements.

For machine tool components, deflection under load is often a very critical requirement. Deflection also affects the resonance capability of the component. This deflection or resonance capability affects the surface finish of the component that is being machined. Excessive deflection or lack of rigidity of a part also affects the accuracy of a machine tool. If the tool under loading conditions deflects more than is normally allowed, machining accuracy will definitely be affected. Hence, to combat such a situation, a thorough deflection analysis of the component under question must be carried out using detailed load, material constants, and part shape. In order to reduce the deflection, rigidity or stiffness has to be increased by selecting the proper material or increasing the size of the component, and a material with a high modulus of elasticity has to be selected. For example, to reduce bed deflection, steel is preferred to cast iron. A weldment will always provide higher stiffness than cast-iron structures.

If the part is very complex to analyze using simple conventional methods, a more detailed and capable analysis tool, such as finite element analysis, should be used. The material constants must be properly investigated before using them in the analysis. Not all cast-iron or steel material properties are similar. In any case, the accuracy of the operating load, material constants, and part dimensions are very important elements for prediction accuracy. In case, these factors change during the design process, so analysis must be carried out again based on the new conditions. The part shapes or forms change very frequently during the design, and the rigidity or deflection of a component is very much influenced by part dimensions and shape. Hence, frequent analysis must be carried out to guide the designer accordingly. A preliminary sketch will be used for preliminary analysis, and a final analysis must be carried out when the part dimensions, materials, and manufacturing methods are finalized.

The calculations and method of analysis must be documented properly for future reference. Justification of any analysis and assumptions must also be discussed and documented properly. The results must be interpreted properly and should be confirmed using proper physical tests if possible. The analysis methods might have to be perfected depending on the amount of deviation of predicted results from actual test results. The results from the actual test are the ultimate proof of any design. The validation and verification of the predicted indices must be made using physical tests. The documentation of all the steps during analysis is very important and will be needed if changes are required at the end. An audit also could be done to make sure proper analysis has been carried out for the component.

During the analysis, the analyst has to include the noise conditions of the component during its use. As mentioned earlier, noise is a variable that cannot be controlled. The strategy of analysis should be to anticipate the noise and represent the effects numerically to make the design robust for the noise factors. The noise components can be as follows:

- Variations such as manufacturing variations, dimensional variations, material variations from part to part, and external and internal usage variations

- Changes in part dimensions and strength due to normal use and excessive wear and tear: thermal or stress aging

- Changes in the customer usage profile such as heavy or high-speed cutting conditions

- External environmental changes such as in temperature, humidity, and corrosion

- Internal environmental changes such as sympathetic vibrations due to neighboring machines, including impact and vibratory loading

The system engineering approach can also be used at the elemental level while performing the analysis. The system environment diagram can be created for the component or the assembly to understand the effects of environmental factors on parts strength and performance indices. One such diagram for index units is shown in Fig. 10.5 below.

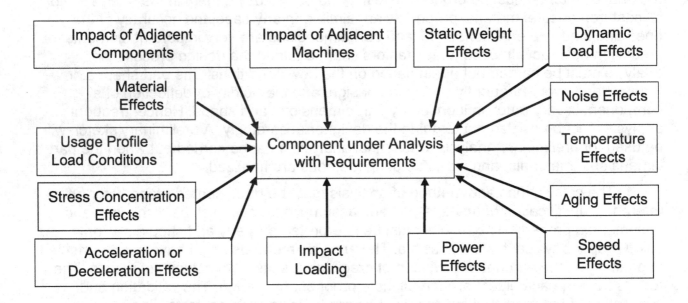

Figure 10.5: System engineering approach for component analysis

10.6 Design Changes and Revisions Management

Normally, any design is changed several times before the final version is firmed up. This final design is then detailed for manufacturing and production. The reason for several design changes could be numerous, as outlined below:

- Manufacturing concerns are the primary reason for changes. They can be done concurrently when design engineering considers the manufacturing inputs. For example, for casting or forging parts, changes are made to make the casting molds, patterns, or forging dies easy to make. The cost of manufacturing is another big concern for making changes. For cast parts, design changes might reflect the reduction of manufacturing complexities. Casting vs. welding is another consideration. If the number of parts to be manufactured is less than a break-even quantity, welding is preferred to casting. In some cases, for high-volume parts, forging can be cheaper than casting. The part design for forging is somewhat different than casting. Availability of casting molds or forging dies could be another reason for design changes. Another concern is the machining of the part. The shape of the part needs to be simple enough for ease of machining. In a nutshell, most of the design changes are made to address manufacturing concerns

- Sometimes part design is changed to accommodate standardization concerns. Standardization reduces the cost of the part. Borrowing the design from other legacy designs is very often done to reduce the cost as well. Parts can also be made in common with other products to reduce the cost. Standardization also affects globalization of design when a company has several manufacturing facilities across the globe.

- In order to reduce wear and tear on a component, contact surfaces can be separated from non-contact surfaces so that maintenance becomes less costly. For example, slide bearings are made in such a way that only bearings need to be replaced and not the whole slide. Contact surfaces are designed in such a way that replacement becomes easier when required.

- Design changes are also made to address safety concerns. Machine sliding door designs are always dictated by safety issues. Covers or shrouds are designed to shield the operator from rotating parts flying out of the workholding. Designs are also made to include safety locks for electrical components. In case of malfunctioning of rotating parts, adequate design changes must be made to protect machine operators or maintenance personnel.

- The next reason for the change could be to reduce assembly time or complexities. The shape or form could be changed to avoid interference with adjacent parts in an assembly. The design has to reflect the assembly concerns, which would reduce total assembly cost.

- Ease of maintenance and servicing the machine must be addressed while designing the part. Total life-cycle cost of the part must be reduced. Ease of putting together a series of parts after dismantling while servicing the machine must be addressed during design.

- Sometimes part designs must be changed to address industrial engineering concerns such as part look, ergonomics, and health considerations. The part might have a pleasing appearance, such as the machine shrouding. The machine sheet metals should be designed in such a way that the machine looks very safe and attractive to customers. The design should be appealing to customers and should reflect safety, robustness, balance, and ergonomics.

10.7 System Design FMEA Considerations

The design failure modes and effects analysis (FMEA) principles should be used to identify the design weakness and the corrective actions to be taken to correct these deficiencies before the product falls into the hands of customers. Fig 10.6 shows some of the inputs required for such an analysis. A similar format is also recommended by the SAE (Society of Automobile Engineers), detailed in SAE J1739 for automobiles. The following discussion stems from the procedures outlined by the SAE but modified for machine tool applications.

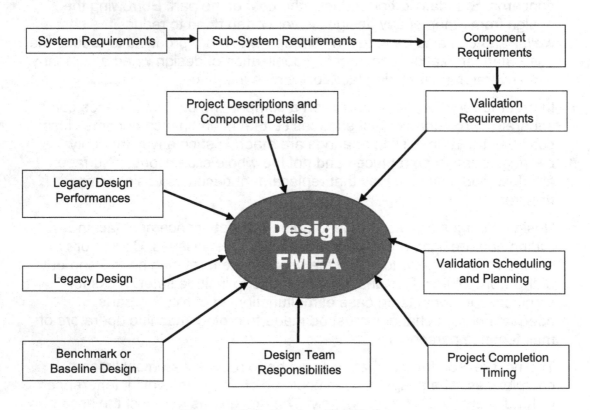

Figure 10.6: Design FMEA for component analysis

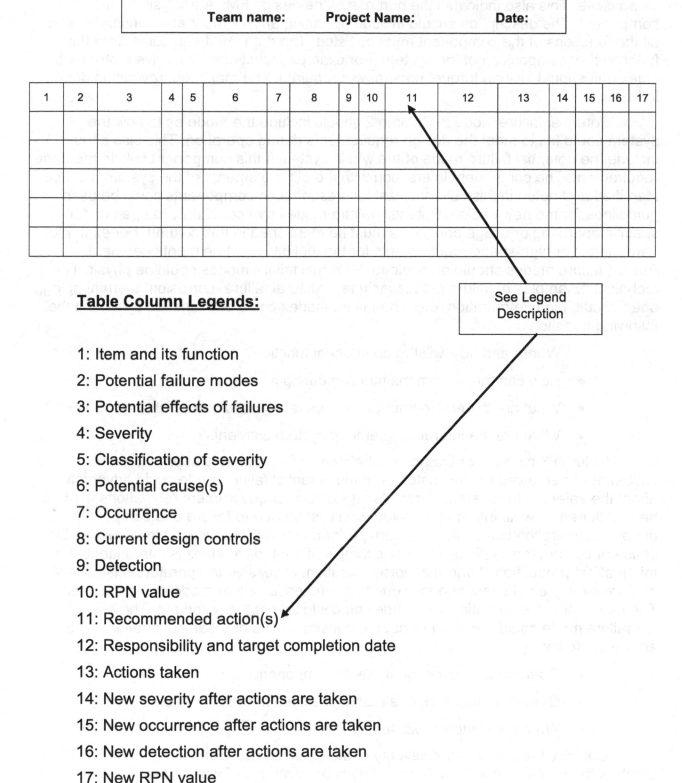

1	2	3	4	5	6	7	8	9	10	11	12	13	14	15	16	17

Potential Failure Mode and Effects Analysis (Design FMEA)

Team name: Project Name: Date:

See Legend Description

Table Column Legends:

1: Item and its function

2: Potential failure modes

3: Potential effects of failures

4: Severity

5: Classification of severity

6: Potential cause(s)

7: Occurrence

8: Current design controls

9: Detection

10: RPN value

11: Recommended action(s)

12: Responsibility and target completion date

13: Actions taken

14: New severity after actions are taken

15: New occurrence after actions are taken

16: New detection after actions are taken

17: New RPN value

Figure 10.7: Design FMEA form format

The entries for requirements for the component in Column 1 should be as brief as possible. This also indicates the purpose of the design FMEA analysis for the component. The description should include operating environment and conditions, and all the functions of the component must be listed. The data input can come from the functional decomposition of the system. For example, functions for a drive motor for the index unit should include torque, horsepower required, and maximum revolution during indexing.

Potential failure mode in Column 2 should include the mode as to how the system could fail to meet the design requirements during operation. This also should include the potential failure mode of the whole system if this component fails to meet the requirement. The component interaction with the other elements of the system must be identified and noted in this column. Past failures for such components could be used as guidelines for the new design. Potential failure modes that could only happen under specific operating or usage conditions must be mentioned in this column. For example, hard metals or high-speed requirements for the tool bits must be mentioned here. Absurd failure modes should be avoided. Potential failure modes could be physical or technical. Examples of failure modes include structural failure, corrosion, thermal aging, open circuit, high deformation, etc. The failure modes could be identified by asking the following questions:

- Where and how will the component function?
- How can the system malfunction during its usage?
- What are the environmental and operating conditions of this component?
- What are the interactions among system elements?

Column 3 consists of the potential effects of failure that is apparent to the customer or perceived by the customer. If the potential failure mode could adversely affect the safety of the system or result in a violation of government regulations, it must be mentioned as well. Intermittent failure of a system due to failure of a component under consideration could cause system performance degradation, which could be the cause of customer dissatisfaction. For example, if the tooling index is intermittent, it might affect production. Abnormal noise, vibration, excessive temperature, instability, etc. are examples of consequences due to the potential failure mode of a component. The machine or tool vibration while indexing could be another example. The effects of the failure mode could determine what the consequences of a particular failure mode are on the following:

- Operation, function, or status of component
- Operation, function, or status of the system or sub-system
- What the customer will feel or see

Column 4 represents the severity of the potential consequences or effects. The death of the machine due to malfunctioning is the example of worst-case severity. Severity applies to the effects only, and it can be measured on a scale of 1 to 10, where 10 represents the worst case. A reduction in severity ranking can be obtained by better design. Hazardous or unsafe conditions without warning can be the highest form of severity, i.e., a 10 rating.

Column 5 needs input for any special process control required for the component. Each item identified in the design FMEA should have an assessment for the need of special process controls and be transferred to the process FMEA (PFMEA) for further consideration. The component characteristics affect machine functions, compliance with government regulations, customer requirements and satisfaction, etc. When the failure mode has a very high severity rating of 9 or 10, the potential critical characteristic is identified, such as hydraulic pressure supplied to the workholding devices. When a failure mode has a severity rating between 5 and 8 and the product affects customer satisfaction, the significant characteristic is identified, and "YS" is entered in the column. Also, a PFMEA is initiated.

Column 6 represents the potential cause of failures, which could mean design weaknesses that lead to failure. Root causes of such failures must be determined by the team members. Then corrective action has to be taken to avert the failure of the design in the future. Potential failure modes that arise due to manufacturing issues should not be included in the design FMEA, but they should be included in the PFMEA. While developing the design FMEA, the team has to assume that a part will be manufactured as per the design and assembled as per the assembly procedure set up by the design team. The component must be manufactured within specifications. For example, the material should be per standards, for example, no porosity, correct hardness, etc. The design must not have inherent design deficiencies that force the manufacturing to produce poor-quality parts. For example, symmetric design, undefined assembly procedures, assembly sequence, and part interference issues must be fixed before the design is released for production. The design should take care of standard manufacturing and material limitations, such as surface finish, mold tolerances, assembly timing, etc. A poor design might lead to material yielding, fatigue failures, creep and aging, excessive wear, etc.

Column 7 represents the occurrence of failures. It is the likelihood that a specific cause or failure will happen. The value is somewhat relative. It is also assumed that removing the cause of the failure mode through design change is the only way a reduction in the occurrence ranking can be affected. The likelihood of occurrence of the failure mode can be represented on a scale of 1 to 10. The ranking must be consistent and justified. The value might depend upon the facts, whether or not the design is brand new, the component application is changed, or theoretical analysis predicts the failure. The probability of failure rate might dictate the relative ranking of occurrence.

Column 8 represents the current design control for the component. The design control should detect the failure mode. This column should contain prevention and design validation, which will ensure the identification of failure modes or failure mechanisms. Controls could include a physical site test, theoretical analysis, lab test, and reliability assessment testing, which is used to identify failure mechanisms of the component. The design analysis could be a prime tool for identifying the weak points of a design. A simulation test could be another significant tool to identify the weakness of the design. Design control can include design reviews, analytical studies, computer modeling and analysis, interference analysis, etc. The ultimate objective is to identify the current design controls that will reduce the occurrence of failures.

Column 9 represents the detection of failure modes. The detection is an assessment of current design controls for the component to identify the failure modes or design weaknesses. The objective is to see if the current design control can identify the failure mechanisms, and if it cannot, then the design control has to be improved. The effectiveness of the design control should also be evaluated. The detection rating should be based on the collective judgment of the team. The effectiveness of a design control depends upon design analysis methods, test development methods, legacy design experiences, test result consistencies, time of analysis and testing, etc. For each control method, the detection rating could be established on a scale of 1 to 10, with 10 being the case of absolute uncertainty and 1 being almost certain to catch the design deficiencies.

The next column, Column 10, represents the RPN, risk priority number, for each failure mode. The RPN value is a measure of design risk. The value could be between 1 and 1000. For a failure with high RPN, the team must try to reduce this estimated risk value as much as possible within the scope of the project. Regardless of the ranking and the RPN value, a failure mode with high severity value must be analyzed thoroughly and mitigated before introduction into the market to reduce the risk. The RPN is defined as follows:

RPN = S (severity rating) x O (occurrence rating) x D (detection rankings)

Column 11 is for recommended action to eliminate failure modes. The objective of the recommended action is to eliminate the failures with the highest severity, occurrence, and detection ratings. The changes in design due to this DFMEA analysis are controlled through design revisions. The potential actions could consist of the computer simulation, bench or simulation test, customer site test, etc. The recommended action should be prioritized on the basis of the following:

- Failures with highest severity ratings must have the highest priority (S ratings)

- Next priority is for the failures that have a higher occurrence (O ratings)

- Last but not least, failures with higher RPN ratings have the next priority

It is also the responsibility of the engineer who initiated the design FMEA for a failure mode to make sure that the follow-up actions have been taken and implemented or addressed before production. The design engineer has to make sure that the design is robust enough based on the requirements and that a special process and quality controls are not necessary for the product. The manufacturing process control cannot guarantee that the failure modes will not occur during use. If a PFMEA is not generated for any failure during the DFMEA process, it is taken for granted that failure mode issues have been taken care of by the design team.

Once the corrective actions have been identified and implemented, new RPN values and rankings must be determined and recorded. In case any improvements, as a result of the newly initiated design changes, have been observed, further action has to be taken to improve the RPN and rankings as soon as possible. Sometimes new design changes might bring up some other emergent issues that should also be resolved (Columns 13 through 17).

The system engineering outputs for the DFMEA process can be numerous, as outlined in the following block diagram, Fig. 10.8.

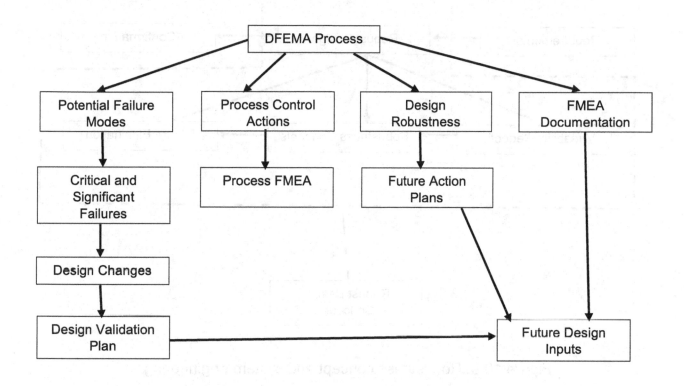

Figure 10.8: Design FMEA outputs

10.8 System Design Robustness

The primary objective of simulation, analysis, and validation of requirements is to make the design robust for real-world usage by customers. The design activities have to interact with technology development, program cost and timing, design variations, and business activities. So, the design has to be a happy medium between customer and business requirements. Core engineering and full-service supplies have to work together to develop the product design and manufacturing methods for the product. The design also has to be flexible enough to include the latest technology without exceeding the cost.

The design for robustness is a new approach promoted by system engineering principles. The robustness of any design allows it to be less vulnerable to usage variations. It controls the rate of decay of system performance over time. The design for robustness starts with the design requirements and ends with the validation and confirmation of those requirements. The recommendation for robust design is shown in Fig. 10.9. The robustness of any design includes functional design and elimination of design failures. The functional design should include variability reduction, reliability enhancements, and quality improvements. The engineering team has to think about the robustness of the product, i.e., its failure rates over time, which should be included as a requirement. Robustness also reduces life-cycle cost.

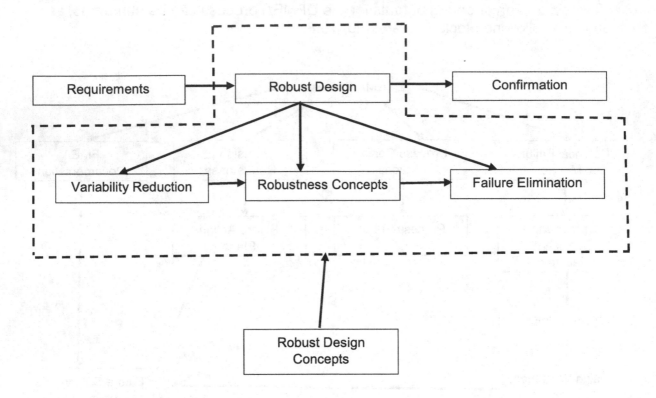

Figure 10.9: Robustness concept and system engineering

The robustness design approach should include product and process improvements, product innovation, reliability enhancements, and integrated manufacturing processes. The approach must include a system engineering approach for the system and sub-systems. The validation method must include evidence of the system performances over time, and the design function must be validated to minimize functional decay over time. The robustness also promotes design flexibility and reproducibility. It also makes the design less sensitivity to noise. Robustness experiments have to be designed and implemented for the verification and validation of the performance of the design under consideration.

Depending on the rate of failure of any design, a robustness candidate for the experimentation must be selected. Identification of new designs with consistent failure patterns must be made beforehand, and design FMEA must also be initiated to identify the necessary actions to reduce the failure patterns. Such priority sub-systems must have a validation plan established. For priority systems, reduction of performance variability and failure rates of new design elements should be the priority. For other systems, reduction of failures over time must be sought. System decomposition should help this process. The system has to be less sensitive regarding the customer requirements over time, and functional degradation over time also has to be minimized to enhance robustness. The customer usage will consist of the noises that cannot be controlled by design functions. Noises have to be anticipated and mitigated while designing the system or component. The concept is displayed in Fig. 10.10.

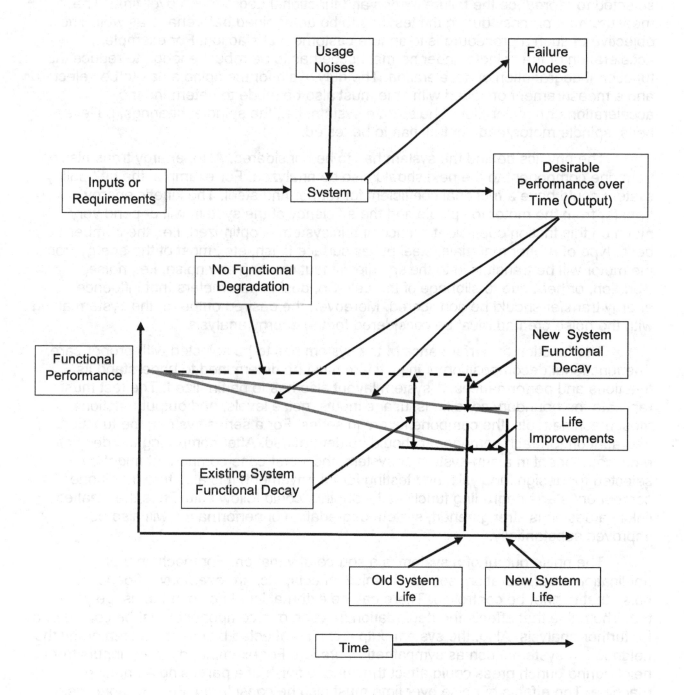

Figure 10.10: Functional degradation and robust engineering

The end objective of a robust design process is to reduce the variability of the desired function over time and eliminate all the failure modes. In order to achieve this objective, the robust design system has to be selected first, and the desired function has to be characterized and analyzed. The real-world usage profile for the system has to be selected next. The anticipated noise conditions or undesirable emergent properties must also be identified. Then a proper experimental method has to be selected to reproduce the failure modes and functional degradation over time. The measurement process during the test has to be determined beforehand as well. The objective of such a procedure is to ensure customer satisfaction. For example, acceleration of the spindle under normal usage has to be robust enough to reduce the functional degradation of acceleration. The duty cycle for the spindle has to be selected, and a measurement of speed with time must also be made to determine the acceleration or deceleration. The spindle system, i.e., the spindle, bearings, pulleys, belts, spindle motor, and control, has to be tested.

The physics behind the system has to be considered. Also, energy transmission from one component to the next should also be analyzed. For example, the belt pulley system could have a frictional coefficient for rubber and steel. The kinetic energy transfer from the motor to spindle and the efficiency of the system will depend very much on this friction coefficient. Hence, if this system is optimized, i.e., the number of belts, type of rubber materials, steel pulley surface finish, etc., most of the energy from the motor will be transferred to the spindle without creating any noise, i.e., noise, vibration, or heat due to slippage of the belt. So, design parameters that influence energy transfer should be considered. Moreover, the desired output of the system along with the noise created must be considered for the energy analysis.

The function or performance of the system has to be selected with proper care. The functional decomposition of the system must be determined to understand its functions and performances. A system layout also has to be finalized. The test must replicate the boundary conditions, usage inputs, noise levels, and output functions. For most machine tools, the components are in series. For a series system, the functional degradation for each component should be determined. After comparing the decay for each component in a sub-system or system, the worst-case component might be selected for design, analysis, and testing for improvement. It is also true that if one component starts degrading functionally, another could follow suit. Once the weakest link in a design is strengthened, system degradation of performance will also be improved substantially.

The noise output of a system is a source of variation. For machine tool application, heat, vibration, sound, chemical effects, etc. are examples of generated noise that cannot be controlled. There can be external input noise in the usage profile, too. The noise that affects the degradation function of a component must be considered for further analysis. Also, the system output can be affected by the noise created by the neighboring system, such as sympathetic vibration. For example, dynamic inputs from a neighboring punch press could affect the surface finish of a part using a grinding machine. The effects of noise over time must also be considered. For example, wear and tear of a component due to vibration should be considered. Experiments or laboratory tests can be set up to determine the nature of decay for such cases.

Tests simulating aging conditions must also be designed. Test results correlation with the analytical prediction must be done before any conclusion is taken up for further consideration. Tests have to be planned, and analytics, hardware, measurement systems, number and duration of tests, cost, etc. must be taken into consideration before they are designed. Steps must first be taken to reduce the variability of function, and then the response has to be adjusted to the desired level. Hence, the test system has to be designed and optimized to correlate the theoretical prediction with the test results. Boundary conditions of a test must match the input conditions of a theoretical analysis for correct correlations.

The selection of noise conditions for a system is a very difficult and tedious job for engineers. To explain the process of selecting noise conditions for a system, let us take as an example a spindle system, which consists of motor, belt, pulleys, workholding systems, actuators, encoders, bearings, and castings. The objective of this step is to develop a strategy that will deal with the most stringent noise conditions that affect machine life. The critical step of this process is to determine, assess, and include the noise effects that simulate the usage conditions for the machine. The noise conditions must be simulated during the tests to correlate the failures and response conditions of the system.

Once the noise conditions are determined, the team has to plan and conduct the tests that will optimize the system and confirm correction of failure modes. Normally, the experiment is a multifactor experiment. During the test simulation process, all the inputs, internal and external noise conditions, and the output conditions might have to be adjusted to create the failure modes for the system. Less effective control factors might be designed to keep the test cost to a minimum. The noise conditions of a component will depend on the following:

- Manufacturing variations: Spindle dimensions, bearing tolerances, belt material variations, motor bearings.

- Component changes in physical dimensions due to usage over time: Bearing thermal expansions, thermal effects on the belt material, temperature of hydraulic fluids.

- Real-world usage profile for the machine or component: Occasional use, frequent use, overloading conditions.

- External environmental conditions: Temperatures, humidity, earthquake.

- Internal environmental conditions: Coolant conditions, sympathetic vibrations and impact due to neighboring machine use, floor vibrations.

- System environment: The components around the spindle system react with the spindle systems during its functioning. Determine the boundary conditions for the system that need to be simulated during the test to create the failure modes.

- Simulation strategy: Inputs, boundary conditions, and noise inputs must be simulated together for the system to create the failure modes. The inputs might have to be adjusted to create the emergent noise conditions for the system.

There are several methods available to identify the cause of variability of a function. One of the most effective is to use the principles of design of experiments

(DOE) for designing the system testing. The other processes are parameter and tolerance design.

- Experimental design of experiments: It provides balanced multiple factors for efficient testing by reducing the number of tests and parameters to be included. It is a filtration process for a test.

- Parameter design: A structured process to reduce functional variation under actual real-world noise conditions. This also determines the signal-to-noise ratio for the component.

- Tolerance design: A way to measure the effects of manufacturing tolerance variations on the functional degradation of the component. It identifies the dimensions of a component that have the highest effect on functional variations. It helps to identify the optimum tolerances for a component and to minimize the functional degradation.

- Failure modes and effects analysis (FMEA): Structured process to ensure. corrective actions have taken and implemented for a specific failure mode.

- Fault tree analysis: Structured process to identify fault causes in a system.

- Block diagrams: Represents system configurations and relationships among interacting or neighboring systems. Helps to understand relationships among interacting system or sub-system configurations. Also identifies system reliability concerns.

A robust design process consists of seven steps:

- Robust thinking approach: Reduction of functional degradation over time of usage.

- System thinking: Include the interacting system components together and follow system engineering principles.

- Noise environments: Identify noise conditions and confirm them by talking to customers.

- Analysis: Theoretical analysis should be conducted to simulate the input and noise conditions and predict the failure conditions before the parts are produced or manufactured.

- Design actions: Include the recommendations from theoretical analysis to change the design. The design should be robust enough to be less sensitive to noise inputs, external or internal.

- Validation testing: Plan the validation testing to confirm the predicted failure modes and then retest the redesigned system to confirm the reduction or elimination of the failure modes.

- Failure Modes: Identify the root cause of the failure, redesign the system to eliminate the concern, and control the process to prevent the repetition of failures. Cross-functional teams must get together to identify the failure concerns, energy transfer paths, and experimental methods to be deployed. The concern is to reduce failures over time of usage. The approach is to "find and fix" to augment customer satisfaction at the end.

10.9 Summary

The application of system engineering fundamentals is a prerequisite to starting any design. Functional decomposition must be completed to understand the functional requirements of each system or sub-system. A system is defined as a set of interdependent sub-systems linked together, and these units work in unison to create a desirable system response and noise. Sometimes "system" and "sub-system" are used interchangeably in the text. The flow of activities should match the "V" flow diagram.

Any design has to satisfy requirements and cost. It should incorporate productivity and efficiency into the design. Design analysis must also be done to ensure customer satisfaction. The design must include manufacturing, reliability, and quality aspects, and it should guarantee quality and reliability. The design should also try to optimize the structure based on customer requirements. The end objective of any design is to satisfy customers.

The analysis of any design is completed to identify failure modes and incorporate design changes to avert the failures. The analysis must take into account customer usage loading, called the "duty cycle," external and internal forces on the system, frictional and inertial forces, dynamic and static cutting loads, and residual forces due to manufacturing. The analysis also should include external or internal noise conditions that affect the system functions or are responsible for failures of the system. The conditions of usage include speed and power consumption, machine design and configurations, external and internal environmental conditions, duty cycle used, etc. The analysis should also predict functional performance.

The analysis steps must start with a proper understanding of the system. The base design should be analyzed first to determine the baseline design performance, which must be correlated first. The system should be analyzed against the requirements. The theoretical confirmation of critical performance is a must to reduce the cost of prototyping and testing. The design should be changed if the analysis prediction is below the required performance. Once the recommendation is implemented, a similar analysis must be conducted again to re-confirm the performance.

After the analysis is completed, detailing the component for manufacturing should begin. The part design is changed to accommodate standardization and manufacturing concerns. Standardization reduces the cost of the part. Borrowing the design from other legacy designs is very often done to reduce the cost of the part. Parts can be made with other products to reduce the cost, and standardization affects globalization of design when a company has several manufacturing facilities across the globe.

The design FMEA should be started after three separate actions are completed: functional decomposition, hardware block diagrams, and reliability block diagrams. The block diagrams are tools used to divide complex systems into sub-systems and components. It graphically represents the relationships among sub-systems or components.

The FMEA team should consist of engineering, manufacturing, assembly, reliability, quality, and safety engineers. The scope of the design FMEA is to determine the stability of the design. It must be stable to start with. Changes must be able to occur as the design FMEA is developed so that recommended actions can be implemented where possible. The FMEA is initiated when the system is new, the old system is being changed, or a carryover design is being used in the new design, and it prevents recurrence of failures.

Robustness, instead of reliability, is a much stronger concept for the machine tool. It is not only a measure of failure, but also the rate at which failures occur in the machine. Robustness, i.e., product reliability over time, can only be determined by conducting strategic and correlated tests. Robustness also represents the function of degradation over time. Validation tests are required for product and process design to create a world-class machine tool. In order to reduce the risk of a new technology, strategic tests have to be performed to prove the application for any particular machine.

This does not mean that tests have to be conducted for all the components or systems of a machine. The critical sub-systems or components must be subjected to such tests. The design approach should be flexible, and test results must be analyzed thoroughly before the design is modified. After implementation, a system should be retested to reduce interference. Only priority and critical systems should be considered for robustness testing. If the cost is an issue, theoretical studies such as fault tree analysis, block diagrams, finite element analysis, etc. can replace physical validation tests.

10.10 References and Bibliography

Cooper, R.G., Edgett, S.J., and Kleinschmidt, E.J., 1998, *Portfolio Management for New Products*, Perseus Books, Reading, MA.

Cooper, R., and Slagmulder, R., 1999, "Develop Profitable New Products with Target Costing," *Sloan Management Review* 40(4), 21–35.

Ford Design Institute Lecture Notes, 1998 Ford Motor Company, Dearborn, Michigan.

Groeneveld, P., 1987, "Road Mapping Integrates Business and Technology," *Research-Technology Management* 40(5), 13–19.

Hatley, D.J., and Pirbhai, I.A., 1998, *Strategies for Real-Time System Specification*, Dorset House, NY.

McGrath, M.E., 1995, *Product Strategy for High-Technology Companies*, McGraw-Hill, NY.

Porter, M.E., 1985, *Competitive Advantage: Creating and Sustaining Superior Performance*, Free Press, NY.

THK America Incorporated, Product Catalog, Schaumburg, IL.

Ulrich, T.K., and Eppinger, S.D., 2011, *Product Design and Development*, McGraw-Hill Companies, NY.

Wheelwright, S.C., and Clark, K.B., 1992, "Creating Plans to Focus Product Development," *HBR* 70(2), 70–82.

10.11 Review Questions

- What are the types of failure modes for a production machine? What are the analysis techniques that can be used to determine the failure modes and their criticality?

- How do you identify the potential effects of failure modes of any machine? Are the failure modes similar for all types of machine tools? Describe the consequences of any failure mode, critical or non-critical.

- For a turning machine, design a severity rating table describing effects, criteria, severity of effect, and relative ranking on a scale of 1 to 10, with 10 as the highest severity rating. What do you have to do for a severity rating of 8 or higher? What are the classification ratings of effects?

- How do you find the root causes of any failure mode? When does the root cause of the failure mode have to be determined and why? What are the assumptions for finding root causes?

- How are design and process FMEAs related? What are the typical failure modes for a machine tool? If a failure mode occurs rarely, should you take any action?

- Design an occurrence rating table for a machining center showing the probability of failure on a scale of 1 to 10, where 10 means the highest possibility. How do you treat the lowest-ranked failure rate?

- What are the design controls required for any failure mode? How do you identify design controls? What are some examples of design control for machine tool failure modes?

- Design a detection rating table for a grinding machine on a scale of 1 to 10, where 10 is the ranking for an absolutely uncertain detection. What are the almost certain failure occurrences of a turning machine?

- What are the recommended actions for failure modes with higher RPN values? What actions do you have to take in case the new RPN value is almost the same as the old RPN value for a failure mode?

- Create a design FMEA for a machining center with all the details required. How does it help to make the machine more reliable and robust? Is design FMEA necessary at all for machine tools?

CHAPTER 11

11 Prototype Building and Testing

11.1 Introduction

The next activity for the project team is to build a prototype to confirm requirements and performance. Warner and Swasey designed a column type of CNC and NC lathe in the 70s. The prototypes were built and displayed in the machine tool show. It became a huge success since the metal removal capability of the machine was exemplary and far ahead of competitive products. The number of castings and forgings produced at that time was very high and needed a machine that could remove tough metals at a much higher rate. However, these models were developed to cater to different requirements. Looking at this success from today's perspective, the machine should not have been a success at all. Now the requirement is high speed and less depth of cut. The amount of metal removed from castings or forgings is much less since the casting process has become very efficient. Instead of molded castings, die castings have come into the picture, thereby reducing the amount of metal required for a similar casting part. The same is true for forged components, too.

Because of advanced analysis tools, components use much less unnecessary material and have a much better life for a particular usage profile. So, current machines are much lighter, smaller, and more productive, with comparatively less power and higher speed, and they have become more sensitive to cost, quality, and speed. New machines have also become much more reliable than older ones of the same capacity. The improvements and efficiencies of new machines can even be improved if a prototype is built and a test plan is in place to verify the requirements. Once the baseline machine is validated against requirements, various models can be designed for higher power and speeds.

The prototype is important for product development. Prototypes can be developed in two possible ways: physically and digitally. A few years back, only physical prototypes were built since the technology for building digital prototypes had not been proven for use. At present, both types of prototypes are built, and they will be discussed in this chapter in detail. This chapter will also discuss the pros and cons of building both types of prototypes and where they can be used for enhancing the efficiency and productivity of the projects. Finally, the chapter will describe a method for developing a test plan for the machines. Physical tests are done only when it is absolutely necessary, since the cost and duration of such tests are extremely high.

11.2 Prototype Fundamentals

A prototype is the first form of a design concept. There are two primary purposes of a prototype:

- Conversion and confirmation of any design concept into its physical or digital form for the first time

- Confirmation, validation, and verification of performances or requirements

The prototypes are normally used for dimensional checks or interferences and verification purposes. It could be used by system engineers and analysis engineers to evaluate the design theoretically before the physical prototype is built. It is cost and time effective. This also helps to formalize the design before it is built. It is also very easy to change the digital prototype since it is a computer model, which can be manipulated much more easily. Industrial engineers use the prototypes for evaluation of safety, ergonomics, and looks. The analysis engineer performs strength and durability analysis on the simulation using various programs, such as finite element analysis software.

The prototype must be built as close to the design as possible, and it should replicate the design intents. Sometimes, instead of a complete full-scale model, a particular sub-system is built to address the issues or concerns with the sub-system. The prototyping consists of various activities, such as developing wood models, foam models, or block modeling for checking interferences. The prototype developing process is basically a step in the design and development process wherein engineers develop an approximate replica of the design concept or sketch. The original prototype is also changed several times to perfect the design using physical or analytical models. The model is approximate, but it should contain all the design intents. Prototype development is similar to developing block diagrams to check the flow of the process. Block diagrams can be used to perfect the process before it is put in place. In a similar way, a prototype is also built to perfect the design. The prototype can also be used to identify the manufacturing process required to produce the design in numbers.

As mentioned earlier, the prototype can be digital or physical. The digital prototype is built in a computer using the design inputs. Physical prototypes are built using tangible materials such as wood, foam, or metals to approximate the design. The physical prototype gives rise to the form or shape of a product, and the prototype system represents physical relationships among sub-systems. It should represent all the degrees of freedom for a system or sub-systems wherever possible.

The physical prototypes can be done in two ways. First of all, a foam or wooden non-functional model is built to check the motions, interferences, and approximate dimensions of the system. Another way is to build a working model that replicates the exact design intent, which can be tested for validation and verification of the requirements. The functional prototype should include all the design intents, power, and motion requirements and materials as per the design. It should feel and look like the actual product and is used as a proof of the design.

The purpose of building these two prototypes is to reduce cost and avoid manufacturing and design defects. The functional prototype should be able to accept instrumentation for validation purposes.

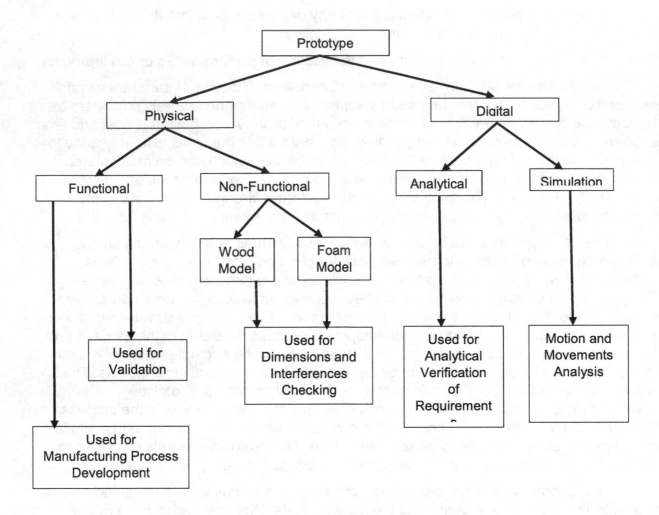

Figure 11.1 Prototype development for production machines

Since most machines are not mass-produced items like automobiles, in general, several prototypes are not made. Normally, to reduce the cost and time of development, only one or two prototypes are made. Moreover, since the cost of prototypes is many times more than the actual selling cost of any machine, it cannot be amortized easily over a limited number of sales, and for design concepts that a market can accept easily, one or two prototypes will suffice. The technology of developing analytical prototypes has matured to the point that the need to develop a physical prototype has gone down substantially.

I also encourage design teams to develop an analytical model, as many would prefer to make instead of developing a physical prototype. The analytical models can simulate motions, material conditions, manufacturing process developments, safety concerns, ergonomics, interference checks, and industrial design concerns.

Unlike automobile prototypes, machine prototypes are very rarely built to resolve focused problems of any unit. For machines, components that are causing problems are changed instead of changing the whole design. In any machine tool, the cost of building is comparatively less than the cost of maintenance of a machine. Reliability and durability concerns are far more important than the production process and methods. Usually, the whole unit does not cause reliability concerns; rather, a few components in a design cause the problem due to overuse, abuse, or insufficient design.

For example, if indexing does not work for a machine, the problem is not with the whole design concept; instead, it is normally the motor drive or encoder or something is jammed. So, the whole unit does not have to be built for evaluation, but some components might have to be changed. In this context, machine tool development is quite different from mass-produced items such as automobiles. For machine tools, customization is more frequently done to satisfy customers. That is the nature of the industry, where most of the machines are modified to satisfy customers' demands and requirements.

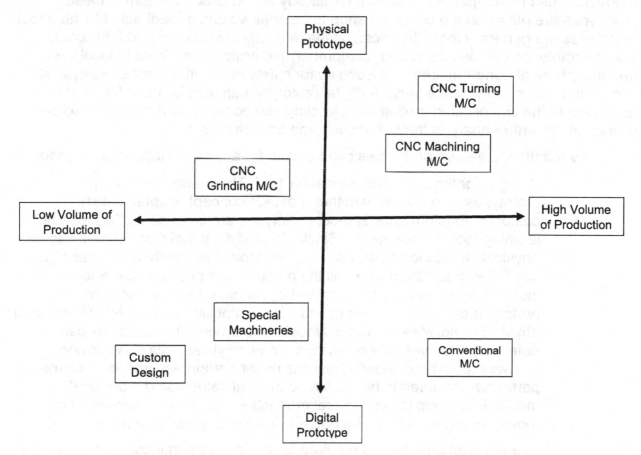

Figure 11.2 Prototype development scheme for production machines

I have also witnessed another aspect of prototype development in the machine tool industry. In case of a high volume of production for machines such as CNC products, more prototypes are built to refine design iterations, including performance improvements and durability concerns. For low-volume machines such as special-purpose or custom-designed machines, production machines are the prototype most of the time since cost and time of production are very high. In this case, quality suffers unless the customization is very simple to implement and the design is based on successful legacy designs. In general, conventional lathes, milling machines, and other standard machines, prototype machines are not built, since the design is well established and durability concerns are very few and far between. Then again, for very costly CNC grinders, a few prototypes are built. For special-purpose machinery, an analytical prototype is the best route to refining the design. Since digital prototypes are intangible, it is much easier to build the prototype for special-purpose, one-of-a-kind machinery, so this is very often done. Analytical prototypes are built for designing, analysis. and simulation. Changes are very quick and easy to make.

For automobiles, comprehensive and focused prototypes are built depending on their purpose. As opposed to these types of prototypes, machine tool prototypes always contain all the modules. Functional prototypes are more common in the machine tool industry. Physical prototypes are full-scale and fully operational replicas of the final product. These prototypes are also built for display and shows. Sometimes these prototypes are placed in the customer shop to get their valuable feedback. Machine tool prototypes are generally look-alike designs that are fully operational and functional. These prototypes can be also called "comprehensive prototypes." In case local laws demand physical performance proof, such as for safety or health reasons, a physical comprehensive prototype is a must. A digital prototype can also be used for a video animation of the complicated motion of interacting components, and this can also be simulated for further analysis through motion and time studies.

As mentioned earlier, prototypes can be used for any of the following purposes:

- Design Confirmation and Validation Tool: The primary use of the prototypes is to find out whether a design concept is apt to satisfy customer requirements. Hence, prototypes are used as verification and learning tools for engineers. Similar to analytical tools, prototypes can be physical verification tools. For machine tools, the weight of the machine has to be determined when all the modules are put together, and the indexing times have to be identified by running the machine. The prototype can also be used to find out the stability of the machine tool, and physical prototypes are the only tools where control interactions can be determined to avert any conditions. The analytical tools only provide answers to some of these questions under certain assumptions, so the performances need to be confirmed and validated using a physical machine that duplicates the real production machine. Confirmation of motion analysis can also be made using the physical prototype.

  The physical prototype is required to confirm industrial concerns or safety hazards under dynamic conditions. The prototype can also be used to get feedback from customers about the look of the machine.

- Feedback Monitoring Tool: The prototype, physical or analytical, also helps the engineers to get feedback from upper management, customers, manufacturing engineers, dealers, vendors, and company employees. The prototype becomes a medium of communication between the design team and everybody else. A three-dimensional physical look-alike product helps everybody to identify deficiencies and good features of the design. I have found that the physical look of a product is quite different than the digital look of the same product. The computer models look much more impressive than the actual machine when physically built. Sometimes a prototype is built to represent the outside periphery of a machine for industrial and safety engineers. Interaction between man and machine can only be had once the prototype is built and an operator is asked to operate the machine under real-world conditions, such as loading and/or downloading the part, maintenance of the machine, ease of handling, etc.

- System Integration Tool: The prototype could be a perfect example of the application of system engineering. It integrates all the functional blocks into a functional working model, a system that contains spindle, power drives, indexing unit, workholding fixture, loading/unloading device and peripheral system, electrical, and controls. This could be called the "system integration" step for the design process. This step ensures all the sub-systems are grouped together to function in unison as a system. Irrespective of the volume of production, the physical prototype is the most effective system building tool. This also identifies difficulties in assembly and sub-system interference issues, which helps manufacturing engineers resolve the issues. The integrated system also helps quality engineers devise methods of inspection for the system, so it enhances effective coordination among engineers of various disciplines. Moreover, this system could be used internally for testing and externally for getting feedback from the customers. For machine tool companies, the most common use of such prototypes is to confirm performance and durability requirements.

- Product Confirmation: The physical prototype can also demonstrate the ability of the project team to function properly and to meet the cost and timing of the project. System and industrial engineers can also demonstrate the ability of the product to satisfy customer requirements. Once a product satisfies the requirements and cost, manufacturing engineers figure out how to produce the system in the most efficient way, manufacture the components, assemble the system, qualify the product, and deliver the product in a stipulated time and volume to satisfy customers. The prototype can also help all the engineers resolve conflicts about technical issues.

- Confirmation of Government Safety Regulations: The physical prototype also acts as a system that can be tested against safety and OSHA requirements. For machine tools, health hazards concerns, such as coolant safety requirements, and operator safety concerns must be demonstrated for the product to be used for government defense projects.

- Prototypes help the team to understand production costs and miscellaneous manufacturing issues. The prototype also helps to streamline the production process and customize manufacturing methods. For example, it could help the design and manufacturing team understand the machining, casting, forging, stamping, and die casting processes.

- A concept design, digital model, or concept sketch cannot be sold to potential customers. With a physical prototype, marketing can show the same to the potential customers, and it is far easier to get a purchase order for the product. The end customer has to like the features and the look of the product before they buy it.

- For a brand-new concept, the prototype might help the team to find the uniqueness of the product for filing patents so that it cannot be copied by other companies. By having a physical prototype, patent attorneys can work with the design team when filing patents. Moreover, the attorney can also find out the patent violations for the product so that the company does not get into trouble after the product is introduced. This reduces the chances of being sued in the future.

11.3 Guiding Principles of Prototypes

In order to develop the physical model of a design, certain principles might be followed to make it effective and productive for the purposes mentioned above. These principles of prototype development could be as follows:

- The prototype must contain as much of the design as possible. Preferably, it should contain the portions of the design that are brand new, unexplored, or open to questions from customers.

- During prototyping, an open mind is better to identify any design discrepancies or deficiencies. Sometimes during prototyping, fresh ideas as to how to design or manufacture better could come up. The team should embrace that and change the design if necessary.

- The prototype helps to question the design principles or critically review the design for manufacturing. It is a good idea not to accept the design as it is and bring up the better way of designing. The first design is not always the best design. So, question the design while prototyping and think up alternative concepts for designing and manufacturing

- In case of discrepancies or difficulties met during prototyping, a logical mind seems to act better.

- Prototypes can be built in stages, starting with the simplest or most difficult sub-systems. Prototypes must be built in the same progression that the assembly technician would use while building the machines. Prototype building can be used for developing manufacturing processes.

- The designers should always interact with other team members, getting their feedback about good and bad aspects of the design so that it can be changed much quicker.

- The prototype must be highly creative during prototype building. The team should think about manufacturing, industrial design, looks, safety, etc. while building a prototype.

- Think about building a prototype with minimum cost but without losing any critical part of the design.

- Keep track of progress on a regular basis.

- Keep a log of prototype development activities on a daily basis. Documentation is necessary to trace back in case of future problems.

- Digital prototypes are easier to build than physical ones and need much less time to model. Sometimes a hybrid approach can be taken. Some less critical portion of the design could use the analytical method, while more complicated ones could use the physical builds. Nevertheless, build the digital prototypes first and then take up the physical build.

- Always build a functional and physical prototype to confirm the design, and there is no alternative to building physical prototypes for confirmation of performance or safety. The physical working prototype is mandatory to identify the undesirable consequences of any design. It could also help the team identify the emergent noise of a design.

- The physical prototype might help the team to reduce the risk of failure during the manufacturing of the product. If the design is not reproducible, it is not a good design and must be changed before it hits the floor. Always keep criticality in mind while building the same.

- Always use the prototype to reduce the cost, time of manufacturing, and complexities of the design. Use it to identify the unnecessary parts of the process down the line.

- Use the prototype to reduce task dependencies and identify the steps that can be done as parallel activities. This reduces the cost and time of assembly.

The success risk of any machine depends on the complexity of the machine. If the complexity is very high, such as for a special-purpose machine where the number of unknowns is too high and the success risk is also very high, physical prototypes must be built. On the other hand, if the complexity is low and the success risk is also low, an analytical prototype could be sufficient. When the complexity is very high but the market success is also low, a physical prototype might be a better choice. The last choice is when both the success or market risk and the product complexity are very low. In that case, an analytical or digital prototype is advisable to keep the cost down.

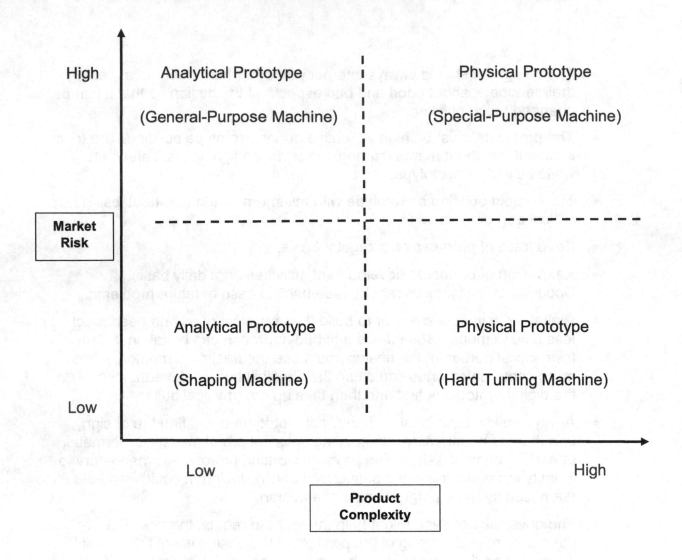

Figure 11.3 Complexity and risk comparison for prototypes

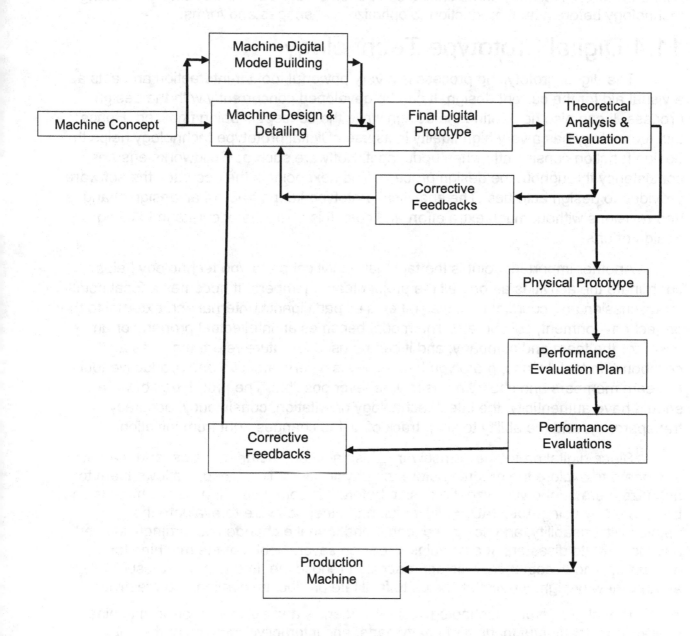

Figure 11.4 Prototype process development

Analytical and physical prototypes require many technologies for completion. Three-dimensional computer modeling and computer simulation are very helpful for early evaluation of the design concepts. Various types of iterative techniques are used to complete the theoretical evaluation. Similarly, manufacturing evaluations of the physical prototype also demand the application of several technologies, such as time and motion analysis, graphical design, man-machine interactions, material evaluations, etc. The prototyping and evaluation process are shown in Fig. 11.4. The corrective feedback must be recorded and implemented before the machine design is released for final production. Recently some critical components have been built using 3D printing technology before actual production to optimize part shapes and forms.

11.4 Digital Prototype Technology

The digital prototyping process is a very powerful tool for interaction and acts as a visual aid for the current design. It can be developed concurrently with the design process. It can also help refine the design while the concept is being finalized. Hence, this technology has a very high fidelity in its use. Digital prototype technology helps the design function consistently when updating it. Software such as Solidworks ensures consistency throughout the design process. The next point is the accuracy the software provides to design activities. The analytical prototype keeps track of all design changes and revisions without much extra effort and cost. It is also very accurate in tracking design efforts.

Another important point is the fact that analytical prototype technology helps coordinate the activities among all the project team members. It becomes a focal point for discussion and coordination. It also helps all participants, internal and external to the project environment, collaborate. The model becomes an intellectual property or an asset for the team and company, and it can be used for future reference or as a collaboration tool. Hence, prototype technology is a very effective and productive tool for team members and should be used whenever possible. The prototype software should have authenticity, the latest technology orientation, consistency, accuracy, transparency, and the ability to keep track of all the changes right from initiation.

Since digital prototypes empower industrial, manufacturing, sales, and marketing engineers to explore the product before the physical form is created, it allows them to optimize, iterate, and visualize the design before it is completed. Innovative models can be created by using AutoCAD, Solidworks, and other software to evaluate the applicability, usability, and design concerns and to make changes accordingly to avert any last-minute disasters. It also helps to design several looks of the machine for discussion and acceptance. This approach can reduce the time to market, especially for a brand-new design. It definitely helps to facilitate product innovation and creativity.

Digital prototyping technology can also identify and optimize man-to-machine interactions, loading/unloading of heavy parts, and interactive motions of the sub-modules of the machine. The design team can use this technology to supplement the physical prototype or for prove-out of performance that does not need a physical machine. For legacy designs, it can completely eliminate the need to build physical prototypes, which are very costly and time-consuming.

The digital prototype helps the team to reduce the design and development time by making it possible to evaluate the concept design theoretically. Consequently, prototype testing also gets reduced. It helps to reduce time to design the sheet metal coverings for the machine. Since this process is performed concurrently and in parallel to design activities, concurrent evaluation of design parameters also reduces the design cycle time and overall design and testing time. Upfront evaluation and identification of design efficiencies help the team to reduce design cycle time as well.

Manufacturing engineers can also validate manufacturing interference issues before the parts are actually manufactured. This reduces the number of physical prototypes and testing fixtures. Some of the dynamic performances can be initially evaluated and re-evaluated using digital prototypes. In a nutshell, a digital prototype has been found to reduce the concept-to-market time, build and test time, and the number of physical prototypes. Overall, it is a win-win situation for the team and the company.

Digital prototypes also help the team in several ways, as outlined below:

- Engineering: Once the requirements for the system are finalized, the engineers start getting into the concept and architecture design of the machine. Instead of using drafting for mechanical and electrical systems and sub-systems, the engineers can start directly sketching the design into the digital software. The design variations can be sketched in the software environments right from the beginning, and several architectures or concepts can be created very easily and in a very short time. Once the design is finalized and accepted, it can be directly used for detailing the components. At the same time, industrial design aspects can also be evaluated and incorporated into the design very easily. The digital prototypes can also be used to find the effects of color schemes, shapes, forms, and materials.

   The mechanical and electrical systems could be designed in the digital prototype at the same time. The system interactions between mechanical and electrical systems could be very easily determined. When the mechanical, electrical, and industrial aspects of a machine's design are baked together in a computer environment using software, the 3D model becomes a full-scale, realistic model of the product. Using it, engineers can not only perform integrated stress and strain under real-world usage profiles, but also determine the dynamic properties of the machine. The stress and strain analysis can help engineers select the proper material for any component to achieve the desired strength. It can also identify the dynamic motions of the design at different usage stages of the machine. Dynamic solutions can find the most optimized motion for the machine. Such evaluations and integrated design can only help the company to speed up the product development process, since computer integration is much faster than the process required to build a physical prototype. Once the design is accepted for further consideration, documentation can be created directly from the digital prototype without any manual intervention. This makes the process of conversion very efficient and productive.

   The digital prototype also helps manufacturing engineers to evaluate the design based on the manufacturing requirements. The manufacturing team works with the design team to design the product so that cost and time to manufacture become as lowest as possible. The digital environment allows manufacturing engineers to work with design engineers to design an optimized machine structure.

Several manufacturing aspects, such as casting considerations, forging process, molding possibilities, injection molding requirements, assembly requirements, and material sections, should be identified before parts are physically manufactured. Design engineers get a chance to shape their design to make manufacturing easier and less expensive. The digital prototype software can also create assembly operations while looking at the 3D drawings instead of mechanically drafted drawings. Three-dimensional drawings are unambiguous by nature and very easily seen in several possible ways. Such interactions and coordination among design and manufacturing engineers before the design hits the floor help to reduce manufacturing difficulties on the shop floor. Manufacturing engineers can design the manufacturing process and shop floor layouts for proper assembly of the machines. They also can figure out the fixturing, tooling, and conveyors required during manufacturing.

For special-purpose, customized machines, customers can get involved during the design process to visualize their product while it is being designed. If required, they might suggest changes to be incorporated in the design to make the machine more suitable for their production purposes. The digital picture can help the customer and the designers interact in a more fruitful way by showing animations, simulations, and safety and material handling possibilities in the design. Customers can also get ready for installation of the machine and help the maintenance engineers understand the product. This increases customer satisfaction for the product.

- Sales and Marketing Team: Using 3D pictures from the digital prototype, with animations, colors, and views of the machine, sales and marketing engineers can create sales brochures, web catalogs, and commercials for sales promotions. Without waiting for a physical prototype to be made, the sales and marketing team can conduct photo shoots and use virtual photography and video animations to create a sales pitch for the product. This is particularly useful for a brand-new and innovative product to be introduced for the first time in the market. This allows these teams to promote the product among potential customers and get involved in the selling process without waiting for the actual product to be manufactured. Potential customers can also visualize the product, which helps them decide whether to buy the product when it becomes available.

- Support Services: The digital prototype can also be used for data management, revisions, change control, release automation, and manage purchasing actions for the bill of materials. Such a data management process connects all the interrelated action teams so they can share the data for their individual requirements. The database remains common to all participants, although the use of data can be different for action teams depending on requirements.

11.5 3D Printing Technology Applications

Another technology, 3D printing, has surfaced in the last two decades and can also be used as an alternative to physical prototyping. In 3D printing, a polymer material, a 3D machine, and software are used to create a three-dimensional physical part. It is also used in rapid prototyping, where a complex part is formed to visualize its form and shape before it is actually manufactured. This process is called "stereolithography" (SLA) or "fused-deposit modeling." Instead of removing metal from a stock during machining, or subtractive manufacturing, a two-dimensional computer-controlled machine deposits material to form the part, and that is why it is called "printing" instead of "machining," The printing needs an electronic file created by design software, such as Solidworks, AutoCAD, etc., and the part is created by adding material layer by layer. This technology adds material to create the shape of the part in 3D, using a fusion process under automated control. This is also called "desktop manufacturing," "rapid prototyping," and "additive manufacturing process." This process is being used for very high-volume parts in automobiles, but it has not found much application in machine tool industries since the machines are not mass produced like automobiles. Nevertheless, 3D printing has become an important part of the manufacturing process to optimize part shapes and simplify the parts before actual manufacturing is done.

The 3D printing process consists of three-dimensional modeling, printing, and finishing. Modeling of parts is completed using design software that creates a 3D solid part file, called an STL digital file, which is downloaded directly to the 2D printing machine. A scanning process is used to create the STL files from the part 3D figure. The software creates 3D geometric data while the part is designed. This process is called modeling. Next comes the printing process. The STL file is corrected for errors by the printing machine before printing. Digital errors could consist of sequencing errors, intersection errors, noise errors, hole errors, etc. The scanning process creates such errors, and they must be corrected for successful printing.

Printing machine software converts this corrected STL file, called a G-code file, into a series of thin vertical layers. The machine slide containing the metal injector moves horizontally and vertically while depositing material to form the part. A G-code file is similar to the files used for machining in a CNC machine. The thickness of each layer, normally about 200–250 dots per inch, is dictated by the resolution of the machine's movements. Although finer resolution creates a better-looking part, the time taken goes up almost exponentially without enhancing print quality. Part construction time depends on the resolution specified, complexity of the part model, and volume required. The time is also dependent on the type and quality of machines being used. For machine tool purposes, injection molding or casting might be less costly. The polymer material in 3D printing is also very costly. Even if this technology helps to visualize the part, it has its limitations for machine tool applications, as most machine tool parts need much higher strengths than what the polymer or plastic material can provide. The 3D printing of machine parts is only meant for prototyping purposes

To produce a part of very fine quality, sometimes the 3D-printed parts have to be machined again. This is especially true if they have to be used for production purposes. This is a hybrid idea of combining subtractive and additive technology to create a part. The 3D-printed parts are often very complexly shaped, with lots of curves and bends. The efficiency and productivity of this process also depend on how the printing software is configured with respect to the base of the part. The orientation of the part with respect to the machine base is very important since the machine only adds material in a linear fashion. The program represents curves and bends with very small linear motions. These might not match perfect bend radius all the time, so there could be some mismatching error in the part. The only way to solve this problem is to enhance the printing resolution, which increases the time and consumes more material. Recent technology allows several types and colors of materials to be deposited on the same part, and this avoids further painting of the part. The parts also need to be cleaned and removed from the base or other reinforcements needed for building it. Nowadays, steel or aluminum instead of polymer is being used for 3D printing. For most parts that are 3D printed, a proper finishing process is also required.

Before a part is created for production and sales purposes, if the concept is borrowed or copied, permission must be given before production use. If the part is for educational, non-commercial, or private purposes, such permission might not be required. Such uses are outside intellectual property rights and might not be a potential legal issue. In machine tool applications, such infringements have not been noticed. In any case, copyright and patent violation issues must be looked into before such technology is used for part production. In general, machining or casting a part seems to be much quicker and less costly, so 3D printing technology has not become a common technology for use in machine tool applications. For some restricted applications in machine tools, such as poppet valves, handless, etc., 3D printing technology could be used in a very limited fashion. Anyway, follow the laws for copyrights and patents while using such technology.

11.6 Design Verification Plan Development

Once the design is firmed up and detailing of the parts is complete, it is time to physically verify the performance of the design. This is almost the final phase of product development before the product enters the production phase. The primary function of this phase is to develop a test verification plan and then conduct the physical tests using the physical prototype. This is the penultimate test of the product. This phase demonstrates the ability of the product to satisfy customer requirements and targets set by the team.

The design verification process starts with the types of components, of which there are basically three: new components, modified components, and carryover components under new loading conditions. These components and the system go through the design FMEA, as explained previously. The test verification plan should be part of the general "plan, do, analyze, and act" (PDAA) cycle for continuous improvement of the machine.

Before the test verification plan for the system or components are finalized, input conditions, such as a real-world usage profile (RWUP), legal and business requirements, engineering requirements, noise factors, and previous test plans, must be considered. The engineers must also have the test acceptance criterion, and timing for the tests must also be determined. With all of these inputs, test schedule, methods, and verification plan are designed and developed. The test results should also be used to correlate the analysis results. There are three types of tests: durability tests, strength tests, and performance parameter tests.

Results from this phase will be used to further refine and optimize the product. The results can also result in the modification and improvement of the manufacturing methods and process selected by the team, and they will help quality engineers identify the critical aspects of the design that must be inspected after or during production. The results will also have to be documented properly for future machine modifications or facelifts required to suit the market.

This phase is done in addition to the analytical steps taken earlier. The physical tests will help to confirm the projections made by the theoretical analysis. Such correlation is absolutely necessary for future reference and use. The physical tests also become a part of a continuous process of improvement, which is recommended by all the quality programs, such as Sic-Sigma, Zero-defect, etc.

Another critical use of this phase of test development is to satisfy the business aspects of product development, i.e., of the customers' voices. The test results will help the company understand how well the design and development team has satisfied engineering requirements and specifications. The tests also help the team to find whether the targets were realistic and reflect customer requirements. It is also important to note that physical test using any hardware or fully comprehensive prototype cannot help the team determine actual customer satisfaction, things gone wrong (TGW), or warranty cost and service issues for the product.

The design verification test plan should be designed and developed to identify product degradation over time (robustness) and correlate the theoretical analysis results, reliability growth curves, manufacturing and quality measurements and controls required during production, process control methods, etc. The verification plan activities flow diagram is shown in Fig.11.5 below. It is self-explanatory. The manufacturing and other services verifications are done normally after performance verification is completed.

To verify the functional performance of the machine, the desired signal inputs, such as material, program, coolant, machining sequence plan, loading and unloading device, measuring instruments, etc. need to be installed in the machine prototype. The output of the machine will consist of the desired and undesired outputs, the latter being "noise" or "error functions." The system also has to be tested with possible noise inputs on the machine as found during real-world usage in the customers' hands. During testing, functional performance has to be measured as per the plan.

It is also true that system reliability cannot be measured due to variability encountered during manufacturing or due to variation in inputs in the customer's shop. Variations can be controlled but cannot be eliminated. Hence, variations from machine to machine will be always there, but the plan should be to reduce their effects on the performance, durability, and reliability of the machine. The system input and output diagram for a system test is shown in Fig.11.6 below.

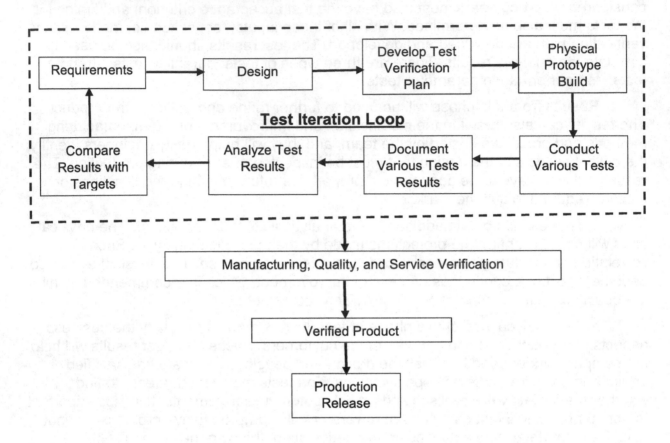

Figure 11.5 Verification phase flow diagram

The verification plan and tests should be so designed that the machine outputs can be measured and compared based on the target requirements. Reliability and robustness depend on product design and process controls. It is also important to remember that changing the design depending on the test outputs is a very time-consuming and costly proposition. Nevertheless, any major discrepancies at this stage create instability among team members, and the project could be in jeopardy. Moreover, changes in one element might bring about unreliability in other components in the long run. Unplanned changes might reduce the reliability of the system as a whole. So, testing should be conducted over some reasonable amount of time to identify performance discrepancies and variations.

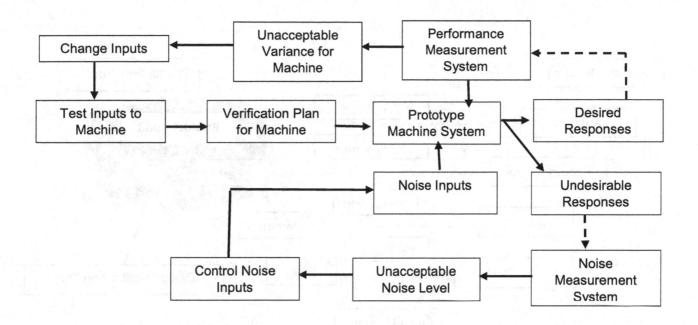

Figure 11.6 Verification of functional performance with noise

It should be remembered that design verification tests are intended to improve the design by iteration, but these are product confirmation tests. Their primary purpose is to establish product robustness, reliability, and durability over a long period. This also confirms performance. For example, by measuring the timing for the index unit under various loading conditions, the index time target is confirmed. By running the indexing over at least a million cycles, the durability performance of the index system is established. Similarly, by running the spindle using a cycle and repeating the cycle over some considerable amount of time, bearing performances can be obtained. The design verification test gives assurance that the product will perform with success in the hands of any customer.

As per the system engineering "V", design verification and confirmation tests are conducted on the right side, where the activities are serial by nature. The system-level test is always done at the end of design and development, after the system is integrated with design-intent sub-system and components. The test phase is shown in Fig.11.7. The responsibility for complete and accurate testing methods and planning falls on the test engineers. Some of these tests can also be done at the end by the suppliers who are supplying major sub-systems or components. The system diagram summarizes the system engineering stages, including the test verification and audit steps. In a nutshell, design verification is done first, and then manufacturing process verification is carried out. Design verification tests confirm the design requirements, whereas the manufacturing process validation demonstrates the manufacturing steps and assembly process so that the product satisfies the budget and time of production without any major issue. After these two steps, production sign-off is completed.

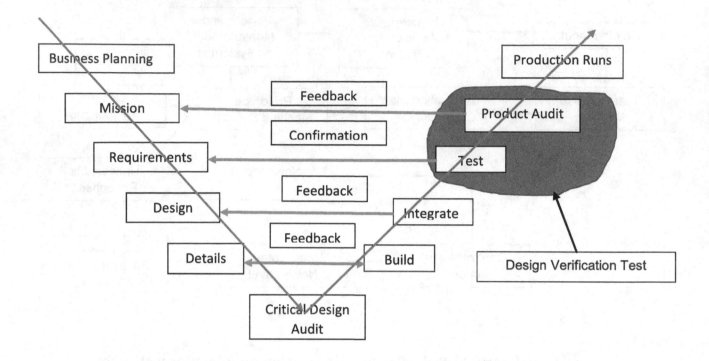

Figure 11.7 System engineering and design verification tests

11.7 Test Levels and Details

Any complex machine tool, such as a CNC turning center or machining center, should be subjected to three types of tests: the test to a predetermined time level (bogey), the test to failure, and the performance degradation test.

Test to a level: This test is sometimes called bogey testing. The loading cycles, failure criterion, and the number of cycles or hours must be pre-defined before the test is started. The failure criterion is pass or fail, i.e., the test is passed if the component has survived the predetermined time or cycles of loading, and is binomial, i.e., either pass or fail. During the test, the function is measured to see if the component has failed to deliver the level required. If the component survives the test, the nature of failure and how long the part can survive until the end are not known. The test does not tell what would happen to the part if it were allowed to run more than the set level of cycles or loading cycles. The loading cycle or the testing time comes from the real-world usage for the component. The number of samples to be tested depends on the criticality and functional aspect of the component. Spindle systems are tested for 4,000 hours at a stretch. The spindle is subjected to cyclic acceleration and deceleration, and it runs for a fixed number of hours or cycles. During this test, failure criterion could be bearing seizure, excessive temperatures over 140–150 degrees F, belt slippage, motor failure, control failure, or stoppage of the test. If the spindle system does not satisfy all the failure criterion, it has failed the test

Test to failure: Normally, critical components or systems are subjected to this type of test. For example, the indexing or hydraulic unit could be subjected to such tests. In this test, the system runs under certain input conditions until it fails. The failure criterion and input conditions must be pre-defined before the test is started. The loading patterns or load cycles must be defined very clearly. Such tests always provide all the information possible. After the test, failure hours, nature of failure, expected life of the component, etc. are known. For a system, if a component fails, the whole system fails to deliver the function. This allows the test engineer to predict the life cycle for the component. After-test inspection of the failed component also allows the determination of the nature of failure or failure mode. Once the failure mode is known, the necessary changes can also be determined to fix the component. The number of failures at any point in time is plotted as shown in Figure 11.8

The system is allowed to run until it fails, so the failure rate, i.e., the gradual loss of function, cannot be known from this type of test, nor can the functional degradation or robustness be determined. If the number of samples tested is great, a failure histogram, like a Gaussian curve, can be determined. The probability of failure can also be learned from such diagrams. This test enables the prediction of failure and mean time to failure for the component or system. Such tests are also conducted to determine the rate of infant mortality, useful life period, and wear-out pattern of a bathtub curve, as shown in Figure 11.9.

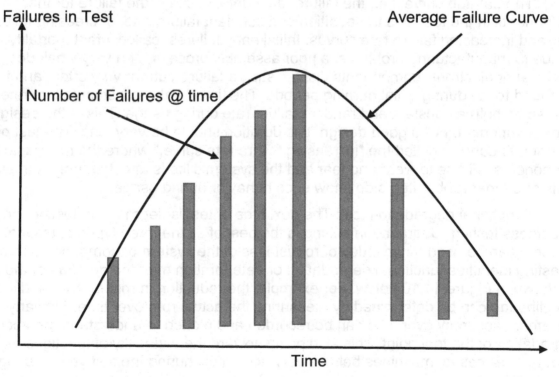

Figure 11.8 Average failure distribution vs. time

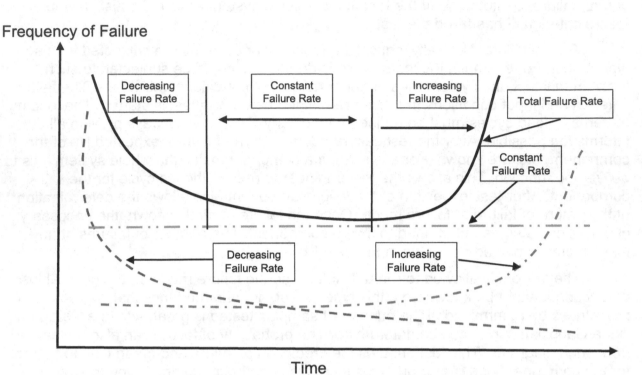

Figure 11.9 Bathtub curve for components

The bathtub curve, i.e., the failure rate curve, displays the failure for the components in the test. It is the addition of a constant failure rate, decreasing failure rate, and increasing failure rate curves. Initial early failures, called infant mortality, can be due to manufacturing problems, a poor assembly process, or a very weak design. Electrical or electronic components display such a failure pattern very often, and they are found to fail during initial burning periods. The time for the constant failure rate represents both a constant and random failure rate during usage. This is the design life of the component. For a good design, this duration should be very long. The last part of the bathtub curve is called the "increasing failure rate curve," where the mechanical components fail due to wear and tear and the system's life is affected. For example, seals and other rubber items do show such behavior during usage.

Functional degradation test: The next type of test is degradation testing, or robustness testing. Degradation testing is the best of all the three types of testing. The outcome can be used as an index of robustness of the system or component. This kind of testing identifies functional degradation or deterioration over the duration of the test, as shown in Figure 11.10 below. For example, the reduction in rpm of the spindle due to belt slippage can be determined by measuring the actual rpm over time. Similarly, axes movement accuracy over time can be recorded and plotted to understand the accuracy degradation of the tool point. This also points toward the variability of functions over time. In most cases, machines behave very accurately during the first year of usage and then deteriorate.

This is very unacceptable for most customers. Degradation, or robustness, testing is often done for high-usage items such as ball screws, belts, motors, conveyors, etc. In such cases, the target for a minimum level of functional performance or rate of decay for the performance over time should be selected and compared to the actuals. This test also gives an idea of preventive maintenance durations. The test data can be plotted, as shown in Figure 11.10, for various components to get an idea of their degradation variability.

The diagram shows the functional decay of various components, systems, or sub-systems under test. All the components are subjected to the same test duration, and during the test, the performance or function is measured at regular intervals, such as index time or spindle revolution per minute (RPM), and plotted as shown. The rate of decay could be constant or variable depending on the type of design, material, and loading. Component D has the worst decay rate, and Component A has the best decay rate. The graph also shows the variability of function at any point in time. Components can be from different design levels, and a comparison of the rate of decay can reveal the best design for the component. Using this data, engineers should be able to predict the functional level of the system at the end of life. The nature of decay can be linear, as shown, or non-linear.

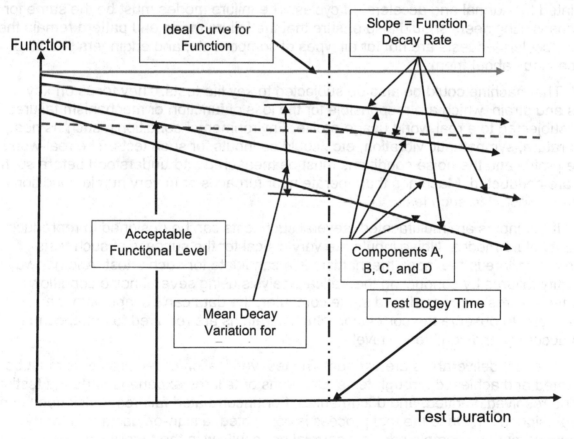

Figure 11.10 Functional degradation test

The information quality level is highest for the decay test and lowest for the bogey test. Moreover, the cost of the decay test is also the highest among all types of tests. For machine tools, normally, a test to failure should be done. In case a rubber seal failure has to be explored, a decay test might be preferred for several types of seals and designs. Depending on the results, seals with the best decay rate can be selected. I have several such tests for machine tool components and systems. For controls and electronic components, test to failure is recommended, whereas, for spindle bearings, a bogey test could be done. Nevertheless, tests should be only done on critical components to reduce the cost and duration of the testing phase.

To investigate failure causes in the shortest possible time, accelerated testing can be performed. The purpose of the accelerated test is to reduce the test duration without sacrificing the quality of data obtained. The failure pattern and duration have to be correlated with field failures. Such tests shorten verification time and developmental time. The accelerated test shortens the test duration by increasing equivalent load inputs. This makes the stress level higher, and that reduces the component life. Another way is to apply only the effective loads, which are high in amplitude and/or frequency. The determination of equivalent higher amplitudes is somewhat tricky, and knowledge of stress levels for failure must be known. Total damage of the normal loading cycle and equivalent load level must be same. The damage comparison has to be focused on during the design of the accelerated test. Moreover, the failure pattern has to be correlated for normal and accelerated cycles. The failure modes must be the same for both cases. Engineers should also ensure that the failure mode and pattern remain the same. Accelerated tests are not for all types of components, and engineers should be very cautious about them.

The machine could be also be subjected to key life tests. They focus on key stress and strain, which are responsible for the loss of function or mechanism failures when subjected to a real-world usage profile. Environmental conditions, such as heat, temperature, sympathetic vibration, etc., could be inputs for such tests. The real-world usage profile and the noise conditions must be measured and understood before such tests are conducted. Machines that operate near furnaces or in very humid conditions can be subjected to such tests.

If the inputs are multifaceted, several such tests can be designed to reproduce different failure modes. Noise inputs are very critical for the success of such tests. Piston ring failure in the index mechanism is a candidate for such a test. Another way is to identify failures by conducting theoretical analysis using several noise conditions. Once the failures are reproduced in the computer, the data can be input into the physical test to observe the correlation. Such iterations are required to enhance the test's accuracy and confidence level.

The test deliverables are various. The test verification of requirements must be completed and achieved through tests. Concerns or failures experienced during testing must be resolved, verified, and documented. For uncorrelated failures, a risk document must be signed off. Once the test process is completed, a sign-off document for the component must be completed. Test correlation results with the theoretical analysis must also be documented and resolved if there are any issues.

11.8 Statistical Confidence and Number of Samples

When the number of samples for the tests is much smaller than the sample population in use, an obvious question arises: what should the number of samples be so that statistical confidence in the prediction of reliability is very high? One of the objectives for sample testing is to predict the probability of failure for the components. The statistical confidence for any reliability prediction is another. Probability of prediction is equal to or greater than the estimated reliability, so the challenge is to match the estimated probability from a test with limited samples with the population reliability in reality.

The accuracy of reliability estimates depends on the number of samples tested. Hence, for each reliability estimate, a confidence level of prediction is also required. Statistical confidence is defined as the probability that population reliability is equal to or greater than the estimated probability predicted by test results. For example, in a test, a part has a 10% failure rate, i.e., it has a 90% reliability of success. The prediction has a 90% confidence level. This means that when a sample of 20 parts is tested, two parts will fail on average (10% failure). Now, if 10 such samples of 20 parts each are tested, nine out of 10 samples will have two failures or less, i.e., 90% confidence in prediction. The confidence level is meant to make sure the reliability estimate is not overstated.

Test planning consists of its goal, its schedule, the number of samples to be tested, the duration of the tests, and the test method. Also, the reliability of a product in the customer's shop is estimated based on the test results. The customer looks for the reliability of the product at the end. Keeping everything the same, rellability is predicted more accurately when the number of samples goes up. More samples mean a more accurate prediction of reliability. The accuracy of the reliability estimate depends on the test methods, data acquisition accuracy, and the accuracy of applied input conditions. Then the test plan and method have to be determined as well.

In order to instill engineering confidence in the test results, an accurate determination of a real-world usage profile is very important. Another concept is to enhance engineering confidence, which seems to be more relevant than the statistical confidence level. The engineering confidence ensures that the components are tested under real-world usage data and that the test methods also simulate the actual running conditions of the machine. The engineering confidence also means that the product will meet functional performance in the field when the machine is used in the customer's shop. A balance has to be made among duty cycle, test methods, cost of the test, and reliability prediction. Reasonable environmental conditions have been considered during tests as noise inputs. The prime objective is to increase customer satisfaction

The number of samples dictates the accuracy of statistical models. Samples should represent the population as accurately as possible. Larger sample sizes provide more accurate results. The worst-case sample size can be calculated using a simple formula if the confidence level and reliability required are given. The worst-case sampling can be calculated as follows:

N = worst-case sample size = log (1 – C)/log R, where R is the reliability desired and C is the confidence level required. Moreover, sampling should not be biased, i.e., random sampling is also a good practice. On the other hand, large sample testing costs more time and money. The sample size and content should represent the population. When engineering confidence is large, the sample size can be small. In order to enhance engineering confidence, testing to failure or degradation testing should be done. More important is to control and include the noise conditions in the test, as mentioned before.

Test methods and test set-ups must be as simple as possible. Complicated procedures lead to many mistakes. Several steps could be taken to make the test simpler:

- Identify critical noise inputs for the test
- Methods must be very clear and simple
- Develop simple test sequence and methods
- Test must include common cause variations and special cause variations
- Mean effects should not include random variations
- Select unbiased samples randomly

The documentation and lessons learned must be completed for all the tests executed. The lessons learned during the development of the product testing, while conducting the tests, or when encountering the difficulties and consequences of noise inputs must be documented properly for future reference. The following should be recorded as well:

- Verification plan
- Test targets and requirements
- Test methods deployed
- Real-world usage profiles
- Test input conditions
- Test set-up details
- Design FMEA severity factor and criticality for the tested component
- Justification for sample size calculation
- Noise control and Inputs
- CAE models, results, and analysis for correlation purposes
- Degradation test, bogey test, or test-to-failure methods
- Functional degradation curves
- Manage variation during testing
- Reliability calculations and robustness values
- Unexpected failure resolution

- Test correlation proof and documents
- Test results in detail
- Recommendations

11.9 Reliability Testing for Machines

The primary purpose of reliability testing is to make sure that the product conforms to the requirements and to identify any weaknesses of the design, which should be fixed before the design hits the floor or the product lands in customers' hands. Depending on the type of testing performed on the product at different stages, it can also determine whether the product will perform without any failure under predetermined testing conditions, which should match customer requirements. Basically, testing enhances the confidence level that the product will perform as intended under reasonable conditions and without failures, i.e., product reliability enhancement. So, the reliability testing for machines can be conducted at various stages, as described below:

- **Tests for Design Refinements and Iteration:** During the design phase, new technologies, materials, and design ideas are introduced to make the design more productive and efficient. In this process, there is always some negative unknowns that need to be identified and fixed before the design hits the floor for production. Hence, some prototypes of critical assemblies, such as indexing units, tailstock units, and spindle units, are built separately and tested under given design conditions. The purpose of such tests is to identify the design weaknesses that need to be fixed. Ultimately, such tests enhance the reliability of the product.

- **Tests for Manufacturing Process Improvements:** When manufacturing has started for the product, its reliability might be affected due to deviations in the process. Once the functional prototype is built, performance evaluation and durability tests are conducted to evaluate confirmation with customer requirements. The performance test evaluates the product's ability to produce parts within the predetermined level of accuracy desired by customers. This is sometimes called a "precision test." This particular performance evaluation test is conducted using customer production conditions in order to make sure the machine produces parts with desired accuracy and precision. The other type of test is the durability test, which is performed to ensure that the machine does not malfunction during actual operation. The machine is subjected to production cycles and programming conditions and runs for 24–30 hours at a stretch to make sure it does not fail during this time period. For most machines, these two tests are performed before they are shipped to the customer. Durability tests also catch the faults of the machine due to poor processes applied during manufacturing, i.e., infant mortality failures.

- **Test for System Reliability:** When a large batch of machines is manufactured, a sample of one or two machines out of the lot is subjected to a durability test under given conditions to ensure that the reliability of the system is maintained at the desired level. Such tests identify failure modes and rates, which can then be used to determine the reliability of the system. It is basically a system-level test. The input or cycling conditions should be designed so that all the subsystems interact with each other in the way that the machine will perform in the customer shop while machining a part. If any assembly shows a typical failure pattern, the failure modes give a clue to the designers as to what went wrong with the design or manufacturing process. If the tests are conducted for a much longer time, it might help the service engineers establish the time and frequency of preventive maintenance for the machine.

- **Tests for Safety Issues:** Some assemblies of the machine, such as a sliding door or workholding systems, are subjected to static or dynamic tests to make sure that the machine does not affect the operator's safety while working. If the workholding system, chuck, or hydraulic cylinder malfunctions, parts may fly through the door and hit the operator who is attending the machine. Tests are performed to ensure such conditions never arise. The machine must not have pinch points so that maintenance personnel do not get hurt while repairing it. The machine must conform to and satisfy health and hazard regulations.

Machine failures can be attributed to design, manufacturing, poor maintenance, overuse, or abusive conditions. The bathtub curve demonstrates the failure rates for three such conditions. Tests have to be designed to identify failures during the infant mortality or burn-in phases and the constant failure rate phase. Failures during the wear-out phase are normally not dealt with here but are taken care of by a robust design phase. Tests should also help the team identify the duration of the constant failure rate stage, which should be as long as possible. Tests have to be designed and conducted to identify the mean time between failures, failure rates, mean time to repair, etc. In a nutshell, the purpose of the tests should be to understand the machine in and out.

The objectives of machine tool testing are numerous. Some of the important ones are mentioned below:

- Durability and Stress
- Failure rate, Failure Mode, and Hazard Rate
- Crash and Safety Considerations
- Performance, Accuracy, and Repeatability
- Conformance with DFMEA and PFMEA analysis
- Weak Design Links

It is also true that the higher the reliability of a product, the higher the cost of the machine is. Once the target reliability values have been accepted, the test has to confirm the product requirements on a consistent basis. The machine has to support customer requirements, company directed quality standards, safety and health standards, and cost requirements. Cost and quality of the machine must be synonymous. Before testing a machine, quality requirements, reliability requirements, safety requirements, and input conditions must be well defined to avoid any confusion later on. Each specification must have a normal value with specified ranges. The failure criterion must be very clearly defined. In case of a failure, the failure mode and failure criterion must be resolved before moving on. In case of any further design or manufacturing modifications, design deviations and actions taken must be recorded for future reference.

Since machine tools are produced in small quantities, the sample size for any test is a matter of great concern. The question is: what is the optimum number of samples for testing? The answer depends on many considerations, such as company product legacy, cost of the product, design complexities, special design considerations, reliability requirements, volume of production, etc. In the case of a machine tool system, using one or two samples is customary. If the system is very complicated, then only one such system is tested in most cases, whereas critical subsystems are tested using several. In the case of electrical components, such as relays, capacitors, or transformers, many units are tested for reliability. The duration of the test in each case is also different. For example, indexing units are tested for at least one million cycles with load and no-load conditions. For bearings, one million cycles are used. For electrical components, normally 20 to 30,000 cycles are the bogey. Nevertheless, there is no fixed standard for such tests, and most companies have their own standards unless the customer contract specifies specific conditions for the tests and input conditions.

In order to characterize the system failure or hazard rate, the test has to be designed to determine the nature of the bathtub curve. It is also true that the characterization test is long and tedious, but that is the only way system improvements are possible. The test should characterize the profile of the system failure, and the plan to enhance the system reliability enhancements or rationing will depend on this hazard rate profile. While testing the system or components, failure or reliability data could be collected for an individual item or the sub-system or system as a whole. It also depends on the objective of the test. If individual data is required, a component test is preferred. If system failure data is required, group data has to be collected accordingly.

11.9.1 Individual Component Test

To characterize the reliability profile or failure rate of an individual component, such as relays, transformers, bearings, etc., similar components are put to the test under similar input conditions. The failure mode and times are noted as the tests continue. The time of failure for similar failure modes must be grouped together and compared for further analysis if required. For example, say n number of similar bearings are tested. The failure times and modes are also noted as the bearings fail during the test.

Since the loading conditions and cycle of load and no load for all the bearings are the same, expected failure modes are also the same. The failure modes could be seizure, ball galling, out-of-round conditions, excessive clearance, etc. Once the failure times for all the failed bearings during the bogey time have been noted, they are arranged in ascending order.

Let us consider a case of testing bearings selected from a production line one at a time. This test would be for individual bearings. There is only one test setup for testing the bearings. We can test one bearing at a time out of all the bearings to be tested. Say bearing one failed in time t_1, bearing two failed in time t_2, etc. The failure times are arranged in ascending order. Hence, we have

$t_1 < t_2 < t_3 <\ldots\ldots\ldots< t_n$, where n is the total number of bearings under test and t_i is the failure time for the ith bearing. As explained earlier, the cumulative distribution of failures of all the failed bearings, F(t), represents the probability of all bearing failures before a given time t. So, F(ti) is equal to the fraction of bearings that failed before time ti or in time ti. Hence,

F(ti) = number of bearings failed/total number of bearings under test

$= i/n$

For example, if 8 bearings out of 16 being tested have failed and test time is equal to 1,000 hours, then

F (1,000) = 8/16 = 0.5, so, the probability of 8 bearings failing out of 16 bearings is 50%, or 0.5. To have an accurate value for this probability, the number of bearings being tested, n, has to be comparatively large. So, a total of 8 bearings have failed before 1,000 hours, and the rest of the bearings lasted longer than 1,000 hours. All the bearings are expected to fail within t_n hours if t_n is large enough. Now the reliability of the bearing can be determined as follows:

R(ti) = reliability of component at time t = t_i = (n − i)/n

$= 1 − F(ti) = 1 − i/n = 1 − 0.5 = 0.5$; thus, the reliability of the bearings at time ti (1,000 hours) = 50%, i.e., 8 bearings have failed before 1,000 hours, and 8 bearings survived 1,000 hours of testing. As mentioned earlier, the probability density function, f(t), is the slope of the cumulative distribution graph at time t.

Hence, $f(t) = [F (t_i + 1) − F(ti)]/[(t_{i+1} − ti)] = [(i + 1/n) − i/n]/[(t_{i+1} − ti)]$

$= 1/[n*(t_{i+1} − ti)]$, where $t_i < t < t_{i+1}$ and i = 1,2,3,4,5……(n − 1)

Now, the hazard rate, h(t), can be calculated also, as given below:

$h(t) = f(t)/R(t) = [(n − i) *(t_{i+1} − ti)]^{-1}$

The experimental data for the test is given in Table 11.1. The values of F(ti), R(ti), f(ti), and h(ti) are also tabulated. The graphs for F(t) vs. failure hours, R(t) vs. failure hours, f(t) vs. failure hours, and h(t) vs. failure hours are displayed in Figs. 11.11, 11.12, 11.13, and 11.14 respectively.

Bearing Number, i	Run Time, ti, hrs	ti+1 – ti, hrs	F(ti)=i/n	R(ti)= (n-i)/n	f(t)	h(t)
0	0	43000	0.00	1.00	1.45	1.45
1	43000	8700	0.06	0.94	7.18	7.66
2	51700	11190	0.13	0.88	5.59	6.38
3	62890	8780	0.19	0.81	7.12	8.76
4	71670	11930	0.25	0.75	5.24	6.99
5	83600	10650	0.31	0.69	5.87	8.54
6	94250	6350	0.38	0.63	9.84	15.75
7	100600	9900	0.44	0.56	6.31	11.22
8	110500	5370	0.50	0.50	11.64	23.28
9	115870	7020	0.56	0.44	8.90	20.35
10	122890	3910	0.63	0.38	15.98	42.63
11	126800	8100	0.69	0.31	7.72	24.69
12	134900	7200	0.75	0.25	8.68	34.72
13	142100	4900	0.81	0.19	12.76	68.03
14	147000	4000	0.88	0.13	15.63	125.00
15	151000	4000	0.94	0.06	15.63	250.00
16	155000	NA	1.00	0.00	NA	NA

Table 11.1 Functional degradation test

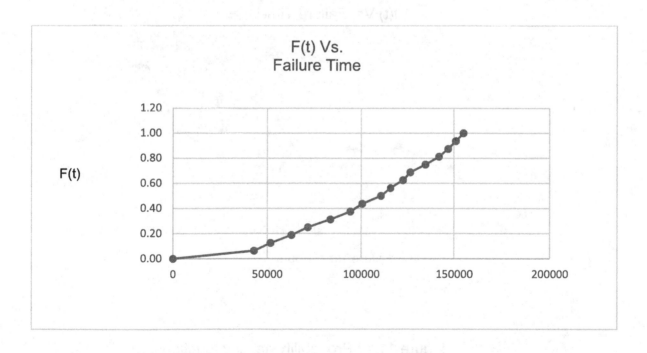

Figure 11.11 Cumulative distribution of failures, F(t)

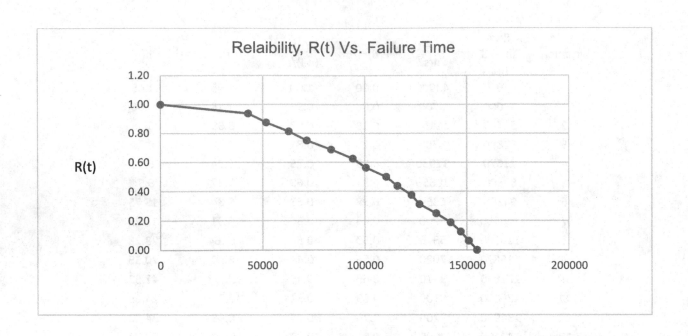

Figure 11.12 Reliability distribution, R(t)

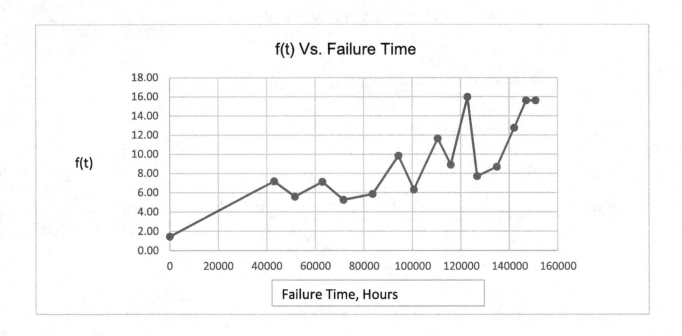

Figure 11.13 Probability density functions, f(t)

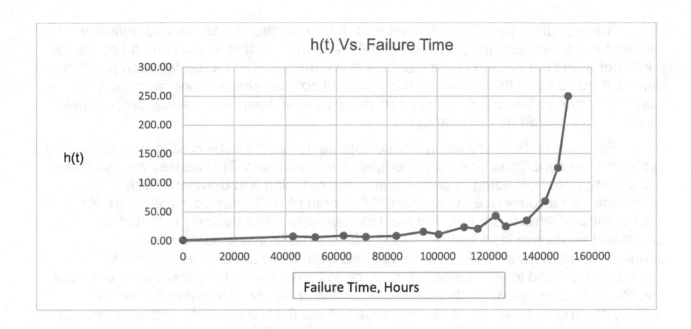

Figure 11.14 Hazard function, h(t)

11.9.2 Accelerated Life Test

The accelerated life test for the product is not very common in the machine tool industry because the product volume does not warrant such tests. Machine tool failures are not that dangerous for the operators and the environment. Such tests, however, are very common in the life prediction for components and assemblies in the auto industry. Moreover, the tests are very costly and time-consuming. The volume of production is comparatively way less than with automobile or consumer-driven products. For machine tool products, the design is very robust, and it takes a very long time for failures. In order to capture the working environment and noise, tests have become very complex and time-consuming.

Hence, key life tests are not often seen in the machine tool industry. Instead of a key life test, a few durability tests are done to make sure that the system works in unison to produce the best possible part and abrupt failures due to manufacturing issues are captured. In order to speed up the test time without affecting its outcome, several input parameters are adjusted to a value that is much higher than those for normal operation of the machine. For example, the cooling fans and electrical components, such as relays, condensers, capacitors, etc., are occasionally subjected to such tests at the manufacturer's end. The key parameters that are adjusted to conduct a key life test could be as follows:

- Higher-than-normal machine cycle speed and severity conditions
- Higher temperatures, hot and cold, than normal operating conditions
- Excessive speed conditions for bearings and rotating elements
- Above-normal safety crash loading conditions

Another difficulty with such tests is to determine the conditions and definition of system failure. For example, the failure of the external loading and unloading conveyor might not affect the cycling of the machine, but it might mean the system can produce parts without loading them. Most of the electrical components, cooling fans, part inaccuracy due to thermal distortion, etc. can be tested using key life test procedures designed and tested by manufacturers.

For such tests, electrical system components can be subjected to excessive cyclic voltage conditions or extreme temperature conditions. This causes excess stress in the components, leading to early failures and reducing test duration. Such a procedure can also reduce the number of components to be tested if the failure is consistent with what is seen during machine operation. The difficulty is to find an equivalent excessive load that will cause a failure mode similar to normal load conditions. The normal load is the design load from the real-world usage profile. The excessive load and lower number of cyclic conditions have to be correlated with actual conditions so that failure or stress damage for the component remains the same. Hence, the engineer has to have a clear idea about the failure mode and frequency of failure for the component. If the load is so excessive that the failure happens far earlier than expected or the failure mode displays excess loading conditions, the load or cycle has to be adjusted accordingly. Hence, a correlation study is required before a component is subjected to such loading conditions in a key life test. The failure criterion should remain the same for both conditions. If tests are conducted using several load conditions having the same failure mode, load and cycles could be extrapolated to find the actual excessive value. Hence, the damage for the part has to be correlated either using a test or a finite element analysis if the component model is available.

For example, say, we are designing a key life test based on damage content for a cantilever long steel beam, propped at one end and loaded at the free end. It is also known that failure occurs when damage is excessive and the material cannot handle the load. So, we have:

- Failure is proportional to damage due to excessive load or temperature

- Damage is proportional to strain induced in the material

- Strain is proportional to stress caused by load

We know that life is proportional to induced stress for the beam. The stress is proportional to the deflection of the beam.

Deflection = δ = P*L/(A*E), where P is the load at free end, L is the length of the cantilever beam, A is the area of the cross-section, and E is Young's modulus for the beam material.

Stress = σ = P/A; so, $\delta = \sigma*constant = \sigma*\alpha$; $\sigma = \delta/\alpha$

LI = Life = $\kappa*(1/\sigma)$, where K is the constant of proportionality between life and stress

LI = $\kappa*\alpha*\delta^{-1} = C*\delta^{-1}$, where C is also a constant

Now, taking log of both sides, we have

Log LI = Log C – Log δ = Log C – Log (P*L/(A*E)) = Log C – Log (PL) – Log (AE)

Log LI = Log C – Log P – Log L –Log (AE) = Log C – Log (L/AE) – Log P

Log LI = Log (C*AE/L) – Log P = Log (D) – Log P

So, the life equation is: **Log LI = Log (D) – Log P**; we can use a known load and life to determine the constant value Log (D). The life equation could be linear or non-linear. The linearized log-log relationship between life and load is shown in Figure 11.15.

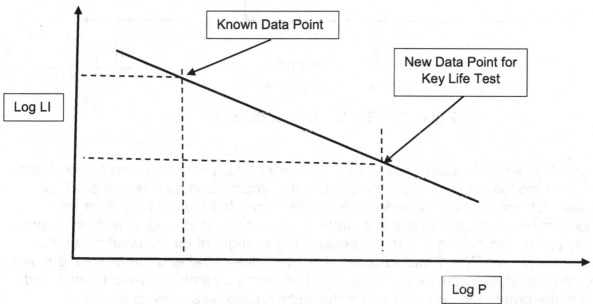

Figure 11.15 Life vs. load log-log graph

Another common application of life-cycle testing is fatigue testing using fixed amplitude and frequency of loading. Many components, such as spindle, steady rest, chuck, and cylinder, can be tested using this technique. The load is normally sine wave, square wave, or sawtooth pattern. The frequency is the rate of application of the load during testing.

N = Number of cycles for fatigue failure

σ_a = Stress amplitude in the part due to loading, lb./square inch, that causes the failure in the part

The values of the stress amplitude are varied, and the number of cycles for failure is noted and plotted as shown in Figure 11.16. This is universally called the S-N curve. The S-N curve is material and design-specific. Since the failure is dependent on stress concentration factors, material factors, notch sensitivity factor, etc., this data is component and material specific. It is, again, a log-log graph for linearization purposes.

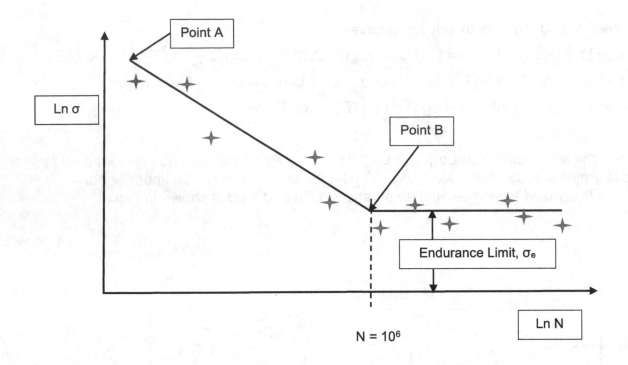

Figure 11.16 S-N curve for the part

For this test, the failure life cycles are obtained using various stress levels. If the amplitude of the load is increased, the stress level amplitude is also increased. The failure test data vs. stress amplitude is plotted as shown in Figure 11.16. A series of tests have to be conducted to plot this curve. The axes are in log-log scale to linearize the data. The endurance limit of the material is the strength of the material when the part fails at a value of N = 1 million cycles. For this value of the endurance limit, the part has infinite life, which is the design objective. For the data points between Point A and Point B in the graph, the general relationship between stress and cycle can be expressed as SN^a = constant = A, where a and A are constants. So,

$$\sigma_1(N_1)^a = \sigma_2(N_2)^a \text{ or } \sigma_2 = \sigma_1*[(N_1)^a/(N_2)^a] = \sigma_1*[(N_1/N_2)^a]$$

$$\text{So, } \sigma_2 = \sigma_1*[(N_1/N_2)^a] \text{ or } N_2 = N_1*[\sigma_1/\sigma_2]^{1/a}$$

Hence, the fatigue life cycles for failure can be adjusted by changing the stress amplitudes, or vice versa. If the value of constant a is 0.10 for any material, we have

$$N_2 = N_1*[\sigma_1/\sigma_2]^{1/a} = N_1*[\sigma_1/\sigma_2]^{1/0.1} = N_1*[\sigma_1/\sigma_2]^{10}$$

If the stress is doubled, i.e., $\sigma_2 = 2*\sigma_1$, then $N_2/N_1 = 0.5^{10} = 0.001$

$$\text{So, } N_1 = 1000*N_2$$

This means that if the stress level is increased twice, the failure life cycles will be reduced by a factor of 0.001 provided the stress level is above the endurance limit for the material and specific design. These values are design and material specific. Hence, if the design is changed, this nature of the graph will change accordingly, and tests need to be run again to re-establish the fatigue cycle values.

There are several other types of tests that are used in the industry, but they are not very common among machine tool builders. For example, sudden-death testing, simulation testing, sequential-life testing, etc. are also carried out. For example, sequential-life testing is done to minimize the number of costly samples for testing since this is a destructive type of test. A sample is collected from a batch of products and tested for failure. The outcome is either bad or good, so the primary objective is to determine the reliability goal of any product, but the test results cannot be used to find mean time between failures, since only one sample is tested at any time. For such tests, a risk is taken to accept the lot depending on the sample result. Hence, population statistics are dependent on the sample result. A question arises as to the confidence level of the result. The batch is accepted or rejected depending on the sample test outcome. There are two types of risks: consumer's risk, called "β risk," and producer's risk, called "α risk." The consumer's risk is defined as when the reliability of the population is less than the tested sample reliability and the batch is accepted for use. On the other hand, the producer's risk is to reject the batch even if the population reliability is higher than the tested sample reliability. In any case, such considerations depend on a high volume of production, such as for computer disks, computers, etc. This method is not very applicable for machine tools due to their low volume of production.

There is one particular application for this methodology for machine tools. If the builder is producing similar machines for a customer in volumes such as 25 to 50, it is very time-consuming to test each one of these products. Normally, a typical sample is selected out of the batch of, say, 50 machines and tested. The whole batch is accepted depending on the reliability value of this single sample. Obviously, the customer or builder takes a risk on whether the batch is accepted or rejected. Without going into further mathematical details, the following analysis can be used in such a situation. Let us discuss the sequential test plan for machines.

11.9.3 Sequential Test Plan

There can be three commonly accepted possible sampling plans for testing:

- Single-Sampling Process: In this plan, the machine is tested for predetermined types of defects and the number of defects observed. This is a very popular plan. Two numbers are defined before testing: d defectives in a sample size of m machines. The criterion is that if there are more than d defectives in the sample size of m machines tested, the lot is rejected; otherwise, the batch is accepted as a lot. If m = 2 and d = 5, then the lot of 50 machines is accepted if the total number of defects is less than five for the two tested machines. The total number of machines in the batch is n, which is much larger than m.

- Double-Sampling Process: In this plan, we select a sample size of m1 (m1 << n) first and test the machines serially. If the number of defectives is less than or equal to d1, the lot of n machines is accepted (similar to a single sampling plan) as a whole. If the number of cumulative defects observed in m1 is more than d2, the lot of n machines is rejected as a lot.

However, if the cumulative number of defects lies in between d1 and d2, a second sample of m2 is selected out of (n – m1) machines. The lot is now accepted if the cumulative number of defects in both samples, m1 and m2, is less than or equal to a new accepted number, d3, which is greater than d2, obviously. It is possible for d3 to be equal to d2.

- Sequential-Sampling Process: The double-sampling process can be obviously extended to include the next set until all the product samples are exhausted. A sequential test process can be described as follows:

  - Select one sample at a time and run the test; count the number of defects in this sample.

  - Another sample is tested, and defects are counted. Then the cumulative number of defects for two samples are noted.

  - The above sequential process is repeated, and every time, the cumulative defects are noted.

  - After each test, there are three possible choices:
    - If the number of cumulative defects exceeds the predetermined accepted defect number, reject the lot

    - If the number of cumulative defects is less than the predetermined accepted defect number, accept the lot

    - Instead of accepting or rejecting the lot at any stage, keep testing the sample until all the available samples are exhausted

Nevertheless, it also has to be remembered that increasing the number of tests indefinitely is not a good plan, since it costs lot of money and time without any added benefits. Improvement of quality has a reasonable limit beyond which the product quality cannot be improved any further without an excessive cost penalty. What the customer requires is the limit of quality for any product. The customer determines the quality standards.

The sequential process for quality testing a batch of machines is an extension of the double-sampling process, as mentioned earlier. In this process, each sampling process generates two definite data points: the number of items processed and the cumulative number of defectives for all the tests conducted. Based on these two data points, the batch is either rejected or accepted or further qualifying tests. Let us define the following:

p_a = Acceptable quality level (predetermined before test/sampling), AQL

p_l = Lot percent defective tolerance, LTPD

p = Percentage of defectives in the lot

α = Producer's risk probability when lot is rejected even if it is a good lot

β = Consumer's risk probability when lot is accepted even if it is a bad lot

m = Number of machines in a given lot or batch, batch size n = Number of machines selected for trial, n < m, sample size

d = Number of actual defectives in a lot (population defectives)

r = Rejection level

x = number of defectives in a sample

The sequential acceptance process has three zones: the acceptance zone, the rejection zone, the and sampling zone, which is in between the acceptance and rejection zones. These three regions are separated by two straight lines, as shown in Figs. 11.17 and 11.18. The line separating these three zones are linear and proportional to the sample number, n. These lines have the same slopes, and they are parallel to each other, as shown. In the sequential plan, when the cumulative number defects exceed Line 1, the lot is accepted. Similarly, when the cumulative number of defects for the samples tested, as shown in Fig. 11.17, exceed Line 2, the lot is rejected. In the sequential acceptance plan, tests are continued as long as the cumulative number of defects is within Line 1 and Line 2. As shown in Fig. 11.17, the lot is accepted, and Fig. 11.18 displays the case when the lot is rejected. The question is how to draw L1 and L2.

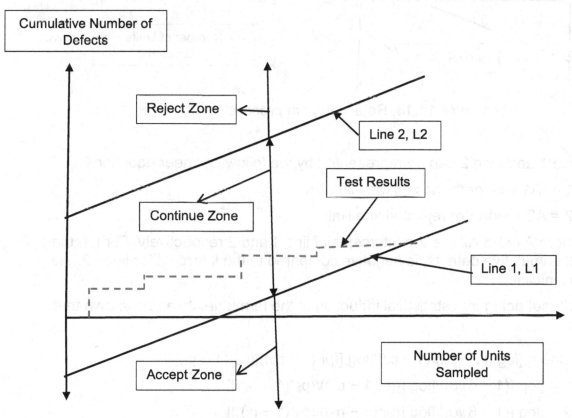

Figure 11.17: Acceptance plan process

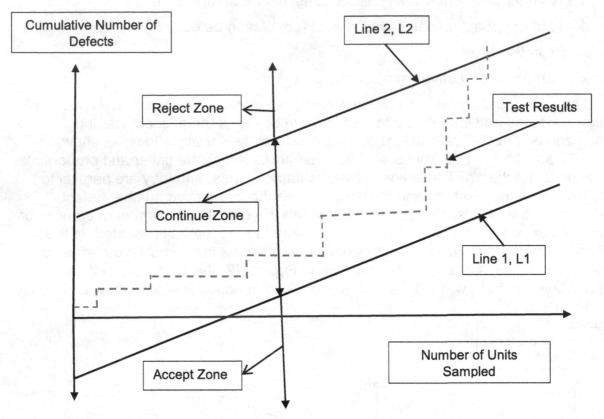

Figure 11.18: Rejection plan process

Line 1 and Line 2 can be represented by the following linear equations:

L1 = -A1 + slope*n; acceptance line limit

L2 = A2 + slope*n; rejection line limit

where A1 and A2 are the intercepts of line 1 and 2 respectively. For L1, the intercept, cumulative defects, is negative compared to the intercept for line L2, the rejection limit line.

Without going into statistical induction of the formulae, it can be shown that

$$\text{Slope} = [\log [(1 - p_a)/(1 - p_l)]/[\log [(p_l*(1 - p_a))/(p_a*(1 - p_l))]]]$$

$$A1 = [\log [(1 - \alpha)/\beta]]/[\log [(p_l*(1 - p_a))/(p_a*(1 - p_l))]]]$$

$$A2 = [\log [(1 - \beta)/\alpha]]/[\log [(p_l*(1 - p_a))/(p_a*(1 - p_l))]]]$$

Hence, in order to draw the limit lines, we have to know the values of the probabilities and the risk factors for consumers and producers. In practice, if the batch number is too low, normally, all the products are qualified individually against the critical requirements.

11.9.4 Average Sample Number (ASN)

The sample size for a sequential sampling process depends on the proportion of defectives in the lot. If the defectives are too many, i.e. bad batches, then the number of samples and tests could be much less, and vice versa. A concept to determine the average number of samples (ASN) curve can be developed when the percent of defectives, p, in the batch is known from legacy data of the manufacturing process. The values of ASN at different values of p are given below:

At p = 0, ASN = A1/slope

At p = 1, ASN = A2/(1 − slope)

At p = p_0, ASN = [(1 − α)*A1 − α*A2]/(slope − p_0)

At p = p_1, ASN= [(1 − β)*A2 − β*A1]/(p_1-slope)

At p = p_2@slope = ASN = A1*A2/(slope*(1 − slope))

With these initial five points, an ASN curve for the product could be obtained as shown in Figure 11.19 below. The ASN curve has a maximum and minimum value depending on the percent defective in the lot size. The maximum value of samples lies in between these two extreme points. Obviously, when the percent of defective in lots is 10%, the average sample number required is minimum. It Is also true that the ASN is a random number and the ASN curve gives an estimated average of the samples, a random variable, for any particular value of p.

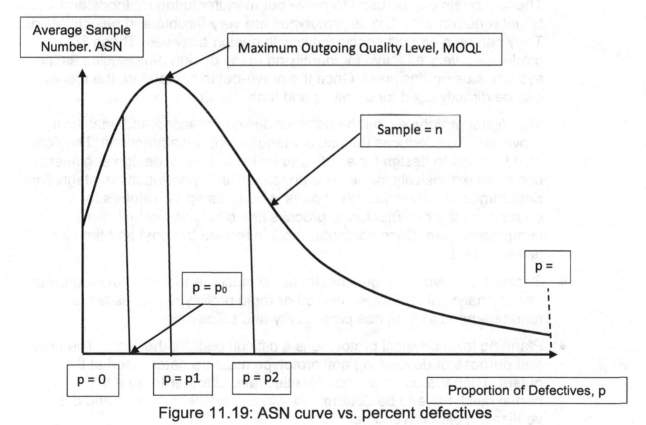

Figure 11.19: ASN curve vs. percent defectives

11.10 Summary

Machine tool development has always needed prototypes for prove-out of the target requirements. A prototype is an approximate replica of the actual production machine. The prototype should have the following characteristics:

- Prototypes can be digital or physical. Digital prototypes are used for theoretical analysis and simulation purposes. Physical prototypes are meant for physical testing to confirm the fact that they satisfy the target requirements. The whole system build is not required if the design is a carryover design with carryover parts. In such cases, prototypes of newly designed sub-systems can be built for testing purposes, and the rest can be done in the theoretical arena. Prototype requirements also depend on the complexity of the product.

- Both digital and physical prototypes can be used for many purposes other than testing and proving out the requirements. Both types of prototypes can be used for showing the product to customers to get their feedback. So, prototypes can be used for coordination, learning, communication, and interaction among team members. They can be the centralized source of information for internal and external customers.

- In recent times, the use of digital and rapid prototypes has gone up substantially due to the fact that they require less cost and time to build. These models can be used to prove out manufacturing methods and target requirements. Digital prototypes are very flexible and easy to build. They can also be built concurrently with design activities. These 3D models are very effective for identifying many design deficiencies as the system is being designed. Once the prove-out is completed, the model can be directly used for detailing and manufacturing.

- The digital prototypes can be used for design iterations and analytical prove-out. This reduces the risk of manufacturing malfunctions. They can also be used to design for safety and looks. Industrial design engineers use them extensively for better ergonomics and man-machine integration. Sometimes the 3D modeling of parts is done using 3D printers to understand the manufacturing process and difficulties in building a complicated part. Such computer models reduce the cost and time for development.

- Because of severe budget constraints in machine tool industries, greater use of analytical prototypes and other rapid prototype processes is recommended to enhance productivity and efficiency.

- Planning for a physical prototype is a difficult task for the team. The goal and purpose of developing any prototype must be determined at the outset. Then the contents, complexities, and whether to do a full build or partial build, have to be determined. Moreover, test methods and design verification plan have to be in place.

- Reliability and robustness must be considered in the product design. Testing the product does not guarantee its success but merely confirms the target requirements as selected by the team. Testing is the last resort to prove out the reliability of a product. Functional performance targets must also be designed into the product, but they should be confirmed by physical testing if required. The targets must be validated and verified wherever necessary to increase confidence in the design.

- The tests must be conducted using a real-world usage cycle and critical noise factors. The test methods and test set-ups must be designed to accept real-world usage profiles and must be capable of measuring critical responses of the system. The test results must be correlated with analytical findings or benchmarked data. The failure modes must be correlated with field failures or analytical failure modes.

- The primary purpose of testing a machine is to enhance the engineering confidence instead of statistical confidence of the machine. The key inputs and noise conditions must be simulated and incorporated in the test to reproduce field failures and fracture modes. A correlated test enhances engineering confidence, which is more important than enhancing statistical confidence. The test quality has to be managed to produce results with minimal cost and time.

- Test execution, methods, and processes must be simple, well-documented, and detailed for floor execution without any problem. The common cause and special cause variation during the test must be controlled and managed so that results are not affected by them. Special cause variations must be eliminated during the test. Test variables must be randomized to get unbiased results. The sample size must be controlled to reduce the cost of test.

- The test should confirm the fact that all the target requirements have been achieved. The concerns and failures during testing must be discussed and resolved before signing off on production. The risk has to be mitigated for unresolved concerns noticed during the test. The test process, results, and conclusions must be documented for future references. Test results must be shared with analytical engineers for their feedback and concerns, and the lessons learned from the test must be documented for future use. The lessons learned should also address reliability and robustness issues, design verification issues and concerns, manufacturing issues and concerns, and correlation issues.

- The chapter also gave some examples of testing plans and the sample number to determine the quality of the lot size. This is an acceptance process plan. The purpose of any acceptance process is to find out if the batch is acceptable based on the test results from a small number of samples.

Single, double, and sequential test plans have been explored. The sequential process plan is a logical extension of the double sequence process plan in which items are tested one at a time and a decision is made whether to reject the lot, accept the lot, or continue with further testing. ASN should also be determined for a sequential process plan.

- Performing an accelerated life test for the product is not very common in the machine tool industry because the product volume does not warrant such tests. Machine tool failures are not that dangerous for the operators and the environment. Such tests are very common in the life prediction for components and assemblies in the auto industry, though.

- To characterize the reliability profile or failure rate of an individual component such as a relay, transformer, bearings, etc., the similar components are tested under similar input conditions. The failure mode and times are noted as the tests continue. The time of failure for similar failure modes must be grouped together and compared for further analysis if required. For example, say, n number of similar bearings are tested. The failure time and modes are also noted as the bearings fail during the test.

- The primary purpose of reliability testing is to make sure that the product conforms to the requirements and, also, to identify any weakness of the design, which should be fixed before the design hits the floor or the product lands in customers' hands. Depending on the type of testing performed on the product at different stages, it can also discover whether the product will perform without any failure under predetermined testing conditions, which should match customer requirements. Basically, testing enhances confidence that the product will perform as intended, performing under reasonable conditions without failures, i.e., product reliability enhancement.

- When the number of samples for the tests is much less than the sample population in use, an obvious question arises: what should the number of samples be so that statistical confidence in the prediction of reliability is very high? One of the objectives for testing is to predict the probability of failure for the components. The statistical confidence for any reliability prediction is equal to or greater than the estimated reliability. So, the challenge is to match the estimated probability from limited sample testing with the population reliability in reality.

11.11 References and Bibliography

Box, G.E.P., Hunter, S.J. and Hunter, W.G., 1978, Statistics for Experimenters: An Introduction to Design, Data Analysis, and Model Building, John Wiley and Sons, NY.

Clausing, D., 1994, *Total Quality Development*, ASME Press, NY.

Duncan, A.J., 1986, *Quality Control and Industrial Statistics*, McGrew-Hill/Irwin, NY.

DeGroot, M.H., 1986, *Probability and Statistics*, Addison-Wesley, Reading, MA.

Ford Design Institute Lecture Notes on Reliability and Robustness, 1998, Ford Motor Company, Dearborn, MI.

Garvin, D.A., 1988, *Managing Quality*, Free Press, NY.

Green, P., and Rao, V.R., 1971, "Conjoint Measurement for Qualifying Judgmental Data," *Journal of Marketing Research* 8, 355–64.

Hatley, D.J., and Pirbhai, I.A., 1998, *Strategies for Real-Time System Specification*, Dorset House, NY.

Kumar, S., and Gupta, Y., 1993, "Statistical Process Control at Motorola's Austin Assembly Plant," *Interfaces* 23(2), 84–93.

Nahmias, S., 2008, *Production and Operations Management*, fourth edition, McGraw-Hill Erwin Companies, NY.

Padhke, M.S., 1989, *Quality Engineering Using Robust Design*, Prentice Hall, Englewood Cliffs, NJ.

Porter, M.E, 1990, *The Competitive Advantage of Nations*, The Free Press, NY.

Rao, S.S., 2014, *Reliability Engineering*, Prentice Hall, NJ.

Ulrich, T.K., and Eppinger, S.D., 2011, *Product Design and Development*, McGraw-Hill Companies, NY.

Wall, M.B., Ulrich, K.T., and Flowers, W.C., 1992, "Evaluating Prototype technologies for Product Design," *Research in Engineering Design* 3(3), 163–177.

Wheelwright, S.C., and Clark, K.B., 1992, *Revolutionizing Product Development: Quantum Leaps in Speed, Efficiency, and Quality*, The Free Press, NY.

11.12 Review Questions

- Can an innovative machine be built and tested completely using a digital process? Do you still need physical prototypes for confirmation of target requirements? Will your opinion change if the machine contains most of the carryover parts?

- Design a verification plan for a CNC turning machine. How would you modify the process if the machine was a very simple one, such as a saw machine? Does the process of developing a verification plan depend on the complexity and content of the machine? Is the verification plan development always necessary?

- What are the primary advantages and disadvantages of building an analytical prototype that looks like a design-intent machine but does not work like a real machine?

- What is the primary use of 3D printing and rapid prototyping? Is it absolutely necessary for the machine tool industry? Does it add to or reduce the cost of the development process? If the machine is not mass produced, does this technology add value to the design and development process for machine tools?

- For a single sampling plan with a sample size of 10, where a given rejection level is also specified as zero, find the analytical expression curve for the ASN curve as a function of probability.

C H A P T E R 12

12 Reliability & Maintainability of Machines

12.1 Introduction

Machine tool manufacturers must have a plan and assessment of their equipment before their machines are installed in customers' workshops for actual production. Sometimes, for special machine tools (SPM), this approach will reduce confusion and bones of contention down the line. The manufacturer and the customer have to come up with this plan together. For most used general-purpose CNC machines, the plan for reliability and maintainability is even more important to gain customer confidence in the products.

A properly designed reliability and maintainability plan will help both users and manufacturers reduce the life-cycle cost for the machine and manage and control equipment failures. This will also help both parties design a preventive maintenance program for the machine to reduce machine tool breakdown time and increase uptime during actual use. Machine availability time has to be increased as much as possible to get the return on investment as quickly as possible. This will reduce the investment cost for the customer and increase customer satisfaction.

The company has to design a plan for this process and has to devise methods to implement it for the equipment. Such a process also helps the manufacturer and user to satisfy the quality programs, such as ISO 14000. The company should have a quality enhancement program for the equipment, and this plan for reliability and maintainability will help the company to maintain and improve the quality programs.

Such a program should outline the goal and objectives of this process. Establishing such a plan will help the manufacturer establish the product among the customer and as a tool to reduce the life-cycle cost for the equipment. This will also help the company and the user define and assess machine tool reliability, failure patterns and rate of failures, cost of operating the machine, mean time to failure (MTBF) and mean time to repair (MTTR), ease of maintenance, machine tool uptime and downtime, effective cost of machining, etc. This plan and process can also help the manufacturer identify the bathtub curve for the equipment and the root cause of failures, such as manufacturing shortcomings, design weaknesses, etc., for the equipment.

The objective of such a plan is to elongate the time for use and reduce the time for infant mortality or failures due to aging. Another primary objective of this process is to establish a standard for the equipment that purchase and quality departments can follow as a guideline to procure the equipment from various manufacturers. This process also helps the company during the procurement process and to keep track of equipment performance for machine life. If the life-cycle cost does not match the predicted values, machine tool designers can also determine the cost of deviation and take steps to improve the design to increase the MTTR and MTBF.

12.2 Reliability and Maintainability (R&M) Process

At the outset, the goals and objectives of the program have to be established for the equipment. Different machines might have a different set of goals and objectives depending on the nature and type of machinery. In general, the goals and objectives of this process could be as follows:

- Optimize life-cycle cost (LCC) of the machine by incorporating proper design and manufacturing methods. LCC is normally commensurate with the complexity and cost of the machine. It has to be optimized to the desired value for the equipment. Life-cycle cost analysis for the machine must be taken up during machine design and manufacturing. Some designs, materials, and assembly might help the company to reduce the life-cycle cost of operation for the machine. Design robustness can reduce repetitive machine failures due to fatigue or other causes. If the manufacturing process or burn-in period is not enough, the infant mortality time could be less than expected, and the process has to be improved to reduce such failures during the initial period of use. The design requirements should have a line item for this index. Eventually, design, analysis, and testing should confirm this requirement.

- The MTTR, mean time to repair, has to be also optimized or reduced to a minimum. In my opinion, if the repair is easier and not too involved, low MTTR can be tolerated. For example, hydraulic valve replacement is much easier and also needs less time to repair. Hence, a lower MTTR can be tolerated, whereas repair of the spindle or indexing unit is very time-consuming, and they are also very complicated to repair. Hence, a very high MTTR is required for such units. Nevertheless, these units or components must have a requirement for MTTR. Normally, design dictates these values for the components. The system could have a different MTTR than components.

- Similarly, the mean time to failure, i.e., the MTTF should also be optimized to the desired level or minimum. Complicated units must have a much higher value of MTTF than the simple components, which are very easy to replace. The more complex the unit is, the higher the MTTF expected from a good system. Again, the system MTTF can be quite different than the component MTTF. Both the MTTR and MTTF might have some cost implications for a system. In any case, the machine must have an acceptable MTTR and MTTF for higher customer satisfaction.

- There should be an established goal for mean time between failures, MTBF, for critical sub-assemblies that affect the cost of production. This is also a critical requirement for the machine. Design and analysis procedures have to ensure the satisfaction of this criterion as the design is developing. Testing of products does not ensure this criterion due to a shortage of testing time.

- The system cost, LCC, MTTR, MTBF, and MTTF are also interconnected for any system. Hence, the system design should have specific requirements for all of these items and must be satisfied accordingly. Since LCC, MTTR, and MTTF provide a good idea about machine quality, the cost is secondary to the satisfaction of these requirements. In the end, the system has to satisfy the cost constraints without sacrificing quality constraints

For purchased components or units, these requirements must be relayed to the supplier so that they can satisfy them while designing their systems. The vendors should satisfy quality requirements by designing, analyzing, and testing their units accordingly. Normally, all machine tool buyers already have these values in their quality system manual, and machine tool suppliers have to satisfy them before the system is purchased. For most customers, the machine tool supplier must demonstrate the design manufacturing and quality control process so that these requirements can be satisfied.

In order to be a preferred supplier of machine tools, the machine manufacturer has to have a quality operating system (QOS) in place, such as QS 9000. The management must be able to demonstrate to vendors the management dedicated to QOS, the procedures for quality, inspection, control, and testing, and the quality enforcement system in place. In other words, the builder has to prove machine tools are built as per the design requirements that will satisfy customer requirements. The builder has to demonstrate the machine's ability and proof of design reviews on a regular basis, as well as capability of design, analysis, and reliability assessment procedures.

The machine builder must have the requirement indices for LCC, MTTF, MTBF, and MTTR. In addition, the machine must be designed for ease of maintenance and safety. The testing and equipment (TE) supplier's responsibility guidelines are also very well described and detailed in the SAE guidelines, "Reliability and Maintainability Guidelines for Manufacturing and Equipment."

SAE guidelines for TE suppliers are as follows:

- Definition of quality objectives, such as statistical process control (SPC), reliability, maintainability, and durability
- Quality control procedures to maintain the QOS
- Guidelines to use DFMEA and PFMEA procedures for design and manufacturing activities
- Procedures to ensure design, analysis, and testing satisfies the design requirements
- Requirement indices, such as LCC, MTBF, MTTF, MTTR, etc., are satisfied
- Inclusion of stress analysis procedures, fault tree analysis procedures, and thermal analysis for the machine design

- Use of predictive tools to confirm reliability data for the machine
- Continuous improvement activities, such as data collection, "plan, do, control, and act" (PDCA) cycle, root-cause analysis, and reliability assessment tools and procedures
- Training program for educational purposes

12.3 SAE Guidelines for Machine Life Cycle

SAE guidelines provide the life cycle phases of any equipment. Following such guidelines is not mandatory for builders or procurement agencies. The machine life cycle consists of five serial and distinct phases:

- Concept Generation Phase
- Product Design and Development Phase
- Manufacturing and Commission Phase
- Run-Off and Operation Support phase
- Decommission Phase

These phases are to be followed by the procurer and builder for machine tool manufacturing and procurement activities. They dictate the processes for procurement, design, build, and quality control. These phases are more suitable for special-purpose machines to suit the production requirements. Also, see Fig.12.1 for the flow of activities for machines. There is an iteration process between the first two phases for product development.

Concept Generation Phase: For custom-built, special-purpose, or modified general-purpose machines, the first phase of concept development is to identify requirements and system development. This phase can be also called a "project proposal phase," where the builder and customer get together to identify the custom requirements required for production cycle optimization or minimization. The development of machine requirements needs feedback from manufacturing engineers, design engineers, customers, service engineers, and quality control engineers.

This phase is signed off on with a clear understating of what the system will do. The selection of a supplier is also identified during this phase of concept development. The builder initiates preliminary research activities, the design action team, and the team leader, who will interact with the procurement team leader. This phase is very critical to avoid future disagreements and discrepancies.

This is also a coordination phase, wherein the teams are formed and procurement process controls are established. Here are the primary activities for this phase:

- Research and data collection
- Proposal development
- Requirements development

- Concept generation
- Team organization
- Team building

Design and Development Phase: In this phase, machines are developed using the initial concepts, requirements, and specifications developed in the previous phase. The requirements and specifications for the product must be frozen before the product design is initiated. The design team develops a machine that will satisfy the plant's requirements. During this phase, in addition to intense design activities, industrial engineers conduct life-cycle cost analysis, which is equal to the total cost of owning the machine for the life of the machine.

The life-cycle cost could include the cost of the machine, the cost of operating the machine, the cost of maintenance and repair, and the cost of commission and decommissioning minus the salvage value. The life-cycle cost is the real valuation of the machine. The design should include ease of maintenance and servicing, safety and ergonomics, ease of operations, ease of loading and unloading, etc. The design team also creates documentation for preventive maintenance requirements for the equipment.

The design has to incorporate all customer requirements into the product so that both the builder and user becomes profitable at the same time. Customer happiness will also ensure the builder's reputation in the long run. During the design phase, several design audits conducted jointly by the builder's team and the buyer's team are also required to make sure that the budget and requirements are aligned. After this stage, the design is signed off on, and the design goes into the manufacturing phase. The audit team could be an independent team whose job is to match the requirements with the designed product. The communication and coordination among team members are very crucial here for the success of the product. The primary activities for this phase are:

- Component design
- Sub-system and system integration
- Design analysis, PFMEA and DFMEA
- Revision Tracking and control
- Communication and coordination
- Feedback management
- Design audits
- Documentation control

Manufacturing and Commissioning/Installation Phase: The primary responsibility in this phase comes from the machine builder and its suppliers. The builder starts manufacturing components and integrates them into sub-assemblies. Then it integrates sub-assemblies into the system to complete the product. In case manufacturing and assembly difficulties arise, the design must be adjusted accordingly to resolve the issues.

Any changes at this stage must be recorded and communicated to the team members inside and outside the builder, and everybody must be in agreement about them. In addition, the manufacturing process or method changes are also to be documented properly for future reference. Design and manufacturing revisions are kept on track by issuing revised drawings and process control documents. Here are the primary activities for this phase:

- Manufacturing process control
- Assembly
- Quality control
- Statistical process control if required
- Process change control
- Coordination and communication

Run-off and Operation Support Phase: For special or modified machinery, the testing is normally done by machining actual production-intent parts after the machine is commissioned or installed in the production line. The machine must be fully operational as based on the requirements for the product. The machine-produced parts must show the machine's capabilities, which are proven by collecting appropriate data and conducting statistical analysis. Data collection and measurements must be done using production-intent gauges and measuring instruments under proper conditions and environments. The data must be recorded, and failure or discrepancies must be recorded and communicated to the team members. Success or failure must be recorded and analyzed properly without any bias, and corrective steps must be implemented before the machine is used for final, continuous production. Here are the primary activities for this phase:

- Prepare the machine for run-off
- Decide the production-intent parts
- Deploy the measuring systems
- Identify the production requirements
- Measure data and analyze
- Publish the results
- Establish and document the procedures and findings
- Incorporate the process changes if required

Decommissioning/Salvage Phase: This is the phase of the machine life cycle. The machine, by now, has reached the end of its design-intent life and could be either scrapped or modified/refurbished for further use. This is called the "reconditioning phase" of the machine. The proper reconditioning of the machine will convert the machine into a production-capable machine with the required accuracies. Economic analysis can be done to find out which is more economical: sale/scrap the machine or recondition it. The return on investment has to be maximized. This is the end of the bathtub curve, where failures are due to wear-outs and component lives are outside the designed life.

The cost of maintaining the product at this stage could be substantially higher than the regular maintenance cost. Failure occurs unexpectedly and hampers production at random. A failure analysis must be performed to understand the failure patterns. The decommissioning of the machine can be based upon the failure and economic analysis, which will depend on the data collected on the failure pattern over time of use.

If the reconditioning is warranted based on the failure track analysis of the machine, the team has to find what design changes are required so that the machine performs almost like new again. Whether the machine is salvaged or reconditioned, previous failure or performance data is required to understand the performance degradation pattern over time. Without the data, unnecessary changes will be done to the components or sub-system. Reliability tracking is also required for the machine. The MTTF and MTBF must be calculated and compared with design intents.

The decision for reconditioning or updating of the machine also depends on the availability of spare parts and the cost of the parts or sub-assemblies. If the cost of replacement is exorbitant, the machine might have to be salvaged or scrapped depending on the end condition of the machine. Sometimes the newly available sub-systems might not integrate well with other existing sub-systems due to design changes made in the new sub-assemblies.

So, proper logic has to be applied before it is decided if the machine should be rebuilt or not. More often than not, reconditioning might not bring back the desired performance for the machine. Moreover, production requirements might have also changed to the extent that the machine is incapable of reproducing the new requirements even if it is reconditioned in the best possible way. Here are the primary activities for this phase:

- Failure report tracking and analysis
- Reliability analysis
- Economic analysis
- Spare cost data and analysis
- Recondition or scrap decision
- Part or sub-system availability

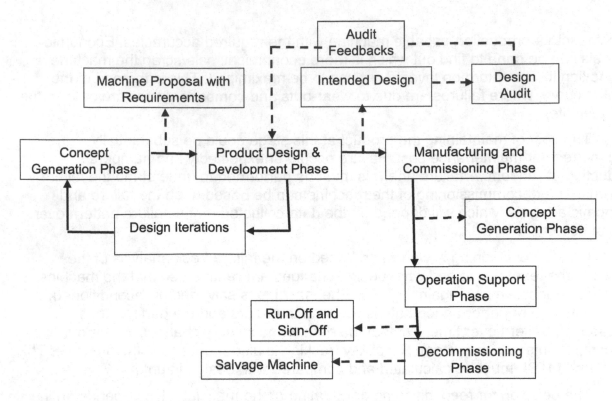

Figure 12.1 Machine life cycle activities flow diagram

12.4 Engineering Tools for Reliability Analysis

The following analytical procedures can be used to enhance the reliability and maintainability of machine tools:

- **Machine Failure Modes and Effects Analysis:** These tools are used to identify the root cause of any failure that is encountered during testing and run-off modes. The root causes for safety issues and performance parameters can be found and eliminated by installing corrective actions. The FMEA process is initiated in case of occurrence of any failure during the equipment life cycle and is used to diagnose the failure mode and identify corrective action. The machinery FMEA, Design FMEA, and Process FMEA can be used to resolve issues regarding machine failures.

- **Thermal Analysis for Electrical Components and Control Elements:** Several analyses, such as heat dissipation rate, cooling and heating rates, and heat generation rate, are performed for the electrical components. Stress analysis for the electrical elements due to heat is also conducted to prevent abrupt failure due to thermal issues. The electrical components are analyzed for their voltage rating, current rating, and power requirements. The control cabinets are also analyzed for heat dissipation rate and cooling requirements. The analysis is performed under severe external and internal environmental conditions.

- **Stress Analysis, Thermal Analysis, and Resonance Analysis for Mechanical Components**: For most of the critical mechanical components, finite element stress analysis and vibration effect analysis using external and internal loading and temperatures are performed to find the durability and static strengths of the component. The analysis will include material properties, real-world usage loading conditions, thermal inputs, and boundary conditions to determine the stress under usage. For some components, such as tool blocks and spindle bearings, thermal analysis is a must.

- **Industrial Engineering Applications:** The industrial engineers take care of the safety, ergonomics, aesthetics, component integration, cost optimization, and maintainability issues of the machine. They coordinate these activities, working simultaneously with the design and manufacturing teams, while the machine is being designed and manufactured. One of their primary functions is to address maintainability issues along with the reliability and test engineers.

  For a machine to be highly maintainable, it should have some specific characteristics, such as ease of ingress and digress, ease maintenance, less preventive or running maintenance schedules, and definitive failure mode correlations. The loading and unloading of the parts must be less strenuous for the operators. Moreover, the design should promote safety features so that the operator can avoid health injuries in case of accidents. For maintenance, components with a high level of failure, such as hydraulic components, must be in easily accessible areas of the machine. The maintenance schedules should be clearly documented and laid out so that technicians and maintenance personnel can easily diagnose the failures and resolve the issues in the minimum possible time. Maintenance manuals should be written with graphics and explanations so that the maintenance technicians can easily understand and implement the necessary steps in case of emergencies. The machine must have an optimized preventive maintenance plan laid out.

- **Reliability, Safety, and Durability Testing:** The machine has to be tested for reliability and durability assessments. The verification plan has to be designed specifically for the machine. The testing plan should adequately determine the reliability and durability requirements. The mechanical, electrical, and software reliability have to be assessed and reported to the buyer for acceptance. All the failures during testing must be documented and reported for further action if necessary. The test plan must include extreme environmental conditions and safety hazards. The environmental conditions must include temperature, humidity, dust, sympathetic vibrations, heat cycling, etc. The test results and failures must be analyzed thoroughly to prevent further failures of the same kind in the future. The test conditions must represent and simulate the environmental and working conditions of the plant. The test results must support the design requirements for the machine.

The deviations must be reported and resolved before the machine is put in operation.

At the end of the machine design, manufacturing, and testing cycle, the following documents must be provided to the buyer for their approval:

- Production run-off data
- Test sequence and results
- Test failure data and deviations data
- Reliability and durability assessment results
- Failure modes and resolution procedures data
- Maintenance schedules
- Safety and hazards test results

12.4.1 Bathtub Curve and Reliability

As explained before, a bathtub curve has three distinct zones for failure rates vs. time of operation. The complete curve is divided into three zones: infant mortality, or burn-in, period; useful life period; and wear-out period, as shown below.

Failure Rate, h(t)

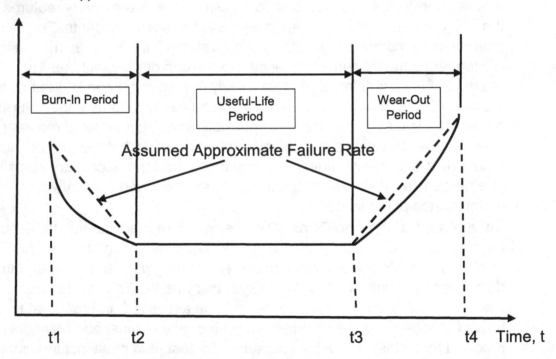

Figure 12.2 Bathtub curve

The Infant mortality, or burn-in, period of the machine consists of failures that occur during a very early period of use, for example, during run-off, initial testing, or installation. Electrical components fail mostly during this period. Failures during this phase might be attributed to the following primary causes:

- Manufacturing defects, such as poor assembly, bolting, or welding connection
- Manufacturing procedures not compliant with design instructions
- Material defects
- Poor installation
- Poor quality inspection
- Improper workholding set-ups
- Electrical component failures
- Wrong machine setting

During the useful life period of the machine, the machine is supposed to operate without much interruption. This is the design-intent life of the machine. Most of the reliability assessments are done during this period. The failures are at random during this phase. During this period, the mean time between failures decreases as the machine ages. In case of a failure, the machine has to be repaired to bring the machine to operable condition. The failures during this period might be due to:

- Excessive load conditions
- High-stress conditions due to external loads
- Improper design considerations
- Abnormal use or overuse
- Misuse of the machine by the operators
- Load and material stress interferences
- Poor preventive maintenance or running maintenance
- Improper part replacement

During the wear-out phase of the machine's life cycle, failures increase substantially over a short period of time. Mostly mechanical components fail during this period due to insufficient life considerations of the design. The machine is almost at the end of its life, and it has to reconditioned for further use. The machine becomes very unreliable and the cost and time of repair will increase substantially. During this period, the team has to decide whether to buy new equipment or recondition the machine. The failures during this phase might be due to:

- Machine beyond its design life
- Machine fatigue due to continuous cycling
- Thermal distortion
- Bearing seizures
- Insufficient design failures
- Material fatigue failures
- Material strength failures
- Load increase beyond design load
- Extreme environmental conditions

12.4.2 Repair or Replace Strategy

A common question is when to procure or replace an asset: when the repair is very frequent or when repair becomes too costly or time-consuming. When the machine is in the wear-out phase of the bathtub curve, this question must be answered. The machine must be in its useful rate life phase, and the failure must not exhibit an exponential rate. The replacement might be justified if the failure takes a lot of time and money to repair or maintain. The answer also depends on the criticality of the machine in the production line. If the machine is not replaceable or available for replacement, the machine has to be maintained without any choice.

This particular analysis does not include the uncertainty of failures during the useful life period or wear-out period. The minimization of cost is the ultimate objective of this approximate analysis. It is assumed that maintenance cost and time for repair increases as the machine gets older and older. This simple analysis is very much applicable to the production machines working continuously or intermittently as long as the hours of operations are kept on track. If the machine needs major overhauling, the reconditioned or brand-new machine will have a similar level of accuracy and performance. The assumptions for the analysis are as follows:

- Failure probabilities are not considered

- Machine operates continuously at the desired level of precision

- No limit on planning time

- New or reconditioned machine is equivalent to the machine to be repaired

- Maintenance, reconditioning, or replacement cost will be considered

- The cost objective function is convex and has to be optimized or minimized

- The cost of maintaining the old machine is proportional to its age in operation

- The machine replacement or reconditioning cost includes decommission and installation costs, i.e., the total cost of replacement

Let's assume the following:

- Replacement cost of the machine including installment and procurement costs = R

- Salvage value of the old machine = S

- Decision variable is the time when the machine is replaced = t*

- Instantaneous cost of operating the machine is proportional to the age of the machine. So, cost = constant rate*age = mu

The objective of this analysis is to determine the time that minimizes the total cost of maintenance and replacement. It is also assumed that the same machine can be replaced as many times as the company wants. Total maintenance cost is shown in Fig 12.3, and the cost curve is shown in Fig. 12.4.

The cycle is between the start and replacement of the machine. The cost has to be determined within one cycle.

Total replacement or reconditioning cost per cycle = (R – S)

Total maintenance cost per cycle. MC = ∫(m*t) dt = (m*t^2)/2 = ½*m*t^2

Average Cost per unit time = C(t) = MC/t = (1/t) *((R – S) + 1/2*m* t^2))

$$C(t) = (R-S)/t + ½*m*t = RS/t + ½*m*t$$

Since the double derivative of the function C(t) is positive, the cost function is convex, with a local minimum with respect to time. The objective is to find the minimum value of time, t, that minimizes the cost function, C(t). Setting the first derivative of the cost function to zero, we have

(-RS/t^2) + m/2 = 0

Optimum time = t* = SQRT (2*RS/m)

For example, a machine is producing a part continuously, and the maintenance cost for the machine is 300 times the age of the machine. So, the maintenance cost for the first year is 0.5*300*1 = $150. Maintenance cost in the second year is 0.5*300*4 = $600. The salvage cost of the machine can be estimated to be $1,000. A new machine or reconditioning cost is $12,000. So, RS = $12000 – $1000 = $11000. The optimum time for the replacement or reconditioning of the machine will be equal to SQRT (2*11000/300) = 8.5 years. So, the machine should be reconditioned or replaced with a new machine at the end of eight years six months.

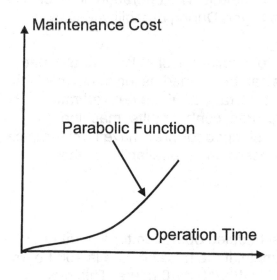

Figure 12.3 Bathtub curve

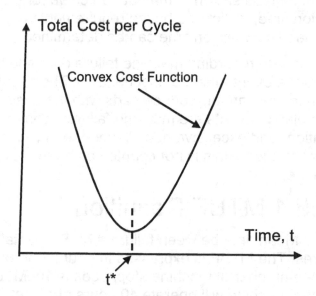

Figure 12.4 Bathtub curve

12.5 Reliability Assessments for Machines

For machine tools, reliability can be defined as the probability of the equipment performing at the design-intent level without interruption or failure for a selected amount of time when the machine is subjected to the designed operating conditions and environments. When reliability is higher, machine downtime is less, production time is up, and the cost of machining is less.

Reliability, by definition, is a statistical measure that represents the ability of the machine to perform without any disruption. It can be measured as the time of operation divided by the available time for operation and is represented by a percentage. For example, if the reliability of a machine is 90%, the probability of the machine operating within a given time is 90%, i.e., there is 90% chance that the machine will operate successfully within the stipulated time. Hence, for machine tool reliability, the following points are to be kept in mind:

- Designed-in time of operation
- Number of disruptions or failures in this time period
- Operational environment
- Machine environments
- Time under downtime

The stipulated time of operation is also called "mission time," during which the machine is subjected to operation. The mission time for the equipment is a design-intent time for the machine, and the machine has to operate in this time without any failures to meet production requirements. The reliability is normally measured when the failure rate is constant, as shown in the bathtub curve for the equipment. The operation time, or mission time, is dictated by the requirements of the buyer. During durability testing of this machine, mission time can be determined.

While recording machine failure data and time of failure, it should be noted that the failure event must be unscheduled. The events can be defined as incidents in which the machine cannot produce parts with the desired accuracy or at the desired rate. Examples of events are machine failures, control failures, control faults, machine vibration, and excessive heat. When the machine is stopped for preventive maintenance or tool changes, this is not counted as an event for assessing the reliability of the machine.

12.5.1 MTBF Definition

Mean time between failure = MTBF = total scheduled operation time/number of failures. The MTBF is expressed in hours. For example, if the mission time is 400 hours and the number of machine stoppages is 10, MTBF is 400/10 = 40 hours. This means that the machine will operate 40 hours at a stretch before experiencing a failure mode. This is an average figure for failures. In other words, the machine will produce at the desired rate and accuracy for 40 hours on average.

12.5.2 Failure Rate, h(t)

The failure rate of the machine is the inverse value of MTBF i.e. machine Failure rate = 1/MTBF = 1/40 = 0.025 failures per hour i.e. the number of failures in a given time in hours. This is the probability of a failure within the unit of time of hours. This means that there is a 2.5% chance or probability that a failure will occur within every hour of operation.

If there are the same elements, such as relays, being used for all of the 1000 machines in a plant, population of relays = 1000. The MTBF of the individual relay has been found to 25,000. Population failure rate = Failure Rate x Number of Relays = (1/25000) x 1000 = 0.04 failures per hour. This means that 4 relays will fail in the plant every 100 hours of operation.

12.5.3 Mean Time to Repair, MTTR

The maintainability of any machine is basically a design function to ease the maintenance of the machine and reduce its cost as well. Maintainability is also expressed as the probability of restoring the machine within a stipulated maintenance time to its original working conditions after it has a failure during usage. The repair procedures must be well defined and followed by the technicians to restore the machine. MTTR also relates to the cost of ownership for the machine, i.e., the higher the MTTR, the higher the cost of ownership for the machine.

The average time to restore a machine to its original working condition is called "mean time to repair," i.e. MTTR. The machine has to be designed so that MTTR is as small as possible to reduce the life-cycle cost of the machine and increase its availability for production. In order to reduce the MTTR for a complicated machine, critical sub-assemblies are kept and replaced as a unit, and then this failed unit is repaired for future use. Such replacements can also be done during preventive maintenance for the machine if the failure trend for the unit is known.

The mean time to repair might consist of several procedures, such as failure reporting and recording, resource allocation, failure diagnosis and resolution method, parts location, actual repair as per procedure, and tryout. All of these need to be done in a very efficient and productive manner so that downtime is as short as possible and uptime as long as possible. Sometimes MTTR is calculated for one type of failure only, but in my opinion, it is better to calculate this index using all the failures of the machine during a window of operation.

MTTR = summation of repair times/number of failures

If there are 10 failures in 1,000 hours of operation and total time for repair time is 100 hours, MTTR = 100/10 = 10 hours per failure.

12.5.4 Machine Availability

Machine availability time is defined as the percentage of time for operation over the design-intent time. So, when the number of failures and time to repair the machine is less, machine availability for production, or uptime, goes up.

One of the primary objectives of preventive maintenance is to maximize machine availability time. The repair is done as per the schedule before actual failure occurs during production hours. The machine design dictates machine availability time, which also depends on breakdown time, repair time, set-up time, tryout time, etc. The machine is considered available for production even if it is standing idle due to, say, shortage of materials.

Machine availability = MTBF/ (MTBF + MTTR), so if the machine has MTBF of 100 hours and MTTR of 5 hours, machine availability is equal to 100/105 = 0.95, or 95%.

12.5.5 Overall Equipment Effectiveness (OEE)

The SAE defines the OEE as a combination of three factors: machine availability, machine performance rating, and part quality rating. Machine performance efficiency depends on the speed and power availability of the machine, i.e., the chip removal rate of the machine. Part quality measures the ability of the machine to produce parts as per the blueprint requirements. During the evaluation of this criterion for the machine, the time for operation time, non-operating time, planned maintenance time, and unplanned downtime should be considered.

Machine Availability %, X% = ((net available time – machine Downtime)/net available time) x 100%; this assumes the inputs are available for the machine to produce parts. Machine downtime includes both planned and unplanned maintenance time, including set-up time. Say this value is 80%.

Machine performance rating %, Y% = idle cycle time per part cycle/(total part produced/operating Time) x 100%. Say this value is 60%.

Part quality rating %, Z% = ((total part produced – total defective parts)/total Part produced) x 100%. Say this value is 90%.

Overall equipment effectiveness (OEE) % = X% x Y% x Z% = 0.8 x 0.6 x 0.9 = 0.432 = 43.2%, i.e., overall, the machine is 43% effective.

12.5.6 Determination of Failure Rate, h(t), for Machine Components

The reliability of a machine can also be defined as

$R(t) = 1 - N_f(t)/N$, where N_f is the number of failures in time, t, and N is the number of parts tested.

$N = N_f(t) + N_s(t)$, where $N_s(t)$ is the number of parts that did not fail during the test duration.

The failure rate, h(t), or hazard function can be defined as below:

$h(t) = (1/ N_s(t)) \times dN_f(t)/dt$, i.e., instantaneous rate of failure

$= (N_s(t) - N_s (t + \Delta t))/(N_s(t) \times \Delta t)$

The failure rate or hazard rate of a component in the time interval (t1, t2) can be defined as the probability of failure per unit time at time t1 given that the component has not failed until t1.

$$h(t1) = (R(t1) - R(t2))/(R(t1) \times (t2 - t1))$$

Let us estimate a failure rate from test data as described below (Table 12.5):

-Total number of components tested under similar test input cycle, N =12,000

-Total duration of test = 2,000 hours

-Test observation is done every 100 hours,

- Calculations:

> Column 1: Observation time, hours

> Column 2: Number of components running at the observation time, Ns(t)

> Column 3: Instantaneous reliability, R(t) = Ns(t)/ N = Col 2/12,000

> Column 4: Rate of change of reliability = $-\Delta R/\Delta t$ = -(R (1) – R (0))/100

> Column 5: Failure Rate, h(t) = $(-\Delta R/\Delta t)/R(t)$

- The similar components are tested against a specific performance requirement. The test conditions, inputs, measurement methods, and output recordings are the same for this test. The test for all components should be started at the same time and should last until failure for each component. The resultant data table and the corresponding failure rate vs. time graph are shown in Table 12.5 and Fig. 12.6 respectively.

12.5.7 Approximate Bathtub Curve Analysis

If the curves for infant mortality and the wear-out phase are approximated by straight lines, the bathtub curves can be represented by an approximate analysis, as shown below. Let us divide the bathtub curve into three zones: burn-in period (t1 to t2), useful life period (t2 to t3), and wear-out period (t3 to t4), as shown in Fig.12.2.

Burn-In Period

The failure rate caused by manufacturing defects during this phase can be represented by a straight line (linear relationship): $h(t) = a_1 t + a_2$, where a_1 and a_2 are constants. Taking the endpoints of the assumed straight line, the constants can be determined. Assuming λ_0 and λ (failure rate for useful life period, constant) and plugging these values in the straight-line equation, we have

$$a_1 = (\lambda_0 - \lambda)/(t1 - t2) \text{ and } a_2 = (\lambda * t1 - \lambda_0 * t2)/ (t1 - t2)$$

$$f(t) = \text{Failure density probability function} = (a_1 t + a_2)*\exp (-a_1 t^2 - a_2 t)$$

$$R (t) = \text{Reliability function} = f(t)/h(t) = \exp (-a_1 t^2 - a_2 t)$$

where a_1 and a_2 are assumed constants, f(t) is the probability density function, and R(t) is the reliability for this zone.

Time of observation	Number of Components Running	Reliability, R(t)	Rate of Change of R(t)	Failure Rate, h(t)
0	N=12000	R(0)=1.00	NA	NA
100	10656	R(1)=0.89	0.00112	1.26E-03
200	9960	R(2)=0.83	0.00058	6.99E-04
300	9501.6	0.79	0.000382	4.82E-04
400	9102	0.76	0.000333	4.39E-04
500	8728.8	0.73	0.000311	4.28E-04
600	8361.6	0.70	0.000306	4.39E-04
700	8001.6	0.67	0.0003	4.50E-04
800	7650	0.64	0.000293	4.60E-04
900	7305.6	0.61	0.000287	4.71E-04
1000	6969.6	0.58	0.00028	4.82E-04
1100	6642	0.55	0.000273	4.93E-04
1200	6322.8	0.53	0.000266	5.05E-04
1300	6013.2	0.50	0.000258	5.15E-04
1400	5712	0.48	0.000251	5.27E-04
1500	5420.4	0.45	0.000243	5.38E-04
1600	5084.4	0.42	0.00028	6.61E-04
1700	4636.8	0.39	0.000373	9.65E-04
1800	4075.2	0.34	0.000468	1.38E-03
1900	3382.8	0.28	0.000577	2.05E-03
2000	2662.8	0.22	0.0006	2.70E-03

Table 12.5 Failure rate calculation

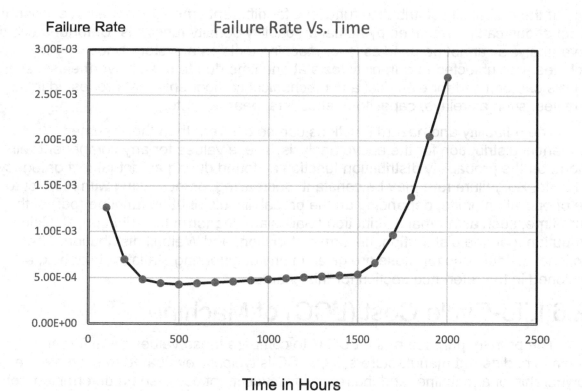

Failure Rate   Failure Rate Vs. Time

Time in Hours

Figure 12.5 Failure rate vs. time

Useful Life Period

Let us consider the useful life period next. The time of operation is in between t2 and t3. During this period, failures occur at random, and they are not due to design shortcomings or deterioration of material strengths for the components due to fatigue, etc. This period is considered to be a design-intent variable. The average failure rate for this period can be assumed to be λ, i.e., the failure h(t) = λ failures per unit time. The failure rate probability density function = f(t) = h(t) * R(t) = h(t)*(1–F(t)), where F(t) is the cumulative failure distribution function over time.

f(t) = probability density function = λ*exp(-λ), assuming an exponential distribution for failures. It does not depend on time t2 and t3; random failures

R(t) = reliability function of the component = exp (-λ * t) = a decreasing function with respect to time of operation

Wear-Out Period

Assuming a linear relationship line in place of a non-linear curve, the failure rate can be represented by an equation of the form h(t) = a₃t + a₄. The average failure rates at time t3 and t4 are λ and λ_max respectively. We have from the assumed linear equation

a₃ = (λ_max − λ)/(t3 − t4)

a₃ = (λ* t4 − λ_max*t3)/(t3 − t4)

If the probability distribution functions for different time zones are not exponential, the equations can be replaced by actual probability density functions. Unfortunately, the above analysis cannot be used as is for machine tools as a system since very few machines are subjected to durability tests at one time due to cost. Nevertheless, such analysis can be used for electrical and mechanical components that are tested in large quantities, such as relays, capacitors, encoders, bearings, etc.

The reliability and hazard functions can be different than the assumed exponential distribution for the above analysis. These values for any component will depend on the probability distribution function as found during an actual test or legacy field tests. So, failure rate may be constant, decreasing, or increasing with respect to time of operation or life, depending on the probability distribution function (pdf) of the failure time, such as normal distribution (Gaussian), lognormal distribution, Rayleigh distribution, gamma distribution, uniform distribution, and Weibull distribution. The functional distributions for these are given in any engineering statistics textbook, as mentioned in the reference section for this chapter.

12.6 Life-Cycle Cost (LCC) of Machine

The primary purpose of the LCC is to compare these values for different machines of different manufacturers. The LCC is typically evaluated to estimate the cost of ownership of a machine, and the return on investment can also be determined from these values, as can the usage life for a typical machine. Performance and cost considerations are equally important and relevant to the selection of machine tools. It is also a very important analysis to estimate the cost benefit of a machine, as well as the design features, estimated life projections, and performance criterion.

Starting from concept generation to the end of life of a machine, there are five phases, as per SAE (Society of Automotive Engineers) recommendation:

- Concept generation using the requirements phase

- Design and analysis phase

- Build and installation/commissioning phase

- Operation and support phase

- Salvage/decommissioning phase

The costs incurred during the first three phases are somewhat-fixed, one-time costs, and the costs during the last two phases could be called "recurring" or "variable" costs. For LCC evaluation, the SAE recommends considering both fixed and recurring costs. The total LCC model includes the following:

- Acquisition costs

- Operating costs

- Scheduled or preventive maintenance costs

- Running maintenance costs

- Unscheduled maintenance costs

- Decommissioning costs

The acquisition cost might include items like purchase price, leasing cost, administrative costs, engineering costs, installation cost, technical training cost, line adjustment costs, shipping and receiving cost, etc. It should include any line item that is required for putting the machine into production.

The operating cost includes direct labor, utilities and consumables, chip handling, part handling, material handling, production loss due to breakdown and maintenance, spare parts storage, acquisition, etc.

The machine maintenance cost includes the cost incurred for both running the machine and preventive maintenance. The preventive maintenance cost includes the cost of spare parts, labor cost per hour for the maintenance, fixed engineering and administrative costs, etc. The running repair cost includes items like breakdown cost per hour, lost production cost, cost of repair, labor cost per hour, spare parts cost, etc.

The decommissioning cost should include items like machine removal cost, shipping cost, salvage cost, dumping cost, de-installation cost, etc.

So, the LCC of any machine is represented by the following equation:

- LCC = acquisition cost (C1) + operating cost (C2) + maintenance cost (C3)+ decommissioning cost (C4) – salvage value (S)

- The LCCs for all the machines under consideration have to be compared. The purchase price of a machine from one vendor could be higher, but the operating cost of the machine could be lower than other machines from different suppliers. Hence, the combined cost is more important than the single-item cost for any machine.

12.7 Mechanical Failure Modes

In order to identify and resolve mechanical component failures due to overload conditions that cause higher static or dynamic stress levels in the component, theoretical stress analysis and failure testing are conducted. Stress analysis is performed under given loading conditions and thermal inputs. A common method of identifying stress theoretically is finite element analysis. Mechanical failures are related to the level of stress, and the reliability of components is also related to the level of stress under operating conditions. The failure rate vs. time for different stress levels of a mechanical component is shown in Fig. 12.6.

When the operating level of stress exceeds the component material strength, the reliability of the component is affected. It can be said that on an average, the higher the stress, the lower the reliability. In order to reduce the chance of failure due to overstress during usage, a margin/safety factor is introduced in every critical design. Since the variation of actual loading conditions in the customer's shop is almost always unpredictable when designing the component, the concept of safety factor is introduced.

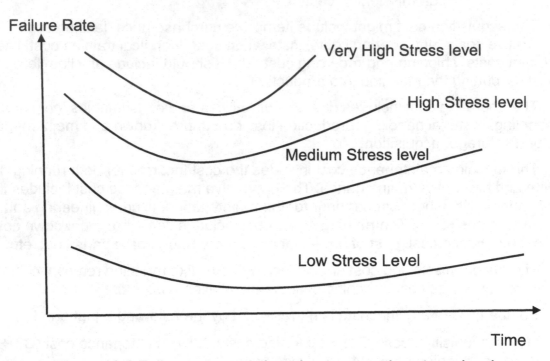

Figure 12.6 Failure rate over time due to operating stress level

Moreover, the material strength also has variability, so there are two variables to be considered: strength variability and load or stress variability, as shown below. If the stress is below the critical level, the failure rate is different than when the component works under a higher level of stress. While designing a component, both the material stress distribution and operating stress due to load variations must be considered, as shown in Fig.12.7

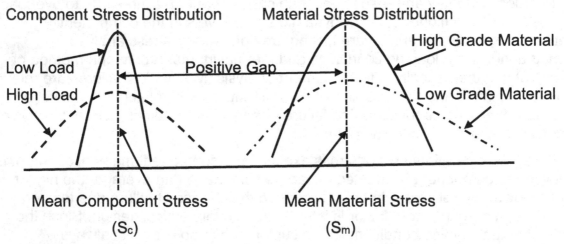

Margin of safety = mean material stress/mean component stress due to load

= S_m / S_c, which should be greater than 1 where $S_m > S_c$

Figure 12.7 Stress-strength relationship for normal distribution

The failure could be due to higher ultimate stress, rupture stress due to excessive external loading conditions, or lower endurance strength for the material. For example, if the cutting load is too excessive due to a harder material or blunt tool, the load stress can become higher than the allowable material strength or stress. The component will eventually fail under such conditions. Also, impact conditions due to loading or unloading of parts could also cause static failures in the workholding devices. Wear-out items such as belts, chains, oil seals, O-rings, etc. could fail due to aging or excessive load or friction. The normal distribution of the load stress and the material strength or material allowable stress is shown in Fig. 12.7.

When the stress due to loading conditions on a component exceeds the allowable stress for the material of the component, a failure mode dictates the failure of the material. As shown in Fig.12.8, where the material strengths of stress due to load interact with one another, a failure mechanism will appear in the interference zone. Even if the mean material strength is higher than the mean load stress (designed-in condition) due to variation of load or aging of material, these two curves might intersect each other to create failure when the material strength is lower than load stress inside the interference zone.

In other words, there is a higher probability that load stress will exceed the material strength at random, and hence, the chance of failure also becomes higher. Under loading conditions inside the rectangular area of interference, the reliability of the component will be much lower, as expected. To avoid such conditions, the S_m has to be much higher than the S_c to avoid interference, i.e., the factor of safety has to be higher. The probability of such interference has to be avoided during the design and selection of materials for the component. Unless the distribution for load and material strengths are well known in advance of design, the factor of safety has to be increased to enhance the reliability of the component

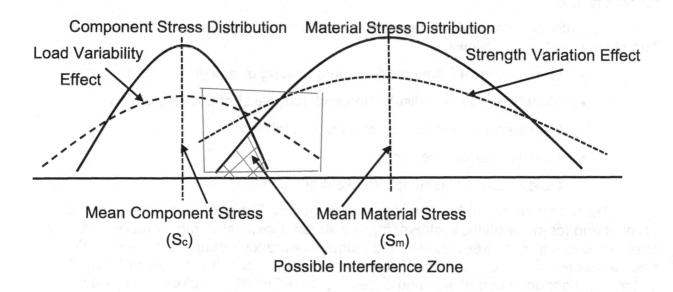

Figure 12.8 Stress-strength interference

Let's take an example for calculating the reliability for a component in the interference zone.

Mean strength of material = μ_m = 30,000 psi

Standard deviation of material strength = σ_m = 3,000 psi

Mean stress due to load = μ_L = 25,000 psi

Standard deviation of load stress = σ_L = 2,800 psi

Normal Variate = $Z = (\mu_s - \mu_L)/ ((\sigma_L^2 + \sigma_m^2))^{0.5}$

$$= (30,000 - 25,000)/ ((3,000^2 + 2,800^2))^{0.5}$$

$$= 5000/4103.65$$

$$= 1.22$$

From a normal distribution, normal variate table in any statistics book, the probability of the interference = 0.1112, i.e., the area of the curve on the left side of this Z value. Hence, the reliability =R = 1 – 0.1112 = 0.8888, i.e., 88.88 or 89%. This means that the chance of failure is 11%. Thus, to enhance the reliability of the component, the load stress variation or load variation has to be minimized since the material is already selected for the design.

The moral of the story is that the designer has to understand that both the load and material strengths have variability associated with them. The interference has to be managed by design only. In other words, both the load stress and material strengths are not constant but are subject to inherent variability. The material strength variability has to be determined by testing only, and the load variation comes from the real-world usage profile or load duty cycles. In most cases, material variability has to be controlled since load variation is hard to control in reality. Failures in the interference zone are random failures.

The interference analysis can be done during the design stage of the machine. The steps could be as follows:

- Identify critical failure modes during testing or analysis
- Identify usage load distribution and material strength distribution
- With same scale and unit, plot both distributions
- Identify interference zone
- Guide design accordingly to avoid any interference

There can be several mechanical failure modes. The failures can be abrupt or gradual and for lower ultimate stress, rupture stress, tensile strength, compressive strength, shear stress, creep stress, and fatigue endurance strength for the material. Failure modes can be due to shear failure, tensile failure, bending failure or thermal distortion, resonance conditions, and excessive vibration. The gradual deterioration of performance is less forgiving than abrupt failures, which can cause safety hazards for the operator and working personnel in the plant.

In order to prevent premature failure and interruption of regular production, the machine has to be under a preventive maintenance schedule, which will also enhance the reliability of the machine.

12.8 Design for Maintainability

Machine maintenance is a critical consideration for selecting a machine tool for a continuous production environment. The design must minimize the frequency of machine failures and increase the time between successive preventive maintenance. The design must also consider accessibility, interchangeability and commonality of critical components, modularity, and standardization to reduce machine downtime. Spare parts should be easily available for critical components. In order to facilitate easy maintenance, a fault diagnostics document must be available, and proper training of technicians is also necessary for complex machinery.

For proper maintainability, accessibility, diagnostics, training, and repairing tools, alignment procedures must be thought out during the design process. For each sub-system, such as hydraulics, controls, electrical, pneumatics, servo drives, workholding devices, and loading and unloading fixtures, properly written documents must be available for technicians for quick repairs. There are several benefits for proper design promoting easy maintainability, such as reduction of mean time to repair and between failures, reduction of spare parts use and availability, elimination of troubleshooting time, reduction of downtime for incorrect repair, increased machine availability for production, lower maintenance cost, reduction of spare parts inventory, etc.

The design for reliability and maintainability should include several review meetings to discuss general design approach and methodology, filed test reports, laboratory test results, machine legacy data and history, design analysis, fault tree analysis, machinery FMEA and conformance with customer design guidelines and requirements, part suppliers and availability, cost conformance, etc. The purpose of any design review is to confirm that the design completely satisfies customer requirements. The relevant issues during the design meetings are as follows:

- Machine safety and ergonomics
- Machine monitoring and diagnosis
- Machine control safety features
- Machine test results and deviations
- Technician and operator training
- Machine FMEA issues
- Measurement methods and devices
- Part and tool loading and unloading procedures
- Maintenance procedures and guidelines
- Part quality Issues and test run-off results
- Spare parts control and availability

12.8.1 Reliability Modeling Examples, Series and Parallel Systems

The machine sub-assemblies can be modeled to determine system reliability while machining parts. Reliability modeling helps to identify the weakest link in components that need to be redesigned to enhance system reliability. This step also helps the team make the machine more productive and reduce costly downtime.

Series System: For example, for a long and slender part (bar part), the operation requires the spindle system, index system, slide system, and tailstock to cut a part. This is called a "system in series," since if any component fails during machining, the part cannot be machined at all. For sub-systems in series, we can evaluate the reliability of the system as shown below:

MTBF of spindle = 1,000 hours = M1

MTBF of tailstock = 1,200 hours = M2

MTBF of indexing unit = 850 hours = M3

MTBF of control elements = 400 hours = M4

MTBF of slide sub-system = 1,800 hours = M5

Next, we calculate, the failure rate, which is inverse of MTBF, as mentioned before. Hence, the failure rates of the sub-systems working together are as follows:

Failure rate of spindle = 1/1,000 hours = F1 = 0.001

Failure rate of tailstock = 1/1,200 hours = F2 = 0.00083

Failure rate of indexing unit = 1/850 hours = F3 = 0.0012

Failure rate of control elements = 1/400 hours = F4 = 0.0025

Failure rate of slide sub-system = 1/1,800 hours = F5 = 0.00056

The system failure rate is the addition of all these sub-system failure rates.

System failure rate = SFR = F1 + F2 + F3 + F4 + F5 = 0.006

MTBF of the system = MSS = 1/SFR = 166.67 = 167 hours.

It can be seen that the lowest-ranking elements dictate the system mean time between failures. Hence, in order to improve the mean time between failures for a system, the system with the lowest MTBF has to be looked into. That is the weakest link. This model can also be used to allocate the reliability of sub-systems, which is called "reliability allocation" by some authors. Another point: sub-system reliabilities affect system reliability. It might be good practice to design all the critical elements with the best reliability instead of rationing reliability among constituents. This could be called "system reliability optimization," and it should be practiced continuously but without exceeding budget allocations.

Parallel System: Next is the parallel system. On example is the spindle and tailstock for a medium slender part where machining is possible without the tailstock. In this case, the spindle and tailstock work in parallel. Another example is the combined work of the spindle, tool block, and indexing unit. One can work without the others working. The system reliability for a parallel system can be determined as follows:

MTBF of spindle = 1,000 hours

MTBF of tool = 100 hours = M6

MTBF of indexing unit = 850 hours = M2

The components are in parallel. In this case, the system MTBF can be determined as follows:

System mean time between failures = MSP = 1000 + 100 + 850 – (1/ (1/1,000 + 1/100 + 1/850)) = 1,950 – 1/(0.001 + 0.01 + 0.0012) = 1,950 – 1/0.0122 = 1,868 hours

Hence, system MTBF for parallel system = 1,868 hours

System failure rate = 1/1868 = 0.0005

For a parallel system, the failure of one element does not affect the system failure rate that much since the system can go on without the other one working simultaneously. In general, parallel systems, such as two spindles, multiple tool blocks, etc., are costly and might cause redundancy for the system. For very critical elements, parallel systems might be necessary so that one can be repaired while the machine is working with the other sub-system.

12.9 Machinery FMEA

As explained in previous chapters, FEMA procedure can be applied to reliability and maintainability activities for machine tools. The machinery FMEA could help the team to identify potential failure modes of the system, their effect on the performance and operation of the machine, and the potential causes of each failure mode and to develop a plan to possibly eliminate and/or reduce their cause. In order to save cost and time, only critical or repetitive failure modes are considered at first. Moreover, an iteration is needed when implementing these corrective measures, since one correction might create other potential failure modes. Hence, after implementation, the team has to ensure that the system is free of any other failure modes.

The machine FMEA (MFMEA) creates a plan for the machine system or sub-systems, whereas the design FMEA (DFMEA) creates the plan while designing the system. The MFMEA is created by the buying company, and the DFMEA or PFMEA for the machine is created by the equipment manufacturer. Once developed, it is the responsibility of the equipment manufacturer to implement the action plan for the machine. The MFMEA is a living document and should be updated as and when changes are made to enhance the reliability of the system. The FMEAs should be well documented for future reference.

Among other things in the MFMEA, the identification of potential failure modes is the most important and critical step.

Any failure that causes the machine not to produce at the desired accuracy level comes under review while the MFMEA is developed for the equipment. Machine failure modes can occur in many possible ways:

- Component failure: spindle stops moving
- Potential failures: part accuracy degrades gradually over time
- Performance deviations from the norm: indexing time increases beyond acceptance level
- Failure beyond repair: scrap the machine

The potential effects of a failure mode could be in terms of machine downtime, excessive part defects, or scrap and safety hazards. The quality inspection of machined parts could demonstrate quality issues with the machine. Safety hazards create a nonworkable condition, such as a machine door not working. The causes should be identified first before developing their effects.

The effects of machine failure can be numerous. For example, machine downtime can lead to functional and economic losses due to breakdown, new set-ups and adjustments to get production going again, trials to ensure the machine is set up right, production of quality parts to ensure the part cycle is restored again, etc. Every one of these has economic consequences, which sets up the priority of actions. Part scrap is another important aspect of machine breakdown. Sometimes machine failure might cause a costly part to be defective due to poor surface finish, chip marks, heat distortion, loss of tooling, etc.

Abrupt machine failure sometimes creates a hazardous condition for the operator; for example, failure of workholding equipment might lead to a part being thrown out the safety doors and hitting the operator, causing a life-threatening situation. Machine failures and their effects and causes depend on the type of machine. Hence, individual FMEAs have to be created for different machines. Safety concerns are the most critical consequences of any breakdown.

Each item must have a severity rating. In order to reduce the severity rating of any breakdown, design changes are necessary. The severity of any failure could depend upon safety hazards, part scrap, downtime effects on the production quantity, and government regulations. The ratings are done in a comparative fashion for all the failure modes, starting from worst to least severity. The severity rating could have a scale of 1 to 10, with 10 having the worst consequence, such as death or loss of body parts. The failure modes with the worst severity ratings are handled first. Operator safety has the highest priority, whether identified or not.

The potential causes of any failure mode could be poor design or design deficiencies, environmental changes, system malfunctioning or non-cooperation, control failures, etc. The data for the occurrence of the failure in failure modes can be obtained from customer concerns, legacy MFMEA, warranty data, maintenance records, field or laboratory testing, etc. The occurrence rating for the failure modes depends on the frequency of occurrence during a stipulated time frame. Each cause must have an occurrence rating.

The next item to consider is the mechanism that is in place to control or eliminate the failure modes. This mechanism identifies potential failure modes or the effects of failures and prevents them from happening. For example, controls should reduce the number of failures and report failure mode occurrences. The documents are used by plant workers to install future action plans in case the failure modes happen again.

As explained earlier, the risk priority of each failure mode is calculated using the following simple formula:

Risk Priority Number (RPN) = Severity (S) x Occurrence (O) x Detection (D), or

RPN = S x O x D

The RPN is a relative value and has no meaning otherwise, and it is used for prioritizing the failure modes for design changes. RPN also helps the team to identify the root causes of any failure and determine the weakest design links of the system. The system redesign should start with the highest RPN item and end with the lowest. Hence, the last piece of action is to identify, for each RPN, a list of actions to counter the design faults. Each RPN must have a list of recommended actions to correct the situation.

When any failure mode has a very high RPN rating, it must be addressed before the machine gets out the door. The design changes are issued using revision numbers. The DFMEA is a living document and needs updating after a change is incorporated. Once the correction is implemented, a revised MFMEA must be issued to understand the effects of the changes made. Documentation must be completed.

12.10 R&M Verification Methods

This is also called a trial run or run-off for the machine before it is shipped to the customer's plant. The successful run demonstrates the ability of the machine to satisfy the customer with respect to part quality, endurance, accuracy measurements, capability indices, etc. The customer signs off on the machine for dispatch to their own plant from the machine manufacturer's plant. The test normally consists of the following aspects:

- Continuous durability runs using cycling conditions for 48 hours
- Dry cycle runs
- Continuous part run-off
- Vibration test
- Statistical quality control test for the produced part
- Electrical component cycling test
- Fault diagnosis
- Control software verification test
- Software reliability

The results of the tests are documented properly, and deviations, if any, are discussed among team members for acceptance or rejection. If necessary, design changes are incorporated to satisfy customers. The intent of this run-off test is to make sure that the machine produces quality parts at an agreed-upon rate at the customer's plant without any hiccups. Moreover, the intent is to reduce or eliminate start-up failures, reduce the effects of thermal conditions on part quality, satisfy customer requirements, and resolve hardware and software issues, control issues, tooling and set-up requirements, etc. Basically, a run-off test is a true representation of actual production conditions in the manufacturer's plant before the machine is dispatched.

12.11 Summary

This chapter deals with the following important points of reliability and maintainability of production machinery. The chapter also deals with various engineering tool and methodologies to determine the reliability of the machinery system. The topics covered can be summarized very briefly as follows:

- Goals and objectives of machinery reliability and maintainability program
- Five phases of the machine life cycle
- Machine requirements, specifications, and conformance procedure
- Reliability and maintainability procedure for machine tools
- Machine reliability
- Mean time between failure, MTBF
- Mean time to repair, MTTR
- Hazard and failure rate, FR
- Determination of MTBF, MTTR, FR, and h(t)
- Reliability of series and parallel systems of machine
- Machine availability for production
- Bathtub curve explanations
- Calculation of failure rates in three zones of bathtub curves
- Development of equipment profile to satisfy customer requirements
- Determination of safety margin and interference between load stress and material allowable stress conditions
- Various failure mechanisms and failure modes for mechanical and electrical elements
- Determination of failure modes and resolution process
- Stress and strain analysis due to loading and thermal inputs
- Tests to verify the reliability and maintainability of the machine

12.12 References and Bibliography

Box, G.E.P., Hunter, S.J., and Hunter, W.G., 1978, *Statistics for Experimenters: An Introduction to Design, Data Analysis, and Model Building*, John Wiley and Sons, NY.

Clausing, D., 1994, *Total Quality Development*, ASME Press, NY.

Ford Design Institute, 1998, Lecture Notes on Reliability and Robustness, Ford Motor Company, Dearborn, MI.

Hatley, D.J., and Pirbhai, I.A., 1998, *Strategies for Real-Time System Specification*, Dorset House, NY.

Lewis, E.E., 1987, *Introduction to Reliability Engineering*, John Wiley, NY.

Netter, J., Wasserman, W., and Whitmore, G.A., 1978, *Applied Statistics*, third edition, Allyn and Bacon, Inc., Boston, MA.

Padhke, M.S., 1989, *Quality Engineering Using Robust Design*, Prentice Hall, Englewood Cliffs, NJ.

Rau, J.G., 1970, *Optimization and Probability in Systems Engineering*, Van Nostrand Reinhold, NY.

Rao, S.S., 2014, *Reliability Engineering*, Pearson Publishing, Prentice Hall, NJ.

SAE Guideline, 2002, "Guidelines for Preparing Reliability Assessment Plans for Electronic Engine Controls," APR5890, SAE, Detroit, MI.

SAE Guideline, 1999, "Reliability and Maintainability Guideline for Manufacturing Machinery and Equipment," M-110, SAE, Detroit. MI.

SAE Guideline, 1999, "Reliability and Maintainability Guideline for Manufacturing Machinery and Equipment," M-110.2, SAE, Detroit, MI.

Wheelwright, S.C., and Clark, K.B., 1992, *Revolutionizing Product Development: Quantum Leaps in Speed, Efficiency, and Quality*, Free Press, NY.

11.11 Review Questions

- Develop an example of the equipment life cycle for a machining center for a production environment. Does it change if the machine is used in a job-shop environment? Identify the critical process steps and the implementation process for each phase of the life cycle.

- Develop a reliability and maintainability matrix for a CNC turning lathe, consisting of tasks during program development, engineering tasks, and continuous improvement tasks in each phase of the life cycle.

- How do you develop an equipment characterization statement for a CNC grinding machine? Also develop an MFEMA for this machine.

- A hydraulic pump has a mission time of 12,000 hours. The MTBF for this pump is rated at 2,000 hours. The MTTR for this pump is eight hours. Find the probability of failure during mission time. What is the failure rate for this machine? What is the availability of this pump? Draw a reliability curve for this component.

- Is the bathtub curve applicable for a CNC machining center? If not, why not? Explain the different phases of the bathtub curve for this machine. How do you propose to develop a plan to reduce the failure occurrence in each of these phases? Can you completely stop failures for a machine? If not, why not?

- How can you improve overall machine effectiveness for a robot loading and unloading parts in a CNC turning center? Can you develop a life-cycle cost model for this machine? The life-cycle costs for two machines are as follows:

  Acquisition cost, operating cost, and maintenance cost for a CNC machining center from Vendor A are $3 million, $10 million, and $8 million respectively.

  Acquisition cost, operating cost, and maintenance cost for a similar CNC machining center from Vendor B are $2.5 million, $11 million, and $10 million respectively. What is the LCC for each machine? Which machine would you recommend buying?

- The MTBFs for Machines A, B, and C are 25, 50, and 38 hours respectively. The machines are working in series to produce a part. What is the MTBF for the total system? if the machines are put in parallel, what would be the system mean time between failure?

C H A P T E R 13

13 System Quality Management

13.1   Introduction

In the previous chapter, statistical considerations for machine testing were highlighted without going into derivations. It has to be borne in mind that, ultimately, the customer judges the quality and the manufacturer just provides the best product possible. For machine tool products, sampling is not that much of an issue since the volume of production is very small and a small number, one or two, is used for qualification requirements. It is also true that delivering quality to the customer, as per requirements, is the ultimate task, which is very hard in most cases. Sometimes no amount of testing or statistical analysis is sufficient to convince the customer of product quality unless it performs as per the customer's expectations.

The next issue is how to define quality. How does the builder define quality and its management for the product? The general indices for total quality management (TQM) were defined precisely by Feigenbaum in 1985. These guidelines need to be slightly revised to suit machine tool applications. He defined quality control as a tool to integrate all aspects of product quality, i.e., quality development, maintenance, and improvement, to satisfy customers. I would like to define the product quality thusly:

Product quality is designed into the product based on customer requirements, and it is verified by quality control methods and manufacturing processes that satisfy customer expectations.

So, the builder satisfies the requirements, and the customer proves the quality. Any product is supposed to do what it was designed to do in a customer environment. Hence, the primary definition of machine quality must include customer satisfaction and product requirements. The quality is the bridge that connects these two end objectives.

Many noted authors, such as Garvin, Crosby, and Taguchi, have tried to define quality in many possible ways. Taking into account all these possibilities, a quality product must have the following six characteristics:

- **Precision:** This is a description of random variability, a measure of statistical variability as a result of machine interactions with random noise and environmental conditions

- **Accuracy:** A measure of differences between the true value and actual value

- **Durability:** Able to perform without deviations and interruptions for an expected long time without significant deterioration in quality or value

- **Robustness:** A product characteristic or attribute that consistently performs according to its specifications under noise over the intended design life

- **Safety:** The ability of a product to be safe to operators and maintenance personnel with intended use, as determined when evaluated against a set of established rules

- **Serviceability:** Degree to which the machine can be serviced with ease with an available amount of resources and within a given time

All these characteristics are required for customer satisfaction. Moreover, product requirements should have some metrics for each of these dimensions. In fact, quality is a multidimensional attribute, and all of these need to be satisfied to define a quality product. The attributes represent specifications that will lead to customer satisfaction.

In order to come up with the metrics for each of these dimensions, designers have to depend on legacy products and feedback from potential customers. The builder's responsibility is to provide the right product to the right customers. It has to satisfy customer needs and expectations. Instead of depending on marketing department feedback, designers have to also directly connect and communicate with potential customers for the machine. Manufacturing is the link between designers and customers. Manufacturing operations have to produce the product as per the design supported by quality control operations.

As mentioned before, designers and manufacturing operations have to collect the data from customers and existing products. So, benchmarking products with a new design is also a must for success. Multiple steps need to be performed to connect the product with customers:

- Product data mining from customers, legacy products, and competitive products

- Interpretation and compilation of customer data into manageable and prioritized requirements for the present product

- Development of design specifications to satisfy customer requirements

- Design

- Manufacturing

- Quality assurance and delivery

Collection of design data from customers is easier said than done. There is not any formal and effective procedure available for collecting such data. The procedure has to be product and customer specific by design. The collection procedure has to be very short and productive. The interview questions or feedback questionnaire must be binary, i.e., answers should be yes or no, and not descriptive. Nevertheless, face-to-face interviews are the best way to accomplish these end objectives. Several procedures such as interviews, market clinic, and focus group discussions can be conducted to collect data. Questions should be designed in such a way that product design requirements can be developed right out of these customer answers. This data collection process is far more difficult than it sounds.

In order to connect customer requirements with the product design and manufacturing features, a quality function deployment (QFD) matrix has been developed and implemented for a variety of products. For machine tools, such procedures can also be used. The QFD is also called "house of quality." The QFD matrix relates customer requirements with product attributes, features, or characteristics. The QFD elements are presented in Figure 13.1. The body of the QFD matrix shows what requirements are most and least related to specific attributes. The hat portion, or house roof, of the matrix shows the cross-connections among product attributes. This matrix shows the positive or negative effects of one characteristic with others. From the main body of the matrix, the attributes that most satisfy customer requirements can be determined by providing some relative weights. This also shows where the design effort has to be placed to satisfy customer requirements.

There are many discussions available on house of quality matrices in quality handbooks. At the bottom of the main body matrix, three values are to be filled in for each product characteristic: attribute target, i.e., the relative value of each product attribute, best benchmark value for each product characteristic, and average or normal value for each attribute. These are all relative values used for comparison purposes. The house of quality matrix displays the design attributes that a product must have to satisfy customers and, at the same time, be competitive in the marketplace. This also shows relatively how one product attribute might affect another positively or negatively. In the end, a visionary strategy and innovation should be used to design a product that will satisfy customers.

The company as a whole has to have quality and product strategies, the product design has to focus on a product design theme and a strategy to achieve that. For example, the machine must have the highest spindle speed or highest axes rapid rate to reduce the time for idle motion. A technology strategy has to be developed to incorporate these features into the design, which has to satisfy the requirements without exceeding the cost that customers can afford. The product strategy must include design, manufacturing, and quality strategy. Many machine tool companies have gone out of business because they did not have a synergized product strategy that satisfied customer requirements.

More often than not, many machine tool builders build machines without satisfying most customer requirements.

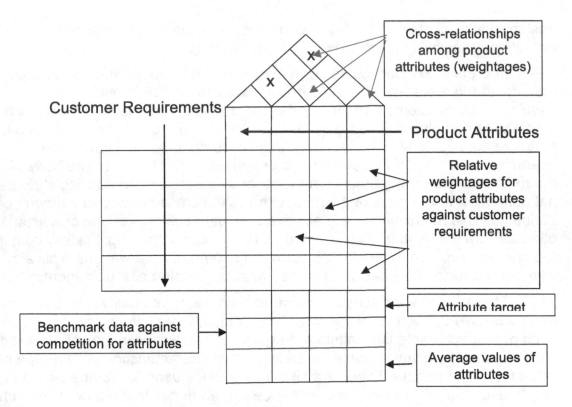

Figure 13.1 House of quality for a machine

The U.S. machine tool industry failed miserably against foreign machine tool products, which delivered more content with less cost. Specifically, Japanese products combined and blended features or attributes with cost, which customers welcomed. To be competitive in the market, the product has to have a combined strategy for precision, accuracy, durability, and robustness as a principal quality direction. Safety and serviceability can be secondary directions. In the real world, it is extremely difficult and unrealistic to satisfy all the requirements simultaneously, but a sincere effort must be made to achieve that goal. When companies deliver a product with a single principal quality direction, they often become vulnerable to competition.

Another point to note is that the actual and perceived quality of a product are two different concepts. In order to have positive perceived quality for any product, the actual quality product has to be delivered for a much longer time in the market, Once the perception of the quality of a product is firmed up in customers' minds, it takes much longer to change it. It has also been seen that some companies would like to be market leaders based on one of these quality dimensions. Such a strategy fails in the long term. The product has to be consistently good over a long period of time, and it has to serve the customers in a very consistent manner. A combined quality strategy is required for any company to be a market leader.

13.2   Strategic Approach for Machine Quality

A successful quality program needs a commitment from all the participants, i.e., management and workers. There should not be any fence in between these two groups of participants. In my opinion, quality management is a bottom-up process and is part and parcel of the system engineering process. Workers must feel the importance of quality, and they should push the idea up the ladder for required training if they need to produce machines that meet requirements. The product's success depends on quality implementation throughout the company's departments and personnel without exception. Feigenbaum and other quality experts have recommended the following:

- Quality does not belong to one department of the company only. It is a discipline for everybody, a practice followed by Japanese concerns. Quality should be the way of day-to-day life inside and outside the company, and the quality of any product should be a matter of pride to all company participants.

- The same quality principles and strategies should also be followed by all companies in the supply chain. The parent company has to administer conformation of the same quality principles throughout.

- Quality principles, management, and control should be transparent to all departments. Quality implementation should not be compartmentalized or customized for any functional and non-functional department. For quality implementation, every department should be given similar importance.

- Management and control of quality function should be the responsibility of all participants, including the lowest level of workers. Everybody is equally responsible for quality functions

- Quality is reflected through products and the people who are involved in making them. Quality should be a cultural norm.

In a quality-conscious company, management creates an environment where even a lowest-level worker maintains discipline. In case of a quality issue, root-cause analysis (RCA) is performed and documented to resolve it. The problem is uprooted from the system so that it does not show up in the future. Instead of focusing only on cost, attention has to be directed to quality, which will reduce the cost indirectly. The company as a whole has to have a quality strategy, organizational structure, quality norms and standards, quality environment, and system-level approach to establish quality in the product. Workers have to empower themselves to enhance the quality of the product.

Quality experts have come up with many ideas, such as zero-defect programs and quality circles, to enhance quality in industrial sectors, and many machine tool companies have incorporated them in factories with limited effect, especially in the US. The primary reason for such limited success is the top-down approach towards quality. It has to be a bottom-up, systemic process.

Many times, a quality circle, quality group, or consortium is formed in a company to discuss and resolve quality issues. Most of the time, this group discusses current problems without paying much attention to preventive quality issues. Prevention is more productive than a cure even in industrial environments. Incorporation of one quality solution should not create another issue in the future. A quality solution includes technical and human issues at the same time. Such groups should recognize the importance of quality training at the worker level to help them resolve quality issues in their own work environment.

In the past three decades, quality enforcement has become a top priority in some machine tool companies. In such cases, quality means the final inspection of the machines to satisfy norms and standards. Quality has become a sales tool instead of a method to improve the quality of the company as a whole. Many such quality tools have become fads and failures in workplaces. Quality only means establishing statistical process control for the products and processes. Quality is product-specific and only a marketing tool to generate revenue. The primary reason for such an upsurge in quality enforcement was numerous, and some of them are as follows:

- Loss of market share
- Stiff overseas competition
- Loss of revenue due to poor quality
- Customer dissatisfaction with poor quality
- High cost
- Very high life-cycle cost for the product

It is a fact that in spite of all efforts put forward to enhance the quality of products, most programs, such as ISO9000, 14000, zero defect, quality circle, etc., have failed miserably to create any significant impact in improving the quality of the product. Companies have lost huge amounts of money incorporating such programs without any productive output. The whole industry has suffered due to its conception of poor quality. The quality program ended up with a "suggestion box" mentality only. The quality program should not become the fashion tool for the company. Quality conception does not change overnight in the customer's mind. Such programs do not change the conception of quality for any product unless the product is designed and manufactured with system quality in mind. Product quality has to satisfy customer requirements.

Hence, quality focus needs a different direction and dimension: in my opinion, quality should be a tool to manage and control variations of each process in a company. A system approach should include quality and its cost concurrently. Quality should reduce the cost of the product in the long run. Every process has to have a quality target with accepted deviation targets. Every operator has to be empowered to manage variability in his or her own environment. For example, a sales department has to have a sales target with a given variance. The company has to establish the target and variance limits for design, sales, marketing, vendors, office practices, etc. That is the concept of system quality, as opposed to product quality.

The shortcomings of U.S. machine tool products entered the limelight after benchmarking studies were completed. It was revealed that U.S. machine tool builders focused on cost, price, and delivery, whereas Japanese builders focused on reliability, features, and product customization without excessive cost. The comparative benchmarking of products between the US and Japan showed the weakness of U.S. products very clearly. Japanese builders gained sizeable ground in the US due to their product strength at a significantly lower cost. Product benchmarking must include quality, features and functionality, reliability, precision, cost, etc. for comparison purposes. All quality dimensions must be considered for the product without any bias whatsoever.

The benchmarking also showed the urgent necessity for quality improvement for U.S. products. The slogan "Made in the USA" took a stronger hit and became secondary to Japanese reliability and cost. Chinese and Korean machine tools were never a threat to U.S. machine tools, since they lacked individuality and precision. This proves that cost with quality is a misnomer and is very much short-lived. I believe that lower cost only creates a local excitement among customers, whereas product quality brings in a universal appeal. Any unbiased benchmarking will point to the weakness of a design, and benchmark data can be used to improve the product and make it competitive. The benchmarking data should be prioritized before implementation to get the most value out of the investment. Benchmarking must be a strategic decision for the company. As mentioned earlier, benchmarking must happen before a system is designed and developed. It should be systemic, i.e., it should include all system characteristics.

In order to find out whether a quality system is helping the company, a quality audit by an independent company should be conducted. The audit should include the following aspects of the total quality process:

- Total quality management and control
- Process control methods
- Productivity and efficiency
- Produce performance and features
- Customer satisfaction
- Perceived quality of product among customers
- Product innovation
- Process documentation
- Leadership style and effectiveness
- Employee satisfaction and involvement
- Continuous improvement strategy
- Quality training program

The ultimate objective of a quality program is to improve revenue by increasing market share. A well-planned quality program will also reduce the product cost. Whether it is the quality program implementation and improvement or SPC, QC, or zero-defect program, etc., product quality should increase customer satisfaction. Quality implementation will require investment and added revenue, and profit should justify such investment on a continuous basis. The quality costs can be direct or indirect. The direct cost is due to finished product scrap, high maintenance costs, after-sales customer service costs, high inspection costs, high rework costs, inventory scrap, etc. Indirect costs can be due to the loss of repeated sales, loss of market share, lawsuits, accidents caused by poor safety of products, etc. In order to evaluate the effectiveness of any quality program, both direct and indirect costs should be included.

The total cost of poor quality is very substantial and recurring by nature. It has been well proven that the cost of poor quality is substantially higher than that of Japanese machine tool products. Quality is never free in reality, but high-quality products bring in higher revenue, higher brand recognition, and larger market share, which are essential for survival and growth on a consistent basis. Quality is the bottom line for a product.

13.3   Design, Manufacturing, and Quality

As mentioned before, quality control meant an inspection department that conducted the inspection procedure following physical measurements of machine movements or statistical process control methods on the part produced by the machines. The first one gave a direct physical measure of accuracy for the machine, and the later method gave an idea about the machine's production capability. Both of these measurement methods were meant to accept or reject the machine for acceptance. This is also to note that inspection methods were done after the machine was designed, manufactured, and assembled. If the outcome of these two inspections were within pre-specified norms, the machine was accepted; otherwise, it was rejected. If the machine was rejected, the root cause of such failure was not known. This post-mortem method was also very dependent upon the way the inspection was conducted. That was the extent of involvement for the quality control department.

Before Japanese machines gained ground in the U.S. market, they were assumed to be of lower quality than American-built machines due to their low perceived quality and precision. It was also assumed that the life-cycle cost of Japanese machines would be far higher than American-built machines. Instead of the idea that the design guarantees the quality of the product, end inspection results proved the quality of the machines. Reliability and durability aspects were completely ignored. Japanese machine builders maintained the low cost and simultaneously worked hard to enhance reliability, durability, and productivity. Ultimately, their efforts paid off, and they gained customer confidence in the USA. Now Japanese builders are at the helm of affairs and enjoying the biggest market share in the USA. They basically focused on design to enhance quality. This approach optimized things to the point that machines were reliable by design and needed very little quality inspection at every stage of manufacturing. Quality wins in the end.

Why did Japanese builders focus on reliability and durability? The answer is somewhat simple. Reliability and durability allow the machine to be operated for longer production time without interference and interruption. Unpredictable failures are few, allowing operators to conduct preventive maintenance. The failure pattern is somewhat predictable, and shut-down can be planned to repair the machines. To enhance the quality of a design, many stalwarts like Taguchi and others have recommended the following steps:

- System Engineering: The product is a system, and its design should be based on systemic engineering, which is a top-down approach. System performance will dictate sub-system performance.

- Robust Design: The design has to be reliable over time. The performance degradation over time has to be within a specified limit that the customer agrees to. The design will be robust against system and environmental noise factors.

- Parametric Design: This is part and parcel of optimized design. The design has to be optimized based on the constraints. The machine parameters that relate the design parameters with machine performance have to be identified by performing design FMEA, etc. These parameters need to be optimized to obtain machine performance that will maximize customer satisfaction. Optimized parameters lead to maximized performance measures for machines.

- Tolerance Design: A suitable range of each optimized parameter has to be determined to make sure such variations do not affect machine performance beyond acceptable limits over the useful life of the machine. Basically, an acceptable tolerance limit of each parameter target has to be specified. The higher the range, the lower the cost of design tends to be, so tolerance limits do affect the cost of the product. Tolerance can be unidirectional or bi-directional as long as it does not affect system performance beyond acceptable limits and the cost is within budget.

Once the design has been optimized, the manufacturing process also has to be optimized to make it less sensitive to manufacturing variables. The manufacturing parameters also need to be optimized with acceptable tolerance levels. For example, for spindle assembly, optimum bearing temperatures need to be determined so that the bearing can be assembled with ease without affecting the shaft and bearing interferences. Assembly temperatures need to be optimized with a tolerance to maintain assembly accuracies. Hence, system principles, robust manufacturing process principles, and process parameters with tolerances need to be identified and implemented for manufacturing processes.

Quality control should be an in-line procedure. It has to be incorporated while the machine is being assembled and parts are being machined. An off-line procedure is basically useless since the damage is already done while the deviations are being identified. Since some processes are irreversible, deviations have to be identified before they are completed. This will reduce scrap and production cost.

Hence, quality starts with the design, and the manufacturing process confirms the design quality of the product. Inspection procedures have to be minimized or eliminated if at all possible. The design has to be economic and compliant with manufacturing resources available at the time of production. The design should be manufacturable with ease. It should also help to produce the machine at the lowest manufacturing cost possible. In the US, the manufacturing process starts after the design is completed, whereas Japanese companies start the design and manufacturing concurrently. A truly innovative process integrates design and manufacturing together so that the design is reproducible with reasonable cost.

The design for manufacturability (DFM) helps designers to integrate the design with the manufacturing of the product. Authors such as Boothroyd and Dewhurst have promoted the cause of DFM to a great extent, but most of these methods are well known to the machine tool industries for a long time. Such authors have stressed the fact that design should promote ease of manufacturing by following some simple design principles while designing the product and not afterward.

These authors have identified the design principles that would enhance manufacturing productivity and assembly efficiency. In my opinion, the design should allow assembly accuracy instead of assembly time reduction only. Manufacturing methods should enhance accuracy since the production volume is limited. Hence, instead of focusing only on assembly time, machine manufacturing engineers should also focus on assembly accuracy. The design should focus on reducing the number of parts, as recommended by these authors.

In addition, I would like to add that machine tool design tends to be very complex for multi-tasked assemblies. Japanese design seems to be simpler. Hence, reduction of complexities and parts are an important task for the design. This task can only be accomplished if and only if design and manufacturing engineers put their heads together to finalize the design. That is why the idea of concurrent design has come into being. In practice, this is effectively not being practiced to the fullest extent. In a nutshell, the design should be simple enough to manufacture with ease and complicated enough to perform the task with consistency. A balance has to be obtained to reduce the cost of the product. A simpler design also enhances the reliability of the product and reduces the life-cycle cost.

Most of the recommendations for parts suggested by Boothroyd and Dewhurst in 1989 are for mass assembly, such as in automobile assembly procedure. In mass assembly procedures, assembly time and easiness are primary factors for reducing fatigue and boredom in automobile applications. Machine tool applications do not particularly belong to such category of assembly. In machine manufacturing, precision, accuracy, and ease of dismantling are the primary criteria. Hence, approaches for part design are somewhat different. The recommendations for part design for machine tool applications are as follows:

- Parts should be assembled in such a way that they can be dismantled and reassembled without sacrificing accuracy; time of assembly is important but secondary

- For most of the assemblies, parts should have adjustment capabilities for alignment purposes; adjustment capability is more important than a self-aligning requirement

- In most cases, part orientation capability is required for alignment accuracy

- Parts require both hands for assembly

- Assembly tools are part and parcel of the assembly; automation is minimal

- Parts can be assembled in any orientation or angle; linear motion is not required

- In order to enhance the rigidity of assembly, parts are secured through proper fastening methods: interference fits, nuts and bolts, etc.

Hence, the advantages of DFM are not very apparent in machine industries. The primary reasons could be various, depending on the type of machines being manufactured. It also has been noticed that in machinery factories, the assembly procedure has a legacy built into it. Technicians gain efficiency by using procedures that have been followed for many years. Experience is given importance. Assembly cost is not a primary factor in machine manufacturing. Accuracy, alignment, adjustment, ease of dismantling, etc. take precedence. Another reason is the low volume of machines. Jigs and fixtures are used for alignment purposes most of the time. The primary focus of assembly is the fact that the machine can be assembled after maintenance with the same accuracy as before and without lots of adjustments.

Machines must be designed in such a fashion that the set-up time and cost are as little as possible. Tool replacement must be made easier. Fixture replacement must be made easier. Part loading and unloading procedures must be made easier. Hence, the part design should be simple without losing functionality and accuracy. Precision is the primary focus in any assembly because accuracy depends on assembly techniques. The design is far more important than procedures, and it should guarantee accuracy and precision. The assembly process becomes more machine-specific when the volume of production is, say, two to three thousand a year, as compared to a target of seven million cars per year. The assembly procedure has to be different in machine tool industries, which are significantly different from automobile companies. Customization is very much acceptable in machine industries.

The design should be the primary source for enhancing product quality. The manufacturing process should be a tool to support design methodology. The quality inspection department should be eliminated, if possible. Cost and quality should be complementary to each other, and design, quality, and cost should be synchronized to satisfy customer requirements. When all is said and done, system quality is an iterative procedure and an honest attempt should be made to enhance the quality and accuracy of the machine at every step. If the question arises as to whether cost or quality is more important, I would prefer quality at a cost that the customer will accept. Ultimately, the customer decides the quality of any product. Quality is first; cost is second.

Moreover, there is nothing new about product quality. It has been discussed and implemented for ages. The only difference has been how to manage the quality and what the optimum quality required by customers is. Customer focus has changed the way control is manifested in a product, bringing statistical quality control into system quality management. SPC should be an on-line instead of off-line process in order to reduce scrap and the amount of rework. At the end of the day, all procedures, techniques, control methods, and implementation have to be geared towards minimizing variability of any design, manufacturing, and processes. If the variability around the mean target is managed properly to satisfy customer requirements, quality objectives have been successfully met.

13.4　Statistical Quality Control (SQC) Methods

Joseph Juran defines quality as "fitness for use." Philip Crosby defines it as "conformance to requirements", "Taguchi defines loss of quality as social loss." The American Society of Quality Control (ASQC) defines quality as "the totality of features and characteristics of a product or service that bear on its ability to satisfy given needs." I think that quality should be defined as "conformance to customer satisfaction"; i.e., the customer defines the quality. So, quality has two primary aspects: design and conformance to customer satisfaction. Perceived quality could be different from manufactured quality. The design quality is geared towards enhancing the reliability or robustness of a product in customer environments. The manufactured quality is geared towards management and reduction of variability of the target. Hence, the product has to be manufactured with design intent and defect-free.

The manufacturing process must be stable, repeatable, and consistent to produce quality goods. These principles are applicable to all manufactured goods and services. If a manufacturing process is not stable, it will produce a bad machine every time. A stable manufacturing process minimizes the variation of target-intent specifications, such as part dimension, assembly variation, assembly time, etc.

One of the predominant uses of statistics and statistical theory in the quality application area is statistical process control, or SPC. Statistical process control is applied to understand the extent and severity of such variations during or after the parts or machines are produced. SPC monitoring can help to identify the root cause of such variations only when the process is stable. SPC can be also used to establish the stability of a process and reduce the variability of the manufactured product.

There is a wide application of SPC in controlling machine tool accuracy. It was invented and applied by Bell Laboratory engineers in 1916. The primary area of use was in the inspection rate of the quality control department. ASQC recommends very strongly that this method be used for production and quality control purposes.

Another proven statistical application is the six-sigma approach for machine tool activities. This approach was applied in the production process by Motorola in the 1980s for its circuit boards to control production defects. The term was coined by Mr. William Smith. This approach can be used in design, manufacturing, and business processes as well.

Many companies, such as GM, Ford, and GE, have used this process to improve their business and products. It has been found to improve the reliability and quality of products. It determines an acceptable number of defects per million opportunities (DPMO), such as 3.9 DPMO. There are many sigma standards such as three sigma, six sigma, etc.

In statistical quality control activities, observations are recorded for random samples drawn from the production batch when the production process is stabilized and the production defects do not belong to assigned causes. Random sampling is recommended for such an application. The observations are plotted in SQC charts to determine whether the process is under control or not. Sample size, frequency of sample collection, etc. have to be pre-determined so that the effects of samples on the outcome is minimized. This a destructive testing procedure, so caution has to be applied for sample selection and measurement methods.

Depending on the outcome of such tests, the batch is selected or rejected. Since 100% inspection is very costly and time-consuming, the SQC method saves time and money on a consistent basis in high or medium production processes. Now, if the accuracy desired or production outputs are too small, SQC might not be suitable for such operations. Hence, for most machine tool applications, SQC is used to accept the machine. Machine accuracy can also benefit from an SQC process, and the results can be used to understand process stability and machine reliability. They can also be used to control the variability of outputs of a machine and as an on-line process control tool to monitor machine tool outputs. The Taguchi method also reduces the variability of product and machine performance. This is an off-line method for quality control, and it can be used to control the robustness factor of any machine. Consistent quality also means higher reliability of production for the machine.

13.5   Dimensional Control and Reliability

The reliability of any production process can be defined as the probability of a part dimension to be within a specified range. There are many variables that can affect the reliability of production. Most significant is the standard deviation of the part dimensions set by the machines and manufacturing process. For any part dimension, the machine is set to the mean target, but because of many in-process variables, the machine produces parts with dimensions that are not exactly the same as each other or even as the set mean diameter. A process can be defined by the standard deviation of the dimensions produced by the machine. This standard deviation depends on tooling, workholding, machine accuracy capability, environmental conditions, material properties, machine rigidity, etc. It is definitely a very complex process since there is no exact answer for any situation and the optimization for the best result depends on the specific situation of machining. Let's explore the effect of standard deviation on production reliability, which basically determines the suitability of a machine to produce parts with the desired accuracy.

It is also assumed that the production process output is normally distributed, which is a very common assumption in the industry. Let's assume the following:

- Normal distribution with two parameters, μ and σ

- Stable process

- Target part dimension = 12 ± 0.02 mm; the machine is set at the target mean of 12 mm, max dimension allowed is 12.02 mm, and minimum dimension allowed is 11.98 mm. UCl = 12.02 mm, and LCL = 11.98 mm

- Mean dimension, μ, and standard deviation, σ, for the process

- Nine samples are drawn from the batch, and the diameter of the part observed is as follows:

Sample Number, n	Observed Diameter, x_i, mm
1	12.05
2	11.99
3	12.03
4	12.08
5	12.12
6	12.02
7	11.08
8	12.07
9	12.02

- The mean of the dimensions: $\mu = \Sigma(x_i)/N = 11.94 = 11.94$ mm, where $N = \Sigma n = 9$

- Variance $\sigma^2 = 0.329^2 = 0.108$ mm $= 0.11$ mm

- Standard deviation $\sigma = (\Sigma(x_i - \mu)^2)(N-1) = 0.329$ mm $= 0.33$ mm

- Hence, the process output has a mean of 11.94 mm with a standard deviation of 0.33 mm

- Let's determine the probability of producing the following dimensions using this standard deviation of 0.33 mm

- Part dimensions are: 12.02/11.98 mm; 12 ± 0.99 mm and 12 ± 1.98

Case 1: Part dimensions: 12.02/11.98 with $\sigma = 0.33$ mm

D_m = mean diameter = $(12.02 + 11.98)/2 = 12.0$ mm; and $\sigma = 0.33$

The probability of producing the part outside the tolerance range is determined as follows:

$P(D < 11.98) + P(D > 12.02) = P(D - D_{bar}/\sigma < 11.98 - 12/0.33) + P(D - D_{bar}/\sigma < 12.02 - 12/0.33)$

$= P(Z < -0.06) + P(Z > 0.06) = 0.476 + .476 = 0.952$

Hence, the probability of producing the parts within acceptable limits of ± 0.02 is $(1 - 0.952)$, i.e., 0.048, or 5%, at best, so very unlikely.

Case 2: Part Dimensions: 12.99/11.01 with $\sigma = 0.33$ mm

D_m = mean diameter = $(12.99 + 11.01)/2 = 12.0$ mm; and $\sigma = 0.33$

The probability of producing the part outside the tolerance range is determined as follows:

$P(D < 11.01) + P(D > 12.99) = P(D - D_{bar}/\sigma < 11.01 - 12/0.33) + P(D - D_{bar}/\sigma < 12.99 - 12/0.33)$

$$= P(Z < -3) + P(Z > 3) = 0.00135 + 0.00135 = 0.0027$$

Hence, the probability of producing the parts within acceptable limits of ±0.99 has improved to a value of $(1 - 0.0027)$, i.e., 0.9973, or 99.7%, at best, so a very likely chance. This is a three-sigma standard.

Case 3: Part dimensions: 13.98/10.02 with $\sigma = 0.33$ mm

D_m = mean diameter = $(13.98 + 10.02)/2 = 12.0$ mm; and $\sigma = 0.33$

The probability of producing the part outside the tolerance range is determined as follows:

$P(D < 13.98) + P(D > 10.02) = P(D - D_{bar}/\sigma < 13.98 - 12/0.33) + P(D - D_{bar}/\sigma < 10.02 - 12/0.33)$

$$= P(Z < -6) + P(Z > 6) = 2 \times 10e-9$$

Hence, the probability of producing the parts within acceptable limits of ±0.02 is 0.999999998, so almost likely. This is a six-sigma standard.

Case 4: Part dimensions: 13.98/10.02 with $\sigma = 0.1$ mm. The process has changed to reduce the variability of the process.

D_m = mean diameter = $(13.98 + 10.02)/2 = 12.0$ mm, and $\sigma = 0.10$

The probability of producing the part outside the tolerance range is determined as follows:

$P(D < 13.98) + P(D > 10.02) = P(D - D_{bar}/\sigma < 13.98 - 12/0.1) + P(D - D_{bar}/\sigma < 10.02 - 12/0.1)$

$$= P(Z < -0.2) + P(Z > 0.2) = 0.84$$

Hence, the probability of producing the parts within acceptable limits of ±0.02 is 16%, so it is somewhat possible. There is an 84% chance that the dimension of the part will be outside the prescribed limits.

Case 5: Part dimensions: 12.02/11.98 with $\sigma = 0.1$ mm (tightened process with less variability)

D_m = mean diameter = $(12.02 + 11.98)/2 = 12.0$ mm, and $\sigma = 0.1$

The probability of producing the part outside the tolerance range is determined as follows:

$P(D < 11.98) + P(D > 12.02) = P(D - D_{bar}/\sigma < 11.98 - 12/0.1) + P(D - D_{bar}/\sigma < 12.02 - 12/0.1)$

$$= P (Z < -0.2) + P (Z > 0.2) = 0.84$$

Hence, the probability of producing the parts within acceptable limits of ±0.02 is (1 – 0.84), i.e., 0.16, or 16%, at best.

So, the production reliability has improved from 5% to 16%, almost three times, by tightening the standard deviation (see Case 1). The above examples show that by adjusting the standard deviation, i.e., the standard deviation of the process or machine, the reliability of production can be enhanced. The variability of the process has to be minimized to enhance reliability.

Process Capability: Any machine must have the inherent capability and accuracy to produce a part that is acceptable for further processing. The capability of any machine is denoted by its ability to consistently produce the part within a specified tolerance range over a specified period of time without any interruption. Hence, the machine builder should ensure that the machine satisfies customer requirements. Process capability is measured by the process capability index, C_p, which is defined as

C_p = acceptable part variation/machine variation capability required

= tolerance range/ $\pm 3\sigma$ = (UCL – LCL)/6σ, where σ is the standard deviation for the machine output and μ is the target mean

The process variation of a machine is considered to be normally distributed, and the machine is statically under control, i.e., a stable process. In order to confirm that the production does not have defects due to assignable causes, control charts are drawn, and the outputs should stay within the upper and lower boundaries. Control charts will be discussed in the next section. The mean and standard deviation for the machine output can also be determined from the measured data or the control chart. For simplicity, the center of the tolerance range is assumed to coincide with the process mean of the machine. This is an ideal situation and does not take into account mean drifts due to variations in tools, cutting conditions, material variations, or wear and tear of the machine over time.

Machines need to be set to take care of such drifts from time to time. This means that they should be set corresponding to the process capability index, C_p = 1, i.e., the tolerance range = 6 times the standard deviation of the process. As shown before, under such conditions, the process statistically produces 2,700 DPMO, i.e., 2,700 defective parts out of 10e6 parts produced, or 270 defective parts out of 100,000 parts produced. When C_p < 1.0, the tolerance range is less than process spreads allow, which means there will be more defective parts than allowed by ±3σ process.

In that case, the machine has to be set tighter or process variables need to be changed so that the process capability index is greater than 1.0, i.e., process variability needs to be reduced to an acceptable level. Hence, C_p = 1 is a minimum. For a continuous production process, the machine should have at least C_p = 1.33, keeping in mind the mean drifts and other process variable changes. For a machine to be acceptable for production, it must have C_p = 1.33 or higher. The higher the process capability index is, the lower the number of defective parts will be.

For example, when C_p = 1.33 and the tolerance range is 0.04 mm, the standard deviation for the machine is required to be:

σ = 0.04/ (1.33*6) = 0.005 mm

z = normal variate = (UCL − μ)/σ = 12.02 − 12/0.005 = 4.0

A, the area under probability density function for z = -infinity to 4 from the normal variate table, is equal to 0.999968. Hence, for the six-sigma process, the oversized part is 0.000032, or 0.0032%, and 32 DPMO will be the rejection rate for this machine under the given cutting conditions, which is more than 3.4 ppm defects.

Mean Drift Effect on Defective Parts: Over the long run, the target mean of the production process drifts to either side of the target mean for various reasons during manufacturing. The mean does not stay at the center of the tolerance range. Hence, the process capability index of the new machine has to be higher than 1, e.g.., 1.33, to take care of such drifts so that the number of defective parts does not exceed six sigma requirements.

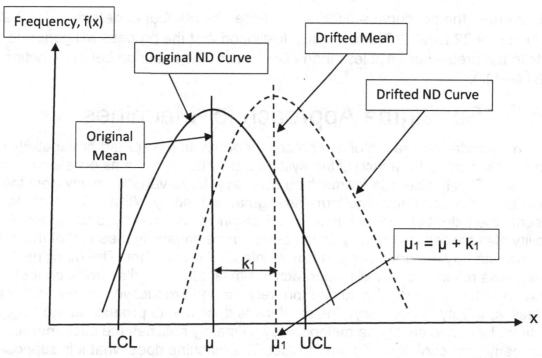

Fig. 13.1 Mean drift for six sigma process (normal distribution)

If k_1 is the process mean drift from the original center of the tolerance range,

$\mu_1 = \mu + k_1$, where $0 <= k_1 <= 1$

When the process is not centered between the end control limits or part tolerance, the process capability index for a three-sigma process, C_{pk}, is given by

C_{pk} = minimum of $[(UCL − μ/3σ), (μ − LCL/3σ)]$ = minimum of $[C_{pk1}, C_{pk2}]$

When k is equal to 1.5σ, and mean drift is to the right, then the area under the curve for a six-sigma process for z = 4.5 is 0.9999966, and the DPMO is 0.0000034, i.e., 3.4 defects per million. Similarly, for a three-sigma process, z = 1.5, and the number of defects is equal to $(1 - 0.93319) = 66,811$ out of a million instead of 2,700 DPMO. Hence, gradual drift over time does incorporate more chances of defective parts to be in the process. Including this drift possibility, the maximum DPMO for a six-sigma process is set at 3.4 defects out of a million opportunities.

For example, let us consider the following case:

Part length per design: 12 ± 0.02 inches

$C_p = 1.33$ for a three-sigma manufacturing process

$C_p = 1.33 = (UCL - LCL)/6\sigma = 0.04/6\sigma$, so desired $\sigma = 0.04 (6*1.33) = 0.005$

UCL = 12.02 and LCL = 11.98 inches

$z = (UCL - \mu)/\sigma = 12.02 - 12/0.005 = 0.02/0.005 = 4$

The area under the pdf curve = 0.999968. Hence, the number of defects = $1 - 0.999968$ = 0.000032, or 32 DPMO. There is equal likelihood that the process will generate defects in the area where it is less than LCL. Hence, the total number of defects = 2 x 32 = 64 DPMO.

13.6   Six Sigma Approach for Machines

The standard deviation of any process represents the spread or variability of the process data. The performance of the system has to be as close as possible to the target mean. This is also true for machine accuracy. More variation away from the target mean means less accuracy. The term "six-sigma," coined by William Smith of Motorola, represents the extent of variation in a manufacturing or business process. More variability also means less quality. In lay terms, more variability means that the machine works fine one day but does not perform as intended on another. The machine becomes less reliable. Six sigma approach is similar to Taguchi's quality control method. In order for a machine to be more reliable and productive, the machine has to have less variability in accuracy. The machine is designed to produce at the target most of the time. Hence, a six-sigma methodology ultimately satisfies the customer and helps the company gain confidence in the product. The machine does what it is supposed to do.

The methodology for the six-sigma approach includes five distinct processes. This approach could be used for either a manufacturing or business process to enhance productivity and efficiency. The implementation of this methodology has five steps, DMAIC, i.e., define, measure, analyze, improve, and control. DMAIC has similarities with Deming's quality control policies. Six sigma process determines the maximum defects allowed in a manufacturing process, i.e., 3.4 defects in one million opportunities, or one defect in 294,000 parts. The machine must operate without frequent failures. Any unreliable machine has a very high life-cycle cost and less productivity. In the six-sigma approach, the tolerance limits for the part from the process is set to six times the standard deviation, assuming a normal distribution of the process.

UCL = Upper control limit= $\mu + 6*\sigma$

LCL = Lower control limit = $\mu - 6*\sigma$

σ_{max} = (UCL-LCL)/12 where UCL-LCL is sometimes called the spread of part tolerance.

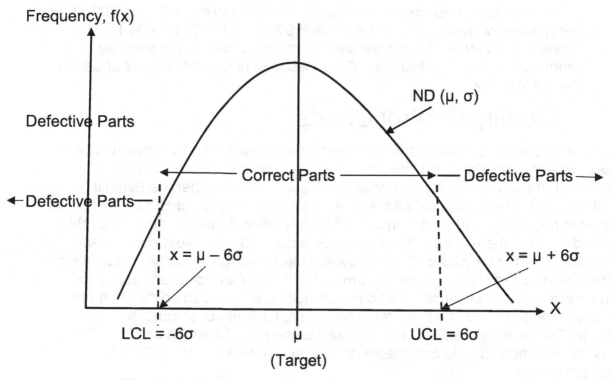

Fig. 13.1 Six Sigma process for normal distribution

The process capability index for a machine is denoted by C_p, which is the capability of machines to produce defect-free parts within a specified tolerance. Let's take an example to explain this:

- Target part dimension = 12 ± 0.02 mm; the machine is set at the target mean of 12 mm. Max dimension allowed is 12.02 mm, and minimum dimension allowed is 11.98 mm. UCl = 12.02 mm, and LCL = 11.98 mm.

- The company buys a new machine to make these parts.

- The company follows a six-sigma approach for its manufacturing process. The machine has to go through a trial to prove its capability to produce these parts within specified tolerances.

- The manufacturing process is stable and follows a normal distribution, and the machine is set at 12.00 mm.

- The process capability index required as per the contract is 1.33, at least to take care of the mean drift up to ± 1.5*σ.

- C_p = process capability index without mean shift = 1.33 = (UCL − LCL)/ (6 σ). Hence, σ = 0.04/ (6*1.33) = 0.005. This is the maximum deviation allowed.

- UCL = 12.02 mm; μ = 12 mm; σ = 0.005 mm

- Z = normal variate = (x − μ)/σ = (12.02 − 12)/0.005 = 4.00

- From a standard normal distribution table, or z table, the area of the probability density function curve from z = -infinity to z = 4.0 is 0.999969. So, the number of oversized parts = 1 − 0.999969 = 0.000031, or 31 DPMO. Since the ND curve is symmetrical, the same number of parts will be produced on the negative side of the mean. Hence, total number of defective parts will be 62.

13.7   Quality Control Charts

The objective of a control chart is to reject or accept a lot as the batch is being produced. This is normally an on-line procedure. A sample is drawn from the manufactured batch, and quality is conducted to determine whether the batch is acceptable or not. This is also called a statistical process control, or SPC. The control parameter is measured and plotted in a chart to determine if there is any assignable cause for defects. If there are, such as operator issue, material issue, coolant issue, tooling issue, etc., the cause must be removed. One of the fundamental conditions of using the control chart is that the process must be stable. Control charts can also be used to determine normal cause variations. Chance variations can be due to noise in the manufacturing process: heat, material hardness, coolant conditions, vibration effects, etc. These are uncontrollable variations of the manufacturing process. The dimensional variations due to assignable causes must be eliminated before test samples are drawn.

The quality control chart was developed by Shewhart and is a graphical tool to find out if the production is statistically in or out of control. If the data is outside the statistical bounds, the process is out of control and needs further investigation before any conclusion is made whether to reject or accept a lot.

There are typically four types of control charts:

- Xbar chart: for mean value of the measured dimensions

- R chart: for ranges of dimensions produced

- C chart: for defects per unit assembly

- P chart: for attributes only, i.e., defective or non-defective

The first two control charts will now be discussed. The primary reason is that these two charts are very frequently used for the manufacturing process. In order to accept a batch of machines, the machine is subjected to production trials, producing the intended part and the primary and critical dimensions, such as diameter, length, ovality, roundness, taper, etc., of the part are investigated. In case of any abnormality, steps are taken to eliminate the causes. If this is not possible, the machine accuracy itself is assumed to be the cause of such deviations. The assumption is that sample means, Xbar, follow a normal distribution if the number of samples is more than five.

The upper bound of control = UCLx = mean of Xbar + 3σ = $Xbar_1 + 3\sigma$

The lower bound of control = LCLx = mean of Xbar − 3σ = $Xbar_1 - 3\sigma$

n = number of samples

m = sample size

r_{bar} = mean value of sample ranges

$Xbar_1$ = mean value of sample means = $[\sum_{i=1}^{n} Xbar i]/n$

σ = standard deviation of the sample means = $\{[\sum_{i=1}^{n}(Xbar i - Xbar_1)^2]/n\}^{1/2}$

$Xbar i$ = mean value of the ith sample

In abbreviated form, the chart control limits are written in the following way:

UCLx = $Xbar_1 + A_2 {}^* r_{bar}$
LCLx = $Xbar_1 - A_2 {}^* r_{bar}$

$A_2 = 3/(d_2 \sqrt{m})$; the value of d_2 depends on the sample size, m. These values can be found in any quality control handbook. Some values are shown in Fig. 13.2 below. The UCLx, LCLx, and the mean sample values are plotted in Fig. 13.3.

Sample Size	d_2	d_3
2	1.128	0.853
3	1.693	0.888
4	2.059	0.880
5	2.326	0.864

Fig. 13.2 Values of d_2 and d_3 for control charts

Part Diameter: 50.0 ± 0.15

Sample Number	Sample Value			Sample mean	Sample Range
1	50.08	50.09	50.07	50.080	0.02
2	50.1	50.09	50.08	50.088	0.012
3	50.05	50	49.94	49.997	0.11
4	49.89	49.94	50.1	49.977	0.21
5	50.15	49.92	50	50.023	0.23
6	49.88	50.06	50.02	49.987	0.18
7	49.97	50.1	50.12	50.063	0.15

m =	3	Xbar1	50.031
d2=	1.693	rbar	0.130
A2=	1.023	n = 7	
UCLx =	50.16		
LCLx =	49.9		

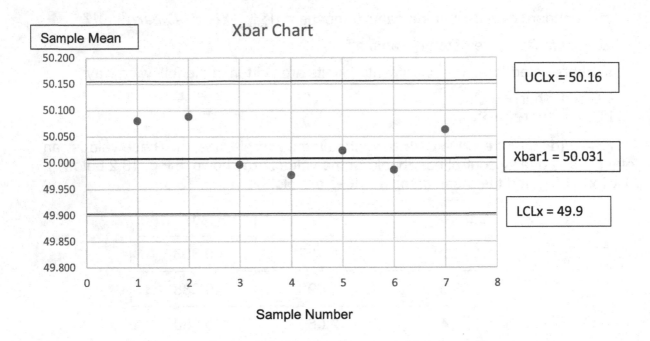

Fig 13.3 Xbar Data and Chart for Process Control

Note: The process is under control since the sample means are within the upper and lower control limits. The next step is to explore the range control chart, as described below.

R-Chart: This is also called the "range control chart" or "control chart" for ranges of the part dimension. The range is the difference between the highest and lowest value of each sample. The R-chart also has upper and lower control limits. This chart basically gives an idea of the variability of the process.

$UCLr = D_3 * r_{bar} = 2.114 * 0.13 = 0.275$

$LCLr = D_4 * r_{bar} = 0.0$

Where

$D_3 = 1 + 3*d_3/d_2 = 2.114$

$D_4 = 1 - 3* d_3/d_2 = -0.114 = 0.0$

The values of d_3 and d_2 are given in Fig. 13.2.

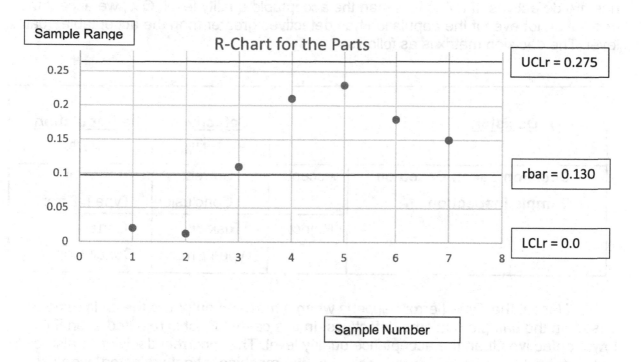

Fig 13.4 R-Chart data and chart for process control

The process of acceptance of the machine for actual production can consist of the following steps:

- Conduct machine position accuracy and repeatability test
- Prepare the machine for run-off test
- Select the methodology, part, and instruments for measurements
- Run an initial test for calculating the standard deviation of the process
- Assume the process capability index required
- Conduct the run-off test and collect data
- Draw control charts for process stability
- Accept or reject the machine
- Document the process

It is obvious that sampling does incorporate some inherent risk when a population is accepted or rejected depending on the outcome of sampling testing. The promise of such testing is that the mean of the sample tends to become the population target when the sample size increases. If this risk cannot be taken, 100% inspection is the only choice to make sure that there is no risk involved in accepting or rejecting a batch. Let's assume that the actual population lot has N% defectives and the sample has n% defectives. If n% is less than the acceptable quality level, Q%, we accept the population lot even if the population has defectives greater than the acceptable quality level. The decision matrix is as follows:

Decision		Defectives in the Population	
		<= N%	>= N%
Decision Taken Based on Sample Inspection	Accept	Correct Conclusion	β Risk or Type II Error
	Reject	α Risk or Type I Error	Correct Conclusion

Hence, the Type I error happens when a machine output or the lot is rejected based on the sample outcome defectives. In this case, the lot is rejected even if it has fewer defectives than the acceptance quality level. This incorrect decision is also called "α (alpha) risk" or "manufacturer's risk," i.e., the machine/lot gets rejected even if it should have been accepted. On the other hand, when the machine or lot is accepted even if the lot has more defectives, the acceptance quality level, this is a Type II decision error. Here the consumer is at risk since he is accepting an actual defective machine. and the probability of making this incorrect decision is called "β (beta) risk." Both of these risks, α and β, are smaller when the number of samples is greater.

Both risks need to be controlled at an acceptable level so that both the manufacturer and consumer strike a balance. Some risk is always there whenever this type of sampling analysis is carried out, but the magnitude s of these risks should be controlled. For most cases, sampling size is predetermined, and only one of these risks can be controlled. In that case, it is reasonable to control the more serious type of risk, i.e., the one that would result in the more costly or undesirable outcome. As discussed in the previous chapter, management and control of these risk factors are possible using an operating characteristics (OC) curve.

13.8   Taguchi Method for Machine Quality

The central concept of Taguchi's method is the idea that quality can be designed in the product and inspection has to be eliminated altogether if possible. The robustness of any product is the idea that it has to work under all environmental noise conditions, which includes material, thermal, cutting conditions, operator treatment, etc. If product quality is more than or less than what the customer desires, it is a loss. If the product has much higher quality than what the customer wants, then all the money invested to make the quality higher than the target is wasted. That is a revenue loss as well. On the other hand, if product quality is less than the target quality, money will be wasted to maintain and service the product, and the customer will not buy it. Moreover, performance around the target quality should have minimum variability. Too much variation is a sign of poor quality, and the customer loses confidence in the product eventually.

Taguchi expressed this concept using a loss function, which is parabolic in nature, or a convex function, which has its minimum at the target quality level. Either side of this target represents a loss and should be avoided. Conventionally, machine performance and reliability characteristics are the two most desirable elements of quality. They should be designed into the product. Inspection, if required, should confirm the fact that the machine supports these design intents. The inspection department should not be a police department; it should be part of the design functions. Readers can consult easily available, quality handbooks to gain knowledge about this important philosophy if desired.

Consequently, product quality has to be designed as per the customer requirements, as mentioned in previous chapters. Similarly, this methodology is to be applied to machines. For a complex product such as a machine, performance is almost everything that it is sold for. Machine performance has to be designed into the product, and performance characteristics must be designed to satisfy customer wants and needs. Machines also must be updated when customer requirements change over time. Sometimes a machine can be designed with future requirements in mind so that changes can be implemented in a much easier fashion.

13.9   Noise Factors for Machine System

A robust design, as mentioned earlier, performs under normal and uncontrollable noise conditions. Robustness points towards the insensitivity of a design to variation of input conditions within reasonable limits. The product must function consistently regardless of input variations, such as environmental changes, material changes, manufacturing variations, etc. These factors are called noise factors since they cannot be controlled. For example, environmental changes, such as temperature, humidity, sympathetic vibration, input voltages, air pressure, etc., cannot be controlled all the time. The machine must perform when the inputs change with reasonable variations. Due to manufacturing variations, two machines might not be exactly similar in their look and performance.

Such variations need to be minimized for a well-designed product. Noise factors can be numerous, some of them can be taken care of during design, and some of them are very difficult to design for. Noise such as vibratory effects could be taken care of by having a machine designed for the highest stiffness and damping. This can help the machine perform better with vibratory inputs. Similar inputs, such as humidity, temperatures, dust, impact and shocks, load variations, etc., are difficult to control. Manufacturing factors can be from other sources, such as dimensional variations, assembly variations, heat treatment changes, human error inputs, premature cutting conditions, etc. These are even harder to control, but machine performance is affected by them to a great extent.

The machine must also perform under wear-and-tear conditions defined by the customers. Any complex, properly functioning machine contains mechanical, electrical, hydraulic, and electronic elements. The noise factors for each of these systems are different. Their interactions create another set of noise condition that also can affect the machine's performance. For a well-designed machine, such interactions must be studied properly, and corrective design steps must be taken to avoid such interactions.

13.10   Machine Inspection Methods

In order to control machine outputs, or even the machine itself, it is subjected to various inspection methods and controls. Several in-process or end-of-the-process inspection steps are designed to ensure that proper manufacturing methods are used and the outputs are well within the desired norms. Most of the machine outputs are checked using statistical process control methodologies (SPC), which consist of physical measurements, control charts, process capability studies, cause and effect diagrams, process FMEAs, etc. These are called "on-line inspection methods" since they are performed as the parts are produced by the machine. Most of the inspection methods and controls are in-line procedures, as they are easy to prove machine quality in a physical sense.

On the other hand, there are some controls that can be taken even before manufacturing. Several design and analytical confirmation procedures can be put in place to make the design perform as intended. These are indirect methods of quality control. The part or assembly is designed and theoretically checked for interference, strength, reliability, and durability in a digital environment. These indirect methods are sometimes referred to as off-line inspection control procedures. This is how the design parameters are optimized to give rise to the best machine performance under given conditions, even with some noise factors, such as load, strength, and thermal variations. The off-line methods are far more effective than on-line inspection methods since changes are much less costly and time-consuming. These are also taken care of by implementing design audits during design procedures. Such quality control is strongly recommended by the system engineering process. Due to the availability of computers with higher memory and more advanced software, off-line inspection methods are gaining ground and replacing on-line inspection methods wherever possible and applicable.

Physical inspections and sampling procedures will never disappear, but their use will be minimized to save time and money as the designs are made possible for higher robustness of the machines. Nowadays, concept designs use parameter and tolerance design to optimize machine performance, as recommended by Taguchi. Design sensitivity is controlled and monitored following system engineering principles. The machine requirements are evaluated under noise variations before the machine is even manufactured.

After manufacturing, experimental designs or design of experience procedures are followed to understand machine performance with respect to design or noise factors. These results are used to design machines with a higher level of robustness and reliability. I hope to see in the near future a complete elimination of on-line inspection methods and controls. Customer use will be the physical verification laboratory without affecting regular machine use.

13.11 Monitoring Machine Health

In order to maintain machine tool quality in the long run, manufacturers must perform at least two steps:

- Monitor machine health by continuing the SPC process applied to the machine output, either in customer shops or in-house
- Collect machine design failure data and perform root-cause analysis to identify the root cause of the failure

These two steps will enhance customer satisfaction and reduce the life-cycle cost of the machines in operation. The feedback will also help the design and manufacturing team improve the design and manufacturing process. If the machine failures are not consistent, the manufacturing process needs to be improved to reduce variations in quality. The failures could be taken as an opportunity to improve the product. Design verification methods might have to be improved to enhance the quality of the machine. Measurement instrumentations might have to be improved as well. The overall objectives of the quality maintenance program should be to analyze the field data or in-house maintenance data to reduce the life-cycle cost of the machine.

In some cases, it has been found that early-stage machine failures are due to poor operator training for the complex machine or the fact that machine manuals are not clear enough for operators to understand how to run the machine productively. To tackle such grave failure situations, the operator has to be trained to keep the machine in proper working conditions, and the operator also has to know how to keep the machine in perfect health by taking preventive steps. The operator has to thoroughly understand the machine complexities and its operations and has to take steps to keep the machine productive and not prone to frequent failures.

The maintenance management team has to keep track of the failure pattern, MTBF, MTTR, etc., and provide such feedback to the manufacturer. The machine operator's training should include education for the following:

- lubrication procedures
- preventive maintenance schedule and its importance
- Collection of failure data
- Hydraulic/pneumatic operations
- Control and electronics components
- Vibration monitoring
- Alignment procedures

- Machine loading and unloading
- Machine overload situations

The machine failure prevention program should include education and training to enhance productivity and machine availability and reduce downtime, etc. Such education and training will definitely improve equipment effectiveness.

13.12 Summary

- Quality starts with customer requirements and ends with their full satisfaction. A quality process supports this migration in a very efficient and productive way. Quality supports the manufacturing and design process, and it starts with the design process. Product quality is an emergent behavior of good design and operation.

- Product quality is built into the product by design, as per the customer requirements, and confirmed by the quality control methods and manufacturing processes, as per the design, to satisfy customer expectations. So, the builder satisfies the requirements, and the customer proves the quality.

- Any product is supposed to do what it was designed to do in a customer environment. Hence, the primary definition of machine quality must include customer satisfaction and product requirements. Quality is the bridge that connects these two objectives. Quality includes precision, accuracy, durability, robustness, safety, and serviceability.

- Good quality reduces scrap and reworking. Quality control and inspection reduce the variation of the manufacturing process and machine outputs. Quality design helps identify the root cause of the problem very quickly. System engineering helps the company to incorporate quality in the product, starting with the design process.

- Defining quality is somewhat difficult and controversial since it depends on many aspects of design and manufacturing. Quality inspection alone is not a definition of quality, and if possible, quality inspection should be eliminated. The design should result in a near-perfect product.

- Quality is product-based and an inherent property of any product. It is defined by the customer and the customer alone. It is both design- and manufacturing-oriented at the same time, and it is the management of system variability of performance.

- Quality has internal and external aspects. External aspects come from the customers, and internal aspects refer to conformance to the design intents. External quality comes from the use of the machine by customers. Internal quality comes from design and manufacturing efforts. Good quality prevents errors and satisfies customer requirements, and it improves the customer-company relationship through the product. A company is known by the quality of the product it delivers.

- Quality inspection uses statistical quality control procedures. The SQC includes acceptance sampling, process control, and design and parameter controls. The SQC helps to identify the uncontrolled process and variations that could be due to assignable or unassignable causes. Unassignable causes lead to random variation, and assignable causes have trends or spikes that need to be eliminated to make the system more controlled. All the SQC processes are controlled.

- Quality focus needs a different direction and dimension: in my opinion, quality should be a tool to manage and control variations of each process in a company. A system approach should include quality and cost simultaneously. Quality should reduce the cost of the product in the long run. Every process has to have a quality target with accepted deviations. Every operator has to be empowered to manage variability in his or her own environment. For example, a sales department has to have a sales target with a given variance.

- A robust design performs under normal and uncontrollable noise conditions. Robustness points toward the insensitivity of a design to variation of input conditions within reasonable limits. The product must function consistently regardless of input variations, such as environmental changes, material changes, manufacturing variations, etc. These factors are called noise factors since they cannot be controlled.

- Taguchi expressed this concept using a loss function, which is parabolic in nature, or a convex function, which has its minimum at the target quality level. Either side of this target represents a loss and should be avoided. Conventionally, machine performance and reliability characteristics are the two most desirable elements of quality. They should be designed into the product.

- Inspection, if required, should confirm the fact that the machine supports these design intents. The inspection department should not be a police department; it should be part of the design functions. Readers can consult easily available, quality handbooks to gain knowledge of this important philosophy if desired.

- The objective of a control chart is to reject or accept a lot as the batch is being produced. This is normally an on-line procedure. A sample is drawn from the manufactured batch, and a quality control procedure is applied to determine whether the batch is acceptable or not. This is also called a statistical process control, or SPC. The control parameter is measured and plotted in a chart to determine if there is any assignable cause for defects. If there are, such as operator issue, material issue, coolant issue, tooling issue, etc., the cause must be removed. One of the fundamental conditions of using the control chart is that the process must be stable.

- The standard deviation of any process represents the spread or variability of the process data. The performance of the system has to be nearest to the target mean. This is also true for machine accuracy. More variation from the target mean means less accuracy of the machine. The term "six-sigma," coined by William Smith of Motorola, represents the extent of variation in a manufacturing or business process. More variability also means less quality. In lay terms, it means that the machine works fine one day but does not perform as intended on another. The machine becomes less reliable.

- The six- sigma approach is similar to Taguchi's quality control method. In order for a machine to be more reliable and productive, the machine must have less variability in accuracy.

- The reliability of any production process can be defined as the probability of a part dimension to be within a specified range. There are many variables that can affect the reliability of production. Most significant is the standard deviation of the part dimensions set by the machines and manufacturing process. For any part dimension, the machine is set to the mean target, but because of many in-process variables, the machine produces parts with dimensions that are not exactly same as each other or even as the set mean diameter.

- A process can be defined by the standard deviation of the dimensions produced by the machine. The standard deviation of the parts produced by a machine depends on tooling, workholding, machine accuracy capability, environmental conditions, material properties, machine rigidity, etc. It is definitely a very complex process since there is no exact answer for any situation and the optimization for the best result depends on the specific situation of machining. Let's explore the effect of standard deviation on the production reliability, which basically determines the suitability of a machine to produce parts with the desired accuracy.

- Quality control should be an in-line procedure. It has to be incorporated while the machine is being assembled and parts are being machined. An off-line procedure is basically useless since the damage is already done while the deviations are being identified. Since some processes are irreversible, deviations have to be identified before a process is completed. This will reduce scrap and production cost.

13.13   Bibliography

Hopp, W.J., and Spearman, M.L, 2000, *Factory Physics*, second edition, Irwin-McGraw Hill Publications, NY.

Eckes, G., 2003, *Six Sigma for Everyone*, John Wiley, Hoboken, NJ.

MIL-STD-105D, 1963, "Sampling Procedures and Tables for Inspection by Attributes."

MIL-STD-414D, 1957, "Sampling Procedures and Tables for Variables for Percent Defectives."

Nahmias, S., 2008, *Production and Operations Management*, fourth edition, McGraw-Hill Erwin Companies.

Phadke, M.S., 1989, *Quality Engineering Using Robust Design*, Prentice-Hall, Englewood Cliffs, NJ.

Rao, S.S., 2014, *Reliability Engineering*, Pearson Publishing, Prentice Hall, NJ.

Taguchi, G., 1986, *Introduction to Quality Engineering*, Asian Productivity Organization, Tokyo, and Kraus International Publications, White Plains, NY.

Taguchi, G., 1987, *System of Experimental Design*, volumes one and two, UNIPUB/Krauss International Publications and American Supplier Institute Press, NY.

Ulrich, T.K., and Eppinger, S.D., 2011, *Product Design and Development*, McGraw-Hill Companies, NY.

13.14   Review Questions

- Can quality be eliminated completely for a manufacturing operation? Is inspection control required for any operation? What is better: on-line or off-line inspection methodologies?

- How do you control quality by controlling variability in the manufacturing process? How do you justify management of variability in a productive tool to reduce machine failure and increase machine reliability?

- Does the SPC process guarantee machine performance throughout its productive life? How do you identify an out-of-control situation for a machining process? Why aren't SPC methods applicable when the process is out of control? What happens to the SPC predictions if the machine output does not follow a normal distribution?

- How are design and quality related? Can quality be enhanced through design only? Does complexity reduce the quality of the product?

CHAPTER 14

14 Industrial Design for Machines

14.1 Introduction

This chapter deals with the role of industrial engineering in machine tool design. It is also proper and timely to mention the fact that I never witnessed any presence of industrial engineers in the design department during my tenure in machine tool design for almost 30-plus years. Even if the application of industrial engineering in the design of production machinery is very critical and worthwhile, very few companies really follow that path while designing and manufacturing machines for production applications. Nevertheless, industrial engineering applications for machines are critical and urgently required to make the machine productive but safe and ergonomically sound.

The application of industrial engineering for automotive and related products are numerous and widely accepted. The success of machine tools normally depends on critical features and performance aspects of the machine. Industrial engineers can play a very important role during machine design and manufacturing. The following critical factors of the machine could be enhanced if industrial engineering principles are applied in a systemic way:

Performance Features: The challenge is to have more performance packed into a smaller machine. The customer always wants to have consistent and reliable features at a reduced cost and smaller size. Important features, like higher spindle power, are required for machining exotic material at a higher production rate. In order to provide higher spindle power, spindle motor size needs to be bigger, as does the control package. So, the challenge is how to fit this larger motor and control into a machine of the same size as the previous. To increase cutting loop stiffness, larger ball screws and stiffer way blocks are required, but the challenge is again how to fit those components into an even smaller size.

Ergonomics: The next critical requirement is to design the machine to reduce operators' fatigue while loading and unloading parts or replacing the tools. The spindle height has to be at a recommended height that is the industry standard. The machine control positions should be at optimum locations so that the operator can approach them at ease in case of emergencies.

The shape of the control box and machine assemblies should be optimized in such a way that the operator and service personnel can access them easily and comfortably. The control box should have a parking and operating position at eye level so that the operator can comfortably operate the machine and service personnel can repair it without much difficulty.

Reliability and Durability: The machine has to operate reliably and without interruptions, increasing production output and accuracy. The machine has to also be robust for performance and durability. The machine has to operate with a reasonable amount of noise. The performance and life-cycle cost have to be optimized accordingly. The machine has to be safe for the operator, and it has to withstand environmental conditions, such as high temperature, humidity, and dust. Durability should not be affected by vibration inputs from surrounding machines. The alignment should not change due to impact conditions.

Size and Weight: Consumers of machine tools have a different view of these two aspects of machines. In the industry, a higher machine weight has been viewed very favorably and everybody thinks heavier machines deliver higher performance and part finish due to higher damping. This is a myth. The stiffness and damping of the machine have to be optimized based on requirements and industry standards. The requirements for stiffness and damping for turning machines are not the same as for grinding machines or machining centers. Machine stability also does not depend on weight, stiffness, or damping. The weight of any machine has to be optimized instead of simply maximized.

Aesthetics: This is not a requirement in the machine tool industry. In comparison to automobiles and cell phones, looks and aesthetics are at a much lower requirement level. As long as a machine has a high production rate, its looks are not important. Very few customers really care about aesthetics or a slick look, and accordingly, builders also do not have a tendency to design the machines with better looks. Nevertheless, the machine's look creates appeal for buyers, and industrial designers normally create that appeal in the design.

Manufacturability: Industrial engineers find out the optimum manufacturing conditions for a particular machine design. They help the team find conditions that help the manufacturing team manufacture the design intents with efficiency and productivity. Optimization of manufacturing inputs is one of the critical jobs that industrial engineers perform for the company. For example, the design of electronics and controls have to be completed with safety, reduced replacement time, and performance in mind. Industrial engineers can also help make it easy for quality engineers to inspect the machine. Overall, industrial engineers can play a critical role in optimizing manufacturing conditions that help the company be more profitable and productive.

Supply Chain Management: One of the most important jobs for industrial engineers is to manage vendors and purchasing functions to reduce input costs. Supply chain management is a critical function for industrial engineers to make the product more economically viable. Proper input materials make the machine more accurate and reduce the machine's production cycle time as much as possible.

Industrial engineers help management control the flow of input materials, storage of materials, and supply of materials at points of production and whenever required. Inventory management reduces and optimizes cost inputs, making machines more profitable. They also design, monitor, control, and execute material movements inside and outside the factory. The primary objective is to supply material just in time for production operations. This effort supports system engineering principles, as mentioned earlier. Another critical operation for industrial engineers is to control and manage the risk management for supplies.

Considering the complications involved in the design of production machines, which includes mechanical design, electrical design, software design, control design, and safety design, to name a few, industrial engineers can play a very important role in optimizing and managing machine design, manufacturing operations, and safety design.

Another aspect of design is to manage technology transfer and management. The team has to find out the readiness of any technology for machine applications. This is not an easy task since the machine tool industry does not give much latitude to using any upcoming technology unless it enhances performance, reliability, and durability without increasing cost. For example, using digital technology instead of analog was adopted with efficiency since it increased the easiness of operation while also increasing production. The ultimate function of industrial engineers is to develop a machine that provides all the desired features without increasing the size and weight at a comparatively reduced cost. The development of product architecture is managed and facilitated by industrial engineers for ease of assembly, manufacturing, and quality control. Hence, industrial engineers help to be innovative and to design a creative and productive machine using proper and proven technology. Their role helps the company be strategic and become successful in the desired market. U.S. machine tool builders should use industrial engineers to be productive and creative in the future and turn around the industry.

The primary job of industrial engineers is to connect customers with the product with respect to aesthetics, looks, styling, safety, ergonomics, and functionality. Actually, in my professional and academic experience, this is one aspect of design that has never been used throughout the world. This is very unfortunate since industrial engineers can help companies optimize their products. This function has to be embedded in the design process right from the beginning and should not be an afterthought. The builders have to keep in mind that the product does not sell well only due to features and performance in the long run and repeat sales in the market also depend on safety, style, and aesthetics.

Hence, industrial engineering functions are part and parcel of the system engineering process. The product is only successful when it is widely accepted by customers due to various reasons that include features and performance. In order to sustain global competition, the machine has to have quality performance along with other aspects that the competition does not provide, such as safety, ergonomics, etc. In other words, the machine has to be healthy in a greater sense to be successful. Imagine a machine having the best features and performance possible but not the sheet metal to protect operators from the dynamic portion of the machine. Such a machine will not sell at all.

If the suppliers are inefficient and inventory cannot be controlled, the product cost will go up eventually for sure. Hence, the product has to blend technicality, manufacturability, and safety. The product has to be ergonomically efficient, too. This means that in order to face the competitive wind in the market, builders have to provide a complete product that has quality performance along with safety and ergonomics to win the customer's mind. Moreover, style, looks, and aesthetics, in general, create the brand aspect of a product, and such considerations foster brand recognition.

14.2 Industrial Design Principles

The Industrial Designers Society of America defines industrial design as "the professional service of creating and developing concepts and specifications that optimize the function, value, and appearance of products and systems for the mutual benefit of both user and manufacturer." Hence, the functions of industrial design engineers are as follows:

- Creation and development of concepts, i.e., design architecture development
- Optimization of functionality, product values, and appearance
- Customer and manufacturer satisfaction

Consequently, the industrial design satisfies the customer by optimizing the product's value. Industrial engineers should be part of the design team, supporting system engineering activities. The industrial design is a part of product design. Industrial design engineers sometimes have to become more than just technical engineers, i.e., they have to be artists as well. They enhance product values through optimization and, at the same time, make it ergonomically safe and aesthetically pleasing. Hence, product design is aimed more towards enhancing the functional value of the product, and industrial engineers join hands with product designers to make it a complete product that satisfies the customer.

Industrial design engineers perform multi-functional duties to help both the designers and manufacturing engineers. These engineers create architecture through shapes and forms using geometry and simplicity. They shape the form of a product to make it functional. Industrial design concepts were originally conceived in Europe after the Industrial Revolution. Engineers produced parts that could be mass produced to satisfy tremendous growth. In the US, the necessity of using these concepts came much later in the industry, and even now, use of such concepts is very limited in the machine industry.

In the US, industrial engineering was used to create product differentiation and brand recognition. It was more of a marketing than a design tool. Very slowly, this primitive idea of using industrial design as a marketing tool is just beginning to change and is being blended with product design to create more product value. Dreyfuss has identified the following primary areas for industrial design engineers:

- **Interface Design:** Linking the product with the users
- **Aesthetics:** Making the product appealing to the customer through forms and shapes
- **Maintainability and Serviceability:** Reduction of life-cycle cost
- **Cost Reduction:** Enhancing product value through optimization
- **Product Differentiation:** Ensuring the product reflects company values and strategies

In addition to the above functionalities, the author would like to add a few more:

- **Supply Chain Management:** Value addition through supply chain controls

- **Ergonomics:** To make the product operator friendly and easy to use

- **Safety Considerations:** Make the product safe as per industrial requirements and safety standards, such OSHA in the USA

- **Manufacturing Support:** Value addition by optimizing the manufacturing process

- **Value Analysis:** Optimization of resources using computer simulation and other technical methods

- **Product Coordination and Management:** Coordination of all the system design activities among different departments through budget and cost analysis

Industrial design engineers get trained in various disciplines, such as basic engineering principles, manufacturing principles and processes, supply chain management, optimization and simulation, system engineering, and project management. Almost all universities around the world conduct undergraduate and post-graduate courses in industrial and system engineering. The scope of this branch of engineering has become wide and complex over time to the extent it has become almost parallel to other branches of core engineering disciplines, such as mechanical and electrical engineering. Anyway, all of these functionalities can be used in machine design, development, and manufacturing areas to make the product more viable and economic.

The importance and application of industrial design have not been fully explored in machine tool industries yet. Even if the application of simulation and optimization is very wide, it has not entered into the machine tool industry. It is somewhat apparent that machine tool industries do not think that such system engineering tools are required for designing and developing machines. There are many aspects of machine tools, such as value optimization, safety aspects, and ergonomics, that could improve substantially if industrial design principles are used.

14.3 Industrial Design Requirements

Historically, industrial design has been applied extensively for either consumer products or aircraft industries. In both of these cases, aesthetics and looks are very important. The automobile industry has also used industrial design to a lesser extent in the areas of safety, ergonomics, and aesthetics. For other products, such as industrial and medical products, the use of industrial design has been minimal. In the construction industries, industrial design has also not found to be of much importance. In all of these products, industrial design has been used by the design team only. A special effort has not been made to explore industrial design as a tool to optimize the product for customer use.

The primary reason could be the amount of interaction between the product and users. When strong interaction and communication are required, industrial design has been extremely useful. For example, cell phones and other home products have used industrial design extensively. To gain fuel efficiency, airline industries have used industrial design principles extensively. For the automobile industry, shape, form, and looks are very important aspects of any commercial design. Human interaction, ergonomics, and safety are also very important for the product to be successful. So, industrial design and system engineering have found use to a greater extent. On the other hand, the machine tool industry has not found any use for industrial design, since the primary purpose of such equipment is production. There is not much "touch and feel" for such products. Another reason could be the small budget for the design and development of such products.

On average, investment in the machine tool industry has been very low, and the idea is to design a machine that performs even if it does not look good or is ergonomically inferior. Moreover, design has given importance to reducing human interaction with the machine. Industrial robots have been used to eliminate manpower. So, one of the design considerations is to reduce man-machine interaction. Nevertheless, industrial design principles could be used to enhance machine value by optimizing resource inputs or simulating the working conditions ahead of the machine's introduction. Another use could be with the technical content of such equipment.

Design in the machine tool industry is pretty much conventional, and all such machines now have a stereotypical and boring look. There is nothing new and exciting in the machine tool industry. The product does not excite users, only giving them a chance to enhance productivity. This concept has to change drastically for a machine to be competitive globally. It has to have its own identity and characterization to gain market. Performance and features are important, but other design aspects such as safety, ergonomics, looks, and ease of use and interaction are also important to design a complete product. The importance of using industrial design principles extensively in such products shows up in their budget allocation. Industrial design is always an afterthought for such products. Historically, the budgetary allocation for industrial design has been at the lowest level for machine design projects. It is also true that the importance of industrial design has not been effectively demonstrated to design and management teams for such products.

The question arises: is industrial design important for the machine tool industry? The answer is definitely yes without any further question. This is a wide-open field for industrial design applications. Even for existing and legacy products, industrial design principles could be used to improve products to make them a commercial success. The products could be improved by enhancing the looks, aesthetics, shape and form, ergonomics, and ease of interaction. Hence, the use of industrial design in machine tool application is multi-dimensional. It can be shown that strategic use of industrial design tools can make the machine more customer friendly and value added. During the design phase, requirements for ergonomics, safety, aesthetics, serviceability, maintainability, etc. could be set up with some metrics.

In the area of ergonomics, the machine tool lacks in many aspects that need further attention to make it more productive and efficient. The ergonomics of any machine must make the machine very easy to use by nature. The control software should be very easy for any operator to use by instinct. In such cases, industrial designers have to ensure that the product talks to the user through shape and form. The communication link should be very natural and instinctive, similar to cell phone or computer use. Another important aspect of machine design is to make the machine easily serviceable so that repair time is as low as possible. Maintenance and diagnostics should be obvious by design. In order to make the machine friendly to service personnel, industrial design for the product must be looked into. In the field of user interaction and communication, machines have to be designed to allow easy access, easy loading and unloading, and higher safety and machine management in case of crashes. Human interaction needs special design considerations to be effective and productive. The method of intervention by the operator while the machine is in machining mode could be improved. Industrial design of a machine should effectively answer the following questions:

- Can the machine be made multitasking similar to computers?

- Can it be serviced while it is producing parts?

- Can the machine select its own cutting path to make the process more productive and accurate?

- Can the machine have its own brain and artificial intelligence to operate efficiently?

- Is the machine safe for operation and service?

- Can the machine be serviced while it is producing parts?

- Does the machine have the unique capability to serve the needs of users in case of emergency?

It has been mentioned before that aesthetics, looks, shape, and form are all important aspects of industrial design function. Before this requirement is included in the system design requirements, the team has to ascertain whether shape, form, or architecture is important for the machine. Does the industrial design help build product branding or differentiation with respect to other similar products in the market?

The utility of this tool needs to be fixed first. For machine tool markets, which are saturated with similar types of machine tools, form or shape is very important to make a dent. Technology for machines is somewhat static and stereotypical by nature. Since everybody is using similar technology and design in an overall sense, can industrial design make a difference? It is apparent that for a machine, a visual differentiation does attract customers in the beginning, before they look into performance features.

The visual appearance of a machine does make a difference among customers as the products stand out from the pack. Industrial design tools can help when the market is saturated with similar products. Sometimes the shape or form does instill pride of ownership among customers. The customer becomes proud to own the machine. It creates a very strong appeal in the customer's mind about the product when performance and features are all the same among machines. The machine has to look robust and safe in the beginning.

An aesthetically streamlined and safe product does create a pride of ownership among design and manufacturing team members. For a machine in general, safety, ergonomics, ease of use and serviceability, aesthetics, shape, and form should be prioritized. Industrial design tools should help the product to establish itself in the market by reducing the cost and enhancing the value of the machine. Each one of these machine attributes should have a requirement metric. If industrial design helps the product sell, this tool has value and should be a part of the design. Management has to push for using such a tool to make a difference.

The customer should be able to imagine and visualize machine performance through its shape and form since function follows them. The machine architecture should communicate to the customer the superiority of the machine over its competitors. Then only the design is successful in totality. The appearance should display a sense of ease of use, ease of maintainability, ruggedness, and robustness. The design must display a sense of customer friendliness and uniqueness. The machine should make users very comfortable and convinced about safety and ease of interaction and communication.

Builders have to find out why customers select some machines and not others. What makes one machine different from another? What are the unique characteristics of a machine? What are the advantages and disadvantages of one machine over others available in the market? Answers to all of these questions can be answered through the use of industrial design tools if applied properly. Builders have to understand the fact that a good product helps build confidence in the consumer. Customers should feel proud to own a particular machine that nobody else has. The product should demonstrate its sellability and usefulness to customers. No amount of convincing and discussion will outperform firsthand experience and the idea the customer forms by just looking at the machine. Before even proving out the performance of the machine, the customer should feel that it is worth investing in it. Japanese machines are slowly doing that in the U.S. market.

For any design program, the economic side of any tool should be evaluated before implementation. The program management team can evaluate the cost of implementation of industrial design tools and processes. Evaluating the cost of this application could have an effect on the cost of the program. Unless there is an economic benefit, it should not be used. For example, the administration of a program should include the cost-benefit ratio for this industrial design tool. The implementation of this program should include all the fixed and variable costs for such implementation. The fixed cost should include the cost of the simulation and optimization tools, salary and benefits for the industrial engineers, manufacturing process tool costs, instrumentation or measurement tool costs, and other facility costs used specifically for this program. The variable cost could include the test program costs, test setup costs, and instrumentation costs required for design evaluations.

Once the total cost for this industrial design program is evaluated, the benefit obtained is determined. Unless the benefits exceed the cost, implementation might not be required. It also has to be kept in mind that some benefits might have advantages beyond the financial, and they should not be ignored. For example, the effect of safety design, ergonomics, and aesthetics on the sale of the machine can be almost impossible to be determined. Another example is the effect of industrial design on corporate identification of machines.

For example, Fanuc, Okuma, and Moir Seiki machines from Japan have definite corporate identities. Their products' shape and color combinations identify which company made them. US products do not have such an identity. Even in such cases, the effects of color and shape combination on revenue can difficult to determine. For many products, corporate identity is established through product shape, size, and aesthetics. Industrial designers can also help the team design logos, packaging, etc. for the products. Visual identity is also very important for many users. For example, all the servo and robotic products from Fanuc Japan have the same color combination and a similar look. They are very visible, and consumers know the company through these products. Industrial designers can definitely help the company and design team to establish such recognition by designing the correct aesthetics, shape, and form of the product. In such cases, a unanimous decision has to be taken by the team members as to whether industrial design is required irrespective of its cost to the program. It is also a fact that product looks and aesthetics can establish corporate identity when the products maintain uniformity and common standards.

The outside appearance of the machines could be the same to establish a sense of visual identification for the company's products. Such effects of industrial design on consumer products have been studied at length by MIT and other universities, but studies on the effect of industrial design on industrial products are almost non-existent. No standards have been established yet for machine tools. Moreover, such studies are carried out by specific companies, and they are not published or disclosed for common knowledge.

It is also a fact that the determination of industrial design effects of products like machine tools is extremely difficult. The determination could be very subjective and arbitrary by nature. In my opinion, a return to the use of industrial design tools and processes for machines will be a net positive, and sometimes also indispensable for certain applications. A customer survey could be conducted to determine such effects for a specific product.

14.4 Industrial Design Process for Machines

Industrial design is part and parcel of the system engineering process, and it should be carried out during design and development. The functionality of the industrial design is almost similar to the mainstream design functions, but it has a different scope. It is an aid to the product design. Side by side with the product design process, industrial engineers take care of the aesthetics, ergonomics, and safety of the product in addition to the supply chain management, project management, and resource optimization. Industrial engineers should do everything possible to make the product safer and easier to operate. There is no hard and fast rule for an industrial design process, but it could include the following steps to be more productive and effective:

- Extract industrial design requirements from the system requirements for the machine

- Identify the metrics and prioritize the requirements that will most satisfy the customer
- Develop system design concepts to satisfy the requirements
- Select designs most suitable for the product
- Complete design details for the selected concept
- Test and finalize the concept
- Develop supply chain and manufacturing requirements
- Integrate the concept with the product design functions

The system engineering process starts with the identification of system requirements that will most satisfy customers. System requirements are then dissected to identify sub-system and component requirements. Similarly, the safety, ergonomic, and aesthetic requirements for machine tools are to be identified and explored for further implementation for the product. Each requirement must also have a metric to satisfy. Some of the industrial design requirements could be subjective, such as decent looks, color combination, etc., but by looking at the legacy company product and competitive products, a definitive requirement has to be developed. Cost for the product could be one of the criteria to end subjectivity. A survey could be conducted among customers to know their preferences with regard to developing a better interaction between the product and the user. The survey could also be conducted among internal customers or departments, which could include, marketing, sales, manufacturing, purchasing, and service. In the end, a metric has to be developed for the attributes and process as to how to satisfy this requirement.

A thorough and deep understanding of customer requirements and the prioritization of such requirements will help industrial engineers develop concepts that satisfy them. Several concepts are developed for comparison, and the concept that satisfies these requirements the most is selected and pursued further. Similar to mainstream product design activities, industrial engineers make sketches that have the potential to satisfy requirements. As a team, the best concept is selected, tested, and integrated into the product. The technicality of the best concept is also simulated and optimized using several industrial engineering tools and software. One such example could be sheet metal design for the machine. The design should allow operators and service personnel to access parts that will need servicing or replacement down the line in the minimum possible time. Hence, if there is a metric for MTTR (mean time to repair), the design could satisfy such a metric. For machine tools, several access doors are provided to satisfy this requirement. Industrial engineers communicate the ideas and integrate the concept into the product design. Such coordination and concurrence are essential for industrial engineers. For the development of supply chain management, they have to work with the purchasing and quality control departments to come up with an effective plan. Such coordination among engineers is very crucial to the success of the product as a whole.

For machine tool application, industrial engineers build models using plastics, wood, foam, and other easily available materials. Such models also help manufacturing engineers develop the manufacturing process for the concept. A feasibility study is also often done for the concept. For the exploration of several concepts, normally, soft models are built, and models with hard metals are avoided at this stage of the game. Such a procedure is highly iterative by nature, and concept development is on the left side of the system "V." Hence, concept development has to be completed using a minimal amount of time and cost.

Similarly, to develop a final color scheme for the product, models with different color combinations are made visible to the participants for their opinion. In addition, cost, time to complete painting, product differentiation, availability of paints, facility requirements, and easy paint maintenance are requirements for color selection. This could be a very lengthy and complex process. Most of such modeling could be accomplished using computers. Computer models are easy and less expensive to develop and build. Computer models can also take care of design integration and manufacturing aspects at the same time.

Once the concept is finalized and selected, the final drawing of the prototype is done. Once the engineering drawing is completed, depending on the nature and complexity of the concept, a hard prototype is built. For crash protection, a physical prototype is built and tested for confirmation. I have done several such testing and development for the main sliding doors that protect the operator. Similarly, the location, opening size, and number of service doors around the machine need lots of engineering effort. Sometimes a complete sheet metal prototype is built to make sure these objectives are accomplished.

I witnessed during my professional career as a design engineer the fact that shrouds or sheet metal designs are done after everything else has been completed for the machine. It is basically an afterthought and does not follow a system design process. Due to this, many times, design changes have to be made to accommodate sheet metal requirements, safety requirements, or ergonomic issues. Such action causes the team to delay some events, and the project cost and duration tends to go up. Another noticeable fact is that the whole design team works in secluded compartments without much communication and interaction with others. During the design phase, product engineers, industrial engineers, manufacturing engineers, and the finance team should work together on a regular and consistent basis.

In recent days, computers have made the design process much easier, and most of the uncertainties and design unknowns can be solved using software and related tools. Several concepts can be built, and digital prototypes can be completed in the computer itself. The color scheme can be developed, and the manufacturing interferences can also be evaluated in a computer. The ergonomics and safety issues for the machine can be solved using virtual software. The design team should make use of such technological advances to minimize project time and cost.

The CAD (Computer Aided Design) or CAE (Computer Aided Engineering), including simulation tools, must be used for the project. Digital prototyping should be done before physical prototypes are built. When the budget and time for developing machines are very restricted, such computer tools become very handy for resolving most issues. The whole system design process becomes more productive and efficient.

From the concept, detailed drawings can be developed directly and in a very timely fashion. Drafting accuracy and quality are also enhanced. Part-drawing management becomes easier since computer software manages all of these internally and can be accessed by many members of the project team simultaneously. This allows the team members with different functions to work on their portion concurrently and without any interference.

Industrial products are very complex and need both the latest technology and users' requirement priorities. Hence, machines are not only technology-push products, but also user-driven products, and the design process has to use methods that will support both of these approaches. The machine has to optimize performance features to be competitive and user-oriented, but on the other hand, it has to be safe and rugged. Since it is mostly a user-guided product and volume of production is also limited due to the low requirements of the industry, budgetary constraints force the team to design the machine in a very time-restricted manner. Hence, most of the time, machine tool projects are not as detailed and elaborate as automobile or customer-driven products. For machine tools, performance measures are very important for the product, and at the same time, user-friendliness and user interface are also very important. This makes the machine design very challenging. The machine has to be cost-effective, performance-driven, user-friendly, ergonomically designed, safe, robust, etc.; all of these aspects have to be bundled into one package. Hence, such products are very difficult to conceive, manufacture, and deliver in the marketplace, which is saturated with similar products.

A common question is always asked about the industrial design process: is industrial design required for all types of products, or is it particularly applicable to a specific type of product. In order to answer this question, the types of products, i.e., consumer-driven products, technology-based products, user-based products, industrial products, etc., need to be discussed as well. Most of the technology-driven products are based on advanced technology, such as production machines, computers, and aircraft. The performance measures for these products are their most important characteristics. Here the product design for the best features becomes the focal point. For these types of products, the management and application of technology are very crucial for its success. Product safety, ergonomics, and aesthetics have to be blended with technology to satisfy the customer. Hence, industrial engineers can help the design team manage the technology in a timely fashion to make the product more attractive to customers. They can also help the product look more sophisticated and be easier to use. Nevertheless, the use of industrial engineering practices could have wide applications, such as inventory management, supply chain management, project management, technology management, cost optimization, value addition, and simulation.

For household products, such as cameras, watches, etc., technology is definitely important, but in addition, a product's ability to communicate and link with customers is very important. The product has to be user-oriented. The competition is very severe for such products. Cost is definitely an issue. Since users are cost-conscious, cost optimization against product features and performance become very relevant for such products. User interaction has to be easier. Here the focus for industrial designers is to design the product for easier communication. In addition, they have the responsibility of managing the product's cost and value in such a way that users can afford to buy it. Even in the case of automobiles, cost management is a very important issue. The product has to deliver what customers look for at a price that they can afford. Here the focus is on product differentiation. The technology is given, but the use of the technology to make the product more customer-oriented becomes more critical. Hence, the primary focus of industrial engineers for this product is to make it more competitive and user-oriented.

It should be pointed out that when technology is used to enhance product performance and features, industrial engineers have to focus on how to do so at a reasonable cost. For such products, industrial design should concentrate on safety, ergonomics, value addition, project management, value management, etc. For consumer-oriented household products, industrial designers have a more involved role to play. They have to focus on customer connection with the product. Here, aesthetics, looks, shape, and user-friendliness become more important for consumers before they buy the product. For example, a chair has to be safe and rugged. In addition, the chair has to look slick and appealing to the customer. The form and shape become very critical. A washing machine has to use the latest technology in such a way that customers can easily use it. Cost management is also critical due to heavy competition.

Overall, industrial design is a sub-process of the total system engineering process and is a part of the product development process. The usefulness of a product development team is almost over when manufacturing activities start. The industrial design process can contribute until the product is delivered to the customer. Industrial engineers can contribute during manufacturing activities, inventory management and control process, quality control process, and value management for the product. In that sense, the industrial engineer's scope is very wide and diverse. Even if the product is delivered to the customer, industrial engineers can help customers optimize their processes for better production. Industrial design activities start concurrently with product design and development activities and end when customer satisfaction is achieved throughout the life cycle of the product. The system engineering process encompasses product design and industrial design activities with equal importance. One compliments the other, and both are equally necessary and important for product success.

14.5 Evaluation of Industrial Design for Machines

In order to identify the importance of the industrial design process in the system engineering of the product, an evaluation procedure for such applications must be established. The goal, objectives, and metrics for the evaluation must be established before the process is included in the design activities.

- Customer feedback on the quality and functionality of the product. The success of the industrial design will depend on the answers to the following questions
    - o Is the control user-friendly?
    - o Does the machine have enough diagnostics?

- o Is the machine safe enough for continuous use?
- o Is the machine capable of controlling the machine in emergency situations?
- o Are the control and emergency switches in easily accessible positions?
- o Does the machine look attractive?
- o Does the machine create a sense of "pride of ownership"?
- o Does the machine create value for the customer?
- o Is the life-cycle cost of the machine optimized?
- Aesthetic appeal of the machine:
  - o Is the machine aesthetically appealing to the customer?
  - o Does the color combination look attractive?
  - o Is it an exciting product?
  - o Does the machine look like a quality machine?
  - o Does the machine cover come off easily when serviced?
  - o Is the main sliding door designed to reduce operator lethargy?
  - o Does the machine look robust and rugged?
- Maintainability and serviceability of the machine:
  - o Are the access doors designed for easy maintenance?
  - o Can parts be moved easily for replacement?
  - o Are the service parts easily and locally available?
  - o Is part removal very easy?
  - o Can all serviceable parts be accessed easily?
  - o Can parts be aligned easily after replacement?
  - o Are the MTTR and MTBF reasonable and competitive?
  - o Is the life-cycle cost optimized and competitive?
  - o Is the design life for the machine optimized for best ROI?
  - o Does the machine exhibit early-life failure issues?
- Manufacturing and inventory issues:
  - o Are service parts easily available without storing in-house?
  - o Does the machine satisfy the production rate?
  - o Does the machine create bottlenecks for production?
  - o Does it consistently produce parts?

- o Is it reliable enough?
- o Is the line balanced?
- o Is the preventive maintenance schedule in line with other machines?
- o Is the machine documentation clear to service personnel and operators?
- o Can the machine be programmed easily?
- o Does the machine add value for manufacturing?
- o Can the machine be controlled remotely?
- o Can it be easily connected to conveyors or robots in the future?
- o Are loading and unloading very easy for the operator?
- o Is there a health concern for the machine, such as coolant management, water leakage, and smoke removal?
- o Is the machine environmentally safe?
- Resource management and control
  - o Are the inputs and outputs for machine manufacturing managed and optimized?
  - o Does the machine have a positive net worth over the life cycle?
  - o Are the inputs easily available?
  - o Has the value analysis completed and optimized for the machine?
  - o Does the machine value satisfy customers?
  - o Does the assembly need well-trained personnel?
  - o Is the project cost under control?
  - o Are the input resources, such as human resources, manufacturing materials, inspection devices, etc., easily available?
  - o Are there enough engineers available to finish the project on time?
- Sales and marketing aspects:
  - o Does the machine create a positive image for the company?
  - o Is it in line with legacy products?
  - o Does the machine create positive product differentiation in the desired market?
  - o Does the machine stand out among its competitors?
  - o Is the marketing cost too prohibitive?
  - o Is the machine a "me too" product?

- o Is the machine competitive performance-wise?
- o Would the customer buy the machine at any cost?
- o Does the distributor like the product's features and attributes?
- o Does the machine value maximize customer satisfaction?

- Optimization and simulation aspects:
  - o Are cost constraints satisfied?
  - o Are the functions optimized?
  - o Have the ergonomics been optimized?
  - o Has the simulation been completed to optimize manufacturing resources?
  - o Have the demand and supply conditions been analyzed effectively for the manufacturing plan?
  - o Have the operational constraints been optimized?
  - o Is it too costly to manufacture the product?

14.6 Industrial Design Applications for Machine Tools

Simulation: The application of simulation tools in the area of industrial design is very useful in solving several types of problems. It is basically a "what if" game for solving industrial engineering problems. It is said: "If everything else fails, try simulation" Simulation can also be used for dynamic problems with respect to time changes. For some applications, a simulation tool can be used as a replacement for physical experiments in a real situation. A simulation tool can be used to solve problems in facility design, queue design in hospitals, machine tool demand situations, etc. The simulation process is shown in the flow diagram in Fig.14.1. It consists of several steps:

- **Define the problem:** Problem definition has to have an objective function, constraints, or constraint equations and assumptions. The problem could be a maximum or a minimum problem. For example, the objective function has to be optimized under a given set of constraints.

- **Model Development:** Decision function, controlled and uncontrolled variables, and constraints need to be defined at this step. The purpose is to develop a mathematical and logical model so these equations can be solved. The value of the objective function has to be determined for various controlled variables. If any variable is fixed, these variables are uncontrolled. If the value of the variable can be changed, it is a controlled variable. Basically, a model function has to be developed wherein the decision variables are linked to the controlled variables. The clock or time of variable change has to be also determined. The clock timing can be fixed or dynamic depending on the type of problem. For inventory management simulation, inventory could be dependent on the amount of consumption, which depends on production and demand. Hence, the clock timing is variable. One other hand, if sales forecasting for a year or two is being simulated, clock timing could be fixed, e.g., every month

- **Program Flow Diagram:** The flow diagram helps to develop the logic of the program, and it helps resolve issues with the program when executed. It shows the program steps in sequential form, and it also shows the mathematical relationship for the simulation equation in the form of $F = f(a, b, c)$, where values of F for different values of a, b, and c (uncontrolled variable inputs) are to be determined. Initial values of some fixed variables have to be specified.

- **Simulation Program:** The programming language could BASIC, FORTRAN, MATLAB, PASCAL, or any other language that the programmer feels comfortable with. Many commercial simulation programs are currently available. For day-to-day use, an in-house program is less expensive, and the time for results is also comparatively less.

- **Collect Data:** Once the program is developed and logic flow is tested for accuracy, the filed data is collected for program inputs. For example, the data could consist of machine cost, machine sales price, machine demand distribution, etc.

- **Model Validation:** The program has to generate realistic data that matches the real-world data as close as possible. The data trend also must be logical and realistic. The generated data depends on the assumptions made. The assumptions should be very clear and simple enough that they can be programmed.

- **Program Execution:** The simulation model is executed for different scenarios. Several runs are made for each of the decision rules, if applicable, and outputs are compared for different conditions. For each simulation model, sensitivity analysis must be carried out to test the sensitivity of the results against input variation.

- **Implementation:** Simulation study is not useful if the results cannot be implemented and validated. It should help in making decisions. The results should help the company to, for example, manage the inventory of incoming materials depending on the variable demand pattern.

Let us take a simple example of predicting production rate based on the past data of demand. The simulation will be for 12 months. The manufacturing cost of each machine is, say, $20,000, and the selling price is $25,000. The problem if production is more than sales is that the inventory goes up, and it costs money to keep the machines in inventory. On the other hand, if the demand is more than what is produced, you lose market share. Hence, a strategy has to be determined for finding the best production rate so that profit is maximized. The simulation procedure described above will be used to illustrate its ability to predict the production rate. The answer will consider two scenarios for production:

Condition One: machines produced this month will be the same as demand in the last month

Condition Two: machines produced this month will meet the expected demand irrespective of demand in the last twelve months

The best decision has to be selected under given cost constraints. We have assumed filling conditions from the past:

- One year is the simulation run, so we have to find what should be the production rate for the next 12 months.

- The demand data for the past 100 months have been recorded and used for simulation.

- The expected demand for the next 100 months has been calculated using demand frequency and corresponding demand.

- The expected demand = $12*0.04 + 17*0.11 + \text{--------} + 40*0.04 = 26$ machines per month.

- Now the midpoint demand data, frequency, and random number allocation for the 100 past months are shown in Fig. 14.4.

- From any random number table, the first two digits of sequential data are taken for simulation. The numbers are shown in Fig. 14.5.

- Twelve random numbers are generated corresponding to 12 months of run time in simulation, and we will convert them to assume demands by assumed 100 random numbers, 00 to 99.

- The random numbers are converted to machine demand as shown in Fig. 14.6 as per two decision rules. Production rate can only be equal to demand.

- Profits are calculated for two decisions, as shown in Fig. 14.6.

- It can be seen that the best decision is to produce machines at a rate of 26 per month irrespective of fluctuating productions every month. The maximum profit for 12 months is found to be $1410000 under decision rule two.

- Now actual data has to be collected to see that simulation results match with the actual data. If it does not, the input data needs to be looked at, and a new simulation should be run.

There are many such applications for the simulation process. Some of the important ones are given below:

- Materials Requirements Planning (MRP): This is used to control incoming materials requirements for the factory. Depending on the simulation output of the production rate, incoming material cannot be adjusted using the simulation procedure.

- Inventory Management: Both static and dynamic, linear and non-linear, simulation procedure can be used to control and manage the inventory of finished goods, in-progress material, scrap inventory, etc.

- Job and Machine Scheduling: In order to schedule workflow for the machines and operators, the simulation process can be also used. The process remains the same, but inputs and decision rules are changed as required.

- Facilities Design: For fast-moving machines, sometimes the outlet design for machine loading and unloading could use the simulation process. Simulations can also be used for complex queuing situations for machines to balance their workload.

The simulation process is not used that often in the machine tool industry currently, but its adoption would make the manufacturing process more efficient and productive.

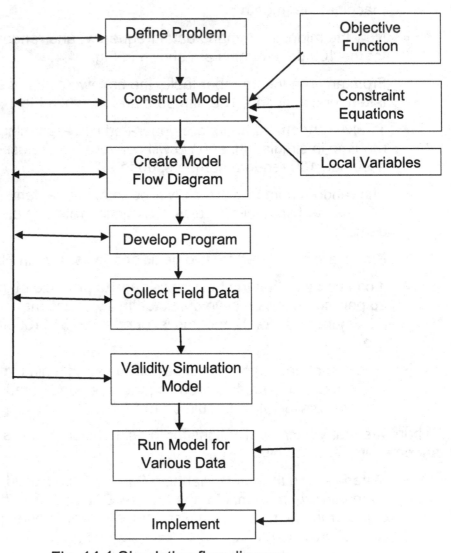

Fig. 14.1 Simulation flow diagram

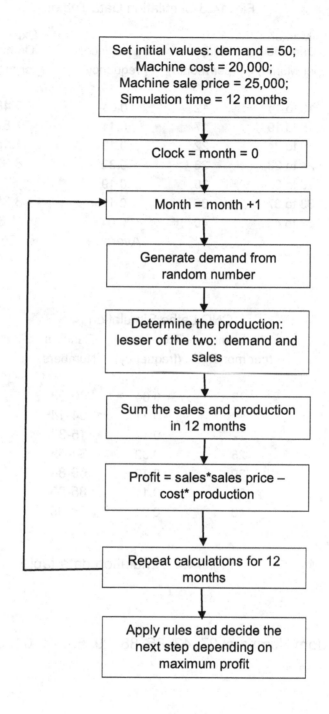

Fig. 14.2 Simulation program flow diagram

Fig. 14.3 Simulation Data Tables

Market Demand (per Month)	Midpoint (per month)	Occurrences (frequency)	Expected Demand (per month)
10 to 14	12	0.04	0.48
15 to 19	17	0.11	1.87
20 to 24	22	0.19	4.18
24 to 28	26	0.32	8.32
28 to 32	30	0.19	5.7
33 to 37	35	0.11	3.85
38 to 42	40	0.04	1.6
		Average	26

CNC Lathe Simulation

Midpoint (per month)	Occurrences (frequency)	Random Numbers
12	0.04	00-03
17	0.11	04-14
22	0.19	15-33
26	0.32	34-65
30	0.19	66-84
35	0.11	85-95
40	0.04	96-99

Fig. 14.4 Simulation data table

Random Numbers: 27, 43, 85, 88, 29, 69, 94, 64, 32, 48, 13 and 14

Fig. 14.5 Simulation data table

CNC Lathe Production Simulation

Month	Random Number	Demand (per month)	Rule 1		Rule 2	
			Machine Produced (per month)	Machine Sold (per month)	Machine Produced (per month)	Machine Sold (per month)
0		26				
1	27	22	26	22	26	22
2	43	26	22	22	26	26
3	85	35	26	26	26	26
4	88	35	35	35	26	26
5	29	22	35	22	26	22
6	69	30	22	22	26	26
7	94	35	30	30	26	26
8	64	26	35	26	26	26
9	32	22	26	22	26	22
10	48	26	22	22	26	26
11	13	17	26	17	26	17
12	14	17	17	17	26	17
Total		339	322	312	282	282

Sales Price: $25000 per machine
Cost: $20000 per machine

Rule 1:   1360000   Rule 2:   1410000

Fig. 14.6 Simulation data table

14.7 Optimization for Machine Inventory

The optimization principle can be used very effectively for inventory management problems. Some companies import machines from overseas companies, store the machines in their facilities, and sell them when the customer orders the machines. In such cases, inventory management is very critical for the success of the company. The questions are how much to order, when to order, and how much to order at one time. The cost structure for the inventory machines must be understood properly for optimizing the number of orders. In order to stock the machine in the facilities, cost structures must be understood properly, and eventually, data must be available for further analysis:

- **Item Cost:** The price of the machines must be paid before the machine is shipped from the overseas company. The cost is called "cost of purchase" and is expressed as dollar per unit multiplied by the number of units ordered or quantity procured. In case there is a discount for a volume of purchase, the discount is applied to the purchase price as well.

- **Ordering Cost:** This cost is associated with transportation, purchase order processing, inspection, receiving, insurance for overseas shipping, etc. Such costs do not depend on the volume of purchase. They are fixed per order or batch of orders. This significantly affects the selling cost of the machine.

- **Stockout cost:** This cost occurs when demand cannot be fulfilled due to a shortage of machinery in the inventory, i.e., running out of stock. Customers want the machine when they need it. They do not want to wait for long, and this cost could include loss of sales since the customer might buy the machines from competitors. It also includes loss of goodwill, etc.

- **Holding Cost:** Once the machines are received, they are held in inventory for a period of time until they are sold. The holding cost is charged as a percentage of the order value per unit time. For example, 10% holding cost means that it will cost 10 cents to hold $1 of inventory for one year. The holding cost might include the following items:

  - **Cost of Capital:** From the time the machines arrive at the facility until they are finally sold to the customers, capital is stuck in the form of machines or inventory. This represents the opportunity cost of the capital invested in the machines since the same amount of money could be invested somewhere else.

  - **Storage Cost:** This includes storage cost or cost of space, in-house insurance cost, and taxes incurred. This is also somewhat of an opportunity cost. If the machines were not kept inside the facility, the storage space could be rented out to earn money. These are fixed costs that might vary with the inventory level.

  - **Obsolescence Cost:** Sometimes machines become outdated and cannot be sold anymore since new and improved ones have come onto the market. Moreover, some machines could become rusty and useless since the customer does not want to buy reconditioned machines. They also sometimes lose accuracy if they are lying idle for a long period. Hence, cost is incurred to make the machines useful again. This is also called deterioration cost due to holding the machine for a very long time. These costs might also include breakage cost associated with holding the machine in inventory and movement of machines inside the factory storage.

The inventory cost is somewhat difficult to assess sometimes. Machine price is easy to handle since it is very well known per order. The ordering cost can be assessed by analyzing and collecting data from accounts. The ordering cost must include all the items that are associated with a particular order, and such costs can also change from time to time. Carrying costs should include the cost of the capital invested in machines to buy them. Carrying costs, such as storage, deterioration, obsolescence, etc., might be obtained from accounts. Stockout cost is the most difficult to assess. Cost of goodwill is not always known. This could be estimated from experience in the past. If repeat orders become fewer and fewer, this could be due to less stock in the facility. Such costs are company and product specific.

Another decisive factor in inventory management is to identify whether the demand for machines is dependent or independent. Independent demand for finished goods or spares does not depend on the company, but it might depend on the market or economic conditions. This demand is independent of company operations and management. On the other hand, dependent demand depends on the amount produced. It is internal to the company. For example, each machine needs one workholding device, so if more machines are sold, more workholding devices must be kept in storage, and they are fitted in the machine just before it leaves the facility. In the following analysis, we will consider independent demand only. For independent demand conditions, a replenishment philosophy is adopted, i.e., when the stock reaches a certain level, more machines are ordered to keep it at the desired level. For independent demand conditions, an economic quantity is ordered to replenish the stock when it drops below a predetermined level.

The famous economic order quantity (EOQ) analysis was developed by F.W. Harris in 1915 and promoted in the industry by Wilson. It is also very widely used in various industries and can be used in machine tool industries as well. The derivation of EOQ depends on the following assumptions:

- The demand rate has a sawtooth pattern with known parameters. In such cases, demand is constant, repeating, and defined from previous data. For example, the demand could be 15 machines per month with no variation, and future demand will remain the same as well.

- Lead time is defined as the time elapsed between ordering and receiving the order in the facility. Lead time is also considered constant and known from past data.

- No stockout is allowed or considered in such an analysis. Since the demand rate and lead time are known and constant, the time to order can be adjusted to avoid stockout conditions

- Machines are ordered in batches or in a known quantity. The lot is stored in inventory before shipment.

- Machine price per unit is constant and depends on the number ordered.

- Carrying cost is linearly dependent on the average inventory level over a period.

- Ordering cost is fixed per order and is independent of the number of machines in an order.

- Only one kind of machine, e.g., lathe, machining center, etc., is in one order. If the prices are the same for all types of machines, then this assumption is not necessary. The inventory level against time is shown in Fig. 14.7.

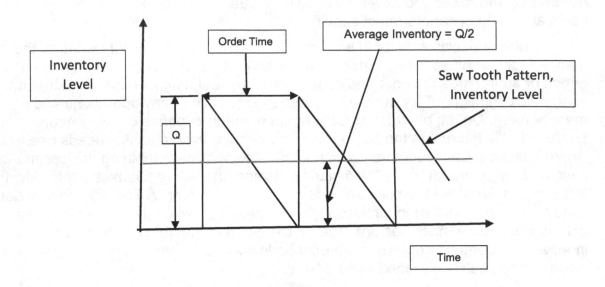

Fig. 14.7 Inventory level vs. time

For the analysis, the legends are as follows:

- D = demand rate, machines per year sold
- Q = machines ordered per batch
- P = machine cost per unit, $/unit
- I = interest rate, % of dollar per year
- TOC = sum total of ordering cost + carrying cost, $/year
- S = set-up cost per order, $/batch

Hence, the number of batches of orders in one year = D/Q = yearly demand/batch size per order.

So, cost per order = number of batches of orders in one-year x S.

Yearly carrying cost = I x P x Q/2; the order arrives just when the inventory level is zero and total inventory becomes equal to Q. The stock is depleted at a constant rate, and the average inventory level is Q/2. Cost of holding one unit per year is I x P.

Total cost per year, TC = ordering cost per year + carrying cost per year = SD/Q + iPQ/2; TC vs. lot size, Q, is shown in Fig 14.8 below.

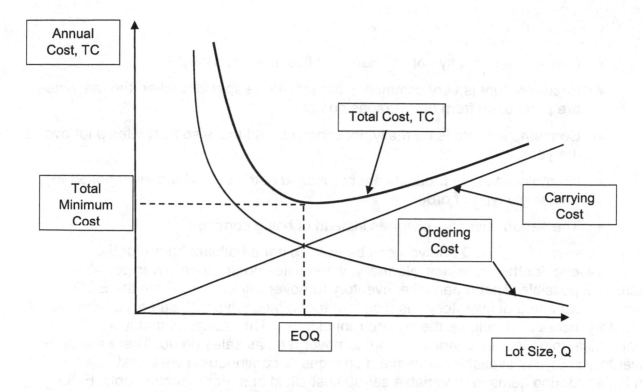

Fig. 14.8 Total cost of inventory vs. lot size, Q

It can be shown that as the lot size, Q, increases, ordering cost goes down since fewer orders are placed per year. Again, as Q increases, the carrying cost goes up linearly. One cost goes down, and the other goes up. When these costs are combined, it shows a quadratic convex relationship with a minimum called the economic order quantity, which gives rise to a minimum cost. The EOQ can be calculated as shown below:

TC = SD/Q + iPQ/2, so for minimum point, first derivative, d(TC)/dQ = 0, then

d(TC)/dQ = -SD/Q^2 + iP/2 = 0

EOQ = (2*S*D/iP) ^0.5, which is also called the Wilson economic order quantity.

For example, D = 200 machines per year

S = $1,000 per order

I = 20% per year = 0.20

P = $12,000

Hence, EOQ = (2*1,000*200/0.2*12,000) ^0.5 = 13 machines per lot, and 200/13 = number of orders = 15 total batches per year, i.e., every three weeks, a batch has to be ordered. The minimum cost of operating this inventory will be

TC = 1,000*200/13 + 0.2*12,000*13/2 = 15,384 + 15,600 = $30,984, i.e., $155 per machine is the inventory cost. Up to the EOQ inventory level, the ordering cost goes down, whereas the carrying cost goes up. At the EOQ point, the minimum cost occurs, where the ordering cost equals the carrying cost component.

This EOQ analysis has many limitations as described below:

- Demand is normally not constant and fluctuates substantially

- Price discount is very common in the industry, especially when the machines are purchased from an overseas company

- Consumption rate is normally not constant, and this also fluctuates a lot over the year

- Normally, multiple products are purchased at the same time as opposed to one single type of order

- The set-up cost also changes instead of being constant

Nevertheless, EOQ analysis can be used to get a ballpark figure for the inventory level for the machines. Normally, when sales go up twice, inventory also doubles if possible. That means the inventory turnover ratio is two. From the EOQ formulae, we see that inventory has a square root relationship with sales, i.e., the rate of inventory increase should be the square root of sales. This indicates that it is not economical to have a constant inventory turnover ratio as sales go up. There are other inventory analyses available for demand changes or continuous review systems, multiple ordering items, and variable set-up cost situations. For machine tools, EOQ analysis is very much acceptable.

14.8 Summary

- The primary objective of the industrial design process is to make a machine safe and aesthetic. It also helps enhance the ergonomic quality of the machine.

- Industrial design connects the product with the customer through its visual look and ergonomic values. Industrial design can help the machine be more competitive and attractive to customers in a highly competitive market.

- Industrial engineers can help the team manage and coordinate the project among all the functional teams, including manufacturing and quality control departments. Industrial design makes the product less costly and easier to manufacture.

- The applications of industrial engineering for automotive and related products are numerous and widely accepted. The success of machine tools normally depends on the critical features and performance aspects of the machine. Industrial engineers can play a very important role during machine design and manufacturing.

- The optimization principle can be used very effectively for inventory management problems. Some companies import machines from overseas companies, store them in their facilities, and sell them when customers order them. In such cases, inventory management is very critical for the success of the company. The questions are how much to order, when to order, and how much to order at one time. The cost structure for the inventory must be understood properly to optimize the number of machines to order.

- Inventory cost is somewhat difficult to assess sometimes. The machine price is easy to handle since it is very well known per order. Ordering cost can be assessed by analyzing and collecting data from accounts. Ordering cost must include all the items that are associated with a particular order, and such costs can also change from time to time. Carrying costs should include the cost of the capital invested in machines to buy them. These costs, such as storage, deterioration, obsolescence, etc., might be obtained from accounts. Stockout cost is the most difficult to assess. Cost of goodwill is not always known. This could be estimated from experience in the past. If the repeat orders become fewer and fewer, this could be due to less stock in the facility. Such costs are company and product specific.

- EOQ analysis can be used to get a ballpark figure for the inventory level for the machines. Normally, when sales go up twice, inventory also doubles if possible. That means the inventory turnover ratio is two. From the EOQ formulae, we see that inventory has a square root relationship with sales, i.e., the rate of inventory increase should be the square root of sales. This indicates that it is not economical to have a constant inventory turnover ratio as sales go up. There are other inventory analyses available for demand, such as change or continuous review systems, multiple ordering items, and variable set up cost situations. For machine tools, EOQ analysis is very much acceptable.

- Industrial design is a sub-process of the total system engineering process and is a part of the product development process. The usefulness of a product development team is almost over when manufacturing activities start. The industrial design process can contribute up until the product is delivered to the customer. Industrial engineers can contribute to manufacturing activities, inventory management, and the control process, quality control process, and value management for the product. In that sense, the industrial engineer's scope is very wide and diverse. Even if the product is delivered to the customer, the industrial engineer can help the customers optimize their processes for better production.

- A common question is always asked about the industrial design process: is industrial design required for all types of products or is it particularly applicable to a specific type of product? In order to answer this question, types of products, e.g., consumer-driven products, technology-based products, user-based products, or industrial products, need to be discussed as well. Most technology-driven products are based on advanced technology, e.g., production machines, computers, and aircraft. The performance measures for these products are their most important characteristics. Here, the product design for the best features becomes the focal point of the design.

- The Industrial Designers Society of America defines industrial design as "the professional service of creating and developing concepts and specifications that optimize the function, value, and appearance of products and systems for the mutual benefit of both user and manufacturer." Hence, the functions of industrial design engineers are as follows:

- o Creation and development of concepts, i.e., design architecture development
- o Optimization of functionality, product values, and appearance
- o Customer and manufacturer satisfaction

- The shape and location of the control box and machine assemblies should be optimized so that the operator and service personnel can access them easily. The control box should be at eye level so that the operator can operate the machine comfortably and service personnel can make repairs without much difficulty.

14.9 Bibliography

Hopp, W.J., and Spearman, M.L, 2000, *Factory Physics*, second edition, Irwin-McGraw Hill Publications, NY.

Industrial Design, Magazine Publications, NY.

Lucie-Smith, E., 1982, *A History of Industrial Design*, Van Nostrand Reinhold Company, NY.

Nahmias, S., 2008, Production and Operations management, fourth edition, McGraw-Hill Erwin Companies, NY.

Ulrich, T.K., and Eppinger, S.D., 2011, *Product Design and Development*, McGraw-Hill Companies, NY.

Schroeder, R.G., 2006, *Operations Management*, third edition, McGraw Hill Publishing Company, NY.

14.10 Review Questions

- Can industrial design add value to machines? If not, how can you modify the industrial design process to enhance the value of machine tools? How do you differentiate an industrial engineering process from a product development process?

- Can industrial design help the manufacturing process be more efficient and productive? Can it also help enhance the quality of machines? Can an industrial design support the system engineering process for a machine tool?

- Discuss the procedure for developing random numbers from a normal distribution using cumulative probability distribution function.

- Discuss the effects of sensitivity analysis for machine tool demands in the market.

- Discuss the advantages and disadvantages of the simulation process for machine tool applications. Is it necessary? Why is such a process not in use in machine tool industries?

- Identify the different types of inventories carried in machine tool industries and machine shops. Identify the various purposes of such inventories.

- Why is stockout cost difficult to assess? How can it be estimated better?

- Why is the item price important in calculating the EOQ for a machine? How can you include the effects of discount pricing?

- What are the differences between replenishment and requirement strategy? Is it important to consider these differences?

C H A P T E R 15

15 Design for Manufacturing & Assembly

15.1 Introduction

At the outset, it is fair to say that a machine tool is not mass-produced. It is a piece of precision equipment that is used for mass production of accurate components. Hence, the machine tool needs special attention during manufacturing to maintain its special class. The pride of any machine tool builder is to build a precision machine, and the product should display that in the purest form. The machine's performance depends on the accuracy of its build. Most theories that promote manufacturing, though, are written for mass-produced components or assemblies, such as for automobile and consumer products, etc. So, the application of methods and processes such as design for manufacturing have to be used with special caution. Machines, in most cases, are produced in batches or, as is the case with special machinery, built one at a time.

Nevertheless, some of the elements of design for manufacturing (DFM), design for X (DFX), etc. could be used to make machine building somewhat easy without sacrificing accuracy. During the design process, some design principles can be used to make manufacturing easier and inexpensive. Any recommendations of the DFM that help the designers make the build easier, accurate, and less expensive should be followed. That is the main point of application of DFM in machine tool design.

The primary idea is to design a machine that can be built easily and accurately to satisfy customer requirements. Product specifications should satisfy these requirements through the design of the product. Instead of following the DFM methodology, I think it is more befitting for machine tool designers to follow the design that will enhance reliability, robustness, environmental and user-friendliness, safety, ergonomic aspects, productivity, efficiency, and accuracy. Such properties of the machine are called "X" in the DFX. This DFX methodology is also sometimes called "design for X." I am of the opinion that a DFX methodology is more appropriate than DFM, particularly for machine tools. This methodology could also be called "focus design." Both of these methodologies are geared towards reducing manufacturing cost through design. DFX and DFM methodologies should help machine tool design engineers in the following areas:

- Development of design methodology that enhances product quality and reduces cost

- Better trade-off policies among design parameters

- Develop and selection of alternate designs to satisfy requirements

- Optimize the design for ease of manufacturing and reduction of manufacturing cost
- Selection of design for higher profitability and brand recognition
- Reduction of manufacturing set-ups and in-process quality inspection
- Quicker assembly process to reduce cycle time
- Accuracy enhancements

When all is said and done, product cost has to be minimized without compromising quality to enable the company to survive and grow. The product has to be financially attractive with respect to similar products in the market. In order to accomplish this objective, the design has to promote better manufacturing methods and processes in addition to promoting and sustaining quality and accuracy. Company revenue depends on the volume of sales and price of each unit. More important is the profitability of the company, which is the difference between revenue and total cost. In order to increase profitability, the cost of the product has to be reduced since there is a limit to the volume of sales.

Reducing manufacturing cost through better design is within the control of the company, whereas sales are somewhat dependent on market conditions. So, performance with reduced manufacturing effort and cost will help to establish the product in the market and, at the same time, increase the profit margin for the company. To be successful in this aspect, the company has to practice and implement the DFM and DFX methodology. As said before, DFX or DFM has to start with the design for the product. If a product can embrace all of these principles concurrently, it will be accurate but with reduced cost. I believe that DFX is better suited for machine tools due to its low volume of production. The design has to promote accuracy, precision, and ease of manufacturing, and only then is the design near perfect.

DFM or DFX starts with the initiation of the design for the product. Design and DFX are concurrent engineering processes. The system engineering process integrates these two parts into an integrated design process for the product. The design for X property needs the same inputs as the mainstream design needs, such as requirements, assembly design and details, etc. Design optimization needs DFX requirements or metrics. DFX also specifies manufacturing requirements, such as durability requirements, robustness requirements, and other properties for the design, and it helps the team reduce the manufacturing cost without sacrificing quality aspects, which are also part and parcel of DFX requirements. Since I believe DFX is more pertinent to machine tools, the discussion going forward will include the DFX aspects of design and manufacturing processes. DFX efforts will combine the design with manufacturing. DFX helps the team to reduce the manufacturing cost by reducing personnel and optimizing manufacturing processes.

DFM or DFX requirements or metrics should be fixed before the design is started. The DFX process should be started right from the beginning and move along with the design process until the product is proven out. DFX efforts should optimize the manufacturing efforts of the design to reduce cost and increase manufacturing efficiency. The primary idea of using DFX or DFM is to reduce the manufacturing cost by modifying the design and manufacturing processes. At every step of the design, cost elements for manufacturing are evaluated. Even if the cost evaluation is somewhat approximate by nature since many manufacturing processes are not known accurately, it is necessary to evaluate the cost of the design from manufacturing perspectives.

Ultimately, the manufacturing process that is most cost-effective without compromising quality is required for manufacturing the product. The cost evaluation might also help the team to make the trade-off for the requirements. In case a requirement turns out to be very costly, it might have to be traded with other similar requirements that are easy to manufacture. Components with excessive weight, such as base, slide, etc., might cost the company, so DFX should be able to identify and suggest ways to reduce that weight. More often than not, the weight of the component has a tendency to grow to satisfy many conflicting requirements such as stiffness, damping, speed, etc. The system engineering process enables the team to optimize the requirement metrics by using DFM or DFX in combination with design activities. Recently, some authors have used the term "DFMA," meaning design for manufacturing and assembly.

Another important aspect of using DFM or DFX for machine tools is that it can improve the assembly technique. In most machine tool companies, the assembly process is not very well documented, and the assembly process is left entirely to the assembly workers. The assembly is mostly manual, and robotized assembly is almost impossible. Hence, the process is not standardized, and machine performance variations can occur due to variations in assembly of units such as slides, spindle, tailstock, loading/unloading fixtures, etc. DFM or DFX principles can be used to automate or improve this process. In order to accomplish this, a metric for assembly time can be put in the machine requirements, and DFX or DFM can optimize the assembly process to minimize the assembly time and reduce its effects on the performance or accuracy of the machine.

15.2 DFM or DFX for Machines

As said before, the primary purpose of the DFM or DFX process is to reduce manufacturing cost and time. There are several ways of achieving this objective:

- To achieve a DFX metric for machines, the number of parts has to be minimized by combining them using casting, welding, etc. wherever possible

- Reduction of the number of parts will end up lowering the cost of inspection, storage, handling, pilferage, obsolescence, etc.

- Reduction of the number parts also lowers the assembly time since fewer parts are to be assembled manually

- Reduction of parts eases manual handling in the assembly process. Ease of handling also reduces the technician's fatigue and frustration

- Having fewer parts definitely comes from the fact that the design is less complex. A simple design is the best design.

- In order to standardize the process to reduce variations, assembly procedure documentation is a must. Hence, documentation gets much easier when fewer parts are required for assembly.

- When fewer parts are required, fewer adjustments are necessary during assembly. In that process, set-up time is reduced as well.

- Having fewer parts tends to enhance the quality and reliability of the machine.

- Having fewer parts also tends to reduce the maintenance or life-cycle cost since the chance of breakdown is less.

- Having fewer parts means there's less chance of variation of machine performance. So, having fewer parts enhances the quality of the machine.

- A simpler assembly means less time to assemble, and that means less manufacturing cost, which enhances the profit margin for the company.

- Parts can be easily procured, whenever necessary, through supply chain management.

15.3 Advantages of DFM or DFX Process

There are several distinct advantages of using the design for manufacturing and assembly in the machine tool arena. The DFM or DFX process uses several optimization tools and methodologies that help to standardize the design for ease of assembly and reduce manufacturing cost and time. The design gets matured through this process. Implementation of this process also helps the design team communicate with manufacturing, assembly, and quality engineers while designing the machine. It also helps to identify the areas that will increase manufacturing cost and time. One of the very critical areas is to manage the tolerances and method of machining for the parts. Most of the time, a very tight tolerance can be avoided without affecting accuracy or machine performance. Since unnecessary tighter tolerances for components affect the way they are machined, grinding costs more money than turning the same part. Casting costs much less than forging or machining parts.

Another area of increasing cost of assembly is using fasteners in the design. Fasteners take most of the time in an assembly. To maintain accuracy, the parts need to be adjusted and positioned accurately. This increases assembly and set-up time, which means more cost. Fasteners must be minimized by design. Moreover, the interface must be prepared for putting together the parts to be fastened, so another machining cost is added.

The DFM process looks for such areas to improve the design. Cost of assembly is far more than the cost of bolts and nuts. More fasteners mean greater cost in the end. Fasteners or riveting should be used only in places when it cannot be avoided at all or is required for ease of maintenance. Let's discuss how these procedures, DFM, DFX, and DFMA, can help to reduce the manufacturing cost for machine tools.

The manufacturing cost includes three major items: cost of parts, cost of assembly of units, and cost of production supports, such as purchasing, supply chain management, in-process and end inspections, loading/unloading devices, measuring instruments, inspection equipment, jigs and fixtures, etc. In general, the manufacturing cost of any machine can be categorized into four sub-categories:

- Finished component manufacturing cost: material, machining, inspection, jigs, fixtures

- Sub-system assembly and inspection cost: sub-assembly set-up, assembly time, in-process inspection, rework

- Machine assembly cost: final machine assembly set-up, assembly time, in-process inspection, rework, assembly fixtures, and set-up tools

- Final machine performance, rework, scrap, and prove-out cost

- Manufacturing support cost: cost of equipment, handling, final inspection, storage, rework, inventory carrying cost, inbound material inspection

Hence, these categories can be summed up in a broad sense: manufacturing cost, assembly cost, inspection cost, and support cost. The idea of using the using DFX, DFM, or DFMA is to identify and improve the design areas that can reduce all of these costs. The costs have to be minimized without affecting the reliability, performance, or robustness of the machine. That is the main idea here. It starts with the design of components, sub-assemblies, and final assembly. The cost is the focus of such an exercise. The design is passed through this cost constraint or cost filter for each one of these categories, so for every possible design, the DFX team evaluates its cost.

This process continues as long as the cost is going down with each subsequent design. When the marginal cost improvement flattens out, the design is optimized for manufacturing cost, and the team can stop looking for further improvements. The only challenge is to make sure the elements of the cost are realistic and decisive. In the beginning, finding the cost can be difficult, but as the design matures and legacy cost data is available, cost evaluation gets easier. At some point, the design has to be frozen and released for production without any further changes.

Once the manufacturing design weaknesses have been identified by the DFX team, alternatives are sought, and the design team incorporates these changes in the mainstream design once agreed upon by all members involved.

The most common evaluation yardstick is to find the manufacturing cost once a design has been completed. If the evaluation does not support the cost metric, the question of design alternatives arises. Obviously, the system design phase is on the left side of system "V"; hence, this interaction process is highly iterative and time-consuming. The design goes through several phases, and finally, it evolves as the best possible design that has optimized manufacturing cost without sacrificing quality. The DFX team also has to keep in mind that cost consideration is not the only constraint and other factors, such as performance measures, safety issues, ergonomics, ease of use issues, etc., must be also considered.

Balancing cost and quality are tricks that need to be learned and practiced to get better at. The optimum has to be determined for each product. When design parameters or requirements change over time, cost evaluation also has to redone to see if any further changes are required for the product. For cash cow products, cost evaluation and design changes are not required since the product is mature enough and does not need them for market penetration. An example is necessary to explain the DFX or DFM process for machine tools further. The headstock of a machine houses the spindle, bearings, and workholding devices. The headstock could be bolted to the base of the machine or the cast. Hence, the decision is to cast the headstock part of the base or separately bolt it to the base. Both design options have advantages and disadvantages over the other. DFM or DFMA have to decide the best design that will reduce the cost of manufacturing.

15.4 DFM or DFX Process Details

The design for manufacturing, assembly, or any attribute starts with assessment and analysis. The ultimate objective is to reduce the cost of the design or cost of assembly or optimize the performance attribute at a reduced cost. Obviously, for any proposed design, an initial cost assessment is required. The initial assessment cost is compared to the budgeted cost for the design. The optimization process starts from this point forward. In order to reduce the cost of the design or sub-assembly, the following points should be considered:

- Reduce the number of parts
- Reduce the cost of purchased components or materials
- Reduce the complexity of design
- Reduce the manual intervention or process during assembly

Once the cost of the proposed design is estimated as best as possible, DFM, DFX, or design for manufacturing and the assembly process starts. The team dissects the design to reduce or eliminate parts that are not necessary or can be combined with other parts to perform the same function. The process of application of DFX in the design is highly iterative, and team members should be in a creative mood to refine the design to reduce the cost without sacrificing performance quality. There are no hard and fast rules about the number of design iterations as long as they are improving the performance at a reduced cost. A note of caution: cost minimization must not reduce the scope, quality, or functionality of any design.

The design scope should not be narrowed down to a bare minimum. On the other hand, time for such iterations should not be a criterion as long as the design is improving. If it takes longer to make the design inexpensive without reducing functionality, the process should be continued until an optimum has been reached. At the end of this process, the design should satisfy the cost criterion for the product as a whole or the cost of the product should be less than the budgeted requirement.

Since the design is in a very liquid stage at this point in time, preparing a bill for materials or manufacturing methods to be applied or assembly process to be applied might be difficult to finalize. Nevertheless, an estimate must be made to go forward. A good way is to compare the present design with the legacy design. The following steps can be performed:

- Prepare a bill of materials (BOM) for the proposed design

- Identify the parts procured from vendors or in-house production

- Identify the vendors that will provide the purchased parts at a minimum cost

- Identify manufacturing methods or process to be applied for the in-house parts and find their cost

- Identify the cost of assembling the parts

- Find the total cost of the assembly by adding the assembly cost, procuring cost, and manufacturing cost

While considering the in-house manufacturing cost for the part, an integrated manufacturing process must be considered. The manufacturing costs should include:

- Raw material cost

- Labor cost

- Equipment cost if new equipment is to be procured to manufacture the part or assembly

- Tooling cost, such as for molds, jigs and fixtures, machining/cutting tools

- Cost of machining time and supplies required

- Cost of waste or waste removal

- Set-up and handling cost

In order to assess the cost of the assembly, normally, variable costs are considered. The appropriation of fixed costs, such as product development cost, finance costs, overhead costs, etc., among the parts can be avoided at this stage since the assessment is a relative measure of the cost. The costs that can be allocated directly to a part should be included. Indirect costs or overhead costs can be avoided at this point in time. For product cost, all the costs, direct or indirect, should be included, but for DFM, DFX, or DFMA, indirect costs should be avoided. In general, the manufacturing cost of any product should include:

- Component cost: For purchased parts, such as spindle motor, belts, cutting, hydraulic components, workholding components, etc., the cost should include the total cost of procurement from the vendors. For in-house-produced parts, the total variable cost of producing a part should be considered. Special consideration should be given to a part that needs special tooling, longer assembly time, special machinery, etc.

- Assembly cost: This includes the cost of assembly time or any special tools such as jigs and fixtures, set-up cost, loading or unloading fixtures, etc. It should also include the cost of inspection and adjustments.

- Inventory cost: If the assembly is kept in storage for a longer period of time, the inventory cost must be included.

- Support cost: Material handling cost, inspection cost, purchasing, shipping cost, and pilferage cost. Such costs should be included only if these parts need special services for production.

Let's take an example for cost assessment: spindle assembly, which consists of spindle, belts, pulleys, bearings, hydraulic cylinders, and chuck.

Part name: spindle

Procurement: in-house manufacturing

Number required per year: 50

Material cost per spindle: $150

Machining cost per unit: $100

Handing cost: $25

Tooling cost for grinding: $5,000

Tooling life: 200 parts

Tooling cost per spindle: $5,000/200 = $25

Set-up cost: $20

Total cost per spindle: material cost + machining cost + handling cost + tooling cost + set-up cost = $150 + $100 + $25 + $25 + $20 = $320

Finished spindle cost: $320

Part name: belt set

Procurement: vendor

Number required per year: 50 sets

Cost of handling: $10 per set

Cost of belt: $100 per set

Cost of inventory: $5 per set

Total cost: $115 per set per assembly

Part Name: pulleys

Procurement: in-house manufacturing

Number required per year: 50

Material cost per pulley: $50

Machining cost per unit: $50

Handing cost: $5

Tooling cost for grinding: $100

Tooling life: 50 parts

Tooling cost per spindle: $100/50 = $2

Set-up cost: $10

Total cost per spindle: material cost + machining cost + handling cost + tooling cost + set-up cost = $50 + $50 + $5 + $2 + $10 = $117

Finished pulley cost: $117

Part Name: bearing

Procurement: vendor

Number required per year: 200, 4 per spindle assembly

Cost of handling: $10 per set

Cost of bearing: $1,000 per set of 4 bearings

Cost of inventory: $5 per set

Total cost: $1,015 per set per assembly

Part name: hydraulic cylinder

Procurement: vendor

Number required per year: 50 sets

Cost of handling: $10 per set

Cost of belt: $2,000 per set

Cost of inventory: $50 per set

Total Cost: $2,060 per set per assembly

Part name: chuck

Procurement: vendor

Number required per year: 50 sets

Cost of handling: $10 per set

Cost of belt: $1000 per set

Cost of inventory: $5 per set

Total cost: $1,015 per set per assembly

Total part cost = finished spindle cost: $320 + belt cost: $115 per set per assembly + finished pulley cost: $117 + bearing cost: $1,015 per set per assembly + cylinder cost: $2,060 per set per assembly + chuck cost: $1,015 per set per assembly = $320 + $115 + $117 + $1,015 + $2,060 + $1,015; total cost = $4,642

Spindle assembly part cost = $4,642

Assembly cost = $250

Inspection cost = $25

Handling cost = $25

Storage cost = $25

Tooling and set-up cost = $50 per assembly

Total spindle assembly cost = $4,642 + $250 + $25 + $25 + $25 + $50 = $5,01

Hence, spindle assembly cost, total = $5,017

Now the question is what to work on first to reduce the cost of the assembly. For the spindle sub-assembly, most of the cost is associated with the purchased component. In this case, the purchase cost is about (115 + 1,015 + 2,060 + 1,015)/5,017 = 84% of the total cost of the assembly. Hence, it would be very logical to focus on vendor management and its cost. In case the in-house cost is proportionately higher than the vendor item cost, then the focus has to be on design, manufacturing, and assembly cost. Most of the time, the design drives the in-house cost. So, focus on the design complexities and number of parts. The process has to be documented properly.

In general, the cost components can include the variable costs specifically attached to components and assembly. The appropriation of fixed costs among units manufactured could be slightly tricky at this point in time, and fixed costs should be avoided since these values are used for comparison only. The overhead cost can be avoided also at this point. Some established companies do have the standard man-hours required for machining, assembly, etc., and the company might have a standard cost per standard man-hour (SMH) as well. These values could be used from legacy data. In case more accurate approximation of cost data is required, legacy data for similar parts can be used, and vendor data could be used as well.

For standard and common parts, such as nuts, bolts, dowel pins, hydraulic valves, springs, bearings, electric motors, hydraulic motors, conveyors, etc. data can be collected from vendor's website. One common procedure recommended for manufacturing is to use common fasteners, springs, rivets, etc. In that way, the volume of such parts goes up to a point where a discount can be obtained from vendors. In case the internal production cost of any special part seems to be higher than purchased parts, it should be offloaded to outside vendors for manufacturing at a reduced cost.

An internal buy or manufacture analysis can also be done. In most cases, for standard industrial parts, such as hydraulic pumps, coolant pumps, relays, encoders, etc., it is much cheaper to buy from outside vendors, who can supply similar parts with a specified technical requirement. For most machines, control and electrical costs are about 60% of the machine production cost. Hence, special cost control has to be exercised for such products by negotiating with suppliers.

Special considerations have to be given when the design calls for special and complex parts. A redesign might be necessary when the cost is excessive and simpler alternatives are available. Multifunctional parts, on average, cost more than single-function items. In some cases, it is better to combine several functions into one single entity unless cost becomes prohibitive. Another common point is the part tolerance specified by the design. If the tolerance is too restrictive and actual performance of the part is not necessary at all, the tolerance should be opened up so that these parts can be machined in a less costly machine or by a less costly manufacturing process.

For example, grinding can be replaced by turning operations, or fasteners could be eliminated by welding joints. For some parts, the manufacturing process could be changed to reduce cost. Instead of manual settings, automated adjustments could be used by employing jigs and fixtures. Other types of machines could be used for enhancing part accuracy instead of costly machines. Hence, for any total part cost, the materials, labor, handling, process, and inspection costs should be considered. The standard machining cost for a grinding machine could be almost two to three times the turning machine operation. The casting process is normally way cheaper than machining, and in some cases, a welding process is as well when the volume is high.

The machining cost is affected by the operator cost, loading and unloading fixture cost, tooling cost, machine idling cost, etc. In a nutshell, a design has to be optimized to reduce the manufacturing cost. Another important consideration is the selection of proper material, In order to reduce the weight, a design might use aluminum instead of standard steel or cast iron. Hence, proper material selection affects the machining cost, handling cost, and material cost itself. It has been seen that 15–20% of the cost can be reduced just by selecting proper material and part design tolerances.

For reduction of assembly time, a design has to incorporate simple and easy adjustment procedures and handling requirements. Parts have to be lightweight and easy for automated assembly to reduce the set-up and assembly time. If a design needs a very accurate adjustment for assembly, an automated assembly process has to be used. The assembly process has to be simple by design. Any complicated assembly process has a tendency to create more rework or wastage in the form of scraps. An easy assembly process needs simpler and lighter parts, with the heavy parts at the bottom, and the assembly should promote a top-down assembly process. Parts should be self-locating with automatic adjustments. Parts should also be symmetrical so that they could be assembled without any special orientation requirements, which increase assembly time. For asymmetric parts, the assembly process should be obvious and naturally aligned. The assembly of parts should avoid the need for minute adjustments if possible.

Interference fits should be avoided throughout. Any complicated design needs more adjustments and alignment procedures. These are the advantages of using steps recommended by design for manufacturing (DFM) or design for manufacturing and assembly (DFMA). For easy assembly of the machines, the design should follow these guidelines to satisfy DFMA or DFM processes:

- Fewer parts
- Top-down assembly process
- Minimum part orientation for assembly
- Minimum manual or physical intervention during the assembly process
- Minimum assembly tools requirement
- One-step or single-step assembly process

- Minimum interference fits in assembly

- Reversible adjustment process

- Minimum in-process inspection requirements

- Easy dismantling

The DFM or DFX process will help the team to reduce the number of parts, which will also reduce assembly time, machining, energy absorption, assembly instructions and lead to higher quality, reliability, and robustness, less life-cycle cost, and eventually higher profit margin for the company. Hence, a design guided by these processes will make the company more competitive, as has been shown by Japanese machine tool products in the U.S. market.

As opposed to Japanese machine tool manufacturers, U.S. builders have shown some resistance to adopting and practicing DMA or DFMA for several reasons:

- Disagreement between design and manufacturing teams: one is superior to the other, a superiority complex

- Design process should be separate from manufacturing process: manufacturing follows the design intents

- First design is the most optimized design: no iteration required for refinement

- DFM or DFMA is required for fast-moving automobile components only, and they cannot help machine tool manufacturing or design

- Japanese follow these principles because their design is inferior and needs help from manufacturing

- DFMA, DFM, and DFX are only suitable for high-volume production and not for low-volume machine tool products

- Customers accept the product as designed

- Finally, inertia against any changes: do not fix it if it works

In spite of the fact that Japanese manufacturers are building better and more innovative products at a much-reduced cost, US builders are slow in adopting these principles. Moreover, the product costing method also a big factor. The effect of using these methods and processes will be pronounced when an activity-based accounting system is used instead of an old and conventional fixed and variable cost method. Fixed cost allocation is a very tricky process, and process improvements might not be accurately reflected unless an activity-based accounting system is adopted. Recently U.S. builders are showing more interest in adopting processes and methodologies that enhance efficiency and productivity.

Moreover, bigger industrial houses, such as Ford, GM, and Chrysler, are also pushing suppliers to use DFM to reduce the cost of machines. It is also true that DFM, or DFX, is a small part of the product design and development process, and any product that has gone through a systemic engineering process and an optimized assembly and manufacturing process has gained popularity among customers and helps the company become profitable in the long run. It is only a question of time before quality machine tool builders in the US discover that these processes and methods are helpful in building a quality company that can withstand competitive pressures and gain market share.

Analysis and methods do affect the cost and performance of the product. The life-cycle cost for an optimally designed machine is always less than products that are designed in a haphazard manner. The system engineering process synchronizes the design with manufacturing, and as a result, the product has higher robustness and reliability, is safe and ergonomic, and has better quality and serviceability. The machine becomes a total product and not a partial product that customers will reject over time.

15.5 Cost Management for Components

Professor Eppinger has mentioned several methods of controlling or reducing the cost of components and assemblies in the automobile field, which is earmarked for high-volume production. I call the process "cost management" since, for most machine tool components, use of lightweight materials, such as aluminum, is very limited due to reliability and durability requirements. Hence, in order to reduce the weight of the components for better acceleration and deceleration, they are heavily machined, and this increases the cost.

An explanation for different types of cost might be in order before any further discussion of cost management. There are several types of cost:

- **Set-up cost:** This cost includes any associated with setting the machine up before it starts producing parts. It might include tool adjustments, tool replacements, workholding adjustments, loading and unloading fixture set-ups, etc. A machine with very high set-up cost is normally not liked by the customer since this is idle time for the workshop and it costs a lot of money. The machine might need special adjustments to produce accurate parts. Part programs are also a part of the set-up cost. At present, Japanese engineers believe that there is nothing called set-up cost since they have designed machines that do not require it, as they are all automatically adjusted. They are designing press machines where a die can be changed in one minute, defined by SMED (single minute exchange of die). It is very necessary that machines be designed with minimal possible set-up requirements. This is a fixed cost.

- **Tooling cost:** This includes all the tooling costs, such as soft or hard die cost, machining center tooling cost, turning machine tools, etc. The tooling cost for machining centers, grinding machines, and turning machines vary depending on the number of jobs produced. Tools need replacement as they wear out. This could be a fixed cost or variable cost. Most factories consider it a fixed cost.

- **Material cost:** This includes all the materials or blank costs for machining purposes. It is a variable cost.

- **Variable cost:** This cost varies with the volume of production: material, labor, supports, inspection, etc.
- **Total cost per unit = sum total of fixed costs + variable cost per unit of production**

So, cost has to be managed accordingly to be within budget without sacrificing quality. Another area is the cost of controls and electrical components, such as relays, capacitors, transformers, cables, etc. To enhance reliability, safety, and durability, components need to be robust and reliable. Control software has to be defect-free.

Servo-control components should be defect-free, and the cost of the purchased component has to be managed to enhance the quality of the machine tools. Here the focus is to maintain the quality of the machines and, at the same time, control cost. Another reason for cost going up is the number of vendors that supply quality parts and assemblies is much less. Due to low volume and fluctuations in demand, the profit margin for the vendors is comparatively low.

Hence, cost has to flow in the right direction for the components and assemblies. For most machine tool builders, management of cost for the purchased component is a severe challenge most of the time. For items such as hydraulic units, control software, control components, electrical components, workholding devices, etc. cost runs anywhere between 60–70% of the machine cost. Consequently, proper supply chain management is the key to cost reduction for machine tools. Identifying a vendor who can supply quality parts at the proper time is the challenge that every builder faces. Hence, the builder has to get involved in the cost management for purchased components by providing DFM or DFX cost reduction ideas to the manufacturers of supplied components and assemblies. Supplier management is a good idea in such cases since this concept is somewhat new to the machine tool arena. The idea has to be implemented in full to reduce and manage the cost of purchased components, which usually control the quality and robustness of machine performance.

The process starts with the design, obviously. If the designer only focuses on the functionality of components and assemblies without any regard for the cost, the cost will edge up for sure. Sometimes the designer has a lack of understanding for cost vs. functionality. The specified part tolerances, heat treatment processes, and design complexities are the primary cost drivers. The design team members have to work very closely with manufacturing, reliability, and safety engineers to specify the tolerance and manufacturing requirements of the components. Design complexities must be managed accordingly. Avoiding grinding machines, machining centers, in-process inspection, etc. must be encouraged while designing the parts.

Open assembly processes are recommended. The design engineers have to understand the difficulty and ease of manufacturing and assembly, and they have to understand the principles recommended by DFX, DFM, or DFMA. The concept could be innovative, but if it cannot be manufactured and assembled in a cost-conscious fashion, the machine cannot be sold competitively. In such cases, the parts might have to undergo a redesign to avoid costly manufacturing procedures wherever applicable. Quality and reliability have to be obtained by managing manufacturing methods and design complexities. A balanced approach has to be used for critical components. This is also applicable for sheet metal parts for machine shrouding and guards. Use of laser cutting machines for such parts turns out to be a very costly proposition and should be avoided unless the volume is very high. On the other hand, a manual process for sheet metal components leaves wavy marks and patterns for machine guards, which deteriorates the look of the machine. So, the design of such parts is very critical for cost management.

Another example is using an electric discharge machine, or EDM, to drill a very small diameter hole in the component. The EDM process is very costly and should be avoided. In such cases, the manufacturing team has to collaborate with the design team during the design phase and hammer out the constraints of the manufacturing process. The manufacturing engineers will guide the designer engineers with respect to manufacturing and assembly of components. The cost management process is a very highly iterative process. As the design matures, it gets harder and harder to change it. So, it has to start with DFM or DFMA principles right from the beginning. Application of these principles at the end of the design process is very hard to manage, and cost will go up as well.

Another example is the use of welding as opposed to machining and fastening the components. Even if welding appears to be cheaper and permanent, dismantling, adjustment, and servicing could be very difficult during the operation of these machines. Hence, proper judgment has to be made as to whether the parts should be welded or fastened together. Even in the case of welding, such as with the base of the machine, length of welding runs, type of welds, quality of welds, butt welds or seam welds, etc. need to be managed properly to contain the cost of the base. The welding might add to the weight of the machine. Sometimes a stitch weld is preferred to continuous weld runs where strength requirement is not critical. For brand-new and innovative concepts, the design sketch needs to be discussed with manufacturing engineers as to the manufacturing viability of the concept.

Most of the time, an innovative design concept cannot be produced at an acceptable cost. In such cases, the concept has to satisfy manufacturing constraints without losing the core principle of the design. This is very often possible for most designs, such as indexing unit design or spindle design. In indexing units, hydraulic clamping of curvic couplings is far cheaper than an electric motor and cam coupling. Belts or chain drives are much cheaper than gear train drives. Casting intricate housings is much cheaper than machining them. Servo control drives for indexing the turret seem to be far costlier than hydraulic motor drives. So, design decides the cost.

As mentioned before, the elimination of the manufacturing process or in-process inspection steps can be avoided to reduce time and cost for assembly and machining. A costly machining process can be replaced with fabrication. Machining can also be replaced with a casting or forging process. The number of fasteners or amount of necessary adjustments and assembly time can be reduced by replacing machining with fabrication, welding, casting, etc. The "net shape" process might not be as useful as it is in the automobile field due to low volume and comparatively larger machine sizes. Another possible way of reducing cost is to eliminate in-process manufacturing and inspection steps and the handling of parts during the manufacturing process. Set-up time can be reduced wherever possible. Instead of dedicated loading and unloading mechanisms, robots can handle loading and unloading. A mechanism can be used to change the tools for turning and machining centers.

Depending on the volume of production and similarities of parts, manufacturing process steps can also be changed. This is also called economies of scale, which is difficult to achieve for machine tool production. Manufacturing variable and fixed costs are reduced when the part cost is amortized over the volume of parts.

The variable cost can be reduced by applying automation or robot applications. When volume goes up, the total part cost goes down due to the amortization of fixed costs over the number of parts produced. The cost of jigs and fixtures, handling fixtures, and other support systems could be also divided among the number of parts.

The primary idea is to produce more parts without changing any fixed costs of the process. Similarly, machining can be replaced by injection molding or welding, forging, and fabrication. In reality, the total cost is linearly proportional to the volume of production. But if a special machine is required to produce a part and a new one has to be purchased, fixed costs go up, and this might not be advisable, especially when the volume of production is comparatively low. The process for making parts should have very low fixed costs and comparatively higher variable costs. In any case, a proper assessment of the part has to be made to decide upon the proper manufacturing process.

Let's discuss a scenario where a part has to be manufactured and the team has to decide whether a new machine has to be bought, a casting process has to be used, or the part has to be machined using an existing machine with a robot loading/unloading device. The casting definitely needs a new mold, which costs money. The volume of production is enough to use an existing machine with a robot and loading fixture. The marginal cost for a machining process is definitely higher than a casting process. If the number of parts to be produced is more than the break-even volume, a casting process is justified to minimize the cost. If the volume of production is less than the break-even volume, the team is better off machining the part using the existing robotized machine. On the other hand, the new machine cannot be justified at all for this case. The fixed cost for buying and installing the machine is a very costly proposition even if the design calls for it. In this case, a design analysis has to be performed as to how to obtain the part machining requirements using existing machines or a casting process. Hence, the volume of production decides the process tools to be used, and the primary objective is to reduce cost without sacrificing quality. For a low volume of production, the idea is to use a process with a low fixed cost and a higher variable or marginal cost.

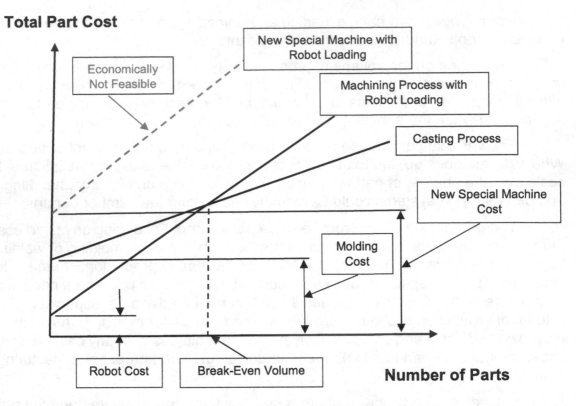

Total Part Cost

Economically Not Feasible

New Special Machine with Robot Loading

Machining Process with Robot Loading

Casting Process

New Special Machine Cost

Molding Cost

Robot Cost

Break-Even Volume

Number of Parts

Fig. 15.1: Break-even volume vs. manufacturing process

Another very feasible way of reducing the cost of the machine is to use standard parts available on the open market. For example, the machine could use standard hydraulic components, fasteners, springs, washers, bearings, etc. to reduce the life-cycle cost. Customers could buy these spare parts for machine maintenance during service. In such cases, the cost of procurement goes down when the volume goes up. The transportation cost might go down, but inventory, obsolescence cost, and pilferage cost might go up also. Inventory carrying cost for purchased items could also edge up when the items are purchased in volume. The volume discounts have to be negotiated against other in-house expenses including cost of purchase and finance.

In case of design changes down the line, too much inventory becomes very costly because the stored components cannot be used anymore. In case of easily available standardized components, items do not have to be stored in-house. Otherwise, the benefits of volume discounts can be offset by finance charges, pilferages, obsolescence, etc. For some costly components, shelf life can be another issue. For example, bearings and rubber components have a limited shelf life beyond which the parts cannot be used anymore. In general, every company has legacy data to support the purchase cost and purchase amount for standard components.

For purchased items from the supply chain, a detailed list of technical and performance requirements can be written for the components by the company and provided to vendors. The company does not say how to achieve these requirements for the product but helps come up with the design.

This is also the same recommendations made by JIT, or the just-in-time process. The supplier is provided with system-level or sub-system-level requirements, and the supplier designs the component to contain the cost and quality as per the requirements. The supplier is given full freedom and responsibility to design and manufacture the component, and they supply the components at a time when the products are required for assembly.

15.6 Assembly Time Management

The following discussion is mostly taken from the concepts presented in Boothroyd and Dewhurst's book and technical publications. As explained earlier, DFMA is an essential part of DFA or DFX. Professor Eppinger also detailed and expanded these principles for automobiles in his textbook, which is listed in the reference section at the end of this chapter. I am going to focus on the manufacture and assembly of machines. The objective of all of these methodologies is to reduce the cost of the product at the end. These concepts mostly have not migrated into machine tool industries because machine tools are not mass-produced items—and they should not be mass produced. Precision equipment should be built with the utmost attention towards quality and accuracy, and the design tends to be more complex by nature. For most machine tools, assembly cost is substantial since more expertise, critical inspection procedures, and assembly time are required to manufacture any precision equipment. Nevertheless, DFA or DFMA principles must be used as much as possible to reduce the cost of assembly.

Application of these procedures can only make the equipment more precise and less costly even if the volume of production is comparatively less than in an automated production system. Another benefit of applying these procedures is to instill consistency of quality and performance in the products. This is an advantage that can be obtained only through these DFA or DFMA methods. Another advantage of these procedures is the reduction of part count and design complexity. A simple design is always the best design. Unnecessarily complex designs can only reduce reliability and durability performance during service. Too many parts will also increase difficulty in maintaining the machine, thus increasing maintenance cost over the life of the machine.

Many consultants and experts of DFA and DFX recommend that teams should keep track of estimated costs, if possible, as the design matures. Cost management is one area builders have to really concentrate on to be competitive in the marketplace. Boothroyd and Dewhurst proposed determining assembly efficiency, which addresses assembly time. In my opinion, the focus should be on precision and accuracy first, and the question of efficiency and productivity comes next. Assembly has to follow a design assembly procedure that supports accuracy and precision. The design will dictate the efficiency of the assembly.

This is different for highly automated mass-produced components, as in automobile and consumer industries. Productivity and efficiency have to be ascertained through efficient design for the machine. For medium- and low-volume machines, the efficiency can be measured using standard man-hours (SMH) required for assembly. From the budgeted system cost requirements for the machine, sub-assembly costs can be obtained with an accepted variation of the cost. Once the sub-assembly cost is obtained, it can be divided into direct and indirect costs. From the direct cost, material and labor costs can be estimated. From the labor cost, assembly cost can be ascertained. If the standard cost for the burden center is known, the standard man-hours (SMH) for the assembly can be determined. Hence, the assembly hours can be determined from the system cost by doing a cost decomposition of the system. The allowed variation for assembly time can also be obtained from legacy data and adjusted for the complexity of the assembly, the number of parts to be assembled, etc. The variation of the assembly hours and cost seems to go up when the part complexity or number of parts to be assembled goes up. The standard man-hours and the variation for the assembly are known and compared against the actuals, as described below.

First of all, an accurate assembly has to be put together using procedures recommended by design, and a highly experienced technician should be asked to assemble the system or sub-system. The assembly process has to be tuned to reduce the assembly time and ease of assembly. Once the assembly process is stabilized over time and the design becomes stationary, several similar assemblies are put together using different assemblers at different times. The performance and accuracy of each assembly should be measured as per the requirements. Once the accuracy and performances are acceptable, the average time required for, say, ten assemblies will be the SMH for the particular assembly and should be recorded.

Let us assume the following:

- Desired quality process = $\pm 3\sigma$
- Desired assembly mean hours with deviations, AT = 4 ± 0.25 hours
- Desired process capability index, C_p = 1.33
- Normally distributed process

Upper control limit of assembly hours = 4.25 hours (UCL)

Lower control limit of assembly hours = 3.75 hours (LCL)

Process capability index required = C_p = (UCL – LCL)/6σ = $0.5/6*\sigma$ = 1.33

Desired standard deviation of the process = σ = 0.0626 hours = 0.06 hours

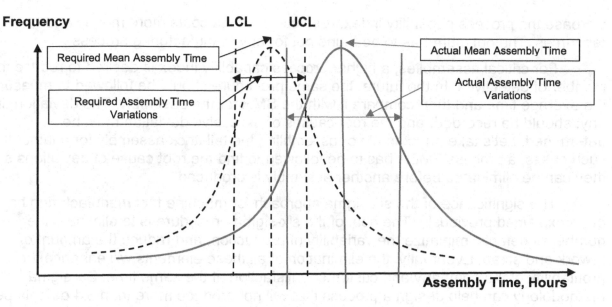

Fig. 15.2: Assembly time comparison

The assembly process is stabilized and is set to a mean of 4 hours ±0.25 hours with a standard deviation of 0.06 hours to follow a ±3σ manufacturing process. The mean of the assembly hours is assumed to be exactly in the middle of the frequency distribution, which is assumed to be normally distributed. The probability of producing an assembly outside the specified tolerance limit is given by

$$P (AT < 3.75) + P (AT > 4.25) = P [(3.75 - 4)/0.06] + P [(4.25 - 4)/0.06]$$

$$= P [Z < 4.17] + P [Z > 4.17]$$

$$= 2*0.00001523 = 0.00003$$

Hence, 30 units will be produced outside the limit per one million assemblies. In case the Cp is changed to 1, σ = 0.5/6 = 0.08 hours. In that case, the probability of producing an assembly outside the specified tolerance limit is given by

$$P (AT < 3.75) + P (AT > 4.25) = P [(3.75 - 4)/0.08] + P [(4.25 - 4)/0.08]$$

$$= P [Z < 3.125] + P [Z > 3.125]$$

$$= 2*0.0009 = 0.0018$$

Hence, 1,800 units will be produced outside the limit per one million assemblies. Hence, the number of defects per million changes significantly when the process capability index is changed from 1.33 to 1.0. In order to reduce the variations of assembly time, assembly procedure, the number of parts, and complexity need to be changed. In order to reduce the mean assembly time and get close to the desired value, the design has to be changed. It is important to remember that design dictates the mean assembly time, and in order to reduce the mean assembly time, the design has to be simple and the number of parts kept as minimal as possible. In order to take care of future shift of the mean over time, the process capability index has to be 1.33 or higher; otherwise, the probability of producing a defective number of assemblies will go up significantly. The manufacturing process has to be tightened appreciably in order to increase the process capability index. A stricter process costs more money and resources. The optimum has to be found out for any manufacturing process.

For critical assemblies, a higher process capability index is desired to reduce the number of defectives. In the future, the same procedure should be followed to measure the average time and then compare it with the SMH for the assembly. The deviation, if any, should be recorded, and the root causes of excessive deviation must be determined. Let's take an example of assembling the tailstock assembly for a lathe. In such cases, a process FMEA has to be followed to find the root cause of deviations so they can be eliminated before another assembly is produced.

The significance of the six-sigma approach for machine tool manufacturing has been explained previously. The goal of the six-sigma procedure is to eliminate the number of defects, minimize the variability of production, and reduce the amount of rework and scrap. Eventually, the elimination of all these elements will enhance the profitability of the company and customer satisfaction at the same time. Six sigma methodology can help design a process that will not produce more than 3.4 defects per million opportunities. To use a six-sigma approach, a DMAIC process has to be followed:

- D means define the problem
- M means measure the extent of the problem: the current sigma of the process
- A means analysis of the data collected: root cause determination
- "I" means improve the process: select the process improvement steps and implement
- C means control the process after implementation: monitor the process to prevent any deterioration over time

The primary objective of using a six-sigma process for manufacturing is to maximize profit and customer satisfaction and incorporate manufacturing consistency and productivity by eliminating defects. It is obvious that before such analysis is carried out, the assembly or manufacturing process must be stable and process error has to be random by nature. The approach should be a simpler design with fewer parts to be assembled. Several design actions can be taken to enhance manufacturing capabilities, as outlined in the next section.

15.7 Reduction of Assembly Time

Simplify Design and Multifunctional Part: It has been mentioned time and again that the design has to be simplified to reduce assembly time. A simpler design will also reduce the number of parts, either by integrating their functionality or eliminating unnecessary parts. Assembly is time-, material-, and labor-intensive when more parts are used. On the other hand, sometimes parts also get very complex when one is designed to perform several functions. Hence, functionality and complexity have to be managed for parts in any assembly. Simplified design also helps vendors supply parts or assemblies in a much quicker fashion and at a reduced cost. Lead time goes down, too. A reduction in the number of parts also lowers purchase cost, inventory carrying cost, and life-cycle cost. Having fewer parts also enhances the productivity and efficiency of any design. Both design and assembly time go down as well.

It also has to be kept in mind that reduction of parts while maintaining the same functionality does not always reduce the cost. Moreover, replacement of parts during servicing might go up since parts cost more. Hence, an optimum balance has to be achieved. Reduction of parts also has other advantages:

- Integrated parts with combined functionality have fewer interfaces and joints. In general, fabrication or welding cost less than machining, but welding reduces adjustment capability. When parts are simple, even if they are integrated, casting molds, forging dies, and tooling costs go down.

- Machining costs for a smaller number of parts go down also. On the other hand, if parts become too complicated, it might cost more to machine them.

- When multiple functions are integrated into the same part, the adjustment necessary for alignment and inspection goes down too. The assembler does not have to worry about their position relationships in an assembly.

- Having fewer parts means fewer design iterations and less communication and coordination among various functional engineers involved in the project. Part detailing time and cost are also reduced significantly. Hence, the design process time is reduced.

- Use of bends, tapers, radii, chamfers, excessive and unnecessary tolerances, and blind holes should be minimized in the part design.

Intuitive Assembly Process: Depending on the design, the assembly of parts might need lots of adjustment, alignment, and reworking to achieve the desired performance and accuracy. Such requirements tend to increase variations of the performance of the assembly and assembly timings. The assembly process should be self-explanatory and obvious by design. Similar software tools could be used to optimize the assembly tools as well. The design should be very easy to assemble. The natural flow of the assembly should be dictated by the design. An assembly software tool can be used to iterate the assembly process. From the concept of simple parts proposed by Boothroyd and Dewhurst, the following can be noted for machine tool components and assemblies:

- Part insertion should always be from the top of the assembly. Vertical assembly is the most normal and is the easiest to do. When inserted in the vertical position, gravity helps the part to position itself in the assembly. Part assembly on an inclined plane is very difficult Especially heavy parts must be assembled from the top.

- Part does not need assembly tools for alignment and location. Parts should be self-aligned. Assembly should have characteristics such as taper pins, dowel pins, diametric guidance, alignment holes, etc. Mating parts should be easily removed in case of refitting.

- Mating parts should not have to be aligned or oriented before assembly. Symmetrical mating, such as with spherical or cylindrical joints, is preferred. In-plane and out-of-plane matching by orienting the mating parts are examples of bad design.

- The assembly should be completed single-handedly. Both hands should never be engaged due to safety requirements. The assembler will also very quickly get tired if both hands are used for assembly operations.

- Use of screws, rivets, nuts, and bolts should be minimized.

- Self-locking mechanisms or methods, such as spring-loaded, toggles, etc. should be used for parts that need to be assembled.

- Reduce the number of interference joints, such as with press-fitted parts.

- Instead of using separate gauges, or go-no gauges, the mating parts should have gauging capability.

15.8 Service and Support Cost Reduction

In order to reduce assembly time and cost, the design could incorporate methods that will reduce the use of support services significantly, as mentioned below:

- Reduce in-process inspection for the assembly

- Parts to be assembled should be individually inspected, as per the part drawing, before assembly is started

- Accurate assembly documentation should be available in detail

- Single-person assembly is preferred to many assemblers working at the same time, if possible

- Design should help to reduce handling time to a minimum

- Reduce overhead charges to a minimum

- Minimize supervision during assembly

- Purchased parts should be inspected and available before assembly is started

- Promote sequential assembly as opposed to a parallel assembly process

- Find consensus among manufacturing, quality, and design engineers before assembly is started.

- Use activity-based accounting for assembly activity

- Accept the help of vendors for the purchased components and assembly

- Use a clean and dust-free assembly environment to reduce scrap and reworking

- For a foolproof assembly process, PFMEA must be used beforehand.

15.9 Foolproof Assembly Process: A Misnomer

There is no foolproof design, foolproof manufacturing process, or foolproof assembly process. These are all theoretical jargon consultants use while giving lectures. All of these processes are highly iterative and subject to change as they mature. Nevertheless, computer tools can be used to refine the process or design. For example, misuse of bolts, washers, springs, etc. sometimes causes some assembly problems and costly rework, and sometimes, parts even need to be scrapped because the wrong screws have been used during assembly by mistake. Such situations can be avoided by using one or two sizes for screws or bolts, such as one small screw and one significantly different size bolt or screw. This will most probably eliminate such confusion altogether.

In general, several design and manufacturing cycles are required to make the design almost error-proof. Another way is to carry out FMEA to identify failure modes and the root causes of potential failures. Neither the design nor the manufacturing process is never final, and both approach an optimum over time. Proper documentation has to be completed whenever a change in the design, manufacturing, or assembly process occurs. So, instead of trying to make the design or process foolproof, improvements should be sought over time to make the design better and better over time.

15.10 Advantages of DFMA or DFX Processes

These processes are to be applied to improve the bottom line of the company by reducing the cost of the machine without sacrificing the performance or accuracy of the machine. All these processes seem to be the same or similar. Even if the process names are different, one complements the other. In the end, any of these will help the team find the optimal design and manufacturing process. When these processes are optimized, the product satisfies the customer, and the company increases its profit.

DFM helps design engineers refine the design to make sure it can be manufactured flawlessly and the end product satisfies the customer with the best possible performance that money can buy. DFM helps save money for the components. The DFM and design processes must be run concurrently. Hence, at the end of the process, the design is optimized to satisfy the customer. Even if the design process might take slightly longer due to several iterations, the overall system process is shorter since manufacturing or assembly process times are reduced significantly.

Hence, the overall process is improved. Total project time from start to finish is way more important than individual process times. The total design cost for the product will also become less by reducing rework and scrap for prototypes. The product will satisfy all requirements including costs. This means that the time spent on DFM or DFMA is time well spent in the long run.

Since development cost becomes higher and higher if the changes have to be made at the end, any change in a mature design is very costly and takes much more time. In Japan, the design takes almost 60–70% of the project time in the beginning, but the product manufacturing and quality process runs very smooth at the end. It is better to change the design in the beginning than at the end of the design. If the design goes through the DFM process, the machine becomes less complicated and has fewer parts. Such a design takes less time to manufacture and inspect the quality of the product. The whole process becomes efficient and productive as a result. DFMEA, PFMEA, and DFM processes also need to be used to identify the failure modes and the root causes of potential failures. The design has to be changed accordingly to eliminate them.

In order to improve the quality and performance of the product, concurrent use of DFM, DFMA, and design must occur during development. Since the number of parts and complexity are reduced significantly, the inspection process becomes less stringent, and in-process inspection is also eliminated, making the manufacturing process more productive. Reduction of parts and complexity will definitely increase the reliability and robustness of the product. It also helps to reduce the life-cycle cost of the product over its useful life. Fewer parts need to be stored for servicing, and less time is required to service the machine. Failure probabilities tend to go down, and they become predictable. Instead of a costly breakdown of machines during production, machines can be serviced on a preventative basis when the machine is idle. This is very helpful for a high-production environment. Design refinement should not be done at the cost of quality or precision. Cost benefits should arise as a result of total product optimization.

Proper use of the DFM process will also help to reduce the life-cycle cost and repair time. When parts are designed for more life, a smaller number of parts are used during operation since components can be reused several times. Due to the standardization of purchased parts, common parts can be used for all machines. This reduces the part storage and handling cost in the long run. Products will have less warranty cost as well.

15.11 Summary

- DFM, DFX, or DFMA are geared towards reducing manufacturing cost without sacrificing system quality and performance. This enhances customer satisfaction by providing the best product possible at a price that customers can afford.

- DFM also reduces the total project timeline by lowering manufacturing and product assembly time. This reduces manufacturing cost through proper design of the product.

- In order to improve the quality and performance of the product, concurrent use of DFM, DFMA, and design must occur during development. Since the number of parts and complexity are reduced significantly, the inspection process becomes less stringent, and in-process inspection is also eliminated, making the manufacturing process more productive.

- DFM or DFX starts with the initiation of design for the product. Design and DFX are concurrent engineering process. The system engineering process integrates these two parts into an integrated design process for the product. The design for X property needs the same inputs as the mainstream design, such as requirements, assembly design and details, etc. Design optimization does need DFX requirements or metrics.

- The most common evaluation yardstick is to find the manufacturing cost once a design has been completed. If the evaluation does not support the cost metric, the question of design alternatives arises. Obviously, the system design phase is on the left side of system "V"; hence, this interaction process is highly iterative and time-consuming. The design goes through several phases, and finally, it evolves into the best possible design that has optimized manufacturing cost without sacrificing quality aspects. The DFX team also has to keep in mind that cost consideration is not the only constraint and other factors, such as performance measures, safety issues, ergonomics, ease of use issues, etc., must be also considered.

- Cost has to be managed accordingly to be within budget without sacrificing quality. Another area is the cost of controls and electrical components, such as relays, capacitors, transformers, cables, etc. To enhance reliability, safety, and durability, components need to be robust and reliable. Control software has to be defect-free. Servo-control components should be defect-free. So, the cost of the purchased component has to be managed to enhance the quality of the machine tools. Here the focus is to maintain the quality of the machines, and at the same time, the cost has to be controlled. Another reason for the cost going up is the number of vendors that supply quality parts and assemblies is much lower. Due to low volume and fluctuations in demand, the profit margin for vendors is comparatively low.

- For critical assemblies, a higher process capability index is desired to reduce the number of defectives. In the future, the same procedure should be followed to measure the average time and then compare it with the SMH for the assembly. The deviation, if any, should be recorded, and the root causes of excessive deviation must be determined. Let's take an example of assembling the tailstock assembly for a lathe. In such cases, a process FMEA has to be followed to find the root cause of deviations so they can be eliminated before another assembly is produced.

- DFM helps design engineers refine the design to make sure it can be manufactured flawlessly and the end product satisfies the customer with the best possible performance that money can buy. DFM helps save money for the components. The DFM process and design process must be run concurrently. Hence, at the end of the process, the design is optimized to satisfy the customer. Even if the design process might take slightly longer due to several iterations, the overall system process is reduced since manufacturing or assembly process times are reduced significantly. Hence, the overall process is improved. Total project time from start to finish is far more important than individual process times.

- Application of these procedures can only make the equipment more precise and less costly even if the volume of production is comparatively less than that produced in an automated production system. Another benefit of applying these procedures is to instill consistency of quality and performance in the products. This is an advantage that can be obtained only through these DFA or DFMA methods. Another advantage of these procedures is the reduction of parts count and design complexity. A simple design is always the best design. Unnecessarily complex designs can only reduce reliability and durability performance during service. Too many parts will also increase difficulties in maintaining the machine and, thus, increase the maintenance cost over the life of the machine.

- Reduction of parts and complexity will definitely increase the reliability and robustness of the product. It also helps to reduce the life-cycle cost of the product over its useful life. Fewer parts need to be stored for servicing and less time is required to service the machine. Failure probabilities tend to go down, and they become predictable. Instead of a costly breakdown of machines during production, they can be serviced on a preventative basis when idle. This is very helpful for high-production environments. Design refinement should not be done at the cost of quality or precision. Cost benefits should arise as a result of total product optimization.

- The proper use of the DFM process will also help to reduce life-cycle cost and repair time. When parts are designed for more life, fewer parts are used during operation since components can be reused several times. Due to the standardization of purchased parts, common parts can be used for all machines. This reduces part storage and handling costs in the long run. Products will have less warranty cost as well.

- If the design goes through the DFM process, the machine becomes less complicated and has fewer parts. Such a design takes less time to manufacture and inspect for quality. The whole process becomes efficient and productive as a result. The DFMEA, PFMEA, and DFM processes also need to be used to identify failure modes and the root causes of potential failures.

15.12 References and Bibliography

Abernathy, W.J. and Townsend, P.L., 1975, "Technology, Productivity and Process Change," *Technological Forecasting and Social Change* 7(4), 379–96.

Bralla, J.G., editor, 1999, *Design for Manufacturability Handbook*, McGraw-Hill, NY.

Boothroyd, G., 1992, *Assembly Automation and Product Design*, Marcel Dekker, NY.

Dewhurst, P., and Boothroyd, G., 1987, "Design for assembly in action," *Assembly Engineering*, January.

Dewhurst, P., and Boothroyd, G., 1989, *Product Design for Assembly*, Boothroyd Dewhurst, Inc., Wakefield, RI.

Dixon, J.R., and Poli, C., 1995, *Engineering Design and Design for Manufacturing*, A Structured Approach, Field Stone Publishers, Conway, MA.

Gupta, S.K., Das, D., Regli, W.C., and Nau, D., 1997, "Automated Manufacturability Analysis: A Survey," *Research in Engineering Design* 9(3), 168–190.

McGrath, M.E., 1995, *Product Strategy for High-Technology Companies*, McGraw-Hill, NY.

Nahmias, S., 2008, *Production and Operations Management*, fourth edition, McGraw-Hill Erwin Companies, NY.

Paul, G., and Beitz, W., 1996, *Engineering Design*, Springer-Verlag, NY.

Ulrich, T.K., and Eppinger, S.D., 2011, *Product Design and Development*, McGraw-Hill Companies, NY.

Ulrich, T.K., and Pearson, S., 1993, "Assessing the Importance of Design through Product Archaeology," *Management Science* 39(4), 429–447.

Urban, G L., and Hauser, J.R., 1993, *Design and Marketing of New Products*, third edition, Prentice Hall, NJ.

Whitney, D.E., 1988, "Manufacturing Design," *HBR*, 66(4), 83–91.

15.13 Review Questions

- Is DFM or DFX suitable for the machine tool industry? What are the fundamental advantages and disadvantages of applying these processes to machine tools? Can DFMA or DFM help reduce the machine's cost?

- What are the advantages of starting the DFM process along with the design process? What are the difficulties of this concurrent process? Does it really save money and time for the product when these two processes start simultaneously?

- How do we justify this process for low-volume machine tools? Is it still necessary to apply such methods for such products? What are the barriers to using DFM for low-volume machine tools such, as CNC grinders or special-purpose machinery?

- Why does the number of parts have to be reduced? Why does the complexity of parts and assemblies need to be reduced? Does this reduce the cost of the machine as well? Can DFMA be used for a very simple high-volume machine, such as saw machine?

- Estimate the production cost for a CNC turning center that you would like to design and produce in the near future. Is there an upper limit to the cost of a machine? How do you justify the cost? What is more important: cost or accuracy?

C H A P T E R 16

16 Project Management and Scheduling

16.1 Introduction

At the outset, it should be said that the discussion on project management and activity scheduling in this chapter is very limited and preliminary. There are excellent reference books on this topic alone. For students and professionals who would like to explore this subject further, I would advise them to study the textbooks mentioned in the reference section. A detailed treatment of this topic is beyond the scope of this book. Project management basically manages, schedules and controls the project activities throughout the project's duration. It helps to control the timing, cost, and principal resources and also guides the team to bring in extra resources or release them whenever required depending on the project schedule. For machine tool projects, I recommend using project management and scheduling tools to control the budget and timing. Normally, a chief engineer or team leader controls the projects. As envisioned in many companies, project management is not a tool for management to control activities, it should be meant for enhancing efficiency and productivity. Project management and scheduling is a dynamic tool for understanding the status of the project and steering the project in the right direction in case of delays or overspending. These tools can be used irrespective of the size of the project or budget.

Many times, I witnessed the cost and timing of a project substantially go up because the left hand did not know what the right hand was doing, duplicate activities were being conducted, or most of the attention was given to the wrong area of the project or to a particular area at the wrong point in time. Project activities need to be scheduled far ahead of when they have to happen to gain efficiency. When a project, big or small, is controlled and managed in a systemic fashion, the project flows smoothly, without any hindrances or hiccups. Success in a machine tool project is even more necessary to integrate project cost with project resources since machine tool companies usually have very limited resources. So, system engineering principles and project management tools are integrated to reduce resource requirements.

Machine tool project failure could be due to various reasons, such as poor system integration, wrong folks at the wrong time, poor control of money, poor time sensitivity, etc. When activities are not sequenced properly, project timing lengthens. Hence, effective and strategic project management is required when the project goes beyond control and needs to be streamlined to gain productivity and efficiency. A project management process is almost mandatory for a machine tool project as an integral part of system engineering. The team also has to have a good idea about project management tools so that the engineers do not feel that they are being controlled unnecessarily and are an environment where they have don't the freedom to invent and design the best machine. The primary objective of any project management and scheduling is not to eliminate or restrict the engineering activities; instead, it helps the project team identify where resources are required to steer the project in the right direction. Project management is definitely not a CEO tool to control the activities of the project; rather, it is for the benefit of the team members so that they do not die from working hard. It is a correctional tool for the team to negate or supplement conventional wisdom.

Sometimes a project fails due to the poor applications led by the legacy experiences of the team leader even if the product concepts or ideas have changed. For example, a team leader experienced in managing turning center developments might have immense difficulty managing a machining center project. The content is different. The project management tool helps the leader identify the project activities and their cost and timing. It is basically a system management integration tool required to manage system development in its entirety. As said before, it is mandatory to run the project as a system that needs project control and integration to complete it on time and within budget. That helps the company introduce the product before its competitors. Instead of personal wisdom for project guidelines, system engineering recommends using systemic project management tools to gain efficiency.

16.2 Machine Project Management and Scheduling

It is true that the success of any company depends on how quickly and efficiently the projects are finished and if they are within budgetary constraints. Project control and management are needed for the following reasons:

- Integrated part of system engineering: customer satisfaction

- Complete project on time and within cost: efficiency and productivity

- Introduction of products within time: competitive advantage

- Human resource management: efficient use of scarce manpower

- Activity scheduling: project efficiency

- Reduction of product concept-to-market timing: market leadership

Most machine tool projects drag on at the end. When the project delays considerably, the product loses its competitive strength in the market, and it becomes a follower instead of a market leader. The product might fail miserably. It is also true that money becomes scarce at the end of the project if not controlled properly. Most of the time, the future success of a machine tool company depends on the success of the product since the company can only afford to develop one product at a time. Project success also depends on the expertise of its team members. Project performance depends heavily on the team's cohesiveness and the understanding of each member of the effectiveness of project management tools.

In many cases, project management is considered a management tool to control activities in a negative sense. It brings in a negative effect on the team's working environment. Most of the time, when a project gets out of control and cost goes up, manpower is added to finish it in a hurry. That is often a mistake for several reasons: wrong personnel, wrong timing, overtime requirement, and significantly added cost to the project. During such confusion among team members, adding manpower indiscriminately to hasten the project activities normally affects the project timing and cost in a very negative way. In turn, project performance is affected. Random use of project management tools creates confusion and misunderstanding among team members. This reduces the innovativeness of team members, which is very essential for the success of the project and company. When confusion arises, project performance suffers, and costly errors are made in the process. Any error made in the beginning of the project gets amplified many times as the project matures.

It is also true that the project team needs proper training and knowledge to implement project management tools for the project. Since every project is somewhat different, the team must have proper knowledge of tools to be employed to solve or resolve any particular issue. For very complicated projects, such training is even more required to make the project successful. Project performance is directly related to business performances as well. The team has to be culturally motivated to use project management tools. System engineering recommends using project management and scheduling to control all series, parallel, or coupled activities. Again, the motivation has to be from the bottom up to make the project successful.

There is also the idea that the project will manage itself if team members are very efficient, knowledgeable, and productive. Confidence is good, but overconfidence is bad in most cases. Again, reactive behavior and actions are inferior to proactive behavior. Assuming the project needs control and management because of its nature, incorporation of project management tools ahead of execution is a good step to start with.

Project management becomes more challenging when the business conditions are averse to the company's natural thinking. For example, when the resources are very limited or scarce, when market conditions are against the introduction of new products, or when the economy is not in favor of new products, use of product or project management becomes even more critical to finish the project as early as possible and with minimum resources. Project scheduling can optimize the time and cost of the project. There are many elements of project management, such as training, active support and encouragement, budgetary control, finance control and availability, communications, etc.

Direct or indirect indications of a poor project management process could be numerous:

- Delayed project
- Overbudgeted project
- Frustration and misunderstandings among team members
- Layoffs and attrition
- High turnover of team members

- Heavy overtime to manage project timing
- Market demand shifts abruptly
- Loss of competitive edge in the marketplace
- Loss of market share among established customers

A delayed or failed project might have serious consequences for the company as a whole. The company might have marketing issues, brand recognition issues, revenue and profit issues, etc. Sometimes the project even has to be abandoned because of the fact that the competition has already integrated the new technology, the technology has changed drastically, the competitive field has changed due to infiltration of foreign competition, etc. Working hard is not necessarily the only criterion for a successful product or project for the machine tool industry.

Productivity improvements and project efficiency have to be brought in by using proper tools and technology. Project management tools identify costly errors during project execution, and they often suggest the actions to be taken. It is very common in the machine tool industry to find talented engineers who can solve machine issues, but it is very difficult to find engineers with all the talents and expertise required to make a machine tool project very successful. It is very difficult to find people with the proper mindset to manage a project technically and administratively at the same time. I have witnessed talented engineers holding a negative bias toward project management due to experiences that they had in the past.

The negative bias of working engineers could be due to improper training, past failed projects, loss of jobs, and improper management controls. They seem to think that project management and scheduling are put in place to keep track of their progress and that project management tools place constraints in their day-to-day activities, making them less creative and productive. It often becomes a big challenge for a project manager to overcome these negative thoughts. Project managers have to demonstrate the value of such project management tools. In the machine tool industry, tradition is a big and significant part of the design process. Traditional management style or design process is often used in machine tool projects. That is why many machine tool projects fail miserably. Project management could appear to be an unnecessary task for a company that has a very stable work environment, dedicated and happy workforce, hierarchical organization structure, creative environment, etc.

Such organization does not exist in U.S. machine tool industries. The situation is very dynamic due to market conditions. The workforce changes very fast, and the company expands and shrinks very quickly to cater to demand conditions. Hence, a conventional project management style might not be suitable in today's environment. The project is highly task-oriented, and a delayed project might cause company failure. Moreover, the development of machines highly depends on what customers want and not what the company wants to design and introduce to the market. The market is totally customer driven, and customer requirements change very quickly, so the project has to be managed properly to deliver the products on time every time with a limited budget and workforce.

16.3 Benefits of Project Modeling

Any project, big or small, simple or complex, has to have inputs, outputs, and a transformation phase that converts inputs into a desirable and productive output. In order to understand the complex relationship between inputs and outputs, system modeling must be done beforehand. The modeling helps the team to understand the scope, constraints, derivatives, and limitations of the project. It also helps the team to understand how the inputs are converted into outputs under given working conditions and system noise. A project model is nothing but a bird's eye view of all the activities and their relationships and interactions, and it helps the team to put all the system elements into a unified platform to visualize the whole picture before execution and also to minimize errors.

Project modeling allows project engineers to understand the project's internal and external environments under which it has to operate. Models also help the team to include forgotten steps and exclude unnecessary steps. Like simulation models, project models help the team to build a reference framework for the project and also to create engagement rules and restrictions. This brings a sense of coordination and motivation to work as a team to make the project a success at the end. Project modeling also helps define the working relationships among various teams and their activities in a timely fashion, which makes it easier to understand the functional relationships among all activities. Modeling also helps the engineers understand how all the pieces of the puzzle have to be put together to resolve the task at hand. It reduces the risk of operation by eliminating errors and optimizing linking relationships among activities in a well-coordinated and synergetic fashion. Finally, models also help to validate, sequence, and optimize activities for more efficiency and productivity.

There are several possible methodologies for building project management models. Most of them follow Henri Fayol's management principles. His five elements for project management are planning, organizing, coordinating, commanding, and controlling. In addition, he also recommended fourteen management directions to build the structure of the model, such as work division, responsibility and authority, unity of command and direction, compensation, line of authority, etc. These are all very general management principles that every project manager should consider for project models.

There are no standard forms or procedures for making a project model. Every company has their own way of doing things. For machine tool activities, the primary focus is on technical content and not on control factions. In most projects, administrative control takes second bench, and as a result, project cost and overruns happen very often. So, some sort of project management and activity scheduling must be incorporated in the project sequence. For most machine tool companies, organization is hierarchical, with a line of authority and reporting established. In order to bring stability to any organization, a chain of command has to be established. On the other hand, a strict line of authority and chain of command sometimes restrict the engineer's ability to create. Hence, a balance has to be obtained.

There are many kinds of project models, such as a circular model, V model, waterfall model, etc. Each one has advantages and disadvantages over the other. Nothing can be universally applied. For system engineering methodology, the waterfall model seems to be very appropriate. This model is sequential in general. It assumes that downstream activity has to be started before upstream activity has been completed. The project flows from top to bottom. In a real situation, this is not true, and such models cannot be followed. The left side of the system engineering "V" model is highly iterative, and the right side of the V is highly sequential, as mentioned before. The waterfall diagram was developed by Dr. Winston W. Royce specifically for project software design.

For machine tool projects, most waterfall models seem to be inappropriate and inapplicable. Their application has a tendency to increase the cost and project duration since there are many parallel or concurrent activities completed in addition to sequential activities to reduce the time (see Fig. 16.1 also). Similarly, another model, called the "spiral model," has also been used in software design. The spiral model was developed by Dr. B. W. Boehm. Software design and machine tool design are two completely different project activities by nature. As said before, project modeling has to be completed depending on the company background, legacy, and nature of machine design project instead of following any such models blindly.

The idea of project modeling and activity sequencing is to control the activity to satisfy the cost and time requirements for the project. Project models should also address risk management. Models should identify the risk of each activity and manage the project with proper risk mitigation methods. Project models should identify proper objectives with constraints, alternative activity path, and critical paths for activities, manage and mitigate risk-oriented activities, etc. They also do not have any universally accepted formats. Companies follow their own path and legacy for project modeling and activity scheduling.

The project model must be applied and validated for any particular project before execution. Sometimes project models are changed depending on the situations that arise after the project has been executed for some time. Due to emergency situations, project paths and activities need to be changed. For successful project management, teamwork and coordinated efforts towards the goal are very important. Team members should follow the common code of conduct without any internal conflicts.

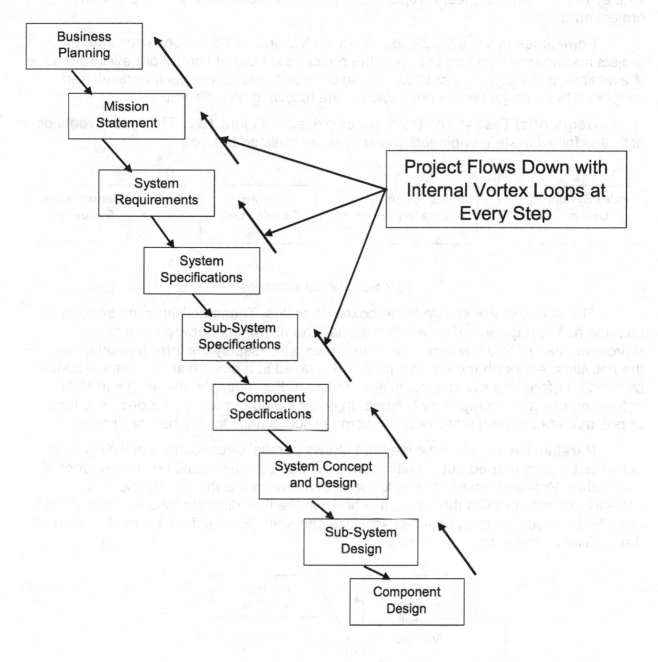

Fig. 16.1 System V and project modeling

16.4 Types of Project Activities

Project management and activity scheduling have three major parts: planning, execution, and scheduling. In addition, it also helps the team to manage resource requirements and risk. Project planning begins with the start of the project, and it evolves dynamically as the project matures. Project execution also controls the activities as they happen and suggests proper steps in order to resolve any hiccups during the project duration.

Sometimes this is also called project control and coordination. The value of project management can be felt when the project goes out of control or new information is available. Basically, project tasks can be of three types: sequential, parallel, and coupled. These project types are shown in the following flow diagrams:

Sequential Tasks: The tasks are displayed in Figure 16.2. The main events or activities follow system engineering practices, as outlined before.

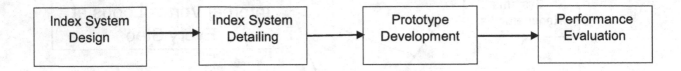

Fig. 16.2 Sequential activities

The activities are shown in the boxes or circles. These activities are sequential, i.e., one has to happen before another starts. The arrow link between activities represents the activity flow direction. The arrows also display the interdependencies of the activities. Although the flow diagram is displayed in a block fashion, some activities can occur before one event is completely finished. For example, detailing can start before the design is completely finished. In order to find out whether a design is feasible or not, manufacturing might ask about some critical details for the part or assembly.

Parallel Tasks: The next diagram shows parallel or concurrent activities, two activities that are carried out simultaneously. These activities must be independent of each other. Pre- and post-activities for parallel activities are the same, and their dependency still remains the same, as shown in the flow diagram below. Pre-activities must be finished before the parallel activities can start. Post-activities can start after all the parallel activities have finished.

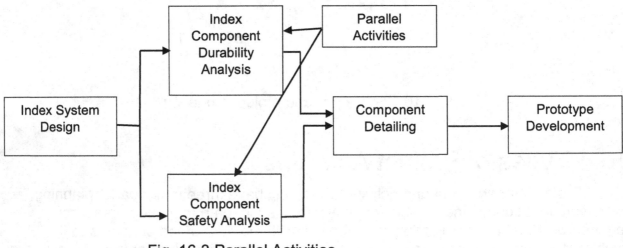

Fig. 16.3 Parallel Activities

The flow diagram for coupled activities is shown in Fig. 16.4. The coupled activities are internally connected to each other. Coupled activities are interdependent. The result of one activity affects the other. Continuous coordination among coupled activities has to happen. For good machine bed development, weight and durability analyses must be conducted concurrently. Once all these activities are completed in an iterative fashion, machine bed detailing starts. The bed concept and design depend on the analysis. The result of this is an optimized bed structure and form for further detailing.

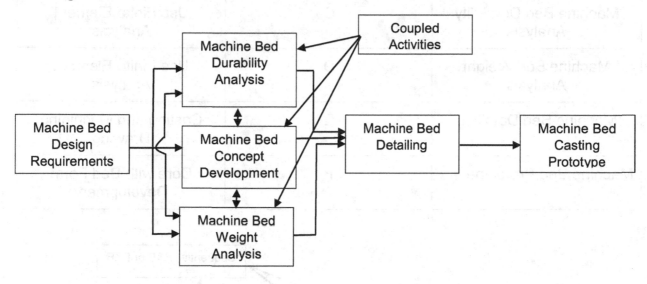

Fig. 16.4 Linked or coupled activities

For some highly complicated projects, the activities are displayed in a design structure matrix (DSM). This is just an extended step to show the relationships among activities or tasks in matrix form. It was developed by Steward in 1981. I have found the block form to be more advantageous and helpful for machine tool projects where the activities are pretty much known by the sequence and the times can be different. DSM displays sequencing and dependencies similar to an activity block diagram. The DSM for a bed prototype development is displayed in Fig 16.4. Before DSM is developed, activity block diagrams must be done. The project tasks are assigned to rows and columns. The diagonal term does not show much. Off-diagonal terms are of primary interest in DSM. Each row displays a task across the table. Depending on the relationship of each task, a dot or X is given for all the boxes where each task is related to other tasks in the matrix. Basically, project managers can use this matrix to coordinate and manage the tasks in the project. Coupled activities are critical with respect to manpower and finance requirements and time. The DSM is helpful when the tasks are sequenced in the order that they have to be executed. That is the reason why this matrix cannot be that useful in machine tool projects. The block diagram shows the same information but in a much simpler fashion. A DSM could be useful for very highly complicated automobile or aerospace projects where the number of linked or coupled activities is too large. A DSM helps to manage them in an efficient manner.

Task	Label	Remarks
Machine Bed Requirements	A	Extracted from System Specifications
Machine Bed Concept Development	B	Three or More Concepts Are Generated
Machine Bed Durability Analysis	C	Use Finite Element Analysis
Machine Bed Weight Analysis	D	Use Finite Element Analysis
Machine Bed Detailing	E	Casting and Machining Drawing
Machine Bed Prototype	F	Core with Bed Form Development

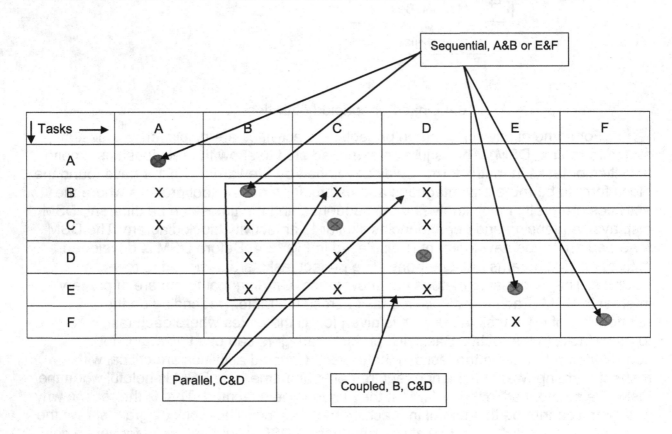

Fig. 16.4 Decision structure matrix, DSM

Gantt chart: This chart is really simple, and it works quite well. It is a convenient means of displaying a series of activities. This chart can also be used for program evaluation and review technique, PERT, and other project management tools. I have used this chart many times and have found it to be very useful for machine tool projects. The chart contains all the identified tasks along the vertical axis, with time as the horizontal axis. The chart is self-explanatory and simple to understand. Anybody can use the Microsoft Project management tool to draw this chart in finer detail. The chart does not show the task dependencies and linking modes or parallel activities. There are ways to import this data into PERT to understand task dependencies. When tasks overlap in time, they can be sequential, parallel, or linked, so sometimes it can be confusing. A typical Gantt chart for bed design activities is shown in Fig.16.5. Even if the graphical form of the activities in a Gantt chart is very useful means of displaying the schedule, it is sometimes difficult to use it as a planning tool since it does not show the sequence of activities.

	Week 1	Week 2	Week 3	Week 4	Week 5	Week 6	Week 7	Week 8
Requirements, A	▓							
Concept Generation, B		▓						
Concept Selection, C			▓					
Bed Detailing Drawing, D				▓				
Bed Analysis, E				▓				
Prototype Development, F						▓		
Prototype Casting, G							▓	
Prototype Machining, H								▓

Fig. 16.5 Gantt chart for bed design

To avert this situation, the activities can be displayed as a network. Networks explicitly display the precedence constraints for all activities. In the network, the events are represented by nodes, and the connecting links between these nodes shows the events. An event is the start or finish of a project task. The network display is the most common way of representing project tasks or activities. The network representation of bed design activities is given in Fig.16.6. For example, event or task A starts at node 1 and ends at node 2. In between nodes 1 and 2, activity A happens. Before H starts, F and G have to be completed. D has to be completed before F starts, and E has to be completed before G starts.

Example 1: Let's create a network diagram for the following activities:

- A has no predecessor
- B has an immediate predecessor: C
- C has an immediate predecessor: B
- D and E have an immediate predecessor: C
- F has an immediate predecessor: D
- G has an immediate predecessor: E
- H has immediate predecessors: F and G
- "I" has an immediate predecessor: H

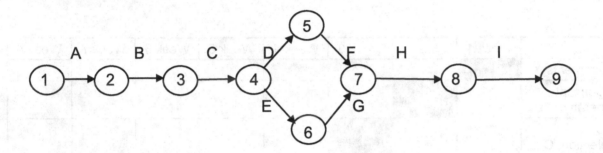

Fig. 16.5 Network path for Example 1

Example 2: Same as Example 1 except task G has immediate predecessors: D and E. in order to satisfy the requirement, activity G has to start after both D and E are completed. In order to represent this situation, a pseudo-activity, S, is inserted between nodes 5 and 6, shown normally with a dotted line. Task S is called a pseudo-activity because it does not take any time to be completed and is a constraint only.

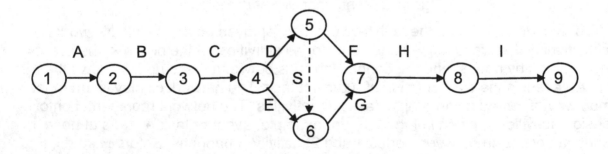

Fig. 16.6 Network path for Example 2

Critical Path Method (CPM): One of the main advantages of a network representation of activities is that we can start doing a critical path analysis for the project. The critical path analysis is a very effective tool to manage and control project activities. It can answer some of the following questions:

- Minimum time for the completion of the project
- Start and finish times for each activity
- Slack time for activities
- Project time optimization

The project critical path analysis must consist of all the activities from start to end. The path follows the sequence of all activities: sequential, parallel, and coupled. Let's consider Example 3 in Fig. 16.7. There is a total of 13 tasks with the same or different times to complete. The tasks are A, B, C, D, E, F, G, H, I, J, K, L, and M. It is also given that D and E have to be completed before G starts. To satisfy this requirement, a pseudo-activity, S, has been added between nodes 5 and 6. The possible paths are:

A, B, C, D, F, H, I, J, M

A, B, C, D, F, H, K, L, M

A, B, C, E, G, H, I, J, M

A, B, C, D, S, G, H, K, L, M

A, B, C, D, S, G, H, I, J, M

The mean times for completing each task are different, with higher and lower ends. In this method, all the task paths from the start node to the end node must be considered. So, the minimum time to complete the project without ignoring any possibilities must be equal to the longest path in the network.

Example 3: Node Network

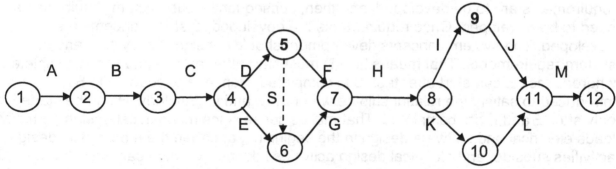

Fig. 16.7 Network path for Example 3

Example 4: A machine tool project has the following tasks:

Task Label	Task Description	Time Required, Weeks	Immediate Predecessors
A	Conduct Market Survey, A	6	–
B	Develop System Requirements, B	8	A
C	Develop Sub-System Requirements, C	4	A
D	Develop System Concepts. D	12	B, C
E	Develop System Layout, E	10	C
F	Design Machine Bed, F	6	C
G	Design Spindle, Tailstock, and Index System, G	14	E
H	Develop Electrical and Control Systems, H	10	E, F
I	Assemble First Prototype, I	16	D, G, H

The machine tool development project has been taken up to design a CNC lathe. The project tasks have been identified, and their times and sequence also have been decided based on past experiences. The team has been asked to follow system engineering principles. The project starts with the market survey and ends with the manufacturing of the first prototype. Once the market data has been collected, system requirements are to be developed, and then, subsequently, sub-system requirements need to be developed. Once requirements are developed, system concepts are developed. The system concepts development should consider the system and sub-system requirements. That means task D must start after tasks B and C are completed, whereas task E can start after task C is completed. Hence, a pseudo-task, S1, is introduced to satisfy the relationship between B, C, and D. Similarly, H can be started only after E and F are completed. That makes sense since mechanical design normally leads electrical and software design in the beginning, and then the mechanical design activities subside, and electrical design activities take off. Activity I can only start after D, G, and H have been completed. It has to be kept in mind that the development of a network diagram can be done after all the necessary tasks have been identified and their times also have to be estimated. The task relationships have to be developed as well depending on the nature of the task. All the tasks have to be completed at the end of the project. Parallel tasks need more resources but reduce project time.

The network diagram representing all these activities for all the tasks is shown in Fig. 16.8. S1 and S2 are pseudo activities to satisfy the constraints, as explained earlier. Since H can only start after the mechanical designs for the bed and other sub-systems are developed, a pseudo-activity, S2, has been introduced between 5 and 6 to satisfy the constraint. For further details on this procedure, the textbooks listed in the reference section should be consulted.

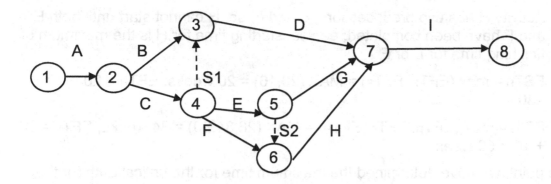

Fig. 16.8 Network path for Example 4

The next task is to identify the critical path for all the activities. The first step is to determine the earliest starting and finishing times for each activity, keeping in mind that pseudo-activity does not any time to finish the task. Next, we have to determine the latest starting and finishing times, without delaying the project, for all the activities. So, four times have to be developed:

- Earliest start time for any activity, n, EST_n
- Earliest finish time for any activity, n, $EFT_n = EST_n + t_n$
- Latest finishing time for any activity, n, LFT_n
- Latest starting time for any activity, n, $LST_n = LFT_n - t_n$
- t_n = individual activity or task time

Now, to determine these times, we can follow this rule:

- $EST_A = 0$ and $EFT_A = EST_A + ACTIVITY\ TIME\ FOR\ A = 0 + 6 = 6$ weeks
- $EST_B = EFT_A$ and $EFT_B = EST_B + ACTIVITY\ TIME\ OF\ B = 6 + 8 = 14$ weeks; B can only start after completing A
- $EST_C = EFT_A$ and $EFT_C = EST_A + ACTIVITY\ TIME\ OF\ C = 6 + 4 = 10$ weeks; C can only start after completing A
- Activity D has two predecessors: B and C, so D cannot start till both B and C have been completed; earliest starting time for D is the maximum of finishing time for B or C
- $EST_D = max\ (EFT_B, EFT_C) = MAX\ (14,10) = 14$ weeks; $EFT_D = 14 + 12 = 26$

In this fashion, early start times and late finishing times have to be completed for all the activities. These times for all tasks are shown in Fig, 16.9 below. The earliest starting time of a task is the maximum of the earliest finishing times of its immediate predecessor. Similarly,

- $EST_E = EFT_C = 10$

- $EFT_E = EST_E + ACTIVITY\ TIME\ OF\ E = 10 + 10 = 20\ WEEKS$

- Activity H has two predecessors, E and F, so, H cannot start until both E and F have been completed; earliest starting time for H is the maximum of finishing time for E or F

- $EST_H = \max (EFT_E, EFT_F) = MAX (20,16) = 20$ weeks; $EFT_H = 20 + 10 = 30$

- $EST_I = \max (EFT_D, EFT_G, EFT_H) = MAX (26,34,30) = 34$ weeks; $EFT_H = 34 + 16 = 50$ weeks

At this point, we have determined the maximum time for the critical path for the project. The critical time is 50 weeks. That is the minimum time required for project completion. In order to find the critical activities of the critical path, we have to determine the latest times for each activity. We have to walk up the stream of activities in the network. The latest finishing time for a task is the minimum of the latest start of the immediate predecessor. Let us start from the last task, I, to explain this process.

$LFT_I = 50$ weeks

$LST_I = LFT_I – t_i = 50 – 16 = 34\ WEEKS$

Now, the tasks D, G, and H end at a node when the activity I starts. So, the latest finish time for D, G, and H has to be 34 weeks.

$LFT_D = LFT_G = LFT_H = 34$ weeks $= LST_I$

$LST_D = 34 – 12 = 22$ weeks

$LST_G = 34 – 14 = 20$ weeks

$LST_H = 34 – 10 = 24$ weeks

Again, F ends when H starts, i.e., $LFT_F = LST_H = 24$; $LST_F = 24 – 6 = 18$ weeks. For task E, G and H are immediate successors. So, latest finishing time for E is the earlier of the latest times for G and H. So, $LFT_E = \min (LST_G, LST_H) = \min (20,24) = 20$ weeks and $LST_E = 20 – 10 = 10$ weeks.

- C is immediate predecessor of E and F; $LFT_C = \min (LST_E, LST_F) = \min (10,18) = 10$

This data is shown in Fig. 16.10. Another column has been added to find the slack time of the tasks. Slack is the time difference between the latest starting time for an event and the earliest starting time. In other words, slack time is the time that an event can be delayed without increasing the total project time. In Fig.16.10, activities with zero slack are the events in the "critical path" for the project. The events in the critical path are called critical events. For this example, A, C, E, G, and I are critical events that need to be completed in order to finish the project within 50 weeks.

Task Label, n	Time to Complete (weeks)	Immediate Predecessor Activity Label	Earliest Start Time, EST_n (weeks)	Earliest Finish Time, EFT_n (weeks)
A	6	Start	0	6
B	8	A	6	14
C	4	A	6	10
D	12	B, C	14	26
E	10	C	10	20
F	6	C	10	16
G	14	E	20	34
H	10	E, F	20	30
I	16	D, G, H	34	50

Fig. 16.9 Computation table for Example 4

Task Label	Time to Complete (weeks)	Immediate Predecessor Activity Label	Earliest Start Time, EST_n (weeks)	Earliest Finish Time, EFT_n (weeks)	Latest Start Time, LST_n (weeks)	Latest Finish Time, LFT_n (weeks)	Slack Time (weeks)
A	6	Start	0	6	0	6	0
B	8	A	6	14	14	22	8
C	4	A	6	10	6	10	0
D	12	B, C	14	26	22	34	8
E	10	C	10	20	10	20	0
F	6	C	10	16	18	24	8
G	14	E	20	34	20	34	0
H	10	E, F	20	30	24	34	4
I	16	D, G, H	34	50	34	50	0

Fig. 16.9 Computation table for Example 4

As explained earlier, a Gantt chart can be generated from the data in Fig. 16.9. The Gantt chart is shown in Fig.16.10. It is much easier to see the overall project tasks with respect to timing. The chart also shows the slack time for each event. It does not help to identify the critical path, but one could say that events with zero slack must fall on the critical path. So, A, C, E, G, and I are critical events as shown in the Gantt chart.

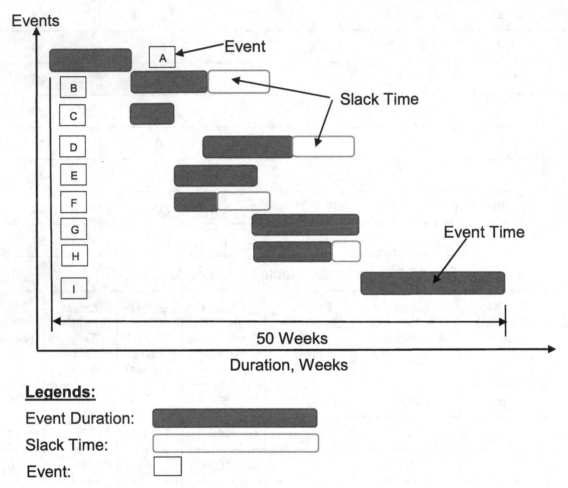

Fig. 16.10 Gantt chart for Example 4

Program Evaluation and Review Technique (PERT): It can be noticed from the CPM analysis that the project times are fixed and given. In reality, project time does vary depending on various situations and bottlenecks during project execution. When project times are subject to change, PERT comes into the picture, and it can provide some effects of time variations, which are also directly connected to direct and indirect costs for the project. PERT can also help to find the effects of time variations on the total time for the project. The outcome of PERT analysis heavily depends on the variation of project times. In PERT analysis, project time is considered to be a discrete random variable (DRV) that follows a random distribution. The project individual times are considered to be independent of each other.

The project time for any event is considered to vary between a maximum and minimum with a mean time for the project. Let us assume:

t_1, t_2, t_3...t_n: Project event timing for event 1, 2, 3......n tasks respectively

t_{min} = activity minimum time required for completion

t_{max} = activity maximum time required for completion

t_{mean} = activity average or normal time required for completion

For example,

t_{min} = minimum time required for completing task A = 4 weeks

t_{max} = maximum time require for completing task A = 8 weeks

t_{mean} = average or normal time required for completing task A = 6 weeks

It can also be said that the project A time has a mean equal to six weeks with plus or minus two weeks variation. Even if a regular normal distribution varies from negative infinity to positive infinity (all values are possible), for calculation sake, it is assumed that a project time less than four weeks or more than eight weeks is not feasible. The frequency is maximum at the mean value of the variable. The maximum value is called the mode, i.e., this is a single mode variation. Another possible distribution for the time could be a beta distribution, where the maximum value can happen anywhere between the maximum and minimum value of activity time, i.e., it is not in the middle as in the normal distribution.

Both of these distributions are frequently used in the PERT calculations. In probability theory and statistics, the beta distribution is a family of continuous probability distributions defined on the interval [0,1] parameterized by two positive parameters, denoted by α and β, that appear as exponents of the random variable and control the shape of the distribution. For α = 2 and β = 2, beta distribution becomes normal distribution. The beta distribution has been applied to model the behavior of random variables limited to intervals of finite length in a wide variety of disciplines, i.e., t_{min} and t_{max}. This beta distribution will be used for PERT calculations. The mean time and standard deviation for the beta distribution are assumed to be as follows:

μ = mean time for an activity = $(t_{min} + 4* t_{mean} + t_{max})/6$. Biased with mean time from experience, this is a 1,4,1 weight scheme, and other weight schemes can be used as well.

σ = standard deviation of activity time = $(t_{max} - t_{min})/6$

v = variance = σ^2

Assuming the time distribution for the tasks follows a normal distribution or Gaussian distribution, which is a special case of beta distribution, the following calculation will be performed. Since the project task times are assumed to be normal, the total project time will also have a normal distribution. The activity times should be independent of each other, i.e., one activity time should not influence or affect other times.

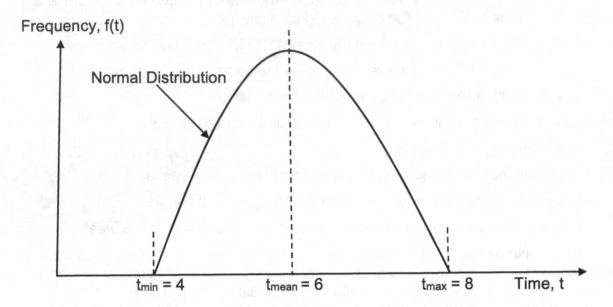

Frequency, f(t)

Normal Distribution

$t_{min} = 4$ $t_{mean} = 6$ $t_{max} = 8$ Time, t

Fig. 16.11 Frequency distribution for project activity time

The activities on the critical path, critical events, will determine the least amount of time that the project will take. In order to reduce the project duration, these activities must be looked into for time improvements by changing how the tasks are performed, for example, using casting or welding instead of machining. The task duration on the critical path are assumed as follows:

Minimum project time = T

$T = t_1 + t_2 + t_3 + \ldots\ldots\ldots + t_n$, where 1, 2, 3, etc. are activities

The mean project time and variance of the project time, assuming normal distribution and independence of task timing, are:

E(T) = expected project mean time = $\mu_1 + \mu_2 + \mu_3 + \ldots\ldots\ldots + \mu$, where μ_1, etc. are task means.

V(T) = variance of project time = $\sigma_1^2 + \sigma_2^2 + \sigma_3^2 + \ldots\ldots\ldots + \sigma_n^2$, where σ_1, etc. are the activity standard deviations.

These values of E(t) and V(t) come from the normal distribution assumption, and the times are independent and random variables. Application of central limit theorem assumes that the activity times are independent of each other and random variables. A random variable is an unbiased variable and not dependent on the special cause variations. Hence the application of PERT for the project could have the following possible steps:

- From legacy data, experience, market available data, or industry standards, the values of t_{min}, t_{max}, and t_{mean} must be determined for all the activities of the project.

- The project times are assumed to be normally distributed with two parameters: a mean and a standard deviation.

- The mean and variance for each activity must be determined after a decision has been taken on the weight factors. Calculate the mean, standard deviation, and variance for each task of the project. These are the estimated values for the project tasks.

- Since mean time for all the activities is known, determine the critical path of the project using these mean values following the procedures outlined in previous sections of this chapter. Also calculate slack values and critical events for the critical path.

- Once the critical path and activities have been determined, calculate the critical project mean time by adding the means of the critical events. The variance of timings for the project can be found by adding the variances of each task. From the variance of the project critical time, the standard deviation for the project time can also be determined by taking the square root of the project critical path variance.

- Using these mean and standard deviation values, a normal distribution equation can be used to draw the nature of the distribution.

Example 5: We are going to use the same data from Example 4. The variation in time for activities is assumed accordingly for calculation purposes. The mean, variance, and standard deviations for all the activities are displayed in Fig.16.11 below. The critical path is A, C, E, G, and I. So, the mean, variance, and standard deviation for the project as a whole can be calculated:

Project time mean, E(T) = 6 + 4 + 10.33 + 14.33 + 16 = 50.66 weeks

Variance, V(T) = 0.11 + 0 + 1.78 + 2.78 + 1 = 5.67

Standard deviation, σ_T = 2.38 weeks

The project timing details can be represented by a normal distribution, ND (50.66, 2.38), with two parameters. So, the total project completion time on average will be equal to 50.66 with a standard deviation of 2.38 weeks. Using these two values and a normal distribution, we should be able to determine different aspects of the project. In this case, the expectation of management is to complete the project within 45 weeks.

Activity	Min Time	Average Time	Max Time	μ	Variance	Standard Deviation
Label	weeks	weeks	weeks	weeks		weeks
A	5	6	7	6.00	0.11	0.33
B	6	8	14	8.67	1.78	1.33
C	4	4	4	4.00	0.00	0.00
D	8	12	18	12.33	2.78	1.67
E	7	10	15	10.33	1.78	1.33
F	4	6	9	6.17	0.69	0.83
G	10	14	20	14.33	2.78	1.67
H	7	10	12	9.83	0.69	0.83
I	13	16	19	16.00	1.00	1.00

Fig. 16.11 PERT calculation table for Example 5

. Let's answer the following questions:

- What is the probability that the project can be completed within 45 weeks? (Standard normal variable Z table can be used for this calculation.)
- What is the probability that the project takes more than 55 weeks?
- What is the number of weeks to complete the project with 0.8 probability?
- What is the probability that the project can be finished within 50.66 ±7.14 weeks; $\mu \pm 3 * \sigma_T$ (six sigma specification for the project completion time)?
- If the process is not centered about the mean but is located at a distance of one standard deviation from the center of the tolerance range for the project time, determine the probability of completing the project within the specified tolerance limits for 50.66 ±7.14, i.e., $\mu \pm 3 * \sigma_T$ (six sigma specification for the project completion time).

Case 1: The probability of completing the project within 45 weeks = P(T < 45)

Normal variable $Z = (t - \mu)/\sigma_T = (45 - 50.66)/2.38 = -2.378$. From the normal variable table, we can find the probability as follows:

P (T < 45) = (0.5 − 0.4912) = 0.0088 = 0.88%, or about 1%. Hence, there is only a 1% chance that the project can be completed within or less than 45 weeks.

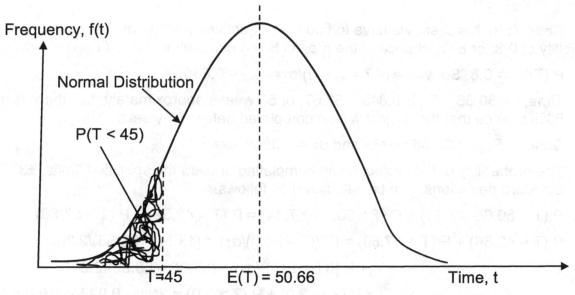

Fig. 16.12 Frequency distribution for Case 1, Example 5

If the team or management wants to finish the project within 45 weeks, they have to add resources or find other means of reducing the project tasks' duration.

Case 2: Now the task is to find the probability that the project will take more than 55 weeks, which is more than the mean time for the project. Hence, the probability is denoted as P (T > 55), which can be calculated in the following way.

Normal variable, $Z = (t - \mu)/\sigma_T = (55 - 50.66)/2.38 = 1.82$. From the normal variable table, we can find the probability as follows:

P (T > 55) = (0.5 – 0.4656) = 0.0344 = 3.5%, or about 4%. Hence, there is only a 4% chance that the project will take more than 55 weeks.

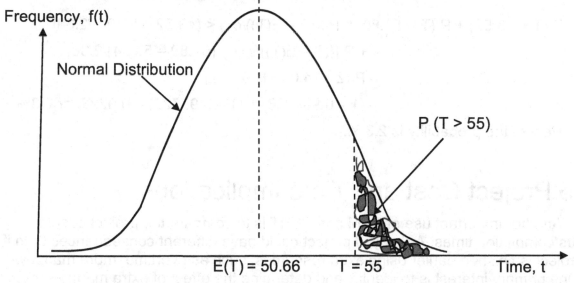

Fig. 16.13 Frequency distribution for Case 2, Example 5

Case 3: In this case, we have to find the project time, which will have a probability of 0.8, or 80% chance of the project being completing. So, $P(T \leq t) = 0.8$

P $(T \leq t) = 0.8$. So, value of $Z = (t - \mu)/\sigma_T = Z_{0.8} = 0.845$

Time, $t = 50.66 + 2.38*0.845 = 52.67$, or 53 weeks approximately. So, there is an 80% chance that the project will be completed before 53 weeks

Case 4: $E(T) = 50.66$ weeks and $\sigma_T = 2.38$ weeks

The probability of the project being completing outside the specified limits, $\pm 3*$ standard deviations, can be calculated as follows:

$P(T < 50.66 - 7.14) + P(T > 50.66 + 7.14) = P(T < 43.52) + P(T > 57.80)$

$P(T < 42.86) + P(T > 57.80) = P[((T - E(T))/\sigma_T) < (43.52 - 50.66)/2.38]$

$$+ P[((T - E(T))/\sigma_T) > (57.80 - 50.66)/2.38]$$

$$= P(Z < -3.0) + P(Z > 3.0) = 2*(1 - 0.9986) = 0.0028$$

- Hence, the reliability of project time completion is 0.9972, i.e., 99.72% chance that the project will be completed within 50.66 ± 7.14 weeks, or $\mu \pm 3*\sigma_T$ (six sigma specification for the project completion time)

Case 5: $E(T) = 50.66$ weeks, and $\sigma_T = 2.38$ weeks

The mean time for this case is shifted from the project mean time, $E(T)$, by an amount of one standard deviation, i.e., 2.38 weeks. The new mean, μ_n, is

$\mu_n = 50.66 + 2.38 = 53.04$ weeks.

The probability of completing the project outside the three sigma limits will be

$P(T < 43.52) + P(T > 57.80) = P[((T - E(T))/\sigma_T) < (43.52 - 53.04)/2.38]$

$$+ P[((T - E(T))/\sigma_T) > (57.80 - 53.04)/2.38]$$

$$= P(Z < -4.0) + P(Z > 2)$$

$$= (1 - 0.999968) + (1 - 0.97725) = 0.0228, \text{ or } 2.3\%$$

Hence, the probability is 2.3 %.

16.5 Project Cost and Time Implications

Another important use of CPM and PERT is to compare the project cost for various completion times. A delayed project could have different consequences than if the project is finished earlier using extra resources, such as overtime, more manpower, etc. One primary interest is to identify and determine the effect of extra manpower on the project completion time. In this case, activity time is considered variable, and project time can be reduced by spending more money. As explained in the previous section, the mean time for the project is the average time that the project requires to be finished. The lower limit of the project completion time can only be had if more money is spent to reduce the project completion time. This could be labeled "reduced time," which is the desired minimum possible time, i.e., the minus three sigma limit for the project completion time if the company is following a six-sigma process. Also, the project task timing should be known beforehand in this case. Assuming a linear cost relationship between the completion time and cost associated with the reduced time, it can be said that a reduction in time is only possible by spending more money. The cost vs. time relationship is shown in Fig. 16.14 below.

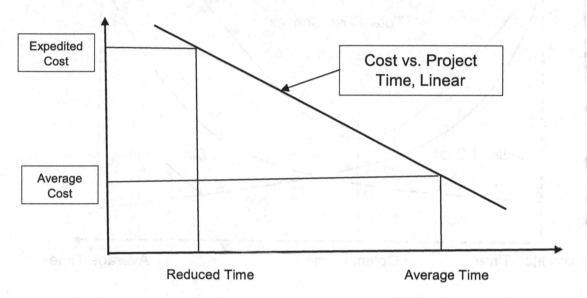

Fig. 16.14 Project cost vs. completion time model

The cost relationship is not always linear. For machine tool projects, human resources are added whenever required and when there is lack of proper resource planning. In such cases, cost could be non-linear. More cost is added when unplanned resources are added at the end of the project and project cost escalates tremendously. Other than resources, project cost might go up non-proportionately when software requirements, such as control software, analysis software, etc., are added at the end of the project. In general, a linear relationship between cost and time is very reasonable for most cases.

As said earlier, in any project, there are two types of costs: direct and indirect. Labor, computers, test equipment, and software costs are examples of direct costs, whereas support costs, financing interest, management cost, etc. are examples of indirect costs. The total cost of the project is the summation of these two costs. Indirect costs increase with project duration, and direct costs decrease with project time. As the project matures, direct costs flatten out, and indirect costs increase at a linear rate. In such cases, as shown in Fig. 16.15, the total cost could be represented as a convex function with a local minimum, which is the lowest cost, and local optimum value, the point at which project time is optimal. Theoretically, this is the most suitable project completion time. The direct and indirect costs with respect to project completion time are usually non-linear, though they can be linear in some situations. As shown in this figure, for the average project completion time, direct costs re less than indirect costs, whereas, for expedited time, direct costs (labor, material, equipment) are substantially more than indirect costs (rent, finance interest, utilities, etc.). With a given task timing, the maximum reduction is possible depending on the amount of slack time for the event.

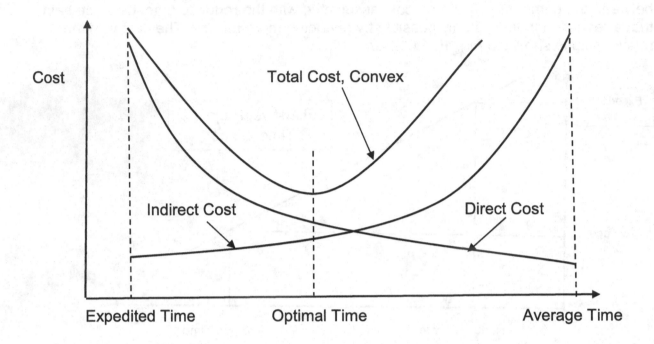

Fig. 16.14 Minimum project completion time, optimum value

The main effort should be to reduce the time for critical path activities. The best possible situation is for all event timing to be made equal to critical path timing. An iterative process can be set up to find the optimum project time. At each step, the total cost is calculated and compared with that of the last step. The process continues until an optimum is reached. Critical path analysis also has to be also performed at every step to make sure that the critical path has not changed due to iterations. Several simulation programs or optimization techniques can be used to solve similar problems with resource optimization or project time minimization.

16.6 Project Management and Organization

The effectiveness of project management also depends on management attitude, culture, style, and organization. The efficiency and productivity of the project depend predominantly on the attitude, knowledge, and expectation of the team members. If the organization has a hierarchical or pyramidal structure, project management efficiency suffers since every line manager is concerned with his or her own efficiency and productivity. The project manager might not have enough authority or responsibility to go beyond his line of duties to get help from other departments to manage the project effectively. When the project needs help from other departments, it might be extremely difficult to receive that help in a timely and effective manner. For machine tool companies, this might not be a major issue due to the small size of the company and loose line of control.

In several instances, I found that having a cohesive and dedicated team is the best possible way to go for machine tool projects. The team is free to act to make the project successful, and a project leader manages the project. The negative side of this approach is the fact that it might be very expensive and time-consuming to go and get special help from other departments whenever required. The team leader has to manage that part of the project along with the project manager, or the team leader becomes the project manager himself. The frequent shifting of employees could create a disruption of team progress and might be more expensive in the long run.

There is another style of organization: matrix organization. In this style of organization, the team leader acts both vertically and horizontally. Matrix organization is a combination of line and project organization. Such a structure is very effective when the organization is extremely large and many projects are running simultaneously. For machine tool companies, this is hardly the case due to their small size and the fact that most of the time, only one major project is taken up. For machine tools, specific knowledge and experience are required to design and manage a project. For example, people designing lathe machine might not be suitable for designing machining centers or grinding machines. So, knowledge and experience could be another problem for running many projects at the same time under one roof.

Instead of a matrix-style organization, it is better to put together a dedicated team that will consist of all the members who have the expertise to carry out the project effectively. For tasks that are required off and on, such as financial activities or legal matters, the team leader can ask for such help when needed without having these members permanently on the team. In reality, such an approach is very productive.

16.7 Project Management Tools and Software

PERT and CPM have been in use since the late 1950s, and CPM was developed in the USA by Remington Rand and the Du Pont Company as noted by Walker and Sayer in 1959. The technique was primarily developed for use in plan maintenance activity. Both cost issue and time management issues were developed. PERT was developed by the U.S. Navy in 1958 as a result of a joint effort by Lockheed, the Navy, and the consulting firm Booz, Allen, and Hamilton. Afterward, PERT and CPM have been used by automotive companies and airline industries to manage complicated projects. There are still some issues with using beta or normal distribution for such analyses to determine the mean and standard deviation for the project. Subsequently, many experts and authors have questioned the use of statistical distributions for PERT programming. I have found CPM to be very useful for machine tool projects. To be frank, though, the use of such tools in machine tools has been sporadic and not very effective for some unknown reason. In addition, advanced methods, such as linear programming and optimization techniques, have hardly been used or explored in detail in machine tool industries all over the world.

Research is still being carried out to refine these methods for network projects. The most difficulty arises when activity timing is decided. If these assumed values are doubtful, the projection will be also wrong. Engineers like to have more time to do a good job, and management wants the project to be completed as soon as possible. This is the inherent conflict. Management wants to control activities, and engineers do not want to be controlled. Nevertheless, project management tools should be used as much as possible to make design development activity effective and productive.

Project management software has been in use for a very long time in some form or another. In the beginning, mainframe computers were used for project management software irrespective of project size and duration. In recent years, most of the software has migrated to a PC environment. This PC software is as efficient as mainframe software, and it is less expensive. Recently, PC-based software has taken the lead, and it is quite suitable for machine tool projects. The following PC-based software is very efficient and can handle almost all machine tool projects.

Heck (1994) and Barney (1995) have given a comprehensive list of all the PC-based software. It is also true that the software environment is so dynamic that the software is changing very quickly. Microsoft Project is an excellent program that can be used for machine tool projects without spending a fortune. It is one of the best programs available and can be effectively used with minimal training for a single project or multiple projects and resources. There are a few shortcomings with regard to report writing, etc. that can be ignored or are not required for a machine tool project. Other programs such as Time Line (Symantec Corporation), Project Scheduler (Scitor Corporation), etc. can also be used, if necessary, in place of Microsoft Project. Similarly, Milestones (Kidasa Software Inc.), Sure Track project manager (Primavera Systems Inc.), etc. can be also used. Microsoft Project, though, is easy to use and menu-driven, and some customization is also possible.

It should be noted that the list of software above is not exhaustive. Open Plan (Welcome Software Company) is also a very good program for use with machine tools. In addition, software companies are coming up with customized project management packages that include CPM, PERT, etc. in one program. These programs can be used for finance, legal, and other business departments, aligning them with technical project requirements, too. They are basically business-based, comprehensive programs that can be used for the whole company.

16.8 Project Planning

Any successful project management process needs proper planning before it is started. The project plan must include a few basic steps:

- What is the primary objective of the project?
- What is the time constraint for the project?
- What are the elements/task/activities of the project?
- What are the resource constraints for the project?
- Is the project brand new and innovative or a repeat of the past?
- What are the validation and verification plans required?
- What should the project implementation plan be?
- What should the operation and maintenance plan be?

- What is the equipment acquisition plan?
- What is the manufacturing plan?

Out of all of these, three basic items of information are definitely required: project activity/task, desired time of completion, and cost constraints of the project. A proper team for executing the project in a very efficient and productive manner can be put together once these three basic elements have been determined. It should also be kept in mind that the planning and execution of any project are highly dynamic and iterative by nature. Market conditions, business risks, and business environments do change over time, and projects need to be adjusted accordingly using proper tools and methodology. Such changes are even more necessary when the project duration gets longer and longer.

The project planning process starts with system engineering practices or system "V" model. The elements of the planning process are to determine the following:

- Project deliverables must be determined first.

- Design and development strategy: a single, derivative, or legacy-based product, etc. For system-based machine tool development projects, a single product development strategy is the best to follow.

- Functional decomposition strategy: Once the system requirements have been developed from the market inputs, system decomposition into sub-system requirements needs to be completed to understand the sub-system- and component-level requirements.

- Risk analysis and mitigation strategy: risk alertness and mitigation for each critical path element or task must be evaluated, and a mitigation plan has to be developed.

- Task or activities determination strategy: the project must be sub-divided into the smallest possible tasks or activities, and the duration and variance of each activity must be obtained as well.

- Network development strategy: once the activities and their possible timing are determined, the network has to be developed. A logical arrangement of all the activities should be determined and put in the network.

- Once the Network, CPM, and PERT analyses are completed, the optimized project completion time, resources required, and cost of the project can also be determined. Optimization or simulation analysis must also be performed to optimize the timing.

- Project activities need to be scheduled as per the network. The activities on the critical path should be done quickly, and slack time for all other activities should be reduced as much as possible.

- Resource requirements (manpower, equipment, software, and finance) need to be assessed and lined up for project execution.

16.9 Work Breakdown Structure (WBS)

In place of functional decomposition of the system, the work breakdown structure (WBS) approach can also be used for project management. For defense-oriented projects, use of system engineering practices is almost a norm. WBS helps the team to divide the project according to system, sub-system, and components (spindle, tailstock, cross slide, etc.) rather than divide the project into functional elements (engineering, financial, legal, etc.) using functional decomposition. In my opinion, either approach is suitable for project management, but WBS can be more appropriate for machine tools. So, we need to discuss this approach a little further. The machine system is broken down into sub-systems and components first. After that, the tasks required to design and develop each sub-system are defined. This is also a top-down approach, as shown in Fig. 16.15.

WBS helps the team build a process to manage the project for the system. WBS is also appropriate for machine tool projects because it helps to assign workforce, manage budget, schedule the work, assess and mitigate the risk, allocate cost to satisfy the budget, and manage the verification plan. Hence, WBS can help the team to manage the project in its entirety. WBS starts with the system and ends with the tasks associated with each sub-system. The following block diagram for a turning center project, Fig. 16.15, explains the WBS process to some extent:

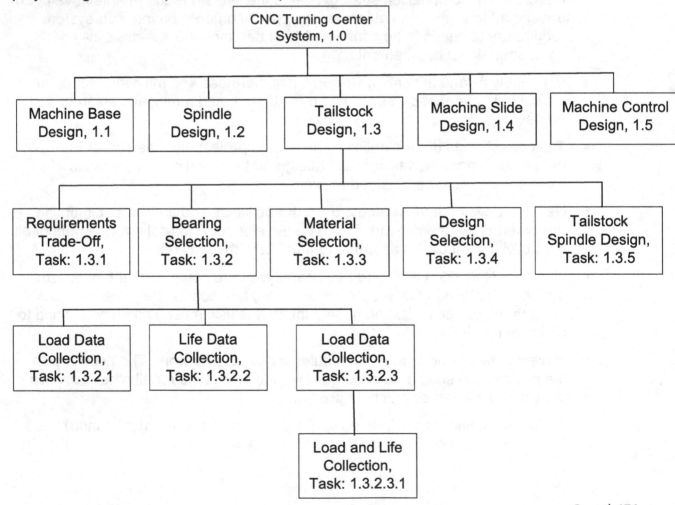

Fig. 16.15 Project task block diagram, turning center project

The block diagram divides the whole project into tasks or activities. Once this is completed, the bottommost activity is finished first, and work flows upwards until it reaches the topmost task for that stream. Hence, the work integration moves upwards, whereas product disintegration flows downwards. Further details on this WBS process can be had from US military standard MIL-STD-881A. This is very suitable for machine tool and defense projects. THE WBS process also supports the system engineering process, and it helps the team develop the system structure by providing:

- Sub-system decomposition
- Sub-system specifications and requirements
- Cost details and management
- Budget allocation
- Activity scheduling
- Time management

The steps necessary to develop a WBS for a machine tool project can be as follows:

- Decompose the product into systems, sub-systems, and components
- Develop the tasks/activities for each sub-system necessary to complete the design
- Keep track of cost for each task and compare against allocation
- Compare each task's timing and budget for proper allocation
- All the tasks must be identified by a suitable numbering system
- Task details should include cost, time, and requirement appropriation
- The disintegration level depends on the project complexity and control requirements
- The task must include design details, validation details, and verification requirements
- Task elements must include the last step of the activity to complete the task

The general flow diagram for developing the WBS can be summarized in the following block diagram, Fig. 16.16:

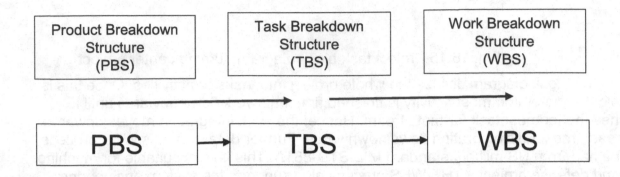

Fig. 16.16 Block diagram for WBS

The step-by-step total-system-engineering action items for the project can be as follows:

- Requirements development
- Product breakdown structure
- Task breakdown structure
- Work breakdown stricture
- Cost/budget appropriation
- Network development, CPM, PERT analysis
- Risk analysis and mitigation scheme
- Project management and control
- Project schedule development
- Resources planning
- Team development
- Work distribution and commitments
- Work initiation
- Plan updates and reporting

Project management and control play an important role throughout the project execution phases: project initiation, project planning, project documentation, planning for manufacturing activities and equipment, supply chain management, product validation and verification, risk mitigation, product launching, and market evaluation. Each of these project activities can be broken down to task levels for further execution and completion. The extended scope of project management is to enhance profit, manage risk and overall activity, identify strategic, technical, and marketing opportunities, etc. Project management also helps the company to analyze the success of the project when completed. In adverse situations, project management might help the team take corrective actions to put the project under control again. The project control function of project management is to monitor the following activities:

- Variance management and control
  - Identify tasks to be controlled
  - Establish control documents and specifications
  - Establish the line of authority for control
  - Develop control mechanism
  - Obtain the result
  - Compare the achieved results with standards
  - Develop variance
  - Management report

Project control activities must be proactive and reactive, but monitoring must be continuous and focused like a hawk. When the project is running very smoothly, project management activity determines if any critical step has been left aside and whether project duration and cost can be optimized by rescheduling activities. In case of adversity, project management focuses on action control to steer the project onto the proper path. Hence, the function of project management is never-ending until the project is completed.

16.10 Project Execution

The next topic for discussion is how to execute a project efficiently and in a productive manner. There are several inherent steps for effective project execution:

- Team coordination and interaction are very important for project success. Difficulties in team coordination arise from conflicts and misunderstandings among team members. Every team member has to follow the instructions laid down by the team leader. Team coordination can be enhanced by proper communication, regular meetings, project timing display, project updates, project incentives, and document updates. In Japan, there is a task for everybody on the team, and everybody is an equal stakeholder in the project. As opposed to American companies, Japanese team members do not get any individual incentives. The whole team wins or loses. The incentives are shared equally by all the team members irrespective of their contribution or position on the team. This is a very unique and proper approach for a successful project.

- Project management is to be considered for the project as a whole. It should include schedule, time, and cost management during product planning, customer requirements, concept development, concept selection and trade-off analysis, system-level, sub-system, and component-level specifications, system-level design and details, industrial design, etc. by creating a task list, task networking, design matrix, Gantt chart, CPM and PERT analysis, risk analysis and mitigation, documentation, and control. Project management is a parallel activity and not a serial activity.

- In case of variations in timings or cost, actions must be taken to put the project under control again. Corrective actions can be numerous, such as scheduling regular meetings and discussing the reasons for deviation, selecting proper team members, putting team members under one roof, resolving team conflicts, requesting extra resource, etc. Special attention has to be given to critical path activities so that project completion time can be controlled. In the worst-case scenario, project scope or timing might have to be changed accordingly.

- At the completion of the project, project evaluation must be done. The evaluation of the project by the project leader must be discussed among team members, and the project's execution style, timing, cost, deviations, conflict, etc. should be freely discussed and documented properly. The strengths and weakness analysis (SWAT) for project execution and management must be completed for future reference. The team discussion can be on the following topics:

  o Project mission and achievements

  o Project performance, productivity, and efficiency

  o Project resources deployment

  o Advantages and disadvantages of project network, CPM, and PERT analysis

  o Risk analysis and mitigation methods

  o Team constitution and management

  o Team leader's involvement

  o Team conflict resolution

  o Engineering activities and effectiveness

  o Negative and positive aspects of project management

  o Project management tools that helped or hindered the project

  o Did the project control tools alienate the project team members?

  o Which part of project management and control was not liked by most of the team members and why?

  o How can project efficiency be enhanced in the future?

  o Is project management and control necessary for the project?

16.11 Summary

- Project tasks are linked to each other, and project dependencies are important for drawing node networks. Tasks can be sequential, parallel, or linked.

- Productivity improvements and project efficiency have to be brought in by using proper tools and technology. Project management tools identify costly errors during the project execution and often suggest the actions to be taken.

- Project modeling allows project engineers to understand the project's internal and external environments under which it has to operate. Models also help the team include forgotten steps and exclude unnecessary steps. Like simulation models, project models help the team to build a reference framework for the project and create engagement rules and restrictions.

- Project management and control play an important role throughout the project execution phases: project initiation, project planning, project documentation, planning for manufacturing activities and equipment, supply chain management, product validation and verification, risk mitigation, product launching, and market evaluation. Each of these project activities can be broken down to task levels for further execution and completion.

- Project control activities must be proactive and reactive, but monitoring must be continuous and focused like a hawk. When a project is running very smoothly, project management activity determines if any critical step has been left aside and whether project duration and cost can be optimized by rescheduling activities further down.

- The work breakdown structure, WBS, helps the team to divide the project according to system, sub-system, and components (spindle, tailstock, cross slide, etc.) rather than by dividing the project into functional elements (engineering, financial, legal, etc.) using functional decomposition. WBS is very suitable for a machine tool project.

- A proper team for executing the project in a very efficient and productive manner should be put together once these three basic elements have been determined. It also should be kept in mind that the planning and execution of any project are highly dynamic and iterative by nature.

- The effectiveness of project management depends on management attitude, culture, style, and organization. The efficiency and productivity of the project depend predominantly on the attitude, knowledge, and expectation of team members. If the organization has a hierarchical or pyramidal structure, project management efficiency suffers since every line manager is concerned with his or her own efficiency and productivity.

- Project management and activity scheduling have three major parts: planning, execution, and scheduling. In addition, it also helps the team to manage resource requirements and risk. Project planning begins with the start of the project, and it evolves dynamically as the project matures.

- The main effort should be to reduce the length of critical path activities. The best possible situation is when all events are made to have the same timing as critical paths. An iterative process can be set up to find the optimal project time. At each step, the total cost is calculated and compared with that of the last step. The process continues until an optimum is reached.

- When project timing is subject to change, PERT comes into the picture, and it can provide some effects of time variations, which are also directly connected to direct and indirect costs for the project. PERT can also help to find the effects of time variations on the total time for the project.

- One of the main advantages of a network representation of activities is that we can start doing a critical path analysis for the project. The critical path analysis is a very effective tool to manage and control project activities. CPM helps control project timing.

16.12 References and Bibliography

Abernathy, W.J. and Townsend, P.L., 1975, "Technology, Productivity and Process Change," *Technological Forecasting and Social Change* 7(4), 379–96.

Bralla, J.G., editor, 1999, *Design for Manufacturability Handbook*, McGraw-Hill, NY.

Boothroyd, G., 1992, *Assembly Automation and Product Design*, Marcel Dekker, NY.

Browning, T.R., and Eppinger, S.D., 1988, "A Model for Development Project Cost and Schedule Planning," MIT Sloan Working Paper 4050, November.

Dewhurst, P., and Boothroyd, G., 1987, "Design for assembly in action," *Assembly Engineering*, January.

Dewhurst, P., and Boothroyd, G., 1989, *Product Design for Assembly*, Boothroyd Dewhurst, Inc., Wakefield, RI.

Dixon, J.R., and Poli, C., 1995, *Engineering Design and Design for Manufacturing*, A Structured Approach, Field Stone Publishers, Conway, MA.

Forsberg, K., Mooz, H., and Cotterman, H., 2000, *Visualizing Project Management*, John Wiley and Sons, Inc., NY.

Gupta, S.K., Das, D., Regli, W.C., and Nau, D., 1997, "Automated Manufacturability Analysis: A Survey," *Research in Engineering Design* 9(3), 168–190.

Kezner, K., 1996, Project Management: A System Approach to Planning, Scheduling, and Controlling, fifth edition, Wiley, NY.

McGrath, M.E., 1995, *Product Strategy for High-Technology Companies*, McGraw-Hill, NY.

Nahmias, S., 2008, *Production and Operations Management*, fourth edition, McGraw-Hill Erwin Companies, NY.

Paul, G., and Beitz, W., 1996, *Engineering Design*, Springer-Verlag, NY.

Smith, R.P., and Eppinger, S.D., "Identifying Controlling Features of Engineering Design Iteration," *Management Science* 6(1), 1–13.

Smith, P.G., 1996, "Your Product Development Process Demands Ongoing

Ulrich, T.K., and Eppinger, S.D., 2011, *Product Design and Development*, McGraw-Hill Companies, NY.

Ulrich, T.K., and Pearson, S., 1993, "Assessing the Importance of Design through Product Archaeology," *Management Science* 39(4), 429–447.

Urban, G L., and Hauser, J.R., 1993, *Design and Marketing of New Products*, third edition, Prentice Hall, NJ.

Whitney, D.E., 1988, "Manufacturing Design," *HBR*, 66(4), 83–91.

16.13 Review Questions

- Is DFM or DFX suitable for the machine tool industry? What are the fundamental advantages and disadvantages of applying these processes to machine tools? Can DFMA or DFM help reduce the machine's cost?

- Is the determination of a critical path for the project critical? How would you use CPM and PERT to improve project efficiency and productivity? Can you use these methods to reduce the cost of the project?

- What are the primary and secondary characteristics of a good team leader? What are the characteristics of a good team member? Is following the team leader's instructions without analyzing the consequences a good approach? What is the ideal relationship between the team leader and a team member? Why should the team be cohesive?

- Is project management suitable for all kinds of projects? Do project management and control cost money for the company? Is it worth it to have a project management team? Is project management disruptive to the project design process? How can you control the project management team?

C H A P T E R 17

17 Reliability Engineering Applications

17.1 Introduction

The reliability of any machine tool must be designed in, and the details of design dictate the reliability and performance of machines. In order to incorporate reliability aspects into the design itself, engineers have to understand the fundamental principles of reliability. They also have to understand the technical design strategy and how to instill reliability in the system and component design right from the beginning. Reliability is a metric that any design has to satisfy, and it has to come from the customers' feedback or requirements. The recent extension of plain reliability is the robustness of the design or the product as a system. System reliability is fundamentally different from component reliability. The manufacturing process also dictates the reliability and robustness of any product. A simple design is normally more reliable than a complicated one. The design and manufacturing have to work in unison to create a reliable product.

In recent years, reliability has become a hot topic for the industries irrespective of their products and services. The industries have come to an understanding that a product cannot be successful unless it is highly reliable. In my opinion, Japan has brought the importance of reliability to this country, and the US is only now waking up. Innovative products without reliability and robustness cannot be successful commercially. Industries should make their products with a reliability metric based on what customers would like to have. The application of system engineering leads to higher reliability for the product.

Reliability is the probability of success and the probability that a system will perform its functions without failure for a desired amount of time and under given environmental conditions with some given level of confidence. This is the design criteria to be accepted for the product. Reliability engineering and practices can be applied through system engineering tools, which can evaluate a product against required reliability metrics; i.e., the system performs without interruption over time at a given confidence level, which is also called robustness of the product. Reliability can also be used as an index of quality for the product.

The reliability of a product is a designed-in safeguard against all types of uncertainties in application and engineering, such as load, material strength, and manufacturing variations. High reliability and robustness ensure a product that will perform smoothly over a design period or the life of the machine. Uncertainties exist in every product. These variations can also be called "noise" for the system or product. The variations of design and application parameters can be as follows:

- Applied load variations
- Environmental variations such as temperature, humidity, dust, etc.
- Material variations
- Manufacturing variations
- Assembly variations

These uncertainties and unknown variations can lead to performance variations of the machine during its desired life. Hence, the reliability of a system varies, and unless these variational inputs are not assessed properly, the product will fail in an immature fashion, which reduces customer satisfaction. The uncertainties of input conditions can be known or unknown while designing the product. The system behaves in a random fashion, and performance outputs cannot be determined. This random system behavior is called "stochastic" behavior. Any system does have inherent randomness that cannot be controlled or assessed beforehand. The failure pattern remains very erratic and random. Another type of system output variation is subjective uncertainty, which comes into the picture due to a lack of knowledge of the system inputs. The system reliability varies due to stochastic variations, and the machine performs or failure occurs in a random fashion.

To control stochastic random variations, the machine system as a whole has to be tested at length and under various input conditions to determine the extent of system variations. Also, the number of machines tested should be more than one or two. Hence, such tests are very costly and consume lots of time, which is not always feasible. If the system availability for tests is very limited, as often is the case for machine tools, the question of confidence level comes into the picture. The prediction of system behavior has a confidence level attached to it to represent the level or degree of uncertainty. For data coming from a very controlled and limited number of machines, the degree of confidence level will be very low, and most of the time, predictions will be erratic as well. Performance variations due to lack of system input conditions depend on the level of system analysis detail or fineness of system simulation conditions.

The assumptions for system modeling have to be as realistic as possible. The system analysis can only be accurate if all the boundary conditions, system inputs, and system modeling are done accurately. If inputs are wrong, output will be wrong. Subjective uncertainty could be present due to several factors, such as operator error, maintenance quality and type of repair, frequency of preventive maintenance, knowledge of maintenance personnel about the machine system, etc. Overall, machine reliability and its enhancement also depend on data collection and analysis. Very often, the analysis includes a safety design factor to take care of the unknown situations or lack of system inputs. Subjective uncertainty depends on the level or degree of lack of knowledge about the system inputs and noise conditions. Stochastic system uncertainty is inherent in the system, and system familiarity is essential to control the stochastic variations.

The objective of reliability assessment for a product is to satisfy the customers who expect the product to survive and perform throughout the intended operation life at a cost that they can afford. The customer is the focus here, and the product should meet or exceed their expectations. This is the primary reason why U.S. products have lost their market. U.S. products have been found by customers to be secondary or inferior to products from Japan or Korea. Innovativeness is good as long as it satisfies customer expectations or requirements. For most customers, reliability is the primary index of a quality product. Now, there are various definitions of reliability of a product. A common textbook definition of reliability is:

- The probability of a product's performance over an intended design life and under specified operating conditions

This textbook definition has several problems described below:

- A product system has aspects other than performance: safety, external looks, aesthetics, etc.
- The operating conditions are defined by customers only and not by management or designers.
- Intended design life is also defined by the customers.
- The reliability or probability of success or performance changes over time.
- It does not specify or address the question of cost vs. reliability.

Another more appropriate and refined reliability definition of a machine could be as follows:

- Reliability is the probability of the successful performance of a product's desirable functions for a pre-defined operating life under customers' operating conditions, meeting or exceeding their expectations at a cost that they can afford.

The probability should include both stochastic and subjective outcomes of the product. The previous definition also addresses cost, performance, and customer requirements simultaneously. In any case, reliability is the success of the product's performance and all other aspects of a system as a whole. If a system performs very well but gets rusted before the desired operating life, then the system has failed. System life, performance, reliability, etc. have to be designed into the system and defined. Reliability should include design, analysis, and verification of the system as a whole. The analysis should include noise and environmental condition inputs and should consider both statistical and engineering conditions.

Reliability analysis should consider operating, manufacturing, material, maintenance, installation, and storage conditions throughout the product's life. Hence, reliability analysis or modeling should include structural analysis, testing, data verification, operating conditions, noise conditions, load analysis, manufacturing variability, material variability, supplier's variability, etc. Reliability analysis must include data verifications and confirmations, root cause analysis, DFMEA, PFMEA, etc.

Hence, reliability or robustness has the following primary characteristics:

- Reliability is a metric to be included in system requirements analysis: field failures or failure during validation

- Reliability is a customer-driven requirement and perception: usage requirement

- Reliability is assessable and predictable with a level of confidence: modeling and analysis of the product managed by engineers

- Reliability is verifiable: engineers' responsibility

- Reliability is also time-dependent: robustness

- Reliability is a property of a system: index of quality for the system and success of a product depends on the reliability

- Reliability is a design function: reliability can be designed into the product

- Reliability cost is part of the life-cycle cost (LCC) for machine maintenance: a design function to keep the LCC as low as possible

The primary objective of a reliability metric is to design and verify the product with a level of confidence. The information sources for this metric can be filed data, supplier's data, and test data. The filed data includes both design input data and customer feedback. The data also gives the failure or success over time.

17.2 Importance of Robustness Analysis

Unplanned system failures cause the most customer dissatisfaction for any product or service. On the other hand, mechanical or electrical systems do fail at some point during their operation. Machines fail just like bridges, vibratory systems, control systems, or any other devices. Hence, reliability assessment and validation of any product or service draw interest from system designers. Reliability is even more important for customers since it is directly connected to cost and profit. In a very competitive market, the reliability of a machine is also a yardstick to establish the creditability of the machine. As the system complexity goes up, reliability requirements also need to be enhanced to keep the system going. For the unmanned operation of the systems in a mass-producing environment, the reliability of a machine is even more necessary to enhance profitability for the company.

Consumers have become more knowledgeable and conscious about system reliability, maintainability, serviceability, safety, ergonomics, and quality. Eventually, engineers will pursue reliability engineering to satisfy the customer. Reliability engineering has to be applied to the design process right from the project's inception, and the reliability engineering process and methodology must be practiced and applied as much as possible. The proper application of reliability tools will enhance the reliability, robustness, and quality of the machine, and only reliable products will survive in the market in the long run. So, reliability engineering is a necessity for the product just like structural engineering or control engineering.

Reliability tools have become more necessary due to the very high complexity of machines. Every sub-system, such as control, servo systems, high-speed spindle system, index systems, etc., has become highly complex, and system reliability is affected by component reliability depending on the configuration or mode of the components. Series systems behave differently than parallel systems. The complexity of the system has become a driver for the application of reliability assessment for machines: as the system becomes more complex, the dependency goes up and up. Non-reliable products cost the customers greatly, and they are becoming more reliability conscious as the system complexity goes up.

In recent years, products have become recognized for higher reliability, mean time between failures (MTBF), mean time to repair (MTTR), etc. Hence, engineers have to design their products to enhance MTBF and MTTR and reduce failure rates during the operating life of the machine. Reliability, maintainability, serviceability, safety, and ergonomics have become part of machine quality in addition to performance. Reliability engineering has to be practiced in every discipline—operation, engineering, finance, services, manufacturing, project management, supplier management, etc.—for the company. Reliability principles can be used to compare the possible designs for the following aspects:

- Cost reduction and optimization
- Quality enhancement
- Redundancy analysis
- Failure mode analysis
- Root cause analysis
- Design analysis
- Failure optimization and avoidance
- Ergonomics and human factor engineering
- Trade-off studies
- Validation and verification

Another aspect of robustness analysis is to understand the difference of treatment between a system and components. A spindle system is different from a bearing in the system. Hence, the reliability analysis of a system is substantially different from component reliability analysis. It is very difficult to analyze the effect of interaction among components with respect to reliability. There is another concept that is relevant: maintained and non-maintained systems. In the case of maintained systems, replacement of a failed part is considered to make the system operational again. For a non-maintained system, the failed parts cannot be replaced. For example, belts in a spindle system are replaced to make the system fully operational again, whereas, for highly vulnerable systems, such as rockets, etc., replacement of parts is not allowed, and a new system is used. For machine tools, most systems are maintained by default. Reliability engineers must also understand the failure patterns of components and systems. Diagnosis and corrections depend on them. Repetitive failures might represent design weaknesses, and random failures could be due to norms of the design and use of the system.

17.3 Reliability Basics

As mentioned earlier, reliability can be used as a metric, and any requirement must have a metric so that it can be measured and tested against a standard. The relationship between reliability, R(t), and unreliability can be expressed as:

R(t) = reliability = number of successes in a test/number of test samples

= 1 – (number of failures in a test/number of test samples)

= 1 – unreliability = 1 – UR(t)

R(t) is normally expressed in percentages. Also, R(t) + UR(t) = 100%, or 1.0.

Failures or unreliability could be also measured in terms of failure percentage, the number of repairs/units in operation, MTBF, MTTR, MTTF, B_{10} life, etc. The reliability of a product is also an index of customer satisfaction. For some customers, small failures can be considered minor interruptions, whereas, for others, they could be considered failures of the product or an indication that the product is unreliable. Reliability is expressed in many forms by the customers, such as a highly durable product, a highly dependable product, a very reliable product, a robust product, etc. Sometimes failures can be less important than the degradation of functional performance. Degradation of accuracy over time is considered to be one of the most important factors for customers. Machines might work, but the parts made out of them are not accurate enough for further use.

Functional degradation over time, i.e., robustness, is even more important for most of the customers. Output variability is another factor that customers do not like. Variability could be in terms of quality parts, accuracy, surface finish, volume of production, etc. A reliable machine is normally robust if it maintains function over time within an acceptable range. Average machine life for production shops is anywhere from six to 12 years. In 10 to 12 years, technology changes drastically, and machine replacement becomes almost necessary to enhance production quality and output. After so many years of use, machines are salvaged to buy more technologically advanced ones.

Customer satisfaction is probably the single most important factor for repurchasing decisions. Customers were not happy with the quality, reliability, cost, and performance of U.S. machines, so they decided to buy Japanese and Korean machines instead. Customer experiences do matter or influence repeat-buy decisions most of the time. Currently, customers are keeping machines longer in use, and the behavior or reliability of machines during this extended usage time is also a very important factor in repeat orders. Extended usage goes beyond the design life of the product. Another reason for selecting a Japanese machine over other manufacturers is the resale value for the product. Reliability enhancements affect the product failure rate during extended usage, and hey also enhance the machine's resale value. Reliability is a key factor for winning customer loyalty for the product.

The company has to have a reliability mission, which should specify customer satisfaction with respect to the reliability of the machine. The mission should be to satisfy customer expectation for reliability and robustness throughout the life cycle of the machine. For example, the reliability mission of the company should be to design and manufacture a machine that will have at least four or five years of useful life without any major interruption or failure with MTBF more than 2,000 hours or more. The failure rate or functional degradation must meet or exceed customer expectations on reliability. For example, useful life, wear-out, or replacement of rubber components should be limited to preventive maintenance schedules. Unwarranted failures should be limited to none, or as minimal as possible to gain customers' confidence. Hence, the satisfaction of customer expectations for performance or failure rates is the primary objective, not what the designer says.

17.4 Continuous Improvement Cycle for Robustness

The continuous robustness cycle for the reliability of any machine includes several activities, as shown in Fig.18.1 below. The objective of this cycle is to improve the reliability of any machine during design, manufacturing, and customer usage. Reliability analysis collects data from several phases of this cycle. Reliability engineering tools diagnose the root cause of the lack of reliability and recommend improvement. The process is continuous and iterative over the life of the machine.

The whole process starts with the customer requirements for robustness and reliability for the machine. The requirements are baked into the design using reliability engineering tools such as root cause analysis, design FMEA, reliability analysis, etc. The previous field data for similar machines is collected to improve the present machine project. Using the legacy data, a reliability mission statement, i.e., reliability metric, is created for the present program. The program reliability objective must support customer requirements. The next phase is to develop a design to satisfy this requirement. Reliability analysis tools are used in every phase to guide the designers to design systems or sub-systems with enhanced reliability. The new and improved system configuration makes the machine more robust with respect to reliability and robustness. The new system improves performance over time and increases design life, MTBF, and MTTR under customer usage or any real-world usage cycle.

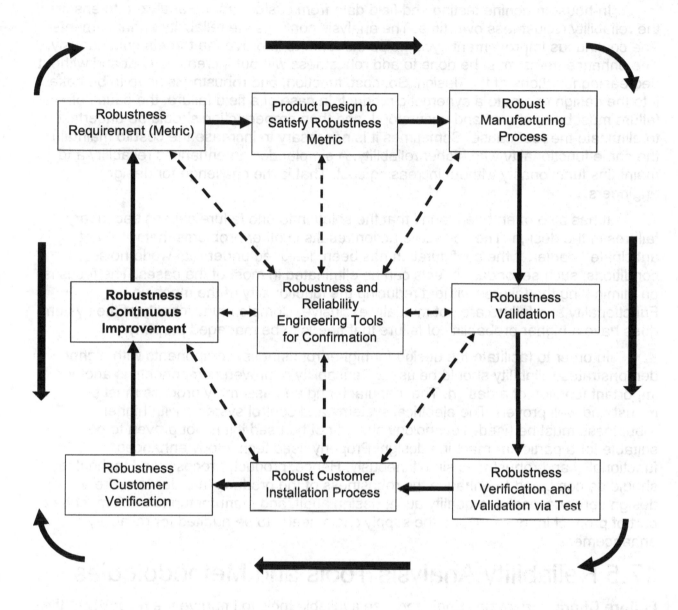

Fig. 17.1 Robustness improvement cycle

Legend:

Data Transfer: ◄- - - - - - ►

Process Transfer: ◄━━━━━━━►

In-house machine testing and field data from customers are analyzed to ensure the reliability robustness over time. The analysis confirms the reliability enhancements. The continuous improvement cycle helps the team to improve the targets continuously. The enhancements must be done to add robustness without increasing cost and without decreasing functions of the design. So, cost, function, and robustness have to be baked into the design to create a synergetic product. In case of a field failure, the cause of failure must be analyzed and understood, and then proper action should be undertaken to eliminate the root cause. Sometimes it is necessary to increase the cost to maintain the same functionality with higher reliability. A simpler design enhances reliability and maintains functionality without increasing cost. That is the challenge for design engineers.

It has also often been found that the solution to one failure causes secondary failures in the design. The corrective action results in other problems that were not anticipated earlier. If the configuration has been designed under real-world noise conditions, such secondary effects can be eliminated in most of the cases. The focus is on eliminating the failures without reducing the functionality of the machine. Functionality and failure are not mutually exclusive. Complex functionality of the system does have a higher probability of failure that needs to be managed properly.

In order to facilitate the design for higher robustness, components with higher demonstrated reliability should be used. Reusability of proven components is another important function of a design. The manufacturing and assembly process must be robust and well proven. The electrical systems and control systems with higher robustness must be used. Technology should not be used if it is not proven to be suitable for a particular machine design. Properly used technology enhances functionality and robustness simultaneously. Hence, product, process, and technology should be combined to enhance the robustness of the product. In order to make a design very successful, a quality audit, design audit, and manufacturing audit must be a part of product innovation, and the supply chain needs to be audited for reliability enhancement.

17.5 Reliability Analysis Tools and Methodologies

Failure Characterization: Engineers use available tools to improve the reliability of the product. They collect the existing product failure data to make necessary changes to the product design so that the failures are eliminated from future designs. They also enhance the product's reliability without degrading functionality and adding cost to the product. To follow system engineering principles, engineers use the following tools:

- Root cause analysis (RCA)
- Failure mode analysis (FMA)
- Stress/strength interference analysis
- Bathtub curve data analysis

- Hazards analysis
- ROCOF plots

Types of Failures: For machines, any failure happens when a product ceases to function in a way that the customer expects. Hence, the quality of any functional performance over time is important. Just delivering a function is not enough; delivering customer expectation in totality is important. Customers define failures. It is also important to note that the customer's definition of failure changes over time. For example, spindle rotation variation within an allowable range might not be a failure in the beginning, but it might be very critical when such variations are not acceptable after some machine usage. For soft metal cutting, horsepower variations might not be critical, but it is considered a failure when the spindle belt slips and fails to deliver proper power for cutting hard metal. Reasonable customers understand that new machines perform better than old machines, but they do have the same expectations for both new and old machines. If a machine makes noise when new, it could be a problem for the customer, but if it squeaks and rattles after 10 years of use, it might not be an issue. Nevertheless, customers have to analyze the complaints and make decisions accordingly. Machine failures can be categorized in several possible ways:

- Random functional variations: If machines behave in an erratic function due to control system malfunction, the failures could be random and might not have any trending characteristics.

- Partial function delivery: This shows partial failure in delivering desired performance. When the machine warms up, the accuracy is good, but it fails to produce an accurate part when it is cold or too hot. Automatic door locks sometimes fail to turn on the machine automatically. Machine leaks occasionally.

- Intermittent but full function delivery: The failures occur intermittently due to electrical circuit malfunction. When the machine is cold, such failures rarely occur, but when it is running very hot, they occur very frequently. Other examples: belt slippage due to heat, control fails to run the machine automatically sometimes, relays malfunction occasionally, etc.

- Total loss of function: Spindle fails to rotate due to bearing seizure, transformer fails to deliver power, servo motors fail, index system fails to rotate the turret, etc.

- Gradual functional degradation: Spindle runout increases gradually, belt slippage over time, hydraulic oil leakage, accuracy changes due to overheating, stick-slip motion of the cross slides, machine noise and vibration, etc.

- Excessive functional performance: Door slams very quickly, spindle stops very quickly, slide moves quickly to collide with spindle, very quick chuck jaw closure, tailstock quill rapid movement, etc.

Failure Severity: It is also true that failures do not always create the same extent of dissatisfaction among customers. The severity of any failure depends on the following factors:

- High life-cycle cost: production interruptions, maintenance cost
- Reduction of design life: complicated design, low durability life
- Very high repair time: complicated mechanism, non-adjustable setting
- Non-availability of replacement parts: obsolete electrical parts, imported parts
- Safety concerns: unsafe design, exposed rotating parts
- Fatality concerns: poor door designs, open movable parts

Among all of these, failures that cause fatalities or unsafe conditions have the highest severity. Failures and functional degradation that reduces the usage of the machines have the second degree of severity. These functional degradations normally help the customers become biased against the machine. There are other failures, such as paint peeling, hour meter malfunction, misalignment conditions, etc., that also cause high concerns among users. To get the most profit out of any machine, its life-cycle cost must be as minimal as possible. When the machine uptime is very high, production volume goes up substantially. One of the biggest concerns of customers is whether the machine can justify the rate of return during the life of the machine. Machines with very low reliability traditionally have very low salvage values. For example, Japanese machines traditionally have comparatively higher salvage values than American-made machines. Life-cycle cost also gets affected adversely due to frequent failures and replacement of wear-out parts, such as rubber belts, O-rings, piston rings, hydraulic valves, etc.

Reliability Measurements: In order to assess the reliability or robustness of a system or component, data must be collected and analyzed. The data has inherent characteristics of variability from measurement to measurement. The data also has statistical properties and distributions. Unless the data is collected properly and consistently for a system failure, reliability prediction and analysis will be almost impossible. The engineering challenge is to characterize a failure by analyzing the data collected from the system. As it has been mentioned before, the probability of occurrence of any event such as failure can be defined by the ratio:

Probability of occurrence of a failure mode, $N = P(N)$ = number of ways failure N can happen/total number of possible failure modes.

The probability measures the likelihood of occurrence of a particular failure, such as indexing failure, spindle failure, ball screw failure, etc. The probability varies between 0 and 1, or 0 and 100%. If the probability is 0 or 100%, the event becomes deterministic; i.e., the system will never fail or will always run without failure. If the probability is 50%, the event might happen or it might not happen, i.e., equal chance.

The reliability of a system can be defined as the probability of a machine performing its desired function over time or design life. Reliability can also be defined as the probability of satisfying a customer. Reliability varies over time or specified life, and reliability or robustness is always expressed as a function of time. Hence, reliability is a ratio that defines the success of survival for a machine delivering desired performance up to time, t, out of the number of machines subjected to an operation or test. The total number of machines under test or operation is also called the population.

Reliability of a system = $R(t)$ = number of success/total number tested = N_s/N_t

$$= (N_t - \text{Number failed})/N_t = (N_t - N_f)/N_t$$

$$= 1 - N_f/N_t = 1 - \text{failure or } 1 - \text{unreliability}$$

Let's take the example of machines with similar indexing systems. Say there are 10 such machines being used in a machine shop. In five years, two indexing system failed or failed to deliver the desired function, i.e., rotation. In that case,

R (5 Years) = 1 – (2/10) = 8/10 = 0.8, or 80%.

Hence the reliability of the index system is 80% within a design life of five years; i.e., 80% succeeded to survive for five years, and 20% could not survive for five years of delivering the desired performance. Here, the entire group of 10 machines is called the population. A sample is selected out of the whole population for a specified test since 100% testing is not feasible. Sampling should be unbiased and selected at random. So, if a company is producing 20 machines in a day and every fourth sample is collected out of the assembly line: population size = 20, and sample size = 5. If a customer says that one machine out of 10 did not satisfy his requirements after one year of usage, the machine has a 10% satisfaction rating after one year. Even if the data is out of a small sample size, it is assumed that all the machines the company produces will have 10% satisfaction rating after one year of usage. Such data provides an approximate estimate of satisfaction rating for the machine, which could be true or false for the population.

Failure Histogram and Distributions: Machine failures can be represented in graphical form, called a failure histogram, which shows the number of failures over the life of the machine. The failure data is discrete and represented by a bar graph. Suppose we are tracking the indexing failures of a CNC lathe machine. The indexing failure for the machine is given in tabular form in Fig.17.2. The failure histogram is shown in Fig.17.3. The horizontal axis represents the life interval, and the vertical axis shows the observed failures in that interval. Most of the indexing failures are observed between 20,000 and 70,000 cycles and the maximum is in between 40,000 and 50,000 cycles. The observations could be made for one machine or set of machines. The frequency of failure could be calculated as:

Frequency of failure, $f(t)$, % = number of failures observed during an interval/total number of failures observed.

The frequency of failure is not the same for the life ranges. If the number of data points collected is very high, the life range will get very small, and the discrete data could be assumed to be continuous when the number of samples is too high. At that point, the bar chart becomes a continuous graph called a "probability density function," $f(t)$, for the failure. It has a normal distribution.

The percentage of failure can also be called the "probability of failure during any life range." Hence, the probability of failure in the life range 40,000–50,000 cycles is 20%, or there is 20% chance that the index system will fail at 45,000 cycles, midpoint of the life span. In the probability density curve, the discrete histogram becomes continuous and smooth. The curve shows the probability of failure at any point in the life cycle for the indexing system. The probability density function, PDF, is a statistical normalized model for the population failure histogram or a failure distribution. Since the graph is normalized, the probability of failure between any two points of the life cycle is given by the area under the curve. The total area under the curve is 100%, or 1.

Life Interval in Cycles	Number of Failures Observed	Frequency of Failures	Probability of Failures	Cumulative Failures	Reliability
0-1000	2	2	2%	2	98%
1000-10000	4	4	4%	6	94%
10,000-20,000	8	8	8%	14	86%
20,000-30,000	11	11	11%	25	75%
30,000-40,000	16	16	16%	41	59%
40,000-50,000	20	20	20%	61	39%
50,000-60,000	14	14	14%	75	25%
60,000-70,000	11	11	11%	86	14%
70,000-80,000	8	8	8%	94	6%
80,000-90,000	4	4	4%	98	2%
90,000-100,000	2	2	2%	100	0%
Total:	100			100	

Fig. 17.2 Indexing failure table

Another form of distribution is called the "cumulative failure distribution," F(t), which displays the cumulative number of failures at any point of life. This is also a continuous and approximate representation of failure bar chart. The graph shows the cumulative frequency of failure through each time interval. Near limit, this also becomes a continuous distribution and is called the cumulative distribution function, F(t), which is shown in Fig.17.3. The cumulative distribution function is helpful to determine the reliability of the system at any point of life.

Reliability, R(t) = 1 – F(t)

Probability of failure between Life 1(t1) and Life 2(t2) = F(t1) – F(t2)

Reliability, R(t), and CDF, F(t), Plot: The cumulative distribution function, CDF, F(t), is very useful for calculating reliability values. The distribution shape is not important. When F(t) is zero, R(t) = 100%, or 1.0, and when F(t) is 100%, R(t) = 0%; i.e., all systems have failed, and reliability is zero. The relationship between CDF and reliability is shown in Fig. 17.5.

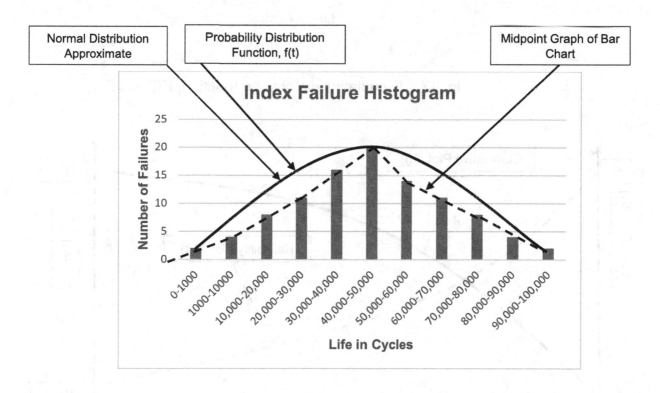

Fig. 17.3 Indexing failure histogram bar chart

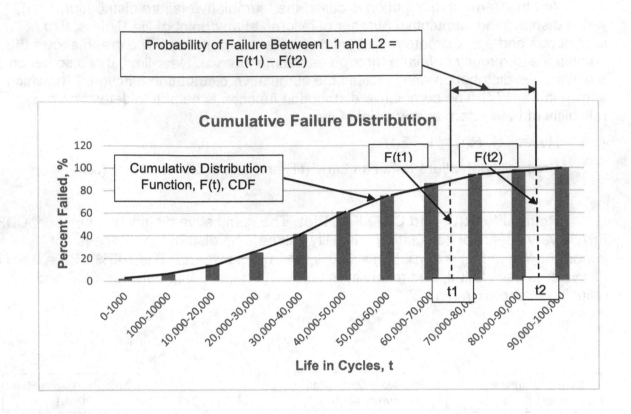

Fig. 17.4 Cumulative distribution function, F(t), CDF

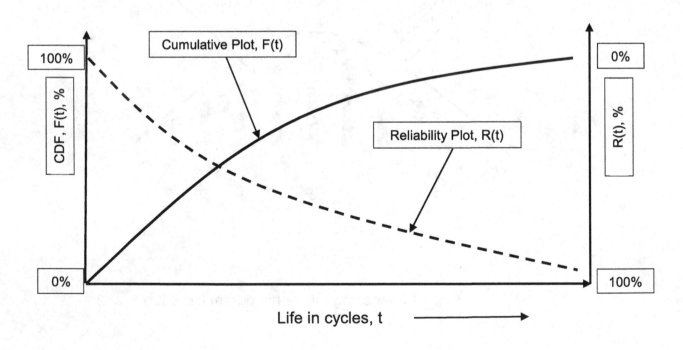

Fig. 17.5 Reliability and cumulative distribution function relationship

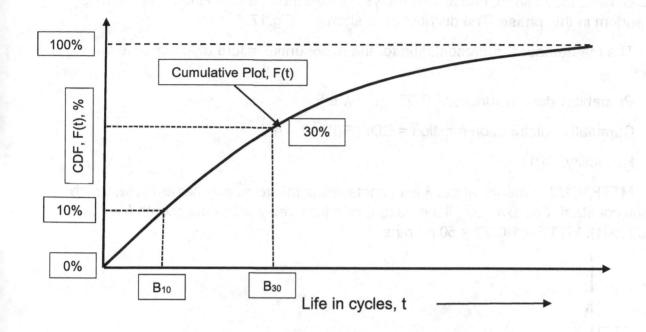

Fig. 17.6 B_{10} Life Plot

Another term in reliability is the B_{10} life for a component. This is defined as the life when 10% of the machines fail; i.e., reliability is 90% at that life. This is particularly used for assessing reliability life of bearings, batteries, lamps, etc. The life is easy to determine from the cumulative distribution plot, F(t). For example, B_{30} life is where 30% of the machines or any system fail. The value of life, t, where 30% of the systems fail under test is called the B_{30} life for the machines. The B_{10} life plot is shown in Fig. 17.6.

MTBF and MTTF: MTTF is the mean or average time to first failure of the machine when under test or operation, and MTBF is the mean or average time between failures. MTBF is used when repair is done to a system and it is used again. MTTF and MTBF are applicable when the failures happen at random and correspond roughly to the horizontal portion of the bathtub curve. To calculate the MTBF, sum the total operation time and divide it by the number of failures during that time.

Normal and Exponential Distribution: Normal and exponential distribution are the most common statistical distributions of failures for machine tools. These two distributions are different in several aspects. Exponential distribution of failures means the failure rate remains constant; i.e., for two equal time intervals, the probability of failure of running machines remains the same. It is applicable to the horizontal portion of the bathtub curve, where the failure rate remains the same up to the desired life of the machine. This is commonly known as "constant failure rate" distribution or "constant hazard rate, h(t)" distribution.

The exponential distribution is applied for components that do not wear out, such as the spindle, index shaft, capacitors, relays, transformers, etc. The failure pattern is also random in this phase. The distribution is shown in Fig.17.7.

The mathematical representation for the exponential failure distribution is as follows:

Probability density function= PDF, $f(t) = \lambda * e^{-\lambda t}$

Cumulative distribution function = CDF, $F(t) = 1 - e^{-\lambda t}$

Reliability, $R(t) = e^{-\lambda t}$

MTTF = $1/\lambda$ = mean, where λ is instantaneous failure rate or hazard rate, which remains constant. For example, if the hazard rate for a relay is 2% per month (i.e., two out 100 fails), MTTF = 1/0.02 = 50 months.

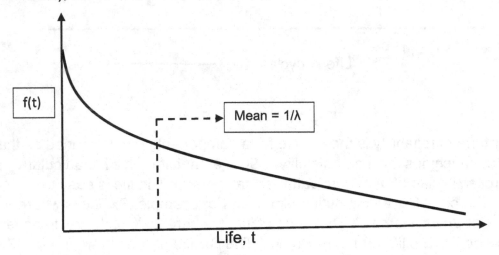

Fig. 17.7 Exponential distribution

Most mechanical components display a normal distribution for failures. This is the best known and most widely used statistical distribution for failures of mechanical components in machine tools. A normal distribution is symmetrical about the mean, and the mean is the same as the mode and median of the distribution. The mode is the highest point of the distribution. In other words, the mean is the expected value of the distribution, and the mode is the value at which the CDF is 0.5; i.e., 50% of the parts fail at the mean point. The mode of the distribution is the value at which probability of failure is the highest. The normal distribution is characterized by two parameters: standard deviation, σ, and mean, μ. The probability distribution for the failure is represented by the following equation:

$$f(t) = (1/((2*\pi)^{0.5}*\sigma)) * e^{-[\{(t-\mu)^{\wedge}2\}/(2*\sigma^{\wedge}2)]}$$

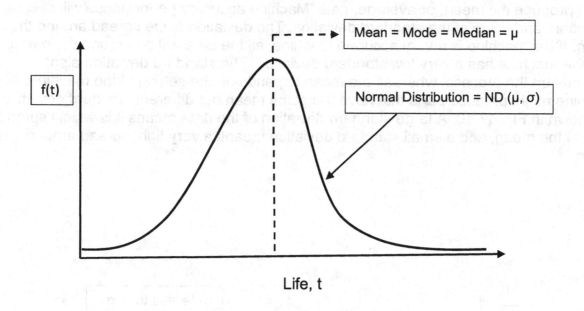

Fig. 17.8 Normal distribution

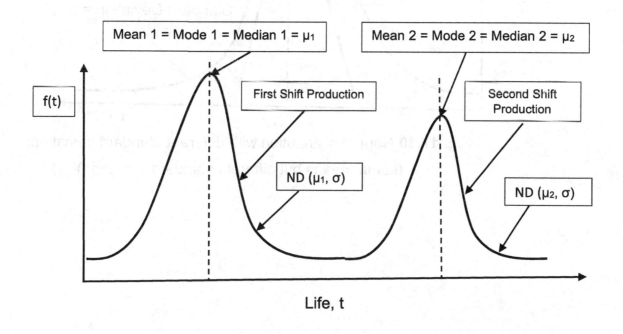

Fig. 17.9 Bi-modal normal distribution for two shifts' production
(different means but same standard deviation)

Any statistical distribution of failure data would show its nature. The machine is set to produce the mean, or average, data. Machine accuracy performance will decide the extent and value of the standard deviation. The deviation is the spread around the mean. If the machine is a very accurate machine, all the data will be around the mean, and the machine has a very low standard deviation. The standard deviation also depends on the process, whereas the mean depends on the setting of the machine. A machine might produce parts that have the same mean but different standard deviation, as shown in Fig. 17.10. A large standard deviation of the data means it is widely spread around the mean, and a small standard deviation means a very tight spread around the mean.

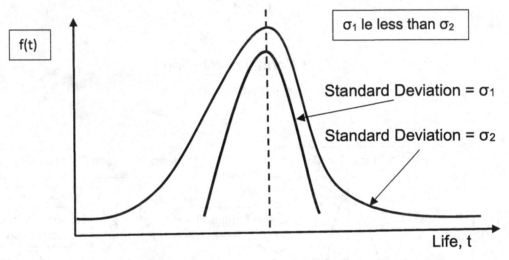

Fig. 17.10 Normal distribution with different standard deviations
(same means but different standard deviation)

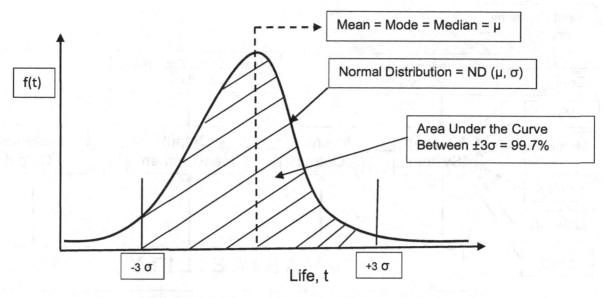

Fig. 17.11 Area under normal distribution graph

Moreover, the area under the normal distribution curve represents the probability as well. For example, the area under the graph between μ ± 3 times standard deviation is about 99.7%, which is the probability of any data within these life values. There are other distributions, such as the Weibull distribution, Poisson distribution, uniform distribution, etc. The Weibull distribution has two or three parameters to describe the distribution. The exponential and normal distribution are two special cases of Weibull distribution depending on the value of the parameters. The Weibull distribution is very often used in reliability engineering.

17.6 Reliability Measurement Process

Before any measure is taken for reliability improvements, the reliability of the present systems must be assessed. Measurements are also required of machine outputs for verification of machine performance. It is also common practice to conduct performance runs before machine approval.

In this section, the effects of the measurement process on reliability assessment will be discussed. Basically, reliability is due to variations, and to control the variability of the machine outputs or field failures, the measurement process has to be controlled. The variation affects machine design. The measurement system is also very important for measuring machine outputs. Variabilities of systems cause variability of outputs. Variability is an inherent property of the system, and it cannot be avoided, but it should be controlled to manage the output quality. Inputs, system interactions, and measurement methods all lead to the variability of machine output. The reliability of the machine also depends on the measurement of the system output. The block diagram showing causes of variabilities is shown in Fig. 17.12.

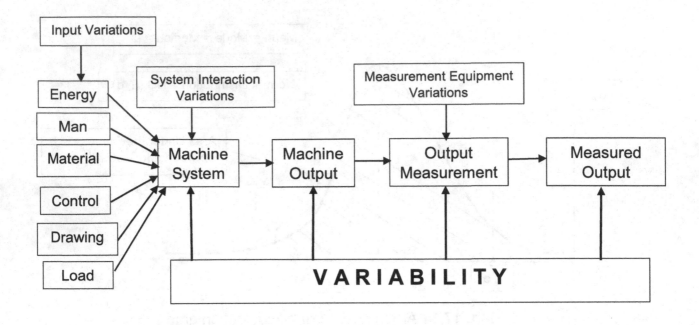

Fig. 17.12 Machine system and variability components

Design, manufacturing, and usage variations cause the system to degrade functionally over time and eventually fail. If an excess load is used for the machine, it will ultimately fail. If the material strength varies substantially, machine stress over the allowable limit might cause the machine to fail prematurely. Noise factors that lead the system to failure might be inherent in the system or external to the system. Inherent noise factors are system interaction variations, material strength, and normal aging due to usage. The system design has to control the extent of these. There is also noise that is external to the system, such as customer usage conditions, load conditions, environmental conditions, external vibrations, operator error, etc. In general, external and internal variations are to be managed to control the system life, performance, and outputs.

Another important factor is the measurement system variation. The measurement system is called gauge variation. The dimensional variations of the output might be caused by gauge malfunctions or measurement methods. Measurement system error must be found before the gauges are used for measurement. The measured part output also might be due to measurement system error and sampling error. Measurement samples must be truly representative of the population, and they must be collected at random after the system is under control. For example, the measurement of vibration using different types of vibration meters will show different vibration levels. The gauges or instruments must be calibrated before use, and capability analysis on the gauges and measuring equipment must be conducted before measurement. The capability will dictate measurement system accuracy and the applicability of the instruments for any particular condition. In addition, sampling accuracy will affect the performance prediction if the sample does not represent the population. A local failure is not necessarily a global failure for any machine.

17.7 Stress and Strength Interference: Probabilistic Design Consideration

It is very well known that material strength, such as ultimate strength, rupture strength, etc., varies depending on material homogeneity and other conditions. Statistically speaking, material strengths and customer usage conditions are both probabilistic and not deterministic. Stress due to loads on a component and material strength for the part have distributions with means and standard deviations. The load creates stress and fatigue in the component, and since load varies, stress varies during usage. Stress and strength must be compared in similar units. The stress of any component of a machine might be due to cutting tool forces, whereas machine strength is the horsepower of the spindle motor. Both of these conditions vary during usage. Voltage fluctuation for a machine could be strength variations, while current drawn by a transformer could be the stress variations. Stress and strength could both be considered to have a normal distribution with a mean value and a standard deviation. If these two distributions overlap to a great extent, as shown in Fig.17.13 below, the reliability of the component will be affected. In order to enhance the reliability of a component, these two distributions must be far apart so as not to cause interference. More interference means less reliability.

Frequency of Occurrence

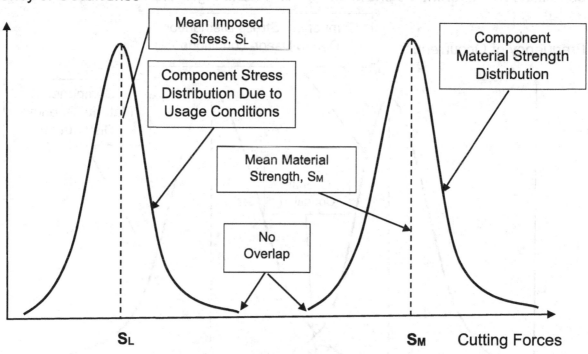

Fig. 17.13 Stress and strength distribution of a component

As shown in Fig.17.13, there is no overlap between stress and strength for the component, and it has the highest reliability. The mean factor of safety, as is very often done in the component conventional design, is:

Mean factor of safety = S_M / S_L

This can be misleading if the distributions overlap each other, as shown in Fig.17.14 below. In the overlap zone of stress and strength distributions, there is a possibility that material strength is less than stress due to external loading. If the usage dictates that the component is subjected to such loading conditions very frequently, the component will fail due to overloading conditions. Hence, the component is safe even if the design factor of safety is greater than one; i.e., the mean strength is higher than the mean stress. This could be an important factor when the design is very fragile. The probability of failure can be determined from the mean and standard deviations of the distribution if the distributions are considered to be normal, which is very common. A Monte Carlo simulation technique can be applied for critical design to know the effects of overlap conditions and avoid unreliable conditions for very critical components. Nevertheless, failures do occur when stress frequently overcomes strength.

As shown in these distribution figures, the frequency of distribution, highest amplitudes, and standard deviations are the same. This also is not always true. Very often, the distributions can be different from each other when the frequencies and tail conditions are different, as shown in Fig. 17.14 and Fig.17.15 below.

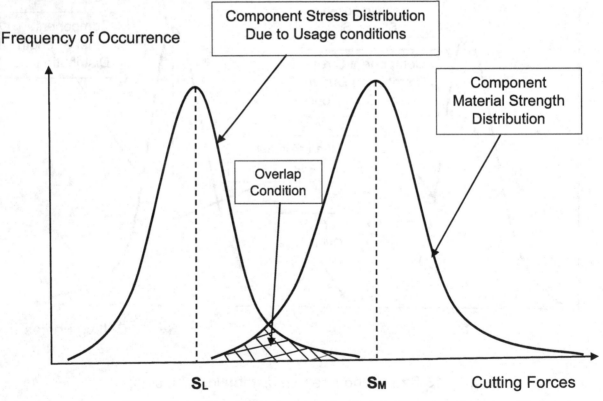

Fig. 17.14 Stress and strength distribution overlap condition

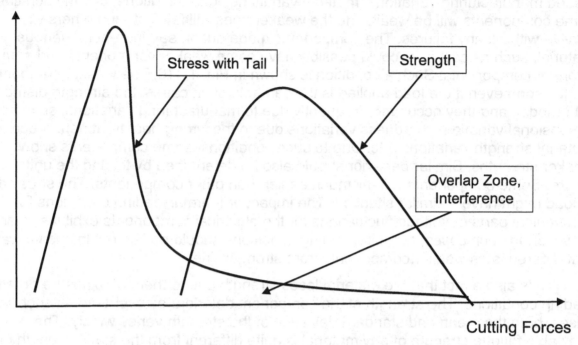

Fig. 17.15 Stress with tail and strength distribution

Frequency of Occurrence

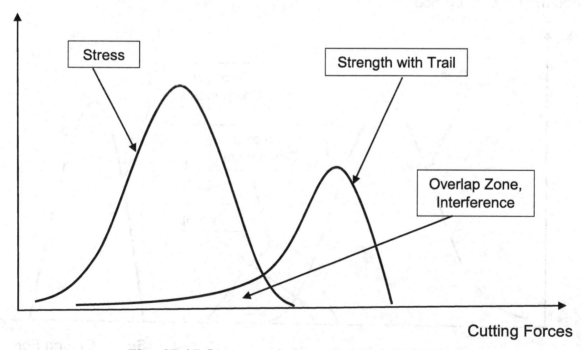

Fig. 17.15 Stress and strength with tail distribution

The tail distribution of stress or strength could be due to several reasons, such as biased manufacturing variations, material variations, load variations, etc. In such cases, some components will be weak, and the weaker ones will fail first, and others will survive without any failures. The components made out of, say, non-homogenous material, such as cast iron, could possibly have some weaker components and some stronger components. Such a condition is shown in Fig. 17.16. The weaker components will fail soon even if the load applied is the same. In such cases, the strength distribution is bi-modal, and they occur very frequently due to manufacturing variations, such as dimensional variations, hardness variations due to the wrong heat treatment procedure, material strength variations, etc. Due to such conditions, some components show weaker strengths. Similar behavior should also be determined by testing the units until failure. Some components will fail much earlier than other components. The stress due to load might have a similar situation. The impact or thermal loading conditions for mechanical parts or voltage fluctuations for the electrical components exhibit similar behavior. In such cases, durability testing conditions should be laid out in such a way that differentiates weaker components from stronger ones.

It is also a fact that the material loses strength due to thermal exposure or fatigue loading conditions. The strength of the component deteriorates over time. In such cases, both the mean and standard deviation of the strength varies widely. The allowable fatigue strength of any material is quite different from the static strength of the material. Examples of material strength degradation are stress concentration due to component design shapes, annealing of material due to exposure to thermal conditions, etc.

Frequency of Occurrence

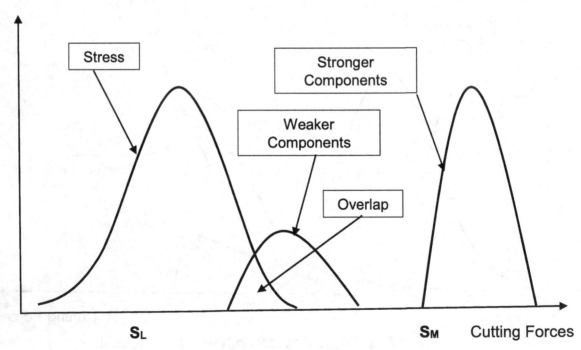

Fig. 17.15 Stress and strength with tail distribution

Another common cause of failure for bearings is the hardening of lubricants, such as grease lubricants. Hydraulic components fail due to the loss of hydraulic oil properties over time of usage. This behavior is shown in Fig. 17.16 below.

Frequency of Occurrence

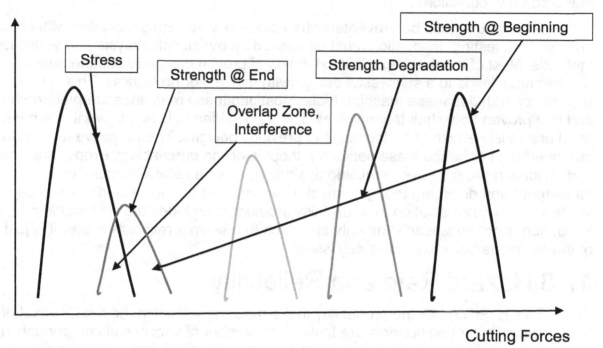

Cutting Forces

Fig. 17.15 Strength degradation of material

Common Failure Examples: Typical failures of machine tools in the customer's shop might be explained using the following explanations:

Machine base bolts are loose: Due to vibration, the machine base bolts are becoming loose, and machine vibration increases to the point that the machine cannot be used anymore. The is called an "early life failure" and is mostly due to production issues and quality inspection failures.

Inaccuracy when the machine is cold: It is often found that machine performance behaves quite differently once the machine has warmed up. The machine's accuracy becomes quite stable when it works for a while. This might be due to the thermal growth effect of the machine due to environmental conditions or it could be due to design weakness.

O-ring fails: Due to continuous usage and heat inputs, rubber components fail prematurely. This could be called failure due to aging or age degradation. Rubber wears out due to thermal conditions or vibrational effects.

The door does not shut properly: Due to the misalignment of the door with the track rails, the door does not shut properly and the machine cannot start automatically. This could be due to the fact that the door seals are worn out, causing the door to jam along the way.

Poor surface finish when the neighbor machines operate: This could be caused by sympathetic vibration due to impact conditions. If a press is operating near a grinding machine, a poor surface results when the press impacts are felt in the grinding machine. This is a poor facility layout problem and has nothing due to machine design or manufacturing conditions.

Most failures can be prevented when creative ideas are associated with proper analysis and testing. Innovation must be backed up by critical analysis and verification methods. Most of the time, a logical extension of proven design to suit the new requirements leads to a successful design with minimal or no failures. This is my experience with Japanese machine tools. Most Japanese machines are platform based and use proven bookshelf technologies. It is a good idea to look into what has been used previously and the shortcomings of previous designs. Then creative ideas can be put together to enhance these designs without injecting unreliability. Proper analysis and validation are the keys to reliable design. To solve a reliability issue, both convergent and divergent thoughts must be entertained to understand the real cause of the failures. A quick solution to a reliability problem often hides the real problem. Validation is not necessarily the only approach to resolve a reliability issue. It is just one of the many methods to resolve any issue.

17.8 Hazard Rate and Reliability

This is defined as the probability that a machine will fail in the next interval of time. When similar components are tested, the number of successful components goes down as time goes up. It is assumed that the component has survived up to time, t. The instantaneous rate of failure, hazard function, or hazard rate is denoted by h(t), which is defined as the ratio of probability density function and reliability at any time, t.

$$h(t) = f(t)/R(t) = f(t) / (1 - F(t))$$

= change in failure rate/machines in operation

= -change of reliability rate/reliability (t)

A common method to measure and analyze reliability data is to use the hazard rate. For example, the hazard rate for an electrical relay for a two-year-old machine might be 20% per month. Hence, there is a one-in-five chance, i.e,.20%, that the relay will fail, malfunction, or burn in the next month. The same relay was working fine for two years of usage. The other way of saying this is that 20% of the relays that have survived for two years will fail in the next month. The hazard rate keeps increasing as the machines get older and older. The same relay might have a hazard rate of 40% per month after four years of usage if it did not fail in the first four years. If the failed relays are replaced as they fail and almost all the bad relays are replaced in five years, the hazard rate will be less.

As explained earlier, the bathtub curve displays the hazard rate or instantaneous failure rate with respect to time or cycles. This curve could be used for repairable or non-reparable items or systems. The bathtub curve, Fig. 17.16, has three phases of failure: infant mortality phase, random failure phase, and wear-out failure phase. Reliability is important in all three phases of a component's life. The failure rate during early life decreases, it increases during the wear-out phase, and in between, it remains constant and random. The maximum design life can be defined as "five years or 10,000 hours of operation, whichever comes earlier," which starts at end of the initial failure life and ends at the beginning of the wear-out phase. In addition, infant mortality rates and wear-out failure rates must also be reduced to gain customer satisfaction. A shorter useful life denotes the design weaknesses of the machine. Early life failures are due to manufacturing and assembly issues or material problems. Wear-out failures are due to material degradation issues.

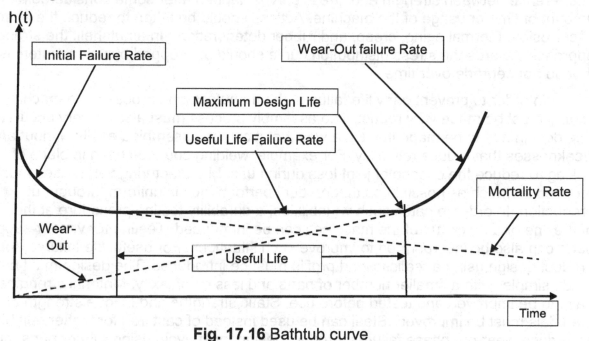

Fig. 17.16 Bathtub curve

The failure patterns are different for the three phases, and random failures can happen in all three phases as well. Infant mortality phase failures grow less frequent over the life of the product; i.e., the chances of such failures are much less after many years of usage of the machine. Moreover, the chances of wear-out phase failures or failures due to aging increase over time of usage. The rusting of sheet metal panels does not happen in the beginning of machine usage but increases very rapidly once at the beginning of the wear-out phase. The general bathtub curve for mechanical and electrical components is shown in Fig. 17.17 below.

For mechanical components, life during the useful phase is very short, and the wear-out phase is very gradual. For electrical components, once the initial burn-in phase is over, components have a tendency to have a very long useful life. Actions for improving the reliability of electrical components during the infant mortality phase might include severe testing and better manufacturing processes for electrical components. Once they are taken care of, they last a very long time on average. The failures must be caught while the machines are still inside the manufacturing shop so that failures in the customer's shop can be avoided for these components. In general, such failures cause a lot of customer dissatisfaction since the component cost is very low but maintenance or repair time can be very high and often interrupts production. This is true for control components as well. For systems, the infant mortality phase is very short, but the wear-out phase starts gradually right after.

Hence, in general, early life failures are due to manufacturing and assembly process issues. Inspection procedures during these phases can be altered to catch these problems. Useful life failures are generally caused by design issues or weaknesses, such as design complexities, load variations, etc. Wear-out failures are mostly due to age or material strength degradation due to continuous usage. Interference between strength and stress curves happen after some considerable amount of time or usage of the machine. Actions should be taken to reduce the effects of corrosion, thermal aging, creep, and rubber deterioration. In a nutshell, the strength's approach towards the stress distribution curve should be stopped by using materials that do not degrade over time.

In order to prevent early life failures, the manufacturing process and machine testing must be made very robust. The assembly process must also be very accurate. The design has to be made in such a way that it can be assembled easily without any weaknesses that reduce reliability. For example, welding could be used in place of bolting to reduce the chance of joint loss during use. Manufacturing variabilities must be reduced as much as possible so that product performance is uniform throughout production. In order to catch such variabilities, a durability testing procedure at the end of the line or a quick run of the machines can be introduced. Testing for vendor-supplied parts can also be incorporated to improve such situations. For useful life failures, robust product design using a realistic load profile must be introduced. The design must be made simpler with a smaller number of parts and less complexity. Material strengths have to be improved and tested before use. Static strengths and fatigue strength of the materials must be improved. Steel can be used instead of cast iron for higher reliability. To reduce wear-out phase failures, product design must avoid using rubber parts, and the design must be robust.

The stress and strength distribution must be far apart to reduce the chance of interference or overlap. Material strengths have to be improved, and resistance to strength degradation must be considered during design. Hence, the design has to be improved for failures during useful life period and wear-out phase, whereas manufacturing and assembly procedures need to be improved for infant mortality phase failures. These are general guidelines for reduction of failures and enhancement of reliability of any product. Before any corrective or preventive action is taken, proper analysis and simulation must be done to find the root cause of failure. The findings must be also supported by verification before the actual use of failed parts in the machine.

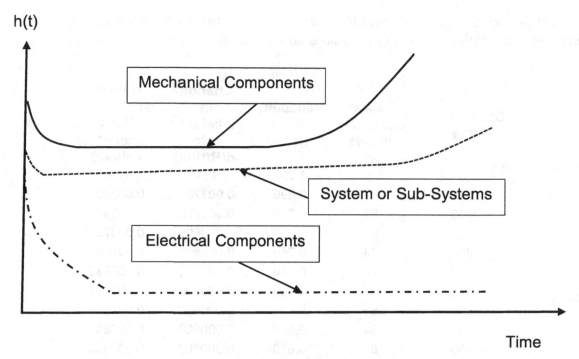

h(t)

Mechanical Components

System or Sub-Systems

Electrical Components

Time

Fig. 17.16 Bathtub curve

17.9 ROCOF Plot for Machines

The ROCOF plot is called the "rate of change of occurrence of failure," which can be plotted when lots of machine data is available for a type of machine. For a lathe machine, data could be collected on the number of failures over the months. The ROCOF plot is a plot of cumulative failures. The main purpose of this plot is to understand and asses the life stage of the product process, i.e., whether the failures are happening in early life or usage life. Let's explain this plot using the example of machines that have been tested and also put into the field. For this machine, we have noted the observation time, t, and the number of components still operating. The observed data is shown in Fig. 17.17 below. For this example, the total number of machines being observed for failures is 100.

The observation time, t, is 1,000 hours of operation. The hazard plot has to be drawn for these machines. Basically, a bathtub plot has to be drawn. The bathtub curve, or hazard plot, is shown in Fig.17.18 below. It helps to visualize the machine failure rate over time. It also identifies whether the problems are at a young age, during adulthood, or in the older days. Comparative plots for all the machine models can help engineers identify whether the failures are due to design or manufacturing issues. Similar plots can also be used to improve the design and enhance the reliability of machines. Machine performance can be also compared using such graphs. Overall, an ROCOF plot helps visualize the rate of change of failures, i.e., whether they are decreasing, constant, or increasing. This graph can also be used to determine the time to buy another machine or salvage the present machine. It helps to identify failure zones or modes and helps to determine remedial measures for machine failures. The chart can be used to identify the useful life period of the design and to determine the warranty period for the machine (constant failure rate zone). When the number of data points is very small, an alternative cumulative hazard plot could also be drawn.

Observation Time, t	Relays operating at t, Ns	Reliability R(t) = Ns/Nt	Change in Reliability Rate, -d(R(t)/dt)	Hazard rate, h(t) = -(Col4 values/Col3 values)
0	100	1.000	N/A	N/A
50	89	0.890	0.002200	0.002200
100	83	0.830	0.001200	0.001348
150	80	0.800	0.000600	0.000723
200	76	0.760	0.000800	0.001000
250	73	0.730	0.000600	0.000789
300	70	0.700	0.000600	0.000822
350	67	0.670	0.000600	0.000857
400	64	0.640	0.000600	0.000896
450	61	0.610	0.000600	0.000938
500	58	0.580	0.000600	0.000984
550	55	0.550	0.000600	0.001034
600	53	0.530	0.000400	0.000727
650	50	0.500	0.000600	0.001132
700	48	0.480	0.000400	0.000800
750	45	0.450	0.000600	0.001250
800	42	0.420	0.000600	0.001333
850	39	0.390	0.000600	0.001429
900	34	0.340	0.001000	0.002564
950	28	0.280	0.001200	0.003529
1000	22	0.220	0.001200	0.004286

Fig. 17.17 Bathtub curve data table

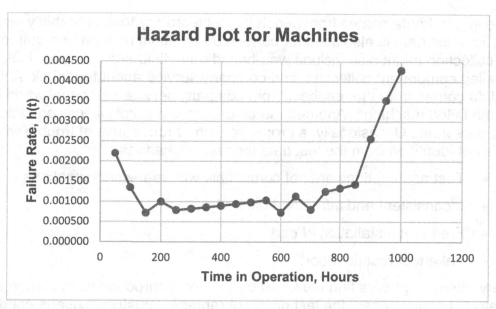

Fig. 17.18 Hazard plot for machine failure

17.10 Reliability Data Collection and Interpretation

Customer data for the machine can be collected from surveys and warranties after machine installation. In-house data can be collected through tests and diagnostics. Hence, data can be collected from external or internal sources. External data comes from dealers, customers, or market surveys, whereas internal data comes from laboratory tests, durability tests, or field tests. Before the data is analyzed, its authenticity must be confirmed for clarity and accuracy. Also, the data needs very careful interpretation. Moreover, it should be collected directly from the machine and should not include customer expectations of the machine. Actual performance is not the same as what customers expect. If the data is collected from the forms that customers fill out when they have a problem with the machine, data extraction is very difficult and should be interpreted accurately.

For example, a customer complaint such as "machine is noisy" is very subjective, and the exact source of the noise must be found before analysis. A common complaint is the "machine does not have enough power," which does not include the metal, feed depth of cut, and speed used while cutting parts. In this case, the machine is being overloaded due to heavy metal hardness, metal's toughness, etc. In some cases, hydraulic oil filters need to be changed, as per the service plan. Instead, the complaint becomes "hydraulic unit is very noisy." Hence, data integrity and description, customer expectations, and machine limitations must be understood properly while scrutinizing the data from filled-in data forms from customers or warranty service forms from dealers. The primary point is that the data has to be understood properly in terms of what the customers or dealers are trying to communicate about problems with the machine.

The internal data comes from bench tests, laboratory tests, durability tests, field tests, etc. The test data is also used for correlation and confirmation of requirements. The data collection might also include web-based data, field service data, dealer data, physical failed component collection, and company service agent feedback. Actually, the best data comes from the feedback from company service personnel when they attend to the failed machines. Another source of data is market clinics conducted by marketing personnel. Occasionally, a proper reason or root cause of trouble with the machine is not identified from the test data for several reasons:

- Test assumptions are not compatible with real-world situations

- Inconsistent and infrequent failures

- Frequent installation of parts

- Fake test results report

Very often, engineers find the failed part did not reproduce failure when tested in the laboratory. In such cases, the test does not represent customer usage conditions. It is to be remembered that data does not represent facts and needs to be interpreted, and this interpretation could be wrong for many reasons, such as misrepresentation, overrepresentation, gauge error, wrong boundary conditions, etc. Test data must be correlated with field data, and environmental conditions and noise must be correlated with actuals. The field data must be understood and diagnosed properly before a decision is made for the root cause. Then root cause analysis must be made using the data before a conclusion is made. Stress interference conditions, bathtub curve, hazards analysis, reliability analysis, etc. must be conducted by engineers to enhance the reliability of the design. Reliability is all about the design of a machine.

The design principles, if properly followed, can only make the design safe and reliable. The variation has to be managed properly for the design process, manufacturing process, quality inspection, measurement process, etc. In order to understand the variation of reliability due to the design and manufacturing process, engineers have to find the effects of piece-to-piece variations, aging mechanisms of the design, customer usage variations, load-cycle variations, environmental variations, and external and internal noise elements for the machine.

Reliability enhancement is recognized as the most important engineering practice for the quality of the machine. Reliability is valued by customers, who use it as a primary consideration when buying machines. Companies also use the reliability yardstick to measure and compare performance against other machines in the market. Reliability must be a part and parcel of the design and manufacturing functions of any machine tool company. System engineering principles integrate reliability engineering and practices into the primary design and manufacturing considerations. Every machine tool company must have a reliability mission. Reliability over time, or robustness, should be a requirement and must have a metric that should be verified by internal and external testing. Reliability should not be a quality signal of any machine for the company, but it should be a living quality standard for the product. Machine performance over time should establish the reliability of the machine in the field. Reliability is an important enabler of customer confidence.

17.11 Summary

This chapter detailed some of the very important and necessary reliability engineering practices that any machine tool company or equipment builder can use to verify the reliability of their products. The technology deployment process for any machine tool company must include reliability engineering requirements for the machine and develop common cause problem resolution practices and a reliability database. The company must focus on building a reliability engineering team that will follow reliability practices and lead process standardization and implementation of reliability engineering in the company.

The design for robustness should start with legacy data, which should include, design, manufacturing, service, chronic problems, and field data for design and comparison purposes. The data can be obtained from customers, services, and other machines in the market. Reliability engineering practices should include a formalized procedure for implementation, real-world usage load profile, verification plan, root cause analysis, PFMEA, DFMEA, analytical models, simulation models to use FEA, and other CAE practices.

This chapter lays out the fundamental principles and applications of reliability. It also describes the overall technical strategy in applying reliability engineering practices for any machine. Reliability tools and methods must be applied to analyze and diagnose reliability concerns for all products. Reliability practice must start with a requirement metric, followed by design and verification. The primary focus for reliability engineering is to develop methods and tools to maintain and improve product and process together.

The customer defines the quality and reliability of any machine. Customer satisfaction and feedback are the ultimate proof of machine reliability. The product must meet and exceed their expectations throughout its life. The reliability of any machine must help customers reduce its life-cycle cost. Proper reliability practice will give rise to the product's ability to perform the desired function over a designated life under an environment that is decided by the customer.

Reliability is definitely a metric that a machine must be verified against. Reliability is not only the customer's perception, but a yardstick for comparing performances of all the machines in the market. Reliability has to be embedded in the product by following reliability engineering practices. Verification of reliability must be obtained from test data, field data, and customer feedback. Verification procedures and tests must comply with the real-world customer usage profile.

Reliability analysis should consider operating, manufacturing, material, maintenance, installation, and storage conditions throughout the product's life. Hence, reliability analysis or modeling should include structural analysis, testing, data verification, operating conditions, noise conditions, load analysis, manufacturing variability, material variability, supplier's variability, etc. Reliability analysis must include data verification and confirmation, root cause analysis, DFMEA, PFMEA, etc.

The objective of this robustness cycle is to improve the reliability of any machine during design, manufacturing, and customer usage. Reliability analysis collects data from several phases of this cycle, and reliability engineering tools diagnose the root cause of the lack of reliability and recommend improvement. The process is continuous and iterative over the life of the machine.

Any system must control and manage variations of the system. Design, manufacturing, and usage variations cause the system to degrade functionally over time and to eventually fail. If an excess load is used for the machine, it will ultimately fail. If the material strength varies substantially, machine stress over the allowable limit might also cause the machine to fail prematurely. The noise factors that lead the system to failure might be inherent to the system or external to it. Inherent noise factors are system interaction variations, material strength, and normal aging due to usage. The system design has to control the extent of this inherent system noise.

Most failures can be prevented when creative ideas are associated with proper analysis and testing. Innovation must be backed up by critical analysis and verification methods. Most of the time, a logical extension of a proven design to suit the new requirements leads to a successful design with minimum or no failure. This is what I experienced in the Japanese design of machine tools. Most Japanese machines are platform based and use proven bookshelf technologies. It is a good idea to look into what has been used previously and the shortcomings of previous designs. Then creative ideas can be put together to enhance these designs without injecting unreliability. Proper analysis and validation are the keys to reliable design. To solve a reliability issue, both convergent and divergent thoughts must be conceived to understand the real cause of the failures.

The bathtub curve depicts failure patterns for three phases of machine life. In addition, random failures can happen in all three phases. The probability of infant mortality phase failures drops over the life of the product; i.e., the chances of such failures are very small after many years of usage. Moreover, the chances of wear-out phase failures or failures due to aging increase over time of usage. The rusting of sheet metal panels does not happen in the beginning of the machine usage but increases very rapidly once it begins the wear-out phase. For useful life failures, robust product design using a realistic load profile must be introduced. The design must be made simpler with a smaller number of parts and less complexity. Material strengths have to be improved and tested before use. Static strengths and fatigue strength of the materials must be improved. Steel can be used instead of cast iron for higher reliability. To reduce wear-out phase failures, product design must us rubber parts infrequently, and the design must be robust.

The ROCOF plot, or the "rate of change of occurrence of failure," can be plotted when lots of machine data is available for a type of machine. For a lathe machine, data can be collected at every failure over the months. If a company has many similar machines in the field, actual field data should be collected for machine failures and performance. The ROCOF plot is one of cumulative failures. The main purpose of this plot is to understand and asses the life stage of the production process, i.e., whether the failures are happening in early life or usage life.

Reliability data needs very careful interpretation. The data should be collected directly from the machine and should not include customer expectations. Actual performance is not what customers expect it to be. Customers always expect more out of any machine. If the data is collected from the forms that customers fill out when they have a problem with the machine, data extraction is very difficult and should be interpreted accurately. Data collection might also include web-based data, field service data, dealer data, physical failed component collection, and company service agent feedback. Actually, the best data comes from the feedback from company service personnel when they attend to the failed machines. Another source of data comes from market clinics conducted by marketing personnel.

17.12 References and Bibliography

Boothroyd, G., 1992, *Assembly Automation and Product Design*, Marcel Dekker, NY.

Box, G.E.P., Hunter, S.J., and Hunter, W.G., 1978, *Statistics for Experimenters: An Introduction to Design, Data Analysis, and Model Building*, John Wiley and Sons, NY.

Dewhurst, P., and Boothroyd, G., 1989, *Product Design for Assembly*, Boothroyd Dewhurst, Inc., Wakefield, RI.

Dixon, J.R., and Poli, C., 1995, *Engineering Design and Design for Manufacturing*, A Structured Approach, Field Stone Publishers, Conway, MA.

Ford Design Institute, 1998, Lecture Notes on Reliability and Robustness, Ford Motor Company, Dearborn, MI.

Lewls, E.E., 1987, *Introduction to Reliability Engineering*, John Wiley, NY.

Nahmias, S., 2008, *Production and Operations Management*, fourth edition, McGraw-Hill Erwin Companies, NY.

Netter, J., Wasserman, W., and Whitmore, G.A., 1978, *Applied Statistics*, third edition, Allyn and Bacon, Inc., Boston, MA.

Paul, G., and Beitz, W., 1996, *Engineering Design*, Springer-Verlag, NY.

Rao, S.S., 2014, *Reliability Engineering*, Pearson Publishing, Prentice Hall, NJ.

SAE Guideline, 2002, "Guidelines for Preparing Reliability Assessment Plans for Electronic Engine Controls," APR5890, SAE, Detroit, MI.

SAE Guideline, 1999, "Reliability and Maintainability Guideline for Manufacturing Machinery and Equipment," M-110, SAE, Detroit. MI.

SAE Guideline, 1999, "Reliability and Maintainability Guideline for Manufacturing Machinery and Equipment," M-110.2, SAE, Detroit, MI.

17.13 Review Questions

- Are reliability engineering practices suitable for the machine tool industry? What are the fundamental advantages and disadvantages of applying these processes to machine tools? Can root-cause analysis help reduce machine failures?

- Is the determination of a bathtub curve for the machine critical? How would you use bathtub curve data to improve machine reliability and performance? Can you use it to reduce the life-cycle cost of the machine? Does stress-strength interference analysis help engineers design machines?

- What are the fundamental characteristics of good field data? Can you use any data available to enhance the reliability of a machine? How do you verify the field data? How does customer usage affect the reliability of a machine? Should you always believe what the customer is saying about the machine? If not, why not?

- Does reliability engineering affect the cost of the machine? Is it worth it to have a reliability engineering team? Is reliability engineering disruptive to the machine design process? How can you control and coordinate reliability engineering team activities?

C H A P T E R 18

18 System Management for Machines

18.1 Introduction

Organizations have to promote system thinking among all their employees. It is not a design function or activity only. This is the most critical step that a company has to take. Company management has to be convinced about the benefits of system engineering and the flow of ideas down the line to the lowest level. The whole company has to be a living system by itself. Most of the time, company management has the idea that system engineering has to be followed by the design team only. That is absolutely not true. System engineering applications have to be taught and promoted among employees.

Many authors, such as Kauffman, Peter Senge, etc., have promoted the definition of system thinking as a collection of components that interact with each other to function as a whole. The same definition has to be extended to the entire company. The human resources, engineering, manufacturing, and quality departments are all part of the system that is the company. The company is a conglomerate of all these constituents, and each one has to function properly in coordination with the others for the company to succeed. Another perfect example of a system is the human body. All parts, bone, flesh, blood, lungs, heart, etc., work in unison to keep us alive and function properly. If one part is not functioning properly, the body becomes sick. Similarly, if the quality department is not working properly or the cleaning department is not functioning properly, the company becomes sick. This is the crux of system thinking. Every part is as important as the other, and every part is required with equal importance for success. The company must operate as an integrated system if it is to succeed in its endeavor to design, build, and market quality and reliable machines.

The coordination and interaction among each constituent of the company have to be based on system engineering principles and requirements. It is a functional requirement of company management to promote system engineering principles for survival as a team or system. Unless system engineering concepts and principles are followed by company management at the very beginning, the company cannot be successful as a system team. Some of the other ideas along this line are as follows:

- A company system is not a linear sum of constituents. It is way more than that due to synergetic or amplification effects.

- Each department has to be as strong as the other. The system will break down as a whole if one department fails to function properly. This does not mean every department has to get involved in designing or manufacturing a product. What it really means is that each department has to be equally strong in delivering outputs as required. The weakest link among the departments will cause the system to fail. For example, in a machine, if the spindle is too powerful and the other elements are not as strong as they should be to accept the power, the machine will break down. A 10 HP machine looks much weaker than a 30 HP machine. A 30 HP motor cannot be successfully put in a 10 HP machine. The system will fail prematurely in a disorderly fashion. Every part of the machine has to be as strong as the spindle motor for it to function. Moreover, the system has to be optimized as a whole. Optimizing one part of the machine does not make the system strong. For example, a strong design team does not help the company to be successful in marketing a good product and making a profit. It needs a sales and marketing department, service departments, quality control departments, human resource department, supply chain management department, etc.

- Productive and strong coordination among departments makes the company strong and successful as a whole. This interaction or interlinking makes a company perform as a system. System efficiency is a result of individual department efficiency. For example, the machine performs best when all the interlinking components or sub-systems work in unison perfectly: tools have to be strong, the spindle has to be strong, the tailstock has to be equally strong, etc. Machine efficiency and productivity depend on the interaction of all these components.

- When a system problem arises, each part has to be looked into. Interaction among the parts needs to be analyzed to understand the real problem. System performance is not a linear sum of the part performances. Interactions amplify the performance of a system. When a system fails, a root cause analysis has to be conducted on the entire system to understand the failure instead of analyzing only a part.

- Any system is always a part of another, larger system. For example, our planet is a part of the solar system, which is a part of the universal system. Similarly, a spindle is part of the machine system. Also, every system is connected to other systems through boundaries, which can be rigid or flexible. The common boundaries are also called interfaces through which systems are connected and interact with each other. The system can be hierarchically integrated.

- For example, a machine system has contexts, such as business environment, marketing environment, legal environment, etc. The machine system also has a shape and form, which lead to the structure of a system. So, a system is not a concept or philosophy but a physical structure. A machine system has a spindle, ways, base, cross slides, tailstock, etc.

- The machine system transforms customer requirements into equipment or a device. System engineering principles help to design, manufacture, verify, and deliver a robust product that is in alignment with affordable cost.

- The system has to have quality, robustness, and cost containment, and it should satisfy development constraints and timing. System engineering can be applied to both simple and complex products and processes. It is quite different from the traditional engineering approach, in which the process starts with the components or bottom. System engineering starts from the top.

- In general, system engineering is process-driven and very systemic and disciplined. It helps the company satisfy the customer's voice. This helps the company to satisfy cost, robustness, and quality and reduce the time-to-market of the product from conception. System engineering is a multifaceted approach towards achieving a worldwide sustained growth of the business, gaining product leadership, achieving product excellence, attaining cost leadership, and taking a lead in customer satisfaction.

18.2   Systemic Management Issues

As mentioned before, system management is very critical for the success of a company. The generic system engineering process has to be curtailed and suited to machine tool companies. The generic system engineering process deals with three major issues: requirements analysis, design, and verification. As a matter of fact, the machine tool system engineering process has to be modified to some extent to suit the machine design and building process. To maintain accuracy, manufacturing and robustness verification are very important factors. The modified system engineering process should consist of requirements analysis, design, manufacturing, verification, and qualification. This quality function is not a policing function, as is normally done in other industries. It is required to ensure the precision and robustness of the product and is a design verification process: the design should be the primary focus. Using electronic tools and methods, this loop has to be iterated over and over again to simplify the design and satisfy all customer requirements. Inspecting quality and incorporating quality into the design are two different aspects of this process. The first one conforms the manufacturing process to the design, and the other helps the design build a quality machine. Quality inspection starts at the design.

Before the start of the design phase, in addition to requirements analysis, the team has to obtain information about benchmarking data, team knowledge capability, manufacturing technology, available technology, resource constraints, cost constraints, etc. These are the systemic inputs required before design starts. The team has to collectively filter all this data to it for their particular system. Even if the machine is a facelift or legacy design, all this data needs to be collected and compared to determine whether another design is required. The requirements analysis should include reliability and robustness, cost, performance, precision, and verification requirements in as much detail as possible.

Once all of the information is baked into an acceptable requirement list for the machine, the design should start. The design should develop possible alternatives, evaluate them against the requirements and cost, determine the manufacturing and verification feasibility of each, etc. Next, a design has to be selected that satisfies most of the constraints. The system "V" model depicts all of these steps. In addition, major functional models can be built by each team for felicitation of their particular process. For example, the manufacturing team could build a manufacturing system V, model as shown below in Fig.18.1. Again, the left side of the V diagram is all about the plan, and the right side of the V is action and confirmation of the plan.

In order to manage the system engineering activities, requirement collection and cascading for the system and sub-system are very critical steps. The cascading process starts with the requirements for the machine as a system and should flow down into the sub-system and component requirements. Also, alternative potential designs must be evaluated against these requirements. A balance between the satisfaction of these requirements and the cost of the system needs to be struck. These critical steps have to be baked into the design at the outset. Requirements lead to constraints of the design, and constraints lead to the cost and performance of the system. The cascading process can also use computer-aided, or CAE, tools for evaluation.

Next, management issues arise in trade-off analysis or studies, i.e., what requirements should be kept in the design and what can be spared or is unwanted in a design. The trade-off studies help the team establish compatibility among conflicting requirements. Trade-off analysis can be considered a filtering tool for the machine or system. It also helps to support customer requirements by creating a must/wants list. Musts are different from wants or wishes. "Wish" types of product characteristics add to the cost of the machine, and the return is less than with the "must" characteristics. These are basically priority lists for design alternatives. Trade-off analysis can also be termed a "balancing" or "optimizing" act among the requirements. The alternative designs must satisfy the "must" requirements at the least and also capture the "want" characteristics as best as they can. Some "musts" create a design risk for the product, but they need to be satisfied.

This process is highly iterative, and the team goes back and forth on decisions to ensure design priorities before the design starts. Iteration refines the selection of priorities, which will guide design considerations. The whole process must be documented very thoroughly for future reference.

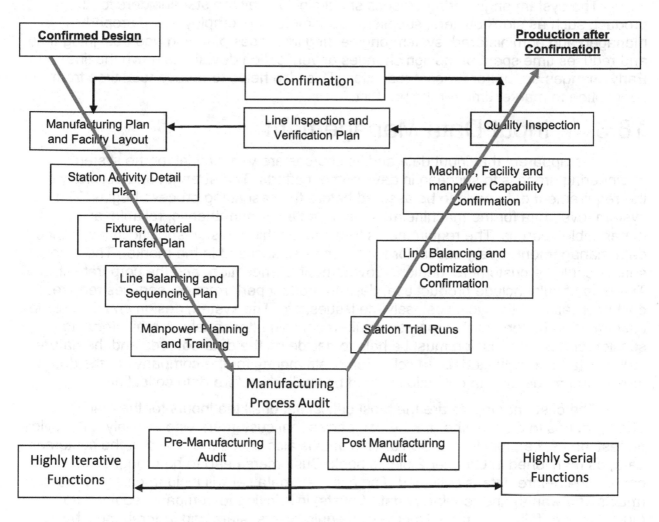

Figure 18.1 System, "V", for Manufacturing Process

This cascading or decomposition process also ties all the system and sub-system or component requirements together in a meaningful way. CAE or other computer tools, such as simulation methods or optimization techniques, can be used to justify the iterative actions. The computer tools also quickly identify the difficulties in satisfying some critical requirements. A thorough feasibility study must be done for each of the alternative designs to refine the design. At the end of the process, a list of the requirements for the integrated system is generated. This becomes the bible for the design from this point on. Each level of the system "V" should be guided by requirements and interfaces. The cascading decomposition process is at the core of the system engineering process, and requirements are connected and linked together to create a synergy of design. Another management issue during this phase of development is to have very strong communication among all the development team members. An open-minded discussion is required during this phase. This process is critical to cost, quality, and development time for the product.

The system engineering process should include all the stakeholders for the product, such as stock owners, suppliers, customers, and employees. If properly managed and administered, system engineering increases planning and designing time and reduces time spent on design changes or verification deviations down the line. Early changes are better for cost reduction. This also helps to reduce total time from conception to market time for the product.

18.3   Input Data Management

It is apparent that input data and its sources are very critical for the system engineering process to succeed in development efforts. The strength and accuracy of the requirement data need to be assured before the cascading process begins. The system-level data for the machine must be collected from authentic, reliable, and dependable sources. The requirements turn into machine design specifications; hence, data management is very critical for the machine to succeed in the market. The data source could be customers, dealers, government, competition, and the corporate library. The data should include product use, desired product performance, services required, cost implications, life-cycle cost, salvage issues, etc. The system design should include customer, regulatory, and corporate requirements simultaneously. A brainstorming session or open discussion must be held to decide on the data sources and the nature of the data to be collected. Data collection costs money for the company, so the data format and mode of data collection should be finalized before data collection.

The customer inputs are the most critical out of all the inputs for the machine. The team has to decide who are the company's real customers who willingly will provide honest inputs, i.e., musts and wants. This data is also called the "voice of the customer" data, as mentioned in Chapter 2 of this book. Customers need to be brought into confidence before data is collected. The complete data set will include customer data (musts and wants) and regulatory data (musts) in addition to company legacy data (successful or failed product). Regulatory requirements, state and federal, must be satisfied. For example, safety, chemical, and environmental regulations must be satisfied by the product. In addition, there might be a local usage trend (wants) for the machine that needs to be looked into. For example, usage of the machine in very hot places could be quite different than cold places because temperature, humidity, etc. affect machine performance. Moreover, the design should also give consideration to local market traditions (wants). For example, some local markets cater to the demands of automobile companies, some satisfy requirements for forging and casting companies, etc. Such inputs can sound trivial in the beginning, but they become very critical when the design is completed.

The next set of important data is corporate data, such as culture, intellectual strengths, standards, financial strength, etc. Human resource management is very critical for the success of the product. The design team also has to know the corporate manufacturing strength and weaknesses. The supply chain management data is also very critical for the success of the product.

For example, if the design uses all the imported parts or assemblies, it costs money and time. It also has a high-risk potential. Another critical set of data is the level of technology to be used in the machine. The applicability of the technology has to be looked into. If the company is trying to design a manual machine, digital technology might not be a matter of consideration. Technology plays a very critical role in the product's success. In a nutshell, the purpose of data collection should answer four primary questions: system performance required, reliability and quality desired, design and manufacturing methodologies to be used, and cost constraints to be satisfied. When all said and done, the requirements for the machine must at least include the following:

- Performance
- Reliability, robustness, and quality
- Cost constraint
- Technology deployment
- Development time window
- Supply chain management
- Regulatory constraints

18.4   Requirement Statement Characteristics

In order to make the requirement document specific and useful to all team members, it must have certain characteristics:

- The requirements must be written in a very unambiguous fashion so that there is not a problem in understanding them. The statements must be very simple, clear, and straightforward. Unambiguity will help to eliminate confusion and incorrect understanding. For example, the statement "the machine must be all-purpose" is vague and ambiguous. The machine could be a turning machine, milling machine, grinding machine, etc.

- The requirement should be one that team members are sure is achievable within the cost, quality, and time constraints. The design should be able to reproduce the requirement. For example, "a turning machine can replace a grind machine function" is an unrealistic requirement. It is not feasible to be designed and manufactured in a real-world situation.

- The requirement should be free of dual interpretation. For example, "the machine should be precise" is subject to various interpenetrations, and precision is not the same for all applications and machines. Turning machine precision cannot be the same as grinding machine precision.

- The requirement must be a real-world requirement and related to the specific product. The requirement must represent the customer's voice and satisfy customers when met in its entirety. For example, "the machine does not need maintenance and lubrication every day" is not a real situation since every production machine or equipment will break down and need maintenance and services to put it into operation again.

- The requirement must have a metric, or it must be measurable in real terms. If it is not possible to have a metric, then it is not a requirement. For example, "the machine must have enough power to cut all steel and cast-iron metals" does not specify the power requirement and specific metal descriptions. This is not a realistic requirement at all since there is no metric around it.

- The requirement must be verifiable if it has a metric. If a requirement cannot be verified, the requirement can be anything possible. For example, "the machine should weigh less than 12,000 pounds when completely assembled with options" is a verifiable requirement. "The repeatability of machining one-inch by six-inch brass rod supported at two ends should be less than or equal to 0.0005 mm" is a verifiable and well-described requirement statement without any interpretation issues. "The machine should have a 20-HP spindle motor" is a verifiable requirement without any ambiguity or doubt.

- The requirement statement should be objective and not subjective. A subjective requirement statement is subject to misinterpretation. For example, "the machine should be painted" is a subjective statement, but "the machine should be uniformly painted yellow on the outside and white on the inside" is an objective statement. Another example: "the machine should have 20 HP and should work in any environment" is not a requirement statement at all because it is very vague and unrealistic. Moreover, this statement is not measurable, not objective, ambiguous, and is also not verifiable. The machine cannot be verified in all possible weather conditions.

- The requirement statement must specify what should be satisfied and not how it will be satisfied, i.e., what and not how. For example, "the machine must have 20 HP" is a good statement, but "the machine must have a 20-HP servo-controlled motor" tells what is specified.

- Last but not least, all requirements must be linked and coordinated properly; i.e., all requirements must be compatible with each other in a sensible and verifiable way. They must also be correlated. A 20 HP machine needs a minimum one-inch square tool shank for proper functioning. The index unit and spindle unit functionalities must be compatible with each other. X-axis and Y-axis movement must be coordinated to create a modulated surface in a lathe.

An example of a well-written and defined requirement statement is: "The machine should have 20 HP available at the spindle and should be capable of working with biodegradable synthetic coolants."

18.5   Requirement Segregation and Management

In order to create an exhaustive list of requirements, they should be divided into attributes. Attributes help the cascading process, requirements analysis, trade-off analysis, and target settings for the machine. The attributes in totality satisfy the requirements for the whole machine, and when summed together, they will represent what the machine is supposed to do. Attributes form a complete list of requirements without showing any priority among them. Basically, for easy management of requirements, attribute lists are formed. For example, customer requirements can be grouped together under the heading of safety, performance, ease of operation, customer friendliness, thermal stability, durability, reliability, electronics and control, life-cycle cost, etc. Similarly, the regulatory requirements can be grouped under the headings of power and energy control, environmental effects, noise and vibration, etc. Under corporate requirements, cost, brand recognition, marketing factors, development time, financing, manufacturing capabilities, intellectual abilities, etc. can be grouped together for further processing.

A series of detailed and critical brainstorming sessions are arranged to create the machine system requirements These teams create the system and sub-system level requirements through cascading and functional decomposition. The requirements detail environmental, performance, reliability, cost, and verification requirements, among others. These requirements are then taken up by each team leader for further action. All the activities, i.e., design, manufacturing, verification, are tied together for the machine system. Machine interface diagrams showing links between machine and external environments are created. Customer usage conditions, such as weight, power, speed, workholding, safety concerns, etc., are firmed up. Machine working conditions, i.e., temperature, humidity, dust, etc., are also considered during interface development. Interface development should consider events throughout the machine life cycle. Requirement targets are arranged by attribute, as shown in Chapters 2 and 3. A document is created for the target of each requirement. Such documents specifically lay out the metrics, targets, and verification methods for each requirement. The cascade diagram for each requirement, as shown in Fig. 18.2, can be hierarchical in most cases.

The primary goal of such an elaborate synthesis process is to strike a balance among targets of each requirement. These requirements help the team to identify design alternatives. This process also helps the team to achieve compatibility. During the design alternative development phase, the team compares several existing company machines, benchmarks competitive machines, identifies technology availability, determines supply chain management and re-use considerations, and then develops design alternatives for further analysis. As said several times earlier, design synthesis is the most critical step since it has to consider performance requirements, regulatory requirements, safety requirements, and technology. This process should be highly iterative, detailed, and well documented.

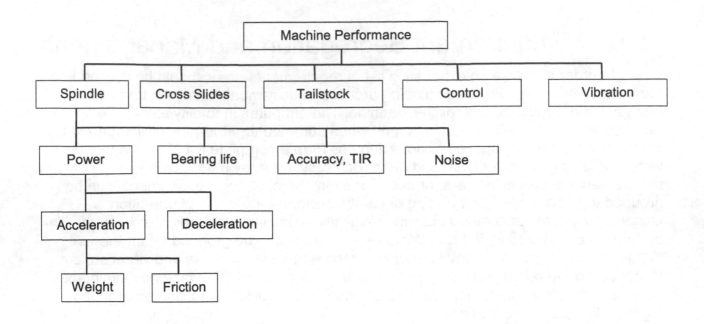

Fig. 18.2: Performance Cascade Diagram for a machine

Once the system-level requirements are confirmed, the next task is to cascade them down to the sub-system level and then to the component level. That is when the component design is begun to satisfy the component requirements. During this phase, several actions are required: the flow of sub-system-level requirements down through the cascading process, functional decomposition of system-level functions into sub-system level, trade-off analysis of the requirements, evaluation of alternative designs, and definition of system boundaries. The importance of developing interfaces, boundaries, design specifications, and verification methods cannot be understood unless this is done completely for a project. Alternative designs need to be developed depending on functional requirements. The problem is that this whole process does not have a fixed and well-defined path. Every company should have its own way of accomplishing this total task.

The primary purpose of partitioning the sub-systems is to create simple manageable interfaces based on the functional requirements, technology to be deployed, optimization and minimization of boundaries, sub-system attributes desired, reusability content, etc. For example, the partitions for a machining center are the spindle, slides, base, tool management system, shrouds, loading fixtures, tooling, etc. Each interacts with the other through boundaries. The boundary is the interface for connecting to neighboring systems. It limits the functions of each sub-system but helps to build compatibility between systems. The boundary also defines the attachments, energy paths, force and momentum transfer, etc. Unless the interfaces and boundaries are broken into manageable pieces, a system diagnosis will be difficult. The design also becomes too difficult.

The sub-system design process includes inputs, functions to deliver, output response, and noise inputs. The interfaces to be built around a sub-system are shown in Fig.18.3 below. The sub-system specifications must include boundaries and interfaces, owner of the system, functions to be performed, design verification methods, system documentation, and a consolidated set of specific requirements that the sub-system has to satisfy. For any design, wastage or noise outputs have to be managed to improve the efficiency of the system.

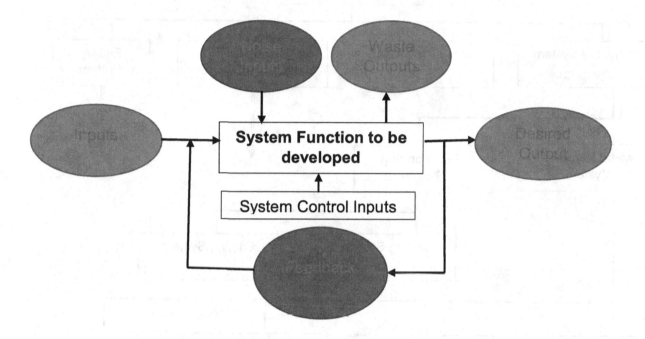

Fig. 18.3: System Input Output Relationship

System design specifications must not include the process to be followed for design or verification guidance. Similarly, organization, team composition, verification procedures, manufacturing process descriptions, etc. should not be included in the specifications. What is desired from a system should be included in the specifications, and how to achieve it will be the responsibility of the system owner. The system specifications are required to gain efficiency and productivity in the product development process, establish communication links and component sharing methodologies, and enhance the team environment. The format for system-level and sub-system- or component-level design specifications should be similar.

It has to be emphasized that the system must have a function to deliver and the system requirements drive or lead the design efforts. The potential alternatives must be driven by function, and the selected alternative for design and development must satisfy the requirements. The validation and verification of the system will confirm compliance between what is required and what is achieved at the end of the process.

Any system is a sub-system of a larger system. For example, a CNC turning machine is part of a metal removal system. The hierarchical block diagram is shown in Fig.18.4 below for a factory automation system that is part of a factory system.

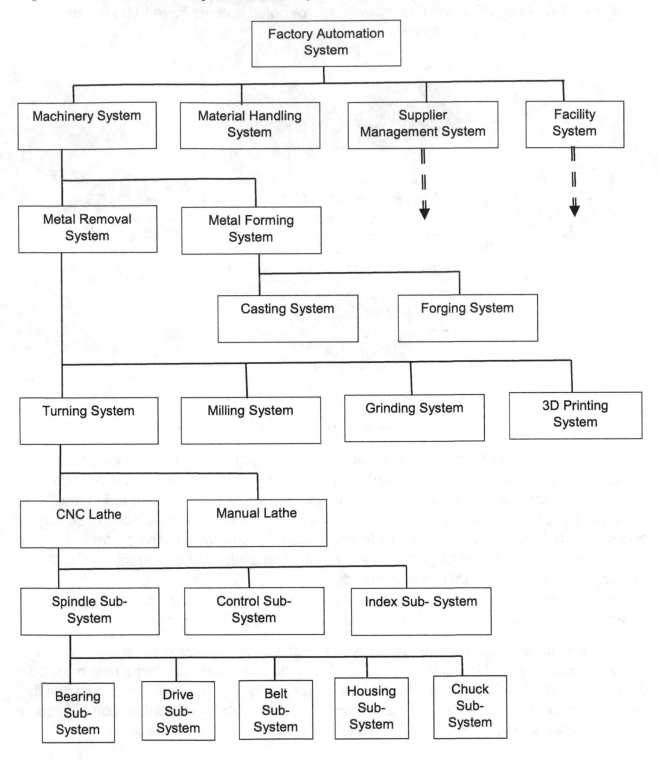

Fig. 18.3: System and Sub-System Relationship Block Diagram

18.6   Management of Verification Process

The verification process is part and parcel of the design and development system. Every system or sub-system has to be designed to satisfy customer requirements. The verification process consists of verifying requirements, methods, testing, and analysis of data, and the verification plan has to be designed and implemented accordingly. The verification can be physical or digital. All systems need to have a physical verification at the end to confirm the machine responses with the desired requirements. Digital verification, such as simulation, layouts, etc., can be used for design iteration and improvement purposes.

The total verification process could include machine system verification, sub-system verification, and component-level verification on the right-hand side of the system "V" diagram of the system engineering process. The machine system is verified against system-level requirements, such as thermal growth, noise, vibration, accuracy, durability, etc. The component-level verification might include strength, vibration resonance, deflection, etc. The sub-system verification for an indexing unit might include accuracy, failure, reliability, etc. Since this sub-process is on the right-hand side of the system "V," these are serial processes. Walking back upstream is a very costly and time-consuming proposition. The whole purpose of the verification process is to confirm that the design satisfies the requirements.

The final verification process should include physical prototypes and should be conducted as per the requirements. It should include requirements for several product aspects: design and development, performance, reliability and durability, robustness, manufacturing and quality control, assembly process, service, and installation. Hence, if there are a requirement and its metric, the verification process must confirm the requirement metric. The verification process and methods to verify the metric will depend on the requirements that are being validated. The design should also satisfy these requirements with the help of analysis and other electronic means wherever possible while the design is being implemented. Physical verification is a confirmation process at the last step of the design and development phase. Similar to design, manufacturing and assembly verifications should also be completed to satisfy manufacturing requirements using the first few confirmation prototypes.

There are several steps in the verification process:

- Identify requirements and the metrics for the system
- Plan verification methods and feasibility
- Conduct verification tests
- Analyze the data collected and comply with requirement metrics
- Pass or revise the design
- Document the process along with findings and conclusions

As mentioned earlier, the verification process could include analysis and physical testing. The analytical verification could include CAE, CAD, Simulation, Finite Element Analysis, and Optimization techniques using digital prototypes, etc. The physical prototypes might include complete machine systems or sub-systems such as the indexing unit, spindle unit, tailstock units, axes, slides, etc. The other part of the physical verification process is the development of test rigs, fixtures, loading devices, instrumentation, etc. Data collection, resolution, and accuracy are very important aspects of the whole physical process. Data analysis will confirm compliance. Documentation and reports are also very important for future reference. In a nutshell, a verification process is required to satisfy the fact that the design meets the requirements.

The whole process includes an integrated machine system, test method planning, comparison process, project schedule, and timing management. The whole machine system is tested for verification of attributes. The selection of attributes for testing will dictate the test process, nature of testing, fixturing, instrumentation, and data collection system. The noise has to be monitored and minimized to conduct a proper. The test assumptions need to be assessed, and a correlation between theoretical or analytical data and test data needs to be established to validate the test process and method. The impact of assumptions on the test results should be assessed before any test is conducted. Test targets, methods, instrumentation, data collection, data analysis, etc. should all be put together to get a good verification process established. Before the tests are conducted, the content of the prototype system or sub-system must be validated.

The advantages of a well-managed verification plan are several. The test planning time is reduced. The manpower requirement is also reduced. Test planning and set-up requirements become easier. All of this boils down to a reduced verification plan cost. Proper planning of test and prototype improves testing efficiency by avoiding test duplication, while increasing test quality, efficiency, and productivity. Proper test assumptions, test inputs, test methods, set-ups, instrumentation, and data collection also help to enhance confidence in the test results and conclusions. The test connects all the stakeholders of the company associated with the development of the machine. It creates a concerted product development team, enhancing coordination and links among team members of different backgrounds, skills, and knowledge levels. It creates a healthy learning atmosphere for all team members.

Verification tests are basically meant to determine whether a system can satisfy customers for a specified machine life under their operating environments and conditions. Various types of performance, durability, or reliability tests are conducted for machines:

- **Tests conducted to confirm the design:** Physical prototypes are built after the design is completed and the trial manufacturing process is followed. The system is tested to identify failure modes and robustness of the system under given operating conditions. The results are analyzed to compare with the desired requirements and record compliances. If necessary, the design might have to be revisited and revised accordingly to enhance the reliability and robustness of the machine.

- **End of line or final system test**: Once the trial manufacturing process is completed and physical prototypes are built, they are inspected to understand the effect of manufacturing methods on system behavior and performance. Most of the time, performance confirmation is completed for the machine to ensure that the manufacturing process followed is a viable method to build the machine. Before the system is delivered to the customer, complex machines are tested for performance and customer acceptance. These are sometimes called acceptance tests, and they are done to satisfy customers before they sign off on the machine for delivery. The machines are tested for capability and quality conformance as well. The primary objective of such tests is to qualify the machine against the customer operating environment and input conditions. Such tests are also conducted to satisfy purchase contracts, which normally include such tests as a precursor to delivering the machine to customer's shops.

- **In-between process, or staging test:** Tests are also conducted in between manufacturing stages to find the effects of specific manufacturing processes on system performance.

- **Reliability or durability test:** These tests are conducts on the system or sub-systems to determine reliability after they are assembled. These tests help to confirm the reliability or robustness of the system under operating conditions at loaded or no-load conditions. They are also useful to establish the servicing conditions or maintenance procedures and frequency of maintenance for the machine. In addition, such tests can be used to understand the warranty life for the machine, what critical parts might fail during operation, and how to modify the design to enhance the reliability of the system to a satisfactory level.

The end of the line test is primarily intended to find failures in the infant mortality phase or burn-in phase, where a high failure rate is observed if the manufacturing or assembly is not done properly. Durability or reliability tests are conducted to find the issues during the operating-life phase, and the extended durability test is conducted to find issues during aging or the wear-out phase, where, again, high failure rates are observed due to mechanical wear of the components. The issues found during the operating-life phase or reliability test are very important, as customers are usually interested in knowing about them. The objective of durability or reliability tests is to determine design deficiencies, mean time between failure, mean time to failure, mean time to repair the system, design life of the system, and reliability or robustness of the machine. Reliability test results also determine the factor of safety margin for components. The other types of tests, such as strength test, safety crash test, impact test, etc., are normally done to find the effects of very high load due to crash conditions.

ISO has standards with regard to machine tool testing for uniformity among the machines. They are ISO 230-1:2012 and ISO 230-2:2006, which suggest the methodology for CNC machine tool testing in general.

Reliability tests are most commonly carried out for new products. The cost of any system is also connected to its reliability. Highly reliable machines normally cost more. Actually, higher cost should be embedded into the cost of the system, and when production volume is up due to heavy demand, the extra cost is amortized throughout the volume of machines sold. Hence, the effect of high reliability should be nullified over time. Since the significance of reliability tests is very high, proper inputs, test set-ups, loading fixtures, instrumentation, data collection methods, etc. should be properly defined before the test. Proper analysis of the data is also required to find the root cause of any failure. The expectation data from the test should have a range of acceptance. The expectations must be realistic and comparable to the benchmarked products.

At last, it should be noted that machine testing cannot be compared to the testing of other consumer products, such as automobiles, household items, etc. since the number of machines available for testing is very small. There is usually not enough money for developing a new machine and testing it. Hence, for machine tools, some risk has to be taken, and risk mitigation is only possible by designing the machine properly right from the beginning. Most of the validation has to be done using digital prototypes. Other engineering tools, such as CAE, CAD, CAM, FEA, simulation techniques, and optimization procedures, have to be used extensively to prove out the product before the design hits the floor. There is no other alternative to this process. This is what I have practiced and recommended throughout my professional career, and I often witnessed similar procedures in Japan.

Due to the small number of components and machines available for validation and verifications, some issues have to be dealt with. One is that applications of statistical tools have been found to be very limited. Hence, the design has to be very robust, and most of the designs should be validated using analytical tools, which will save time and money for the company. A design of experiments (DOE) process will reduce the amount of testing required for validation. Parametric studies can also be done using a limited number of test samples. Moreover, testing of the complete machine system is preferred to component-level testing when the number of samples to be tested is very small. Other test procedures, such as accelerated life-cycle tests, testing until partial failure, magnified load cycle, sudden-death testing, sequential life testing, etc., can be performed when a limited number of samples are available.

18.7   Machine Safety Engineering

Safety engineering for machines has not been talked about much. OSHA (Occupational Safety Health Association) reports that every year, over 18,000 American machine tool operators get injured while working. The injuries include death, amputation of limbs, pinched body parts while working on a machine, etc. The primary reason for such injuries and accidents is the poor and unsafe design of machine shrouds. OSHA and ANSI (American National Standards Institute) have clear guidelines as to the machine guard requirements to enhance the safety of machine tools. In recent years, the number of injuries has come down, but they have not been eliminated. It has also been reported that there are still a lot of machines that violate safety guidelines with respect to the machine guards. Even if the company gets sued and injured workers get compensation, machine designers should try to follow the machine cover guidelines very strictly and religiously to make the machine safer and operator friendly. Beyond that, loss of an operator is a loss to the intellectual ability of a company, and it is very hard to replace a good and sincere operator in the workplace. Hence, the company loses on both counts when an accident happens. Machine tool safety should be a matter of great concern in machine shops and customers should demand the safety compliance of all the machinery that they buy. In addition, when an injury happens, work stops and loss of production becomes a profit loss for the company. Such losses should be counted as a part of the life-cycle cost for the machine.

OSHA also gives some guidelines with regard to the safety maintenance of machinery:

- Check the machine guard conditions once a year as part of preventive maintenance to ensure that safety buttons are in working condition and the shrouds are in good shape to prevent accidents. Find the missing guards and replace them, if necessary, with equivalent safety guards.

- Every machine has some dangerous elements required for delivering the intended functions. For machining and turning centers, tool bits flying out of the tool holders is a very common phenomenon. For lathes, chuck jaws fly out very often, causing serious injuries. The grinding wheel is another part that causes many injuries. It develops cracks and shatters while working, and the chuck flies out. The machine shrouds, grinding wheel covers, etc., if designed properly, can prevent such injuries. Such safety concerns must be addressed while designing the machine. The proper analysis must be performed to resist such impact conditions.

- Currently, all automatic and manual machines are provided with door switches, and they prevent the machine from starting when the door is open. Many times, doors jam and do not contact the switches properly, so operators have a tendency to eliminate such switches to keep production going at the desired rate. Door rails also do not work properly, causing a door jam. Such malfunctions must be looked into during preventive or running maintenance checks and eliminated to enhance safety conditions for the machines.

- For forging presses, hot rolling presses, and bending machines, the design has to ensure that the machine can start operating when any part of the body is inside the working envelope. Both the foot and hand switches must be in working condition. The machine must stop moving as soon as one of these switches in not in the operating condition. Both switches must work simultaneously.

- Another design point could be to use automation of machine working conditions. Mechanical devices can be replaced with hydraulics and electrical devices to prevent unsafe working conditions.

- Eliminate unsafe elements from the design, such as flywheels of a press by using counterweights, for example. The design has to take the lead in making the machine safe for use. Move from mechanical design to hydraulic or electronic design. The maintenance department has to maintain safety by replacing what goes bad during use. It is a joint effort to keep the workplace safe and healthy.

- Safety guards must be analyzed against impact loading or dynamic loading that exceeds the cutting loads by a great margin. Safety has to be started at the initiation of design and incorporated into the design. Machines have to be shielded against unwarranted events that might happen during operation, installation, or movement. Sheet metal design has to ensure that coolants won't leak and cause health hazards and unsafe conditions in the workplace.

- More often than not, fixtures, loading and unloading devices, tool loading devices, etc. cause severe safety issues. Overhead cranes and material handling devices cause unsafe conditions, too. These external work aids must be taken into consideration as part of the machine system and designed accordingly to maintain the safety of the operators and working personnel.

- The service manual for the machine must list the unsafe and unhealthy conditions that might cause injuries for the operator and maintenance personnel. Proper training for the operators must be rendered before the installation and operation of the machine. Safety should be a concern for the management of working personnel as a whole.

- Before buying any machine, buyers and responsible personnel must consult OSHA guidelines for machine safety. They must ensure that the machine follows the guidelines. Both old and new machines must be safe. Machines must be designed for safety and healthy operation.

- It is also a fact that operators use unsafe conditions and responsible people do not enforce safe working conditions in the workshop. Operators must ensure that all safety equipment, such as guards, switches, locks, etc., are in place before the machine is turned on every day. It is the joint responsibility of management and operators to keep the workplace safe and healthy. The buyer has to be aware of the safety working conditions for the machine. They have to be aware of all OSHA guidelines about the machine's safety when buying it. Moreover, maintenance personnel must follow the supplier's maintenance guidelines to keep the machine in a safe condition. A joint effort among management, the supplier, and the buyer has to be in place to make a machine safe and the workplace healthy. The supplier is not the only entity to be held responsible in case of an accident. If the injury is due to poor maintenance, the buyer is totally responsible for any injury.

- The users must be knowledgeable, smart, and aware of machine safety to start with. The machinery supplier designs the machine for general safety conditions. Designers cannot take care of all the user's working conditions to make the machine safe to operate. The company has to follow that process and use common sense to ensure healthy working conditions. No shortcut must be followed to increase production at the cost of safety.

Safety awareness must be promoted throughout the organization to create a safe working place. There is no alternative to applying common sense to make the machine safer for workers.

- The design should take care of pinch points. Crash protection should be designed in case the machine behaves erratically during operation. Operators should not wear loose clothing, have untied hair, or be unmindful while operating the machine. They must also be sufficiently trained to operate the machines before the actual operation is started. Let the operator know about safety while working on the machine. Management must hire operators who are knowledgeable about the machine and its operation. Pinch points must be identified, and operators must be trained to avoid them.

- Common-sense safety rules must be followed: safety glasses must be worn while operating machine; emergency contact numbers must be displayed: loose clothing or loose hair must be avoided; electronic gadgets, such as cell phones, music videos, radios, etc., must not be used while operating a machine; no horseplay in the workshop; etc.

- Guards or covers must be used to protect the operator and maintenance personnel from rotating parts, cutting tools, rotary tools, running fixtures and part loaders, belts, gears, chains, flowing chips, debris, etc. The machine must be anchored and installed as per the installation procedure provided by machine builders.

- Any machine damage or accidents must be logged and reported to the proper authority as soon as possible. Machine maintenance must not be performed while any part of the machine is running.

- Accidents do not just happen; they are caused either by design or due to faulty operation. Accidents do not give notice before they happen, and they occur in the blink of an eye. Common sense, knowledge, and preventive action reduce or eliminate unsafe conditions for the machine. Maintenance and design personnel have to learn from previous mistakes or accidents. Mistakes must not be repeated.

- In addition to OSHA machine tool safety guidelines, ANSI B11 provides similar guidelines for enhancing machine tool safety. This standard provides documentation for design, manufacturing, set-ups, operation, risk assessment, and maintenance.

- ISO (International Standards Organization) and SAE (Society of Automobile Engineers) also provide guidelines for machine tool design, construction, operation, and maintenance, including test methods and measurement systems. SAE primarily focuses on the use and operation of machines specific to industrial applications.

- Various machine tool organizations such as OSHA, ISO, SAE, ANSI, etc. provide safety, design, maintenance, and operation guidelines that will enhance machine tool safety and increase productivity. For automatic factories, there are guidelines as to how to connect the machinery through conveyors, automatic guided vehicles, factory automation, etc. so that factory safety is increased as a whole. The ISO 230 standards series addresses procedures and design guidelines for machine tool accuracy, noise emissions, thermal distortion, vibration effects, risk assessment, and risk mitigation methods. OSHA also provides clear guidelines for various types of tools used in the workshop. Hence, there is no dearth of safety guidelines and documents available to enhance the safety of machines and workshops. These guidelines must be followed by management, operators, designers, and other stakeholders to create a safety environment where everybody is happy and healthy.

- ANSI B11 safety standards specifically provide guidelines for the safe operation of machine tools in general or for specific types of machines. They give guidelines for controlling and managing safety hazards for turning machines, machining centers, electric discharge machines, presses, drilling machines, and forging machines. Design engineers should consult these standards before they start designing machines. Safety standards must be part and parcel of system engineering requirements.

- ISO 16089:2015 standards specifically address machine tool safety considerations, including the risks and hazards associated with manual and automated grinding machines. It covers significant hazard conditions for workpiece holding, loading, devices, etc. It also covers applications, risk management, and mitigation methods for machines when connected with other machines used for automatic production lines. ISO 12100:2010 details the information to be provided by machine tool builders to the buyers. It illustrates guidelines and responsibilities for the user community as well.

Recently another idea has cropped up with regard to the energy consumption of machine tools. The design is asked to make the machines energy efficient since this affects the life-cycle cost of the machine. ISO has published a few papers to provide energy guidelines for machine tools. The performance of any machine depends on the energy consumed and efficiency of the machine to convert machine available power into chip-cutting for metalworking machines.

The energy consumption of any machine depends on various factors, and the combination of all the interactions among machine elements in the cutting loop determines the energy efficiency of the machine. One of the most important requirements for any machine tool is to make the machine as energy efficient as possible without losing productivity and performance. The machine should be environmentally friendly, too. ISO has some guidelines and suggestions to that effect. These are finer points of machine design that have not been optimized yet. Standards for environmental friendliness and requirements must be followed as per the ISO guidelines 14955-1 and 14955-2. These standards can help the builder to measure the energy requirement without mitigating performance requirements. Energy and environmental compliance will help the machine be productive and efficient. The problem is that machine tools are diverse in every sense of design and use. Maintaining the same requirements among all types of machinery could pose a problem for both builders and users. ISO 19455-1 addresses the methodology of designing machines to enhance their energy efficiency. The standard looks into various machine tool components and processes that require energy to function properly. These components should be designed to minimize the energy requirement. Machines should also be able to operate using solar energy in the future.

The idea is to design machines to be energy efficient without losing capability, quality, and reliability. There is no risk in using energy-efficient technology for machines. Mr. Reines has developed these standards in very well-written documents that designers and users should take advantage of. For example, spindle motors, axes motors, hydraulic units, etc. can be designed to minimize energy requirements during use. This is a new challenge for designers since factory automation needs hundreds of machine tools to satisfy automobile demands. Hence, energy efficiency will help the user conserve energy and reduce the cost of production. ISO has now published the more recent document ISO-19455 series, publicizing and recommending energy efficiency while designing machines. This is going to be state of the art for machines for a long time to come.

OSHA standards for machine tools are listed below for future reference (taken from OSHA website):

- 1910 Subpart O, Machinery and machine guarding. Includes definitions, general requirements, and different kinds of machinery requirements.
    o 1910.211, Definitions
    o 1910.212, General requirements for all machines
    o 1910.213, Woodworking machinery requirements
    o 1910.215, Abrasive wheel machinery
    o 1910.216, Mills and calendars in the rubber and plastics industries
    o 1910.217, Mechanical power presses. Includes general and specific requirements for construction, safeguarding, dies, inspection, maintenance, modification, operation, injury reporting, and presence sensing device initiation (PSDI).
    o 1910.218, Forging machines
    o 1910.219, Mechanical power-transmission apparatus

18.8   Summary

The coordination and interaction among each constituent of the company have to be based on system engineering principles and requirements. It is a functional requirement of company management to promote system engineering principles for survival as a team or system. Unless system engineering concepts and principles are followed by company management from the beginning, the company cannot be successful as a system. Organizations have to promote system thinking among all employees; it is not just a design function or activity. This is the most critical step that a company has to take against. Company management has to be convinced of the benefits of system engineering and the flow of ideas down the line to the lowest level. The whole company has to be a living system by itself. Most of the time, company management has the idea that system engineering only has to be followed by the design team. That is absolutely not true. System engineering applications have to be taught and promoted among all employees.

Any system is always a part of another larger system. For example, our planet is a part of the solar system, which is then a part of the universal system. Similarly, a spindle is a part of the machine system. Also, every system is connected to other systems through boundaries, which can be rigid or flexible. The common boundaries are also called interfaces, through which systems are connected and interact with each other. In order to manage system engineering activities, requirements collection and cascading for the system and sub-systems are very critical steps.

The cascading process starts with the requirements for the machine as a system and should flow down into the sub-system and component requirements. Also, alternative potential designs must be evaluated against these requirements. A balance between the satisfaction of these requirements and the cost of the system needs to be struck. These critical steps have to be baked into the design at the outset. Requirements lead to constraints of the design, and constraints lead to the cost and performance of the system. The cascading process can also use computer-aided tools, or CAE tools, for evaluation.

Even if the company gets sued and injured workers get compensation when accidents occur, machine designers should try to follow the machine cover guidelines very strictly and religiously to make the machine safer and operator friendly. Beyond that, loss of an operator is a loss to the intellectual ability of a company, and it is very hard to find and replace a good and sincere operator. Hence, the company loses on both counts when an accident happens. Machine tool safety should be a matter of great concern in machine shops, and customers should demand the safety compliance of all the machinery that they buy. In addition, when an injury happens, work stops, and loss of production also becomes profit loss for the company. Such losses should be counted as a part of product life-cycle cost.

The idea is to design machines to be energy efficient without losing capability, quality, and reliability. There is no risk here in using energy-efficient technology for machines. Mr. Reines has developed these standards in very well-written documents that designers and users should take advantage of. For example, spindle motors, axes motors, hydraulic units, etc. can be designed to minimize energy requirements during use. This is a new challenge for designers since factory automation needs hundreds of machine tools to satisfy automobile demands.

Due to a very small number of components and machines available for validation and verification, another issue is that applications of statistical tools have been found to be very limited. Hence, the design has to be very robust, and most designs should be validated using analytical tools, which will save time and money for the company. A design of experiments (DOE) process will also be very applicable for machine tools, helping to reduce the number of tests required for validation. Parametric studies can be done as well using a limited number of test samples.

Various machine tool standards provided by organizations such as OSHA, ISO, SAE, ANSI, etc. provide safety, design, maintenance, and operation guidelines that will enhance machine tool safety and increase productivity. For automatic factories, there are guidelines on connecting machines through conveyors, automatic guided vehicles, factory automation, etc. so that factory safety is increased as a whole. The ISO 230 standards series addresses procedures and design guidelines for machine tool accuracy, noise emissions, thermal distortion, vibration effects, risk assessment, and risk mitigation methods,

Durability or reliability tests are conducted to find the issues during the operating-life phase, and the extended durability test is conducted to find issues during aging or the wear-out phase, where, again, high failure rates are observed due to mechanical wear of the components. Customers are usually interested in knowing the issues that are found during reliability tests. The objectives of these tests are to determine design deficiencies, mean time between failure, mean time to failure, mean time to repair the system, design life of the system, and reliability or robustness of the machine. The reliability test results also determine the factor of safety margin for the components.

Other types of tests, such as strength, safety, impact, etc., are normally done to find the effects of very high load due to crash conditions. The machine should be environmentally friendly, too. ISO has some guidelines and suggestions to that effect. These are finer points of machine design that have not been optimized yet. Standards for environmental friendliness and requirements must be followed, as per the ISO guidelines 14955-1 and 14955-2.

These standards can help the builder measure the energy requirement without mitigating performance requirements. Energy and environment compliance will help the machine be productive and efficient. The problem is that machine tools are diverse in every sense of design and use. Maintaining the same requirements among all types of machinery could pose a problem for both builders and users. ISO 19455-1 addresses the methodology of designing machines to enhance their energy efficiency.

18.9   Reference and Bibliography

Box, G.E.P., Hunter, S.J., and Hunter, W.G., 1978, *Statistics for Experimenters: An Introduction to Design, Data Analysis, and Model Building*, John Wiley and Sons, NY.

Dixon, J.R., and Poli, C., 1995, *Engineering Design and Design for Manufacturing*, A Structured Approach, Field Stone Publishers, Conway, MA.

Ford Design Institute, 1998, Lecture Notes on Reliability and Robustness, Ford Motor Company, Dearborn, MI.

ISO Standards, International Standards for Organization, www.iso.org.

Kidder, T., 1981, The Soul of a New Machine, Avon Books, NY.

Lewis, E.E., 1987, *Introduction to Reliability Engineering*, John Wiley, NY.

Machine Tool Safety Standards, Wikipedia and internet references.

Nahmias, S., 2008, *Production and Operations Management*, fourth edition, McGraw-Hill Erwin Companies, NY.

Netter, J., Wasserman, W., and Whitmore, G.A., 1978, *Applied Statistics*, third edition, Allyn and Bacon, Inc., Boston, MA.

Osha Standards for Machine Tools, Department of Labor, United States Government, www. osha.gov,

Paul, G., and Beitz, W., 1996, *Engineering Design*, Springer-Verlag, NY.

Rao, S.S., 2014, *Reliability Engineering*, Pearson Publishing, Prentice Hall, NJ.

SAE Standards, "Hydraulic Power Units Used on Machine Tools and Equipment for the Automotive Industry," J1689-199411, SAE, Detroit MI.

Tooze, A., 2009, *Machine Tools and the International Transfer of Industrial Technology*, "The Global History of Machine Tools, Knowledge, Narratives and Fiction," 30-31, King's College, Cambridge.

Ulrich, T.K., and Eppinger, S.D., 2011, *Product Design and Development*, McGraw-Hill Companies, NY.

Wheelwright, S.C., and Clark, K.B., 1992, *Revolutionizing Product Development: Quantum Leaps in Speed, Efficiency, and Quality*, Free Press, NY.

18.10   Review Questions

- What is system engineering? How is it different from a conventional engineering process? Does it cover design, manufacturing, quality, and safety without affecting the cost? Does it pose any risks for machine development and manufacturing? Does an unsafe machine cost more to the users?

- How do you apply a system engineering process to machine tool safety? How can design and analysis be used to enhance the safety of machine tools? Why safety engineering practices must be followed in the workshop environment? Are machine tool standards constraints for machine development? Are machine tool standards important for machine tool development?

- Can you identify the safety practices to be followed for grinding machines? Who is more important to keep the machine safe: users or designers and builders? How do you justify using standards for designing machines? What are the negative aspects of using safety standards for machines?

- Are safety standards and technology connected? Can technology make a machine safer? What are the effects of a technology on the safety of machines?

- How do you design verification process and method to enhance machine tool safety? Is it necessary to spend money and time to design verification methods for a machine? Why do we need safety verifications? Can we have the same verification process for all types of machine tools?

C H A P T E R 19

19 Functional Analysis and Applications

19.1   Introduction

The application of system engineering principles started in the late 1950s with projects such as ballistic missile development, etc., which were very complicated. Conventional engineering processes could not manage the complexity associated with such projects. In addition, several companies, such as General Electric, AT&T, TRW, Lockheed, and other defense contractor followed suit. New product development combined the complexity with new technology to solve new, challenging requirements. Both system and process performance needed to be satisfied for a successful product. It was a new beginning for engineering process development. System engineering, modeling, and analysis were also supported by Mr. Art Hall of AT&T and Mr. Harold Chestnut of General Electric for this new engineering process initiative.

A new methodology, system engineering tools, and process development were initiated and developed by these engineers. Eventually, military standards MIL-STD-499 for system engineering was released. Also, commercial standards IEEE 1220 was published in 1998. The system engineering society, INCOSE, was formed in 1992. INCOSE's systems engineering handbook was published in 1998. Moreover, DSMC's system engineering management guide was published in 1989. NASA also published the system engineering handbook in 1995. Several NASA project failures, such as the Challenger space shuttle accident, MARS Polar Lander failure, etc., created an urgent necessity for system engineering processes to be used for all such complicated projects so that mistakes were not repeated. It was recognized that individual element problems are different from system problems due to interaction among those elements. Hence, system risk management has to be dealt with separately.

System problems are not a linear sum of the individual element problems. System problems have to be understood in a different context. For example, a servo motor problem, i.e. failure mode, root cause, etc., will be completely different than when it is connected to an axis system due to cross-linking and interactions. The system problem has to be analyzed in a system environment, and system failure root cause and failure mode have to be understood when a system is integrated.

System engineering principles should help the team understand system problems, and the solution to system problems can only come from system engineering applications. For example, system reliability changes drastically when individual elements are integrated in terms of function as a system. System complexity is different from individual sub-system complexity. The boundary interactions of the sub-system make a different situation for the system failure modes. More often than not, system failures are very hard to find, and they cause most of the damage. A rubber O-ring might not fail under specified loading conditions, but it might fail when it is used in a system due to system interactions, as happened in the case of the Challenger space shuttle accident.

19.2   System Complexity

The common cause of system failure or lower system reliability is system complexity. More complexity leads to loss of reliability. By definition, complexity is when elements are integrated together and system behavior is substantially different or cannot be predicted from element performance. System performance is difficult to define and depends on the performance of the individual element performance in a non-linear fashion. System knowledge and behavior need to be understood properly to predict system behavior or failures and successes. System complexity increases substantially when a system has many attributes or functions. In other words, a complex system is not a simple linear system. A system can be defined as many ways as somebody can think. One such definition is that a system can be thought of as a logical combination and conglomeration of individual elements interacting together to generate system functions that are more than the individual element functions.

For example, a turning center has individual elements such as indexing unit, spindle unit, tailstock unit, cross slides, shrouds, controls, etc. Each of these individual elements can perform some functions when they stand alone and are separated from each other. When they are interlinked together in a system, they can machine a part into the desired form. An indexing unit or control by itself cannot machine a part, but when acting together, they become capable of generating a machined part. The emergent function of a system is quite different from individual functions. More is different in many possible ways.

System complexity is increased due to a large number of attributes and functionalities. The system is integrated with lots of similar or different parts in many possible ways. The linking of individual elements determines the system performance. The system design also depends on the intellectual and cognitive abilities of system designers or integrators. System complexity is very hard to define most of the time. The complexity of a system arises because the human mind cannot completely apprehend or predict system behavior due to insufficient knowledge or poor understanding of the system. The human mind is limited in thinking about system behavior as the system grows more and more complex. As Dr. Philip W. Anderson, Nobel laureate in physics, 1977, explained, "More is different." A conventional design process converts requirements into a physical system.

For system design, the designer has to consider performance, attributes, requirements, hierarchy, constraints, cost, reliability, interlinking, interfaces, etc. When a hammer is designed, the designer thinks about a wooden handle connected rigidly to an iron head to create the functional capability of delivering an impact condition. When the same principle is applied to design an impact forging machine, the situation becomes quite different and complicated. A complex system has to deliver multiple functions simultaneously. The complete behavior of a system might not be known for practical purposes. Unknowns always remain unexplored for a complex system.

In order to understand system behavior even partially, synthesis, analysis, simulation, prototype building, testing, data interpretation, noise analysis, etc. might have to carried out again and again. The mode of integration and use dictate the performance quality, reliability, cost, user-friendliness, etc. Human intervention and perception of a system depend on the intellectual ability of a person who is trying to understand the system. Moreover, this is not restricted to technical systems only. It could include technical, organizational, political, and human systems as well.

System validation is critical for identifying system behavior, performance, reliability, and noise outputs. The more complex a system is, the more validation and verification is usually needed to understand it and its behavior. System validation is a complex function of system complexity. The validation process should include target requirements, validation and verification planning, physical and digital prototypes, technology verification and selection for testing, simulations, analysis, process validation, field testing, and laboratory testing. It is quite a detailed procedure. Validation provides different insights into a complex system. Physical testing is not sufficient alone since testing assumptions become a critical issue. For system identification and verification, both physical and analytical verifications are necessary. To make a system highly reliable, various types of validations are required.

It has also been found by many researchers that the type of organization, open or close, affects communication complexities. Hence, interpretation of test results is sometimes very subjective depending on cultural and organization structure. The communication system must be open in the organization. Failure should be considered a gateway to better opportunities. For machine tool builders, this sometimes can be an issue if machines are designed and built using various suppliers and vendors. If all of these suppliers are not in tune with the buyer's organization and culture, the ultimate result could be unsuccessful. In other words, both buyers and suppliers should be part of a system working in unison to create a product. If the company is profit oriented, bureaucratic, legacy dependent, or culturally conservative, communication is hindered and broken very often. For such companies, a complex system design could be at fault.

Product complexity arises from many factors. One of the common reasons is the multifunctionality of a product. The design also becomes very complex when a product has to consider platform sharing, legacy, hastened project schedule and delivery time, severe cost constraints, global factors, new technology, and various customers across the globe.

Product complexity is also dependent on the technology to be deployed in the product. Technology obsolescence is an important factor when the product development time stretches on for too long. If a product design is happening globally, coordination becomes too difficult when a product has to satisfy requirements that are not compliant with each other: electrical requirements in the US are not same as in the UK. Team location is also a very important consideration for product design since the design philosophy changes from country to country. Software commonality across all products and countries makes the system design very complex and sometimes very difficult to achieve. Language can be a major issue in such cases. I have experienced severe complexity arising out of the cultural differences between the US and Japan. The perspectives of design, design philosophy, design approach, system validation approach, etc. are all quite different.

Organizational complexity also adds to the product complexity if two companies do not have a uniform product focus. In any case, quantification of the complexity of any system or product is very difficult in general. A system engineering approach might help to reduce the design risk for a complex product. Nevertheless, the system complexity must be assessed before a project is started. Hence, the system complexity depends on the number of components in the system, their relationships with each other, the mode of interactions between them, and the operation states of each. Obviously, as any of these factors increases, the complexity goes up. For example, let's take as an example the spindle unit, which consists of a spindle, bearings, and housing, i.e., three elements or components. Now, the arrangement of these components is that the spindle is supported by bearings, and the housing holds the bearings in place. Each element can interact with the others while working or when the spindle is rotating and taking external cutting and gravity load. So, we have three interacting elements. The number of interactions: 2 x 3 = 6. The number of relationships among elements is three; i.e., the shaft, bearings, and housing are related to each other, and one cannot survive without the other. Next, consider the possible state of, say, the spindle, either working properly or not. Each component has two binary states: working or not working. Hence, the number of states of all the elements, i.e., the system, are eight (permutation of three component individual states). So, we have:

Components: 3

Interactions among components possible: 6

Relationships between three components: 6

Maximum possible system states: 8

The system design complexity also affects the design of a feasible system architecture. In order to satisfy all the above constraints, the system architecture has to support the requirements, as outlined above. Another complication comes into the picture when a system interacts with external environments. A spindle unit shares an interface with the base of the machine. Force and momentum act through this boundary. The machine is a closed system if it does not interact with the outside boundary. When the machine acts with external energy sources, such as impact, temperature, and humidity, the same system becomes an open system. For an open system, one of the elements or the whole system interacts with the outside boundary elements or environment. Energy flow is another example of an open system.

An external energy source, such as electrical inputs, makes the system an open system. If the system can run with an internal battery system and does not need an external energy source, it is a closed system. The external inputs and system response will add to the complexity of the system. The management of inputs from external sources and of the system response to the external source will affect the system complexity since three interrelated entities need to be managed. For example, the interaction between a machine and operator adds to the complexity of the system. Management of a complex system needs a different mindset since the emergent behavior, inputs, response, interactions, relationship management, etc. add to the complexity.

A complex system design needs a different mental model to deal with the complexities. Due to dynamicity, system behavior changes, which triggers the changes required in input and output management, noise management, environment management, etc. For machines, the complexity can be managed very well. The same machine, though, when used with other machinery, conveyor systems, control systems, external computer system inputs and outputs, becomes unmanageable. The system behavior prediction becomes almost absurd, and failure does happen for such a complex system. It takes a long time just to understand the system's internal and emergent behavior.

For control system software, it becomes very difficult to ensure accuracy. Fortunately, a machine system or an automobile system is pretty much defined. They all have similar subsystems and relationships, and links are well defined. Management of such systems is comparatively easier than managing a factory automation scheme. Similarly, management of a flexible manufacturing system design is a completely different task, one that needs a much higher mental level. That is the reason a stand-alone machine seems to be far more reliable than a flexible manufacturing system using the same stand-alone machine. The dynamics of a flexible manufacturing system are quite different.

When designers explore new unchartered areas, such as designing for a factory of the future, a completely new flexible manufacturing system, or automated factories, requirements for such systems might not be known beforehand. These can be called "unprecedented systems." For example, designing machinery for the moon or Mars could be a completely new and challenging area. Some requirements could be thought of, but not all can be charted for sure. Since the requirements are not known properly, the goals and constraints for such a new system will also be unknown. The system architecture will be different, and system complexity will be increased substantially to satisfy all the desired requirements. The efforts and intellectual abilities for designing such a system will be at a different level than normal machine design activities.

19.3   Steps to Reduce System Complexity

Reduction of system complexity without mitigating performance and reliability is a great challenge for the design team. Conventionally, it has been said that greater performance leads to higher complexity and, eventually, cost.

System complexity also increases the life-cycle cost of the product. There is an expected performance for every product, and a definite measure has to be taken to achieve it. This is the conventional approach to system design. System engineering is a different approach towards a design that contains both the cost and system requirements in a much easier fashion. Intellectual ability and design experience of the team members help the company design a product that does not compromise performance and increase the cost. This is due to the fact that system requirements include all the constraints for the design, and the accepted design will satisfy the requirements. There is no surprise at the end of the project. System engineering also optimizes the manufacturing process, which reduces the manufacturing cost for the product.

Moreover, cost is not a static number, and it changes as the design progresses unless strict control is kept. The cost of the design depends on how the components are connected in a system. It is also true that the cost of a system is affected by the reliability, robustness, quality, design diversity, life-cycle cost, etc. Cost accounting methods do not address the dynamicity of the design and manufacturing process. Instead, it gives the cost of a system at a single point only. Conventional engineering processes focus on system design and the functionality of the design satisfying requirements but not the cost on a running basis. The performance of a machine is the primary criterion of a design, and cost comes at the end.

System engineering helps the team keep a track of the system cost in addition to satisfying performance requirements. System engineering practices need the cost accountant to be part of the design and development team. The cost of any design is determined as soon as a new design or variation is proposed. Cost becomes an issue for the project team, which includes, design, manufacturing, industrial engineering, quality, service, and marketing personnel, and not for management only. At the end of the day, a project is successful only when it makes a profit for the company. Cost should be a primary index of success of the project in addition to the performance and reliability of a machine. The total program cost starts going up right from the start of market identification. The design and development phases cost the company the most. The requirement development phase and concept development phase cost comparatively less than the design and development phase if controlled properly. As mentioned before, the design phase defines the material cost, manufacturing cost, delivery time, service cost, quality reliability cost, etc.

The cost of a product also depends on the uncertainties, both external and internal, of the company. Internal uncertainty is the primary risk of the design. The probability that the design will satisfy the requirements determines the design risk. The product life and life-cycle projections are the primary risk factors. Customer satisfaction is well connected to such internal uncertainties. In most cases, a product fails because of very high risk with these two factors. Then there are external risk factors, such as market conditions, economic conditions, global competition, political unrest, etc. Even if the internal risks can be controlled, external risks are extremely difficult to predict or control. External risks also affect product introduction timing and cost factor. Machine tools are strongly affected by external risk factors since these are open systems.

System complexity affects the product in a great way. Most of the machine tool fails when the complexity and design risk are very high. Any new design with new concepts or unproven technology also affects the system complexity and eventually the product success rate. It is always necessary for the design team to be aware of the system complexity issues. If the design is too complicated to maintain and operate, the product might have very limited success. The complexity also affects reliability, robustness, and quality of the system, and a complex design is also a cost driver for manufacturing. The simpler is always the better design. System engineering practices can help the team control complexity and the manufacturing process. For machine tools, the general system engineering process should be curtailed or suited to the design system as necessary to keep the project time and cost as low as possible without affecting reliability and robustness.

19.4　System Engineering Standards and Process

As mentioned in the last section, design projects can tailor the system engineering process into specifics that help streamline the activities and enhance the customer satisfaction index. The customization of the generic system engineering process might make the process much simpler and time-effective. The tailored system engineering process might include the following steps:

- **Market Analysis for a Specific Machine**
  - Market Survey
  - Customer Needs and Wants Data
  - Customer Feedback
  - Benchmarking Data
  - Quality and Reliability Data
  - Price Data

- **Requirements Development for Machine**
  - Customer Needs and Wants Analysis
  - Product Metrics Development
  - Trade-Off Analysis
  - Cost and Time Analysis
  - Cost Constraints
  - Project Outline Development
  - System and Sub-system Requirement Development
  - Requirements Rationing Matrix
  - Constraint Development

- **System Desegregation and Synthesis**
  - System Architecture Development
  - Functional Decomposition and Analysis
  - Concept Development
  - Preliminary Analysis and Synthesis
  - Cost/Profit and Payback Analysis
  - System Development Time

- **System Design and Analysis**
  - System Alternative Design
  - System Analysis and Simulation
  - System Trade-Off analysis
  - System Optimization
  - System Selection
  - System Requirements Matrix Development
  - System Selection
  - Sub-System Design
  - Cost and Time Control
  - Break-Even Analysis
- **Sub-System and Component Details**
  - Sub-System Design
  - Component Design
  - Component Analysis and Optimization
  - Cost Optimization
  - Design Audit
  - System Risk Management and Mitigation
  - Manufacturing Issues and Process Development
  - Quality and Reliability Assessment
- **System Integration and Manufacturing**
  - Prototype Development
  - Sub-System Integration
  - DFMA Development
  - Manufacturing Process Optimization
  - Verification and Validation
  - Supply Chain Development and Management
  - Reliability Analysis
  - Life-Cycle Cost
  - Warranty and Service Considerations and Confirmation
  - Requirements Confirmation and Deviation
  - System Safety Confirmation and Product Introduction

There is no established standard for following the above tasks for machine tools. The team has to follow as much as it can to make the project successful. Actually, the system engineering standards started with military standard MIL-STD499. A new version of this original standard, MIL-STD-499A, was published in 1974. The next updated version of this latest standard came in 1994 as EIA/IS632. which was then updated as EIA632, which came into being in 1999. This latest standard was updated in 2002–2003 and was called ANSI/EIA632. IN 1998, IEEE1220 was published. These standards help the users follow a structured, disciplined approach towards system engineering practices. The standard covers both the product and processes. It also helps resolve system problems during its use. EIA-632 provides guidelines for system designs and methods as to how to use the system engineering principles for various types of projects. IEEE1220 also helps guide the project team to adopt system engineering processes for all phases of design and development. It is also true that standards cannot always be followed to the least detail for the entire project. Still, it provides guidelines as to how to tailor the project in a system engineering context. No standards can be or should be followed in their entirety, but general guidelines will make the design and development process very efficient and productive. Any standard gives overall process detail but not specific details for all projects. It helps the team pick up the process tools to make the design and validation more efficient.

There are military standards available for various project management and operations. For example, for system life-cycle processes, ISO 15288 can be followed. For product development support, MIL-STD-337 provides cost guidance for the design. For drawing practices, MIL-STD-100 can be consulted. For information systems, MIL-STD-7935 can be used. For human factors and ergonomic considerations, MIL-STD-1800 can be used. For manufacturing guidelines, MIL-STD-1528 can be consulted. For quality considerations, MIL-HDBK-50 and ISO 14000 can be used effectively for machine tool design projects. There are various standards and handbooks for maintainability guidance, such as MIL-HDBK-791, MIL-HDBK-472, etc. For reliability, MIL-STD-785, MIL-STD-1530, MIL-STD-2164, etc. can be consulted. For system safety engineering practices, MIL-STD-882 can be used. These standards are not specifically meant for machine tool design and manufacturing but give a general perspective for design, cost, manufacturing, and quality. Another good standard/handbook for heat management for machines is MIL-HDBK-251. For defense projects, these standards are almost a must, and other standards, such as MIL-STD-881, MIL-STD-1771, MIL-STD-1521, etc., must be consulted.

In general, EIA-632 and IEEE-1220 explain the system engineering process, and most of these standards can be applied to the machinery business to a great extent. EIA-632 describes the connections among many aspects of the business, such as management, product evaluation, system design, and supply chain management. These processes interact with others, from product inception to when the product is delivered. This is somewhere analogous to the system "V" process. The standard shows how technical management is connected to technical evaluation of the product by interacting with different processes such as supply chain management, system design and development, product manufacturing, etc. The relationships of processes are shown below in Fig. 19.1.

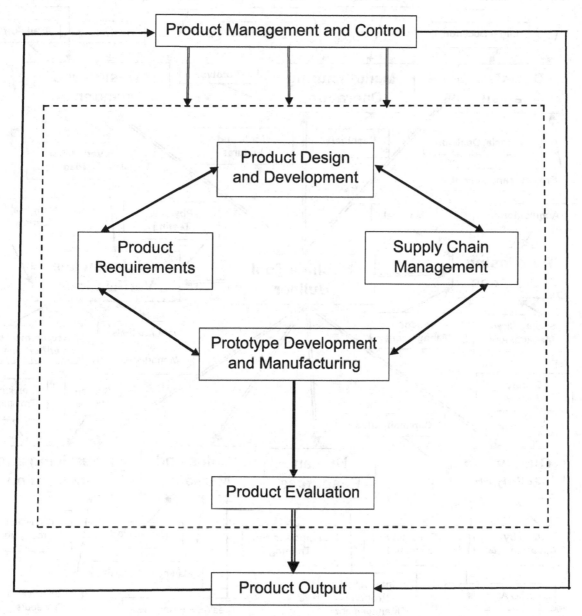

Fig. 19.1 Product process relationships

EIA-632 also provides a relationship model for all types of business entities and functions. This model connects all the business functions for the manufacturer and helps the manufacturer to develop business links for the product's success. The model is shown in Fig.19.2 below. Actual standards recommendation and requirements have been modified to suit machine tool manufacturing.

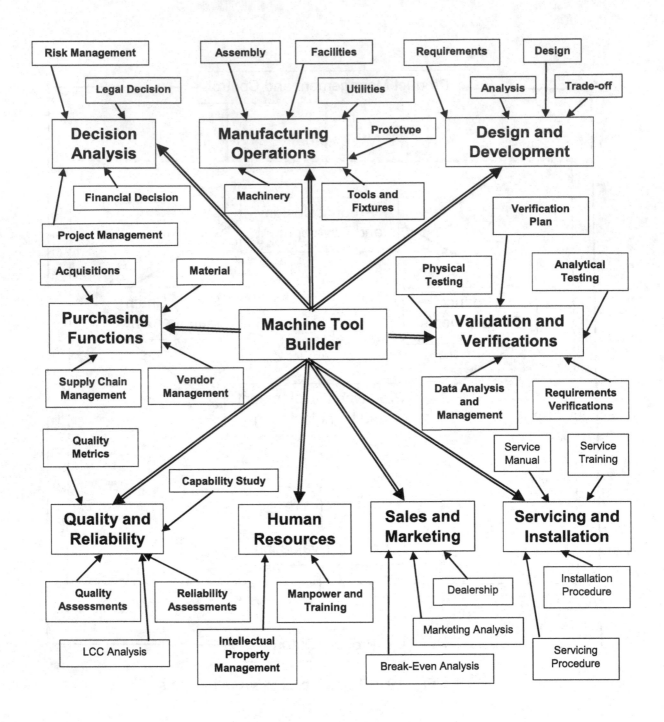

Fig. 19.2 Relationship-activity model

The system engineering process recommended by IEEE 1220 is slightly different from the system "V" approach. In this approach, the process inputs start with requirements analysis. This analysis leads to the requirements verification phase. After the requirements are verified, functional analysis is completed. Then functional aspects are verified. Next, design synthesis is completed, and then the last stage is design verification. In each stage, analysis is carried out to verify stage functions and theoretical confirmation. This particular process can be followed for theoretical verification of the design, as shown below in Fig. 19.3. The standard recommended steps will be modified to suit the machine tool environment, i.e., digital or analytical prototype development.

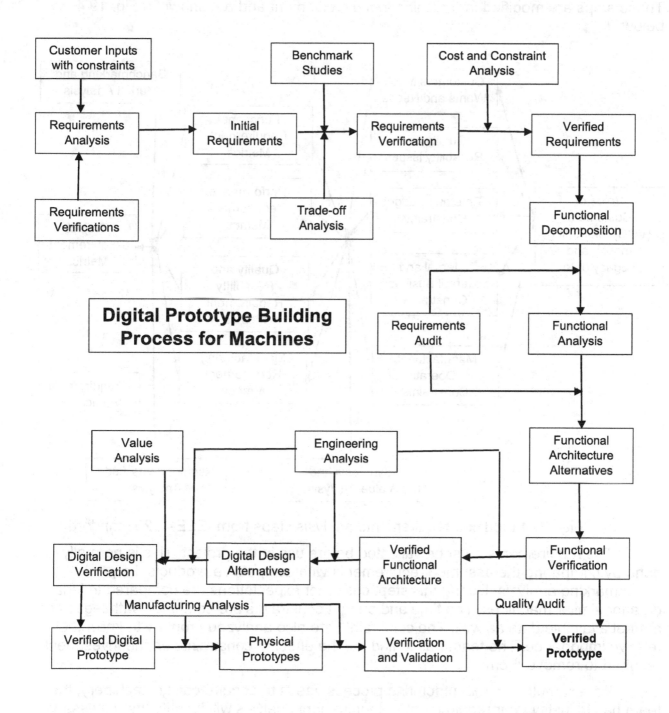

Fig. 19.3 Modified action steps from IEEE-1220 standard for machines

The standard IEEE-1220 also details the requirement development process. These steps are modified for machine tool development and are shown in Fig. 19.4 below.

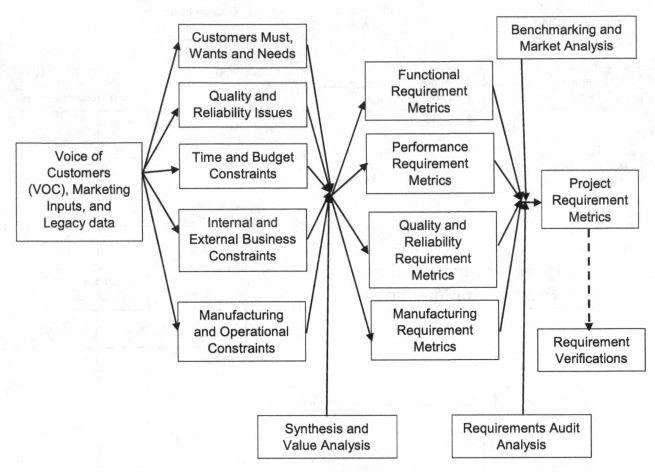

Fig. 19.4 Modified requirements analysis steps from IEEE-1220 standard

The requirements must be validated before use in the project. This is normally done by comparing the assumed requirements with competitive products, i.e., benchmarking analysis. During this step, customer expectations are compared to legacy data for similar machines. The time and budget constraints are compared with legacy or market available data as well. The constraints are also analyzed properly to know their relative importance. The team has to find out the effect of constraints on the investment required to remove them.

For example, if a manufacturing process needs to acquire costly machinery, the team has to justify their requirements. Requirement analysis will identify the necessary adjustments for the requirement list. The refined requirement metrics become the project target, which needs to be satisfied to enhance customer satisfaction. The requirements validation process is very critical to the success of the project within budgetary constraints. For some projects, a trade-off analysis for the requirements might be necessary to decide their priority and establish the superiority of the new machine over available machines in the market. The validated requirements become the project constraints that need to be satisfied and well documented.

19.5   Functional Decomposition and Analysis

After the requirement analysis is completed, a functional analysis of the machine needs to be done as per IEEE-1220 standard. The functional analysis consists of three steps: functional context analysis, functional decomposition, and functional architecture development. The input for this process comes from requirements analysis and functional verification data. The context analysis consists of several steps: functional behavior analysis, functional interface development, and performance requirement analysis. The functional decomposition analysis consists of subfunction development, functional development timing, functional failure modes development, functional safety development, etc.

In simple terms, functional decomposition helps the team divide the system functions into sub-system functions and then to component functions. This is generally done to define the functions or tasks each component or sub-system is supposed to do to generate the system functions. Instead of looking at a system problem, functional decomposition helps to break the functions into more understandable sub-functions. There are several sub-functions for a machining system. as shown in the block diagram in Fig.19.5. The functional decomposition of the turning machine is not fully developed in the diagram. The process is partially developed. Hence, it can be seen that when the machine system is very complex to deal with, the functional decomposition of the system into sub-systems and then into components will help identify functions to be performed by each component or sub-system. This helps the team to design the components.

For very simple designs, or if the design is team is very familiar with the products designed earlier, such detailed decomposition might not be required. For example, a turning machine design team might be very familiar with the design of the turning center, but when the similar machine has to interface in a flexible fully automated manufacturing system, a lot of complexity arises, and at that time, functional decomposition will help the team design the system as a whole. Functional decomposition simply means dis-integrating the system functions into manageable sub-functions to be performed by each system component. For example, as shown in Fig.19.5, the machine is composed of a spindle, cross slide, longitudinal slide, tailstock, loading/unloading fixture, etc. For each sub-system, there are functions to be performed. Then these functions are sub-divided into component or unit functions as shown. In a similar way, other sub-system functions need to be developed.

It should be kept in mind that there is no standard way of developing such a block diagram for a very complicated system. Some guidelines are provided by IEEE-1220 standard. Moreover, functional decomposition of any system cannot be absolutely correct, and many versions can exist simultaneously. It is always better to start from a legacy product and then modify it for the new product. Most machine tool design follows the same guidelines. Functional decomposition can also be developed from the energy flow or job flow sequences. Decomposition can also be developed from a customer's list of actions during the machining process. The functions of the machine must match the ways that the customer is going to handle it.

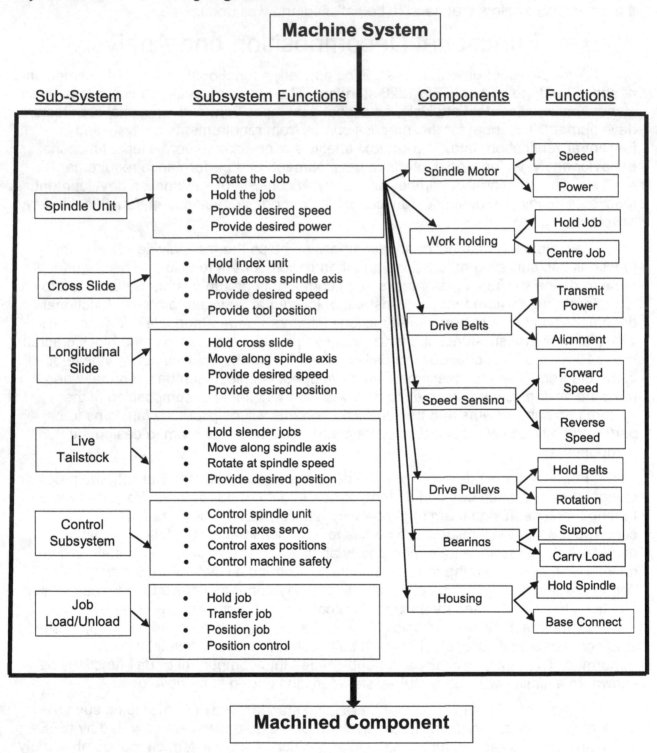

(Note: The functional decomposition is not complete for the whole machine)

Fig. 19.5 Functional decomposition for a CNC turning machine

A very heavy job cannot be handled very accurately or precisely. The loading and unloading devices have to be handled differently. The machine positioning system has to be different and has to be designed accordingly. The International Council of System Engineering (INCOSE) recommends system engineering practices for all technical and non-technical business. The following block diagram for requirement analysis, Fig.19.6, is recommended by INCOSE. The theme is similar to other standards, but the outline and terminologies are slightly different.

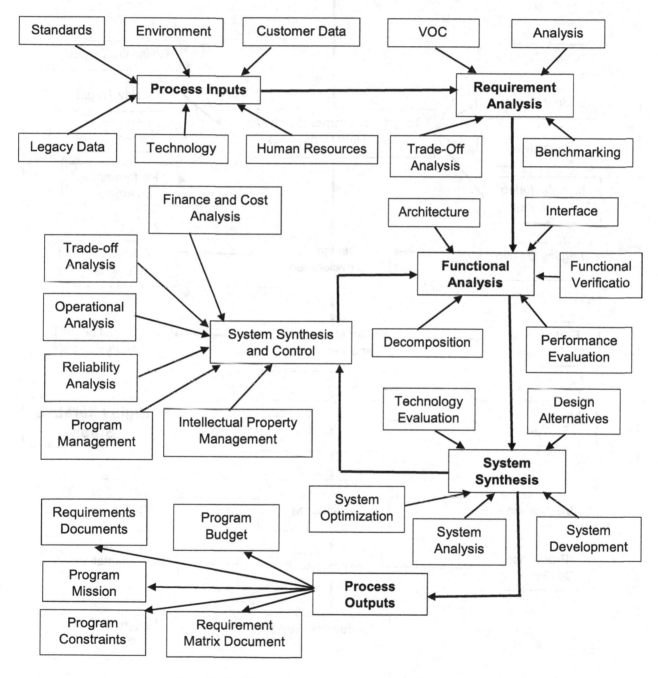

Fig. 19.6 Requirements analysis process (INCOSE)

To conclude this section of the chapter, I recommend the following steps for machine tool design and development. These actions are in line with whatever has been discussed so far but simplified to the point that the project can be completed successfully within budget and time. The simplified steps are shown in Fig.19.7 below.

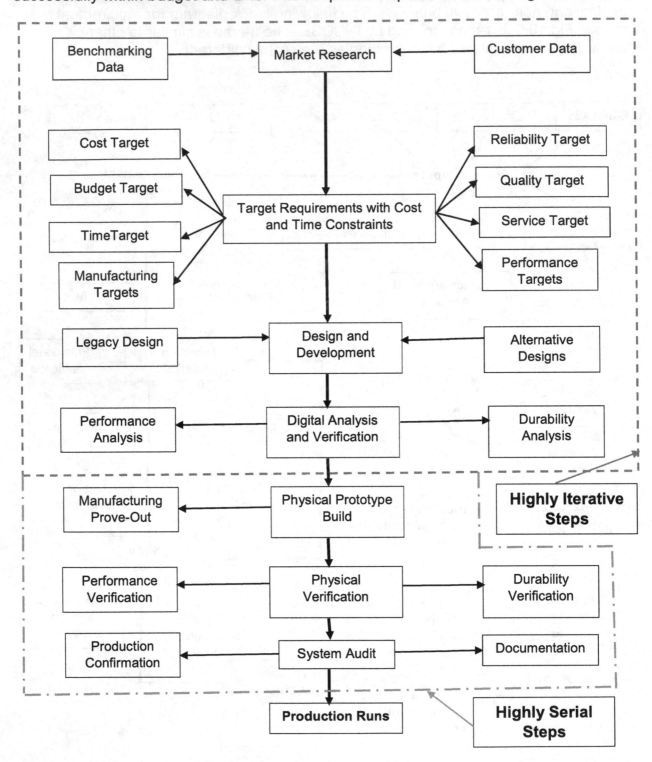

Fig. 19.7 Recommended machine development steps

The first few initial steps, as indicated in Fig. 19.7, are highly iterative in nature. Everything must be designed, analyzed, and tested using digital prototypes. This might be called digital or analytical verification, and it saves time and money. Machine tool projects are always under severe budget restrictions due to a lower volume of production. Iterative technique is a very powerful way to refine a design or any concept. In Japan, iteration of design is a must, and iteration procedure takes a sizable amount of time during design and development. On the left side of the system "V," all the steps are highly iterative, but the right-side actions are highly serial and non-iterative. During iteration, several actions are required:

- An iterative loop requires reconsideration of any decision against a norm. The decision is questioned repeatedly to justify the action taken.

- The design architecture is discussed after an initial decision and reiterated to refine it.

- The team dynamics are changed, and coordination and brainstorming are required to be successful.

- Target specifications are given values based on cost and competition to arrive at the optimum stage of any decision made.

- Iteration is also required to justify business decisions, project management, and cost issues.

- An iterative loop is required for understanding customer needs, target requirement analysis, value analysis, manufacturing processes, functional analysis and decomposition, system design, etc.

- During the system design phase, any design is iterated for design considerations, design values, design optimization, requirement confirmation, etc. before a final design is selected

The project management and iteration technique add value to the system and eventually promotes customer satisfaction. The idea is to reduce the potential risk of any design or decisions made during development and manufacturing. Since the design is iterated over and over again, it reduces its risk and cost. In some cases, full-fledged system engineering needs to be tailored to the machine tool project where design risk is comparatively less. Cost and quality are major concerns. Similarly, an iterative process could be used for manufacturing development since there are many possible solutions available to manufacture a part or assemble a unit. Iterations provide refinement and optimization of any design or process. The extent of iteration depends on the complexity of the project, uncertainties associated with the outcome of the project, costs involved in the process, etc. If the design concept is innovative and absolutely new, an iteration process is required even more to identify the real value of the concept in the customer's mind. Any potential design must add value for the customer at the end. Iteration steps must be taken during process evaluation and methodology adaptation for any design. Any accepted tasks must be justified and optimized before any action is taken. Without any question, iterative technique must be applied to very large, complicated, high-cost, and high-risk projects, such as machines designed for flexible manufacturing systems.

The team has to ascertain the right use and application of system engineering for the project at hand. To add functionality to an existing design, a full spectrum of system design might not be required since the risk and cost associated with such an application are not enormous. Frankly speaking, full-blown machine design and development starting from scratch might justify the use of system engineering (SE) processes as per the standards. The team has to decide that. Nevertheless, the application of SE recommended practice will reduce the risk of the project. There are no hard and fast rules for SE applications, but they will enhance the value of the design by reducing cost, time, and risk associated with the project. If any machine design project is done at random and in a very haphazard way, the product will have a very high risk of being non-competitive. The cost and time of machine development will be too high to compete in the market place. A step-by-step procedure is required for a complex project to understand why a decision has been taken and adopted. For detailed knowledge of SE processes, several handbooks and standards from INCOSE can be consulted.

For a very complex project, such as the design of a flexible manufacturing system or highly automated machine, very often the team does not know the steps to be taken to make the project successful. When the understanding of a project mission or requirements is vague and ambiguous, as disciplined SE process application is a must for guidance. A systemic approach will help the team understand the width and breadth of the project. In any procedural method, a decision can be traced back very easily for modification or elimination if necessary. Even the engineering tools to be applied for evaluation and optimization must be evaluated before use. For machine tools, applications such as simulation, statistical methods, optimization tools, etc. might contradict each other and confuse the team. Hence, any process or methodology to be applied must be properly designed for success.

19.6   Functional Block Diagram

In order for a system to be successful in a competitive marketplace, it must be fully functional, providing value, performance, quality, reliability, robustness, ease of maintenance, and expandability. The system must have a minimum life-cycle cost, and its value must be appealing to users. Only functionality is not important; all the functions of the machine must have acceptable performance measures that a customer can live with. Functionality and performance must justify the cost associated with the machine. Robustness must be more than adequate for any field application. The MTTR and MTBF should be as high as possible. The repair time must be as minimal as possible to reduce the cost of any downtime. The system should be adaptable to a new requirement without much difficulty. For example, it should be comparatively easy to add a loading and unloading fixture, tool management system, or a robot down the line. In order to reduce the initial investment, the user should be able to add such devices in the future to optimize the new process or requirement. An optimized design should be able to provide these system characteristics.

An SE process will help the team design such a machine. The requirements analysis, functional analysis, and system design must be done in a systemic way. In the requirements analysis, as mentioned before, mission statement, business requirements, project requirements, technical requirements, performance requirements, project constraints, etc. must be identified and documented properly without any ambiguity. The requirements must have metrics as well. In the functional analysis step, functional decomposition, functional and boundary interfaces, functional architecture and confirmation, and functional analysis must be carried out to justify each function. Next, functions and functional architecture must be converted to design shape and form, which should be analyzed based on the requirements. In this phase, all verification must be done using proper engineering analysis and synthesis tools. Digital prototypes must be built and tested for refinement and optimization during this phase. The system design phase is very critical for the success of the project. Alternative designs and thorough analysis and justifications to select the best potential design is very critical to reducing the cost and time of the project. The design phase dictates which manufacturing methodology and manufacturing process are suitable for the design. These three steps are almost mandatory for the system engineering process to be successful.

Functional analysis is fundamental to a systemic approach towards the design and development of a machine. This process has been followed to some extent in several different forms throughout the industry, but a concerted effort has not been made yet. That is the reason why the US has lost competitiveness in the machine tool industry. Functional analysis converts architecture and specifications into performance and design specifications. It helps the design process flow very easily towards a successful end. Functional analysis includes a functional decomposition process, as mentioned earlier in this chapter.

The decomposition process breaks down system functions into sub-system- and component-level functions. It becomes easier for the designer to understand the context, functionality, performance requirements, and constraints of any design. When the system requirements and functions are broken down and passed to lower levels, the lower-level design functions, when integrated together, generate a system function in a concerted way. All the functions become complementary to each other to give rise to the desired system. The time, budget, safety, quality, reliability, etc. become elements of the system and sub-system designs. The functional architecture is developed depicting the boundaries and interfaces among the sub-systems. The interactions across the boundaries and interfaces are logically developed. The functional architecture is a conceptual design that satisfies functionalities. The design is not final at this stage. The functional architecture only shows the initial form of the final design. The functional analysis flow diagram is shown in Fig.19.8 below. It extracts the functional requirements of a machine from the target requirement matrix. These functional aspects are then assembled in a timely block diagram as a sequence or as parallel events. Parallel events occur simultaneously, and sequential events happen one after the other. Moreover, functions need to be prioritized, and trade-off analysis needs to be done to finalize the functional requirements. Functional requirements can be combined together to reduce the numbers.

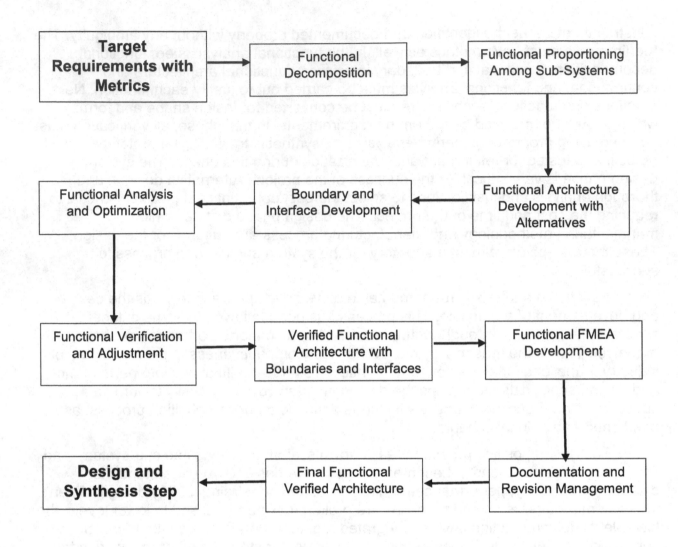

Fig. 19.8 Functional block and flow diagram

The functions of a machine must be measured against the competition. The functional performance and the cost must be competitive, and it must be prioritized; i.e. reliability has priority over functions, etc. All the quality functions can be grouped together for the block or flow diagram. Functions must have a metric to measure, and they must be compatible with customer requirements and enhance customer satisfaction at the end. Functions are what the machine has to do, but not how to do them. Functions do not represent any methodology or action items. A functional block diagram must include all the necessary functions required for the machine throughout its life cycle, such as life-cycle cost, maintenance requirements, diagnosis and ease of maintenance, functional modes, failure modes, etc.

All operational, manufacturing, and administration functions should be included in the functional block diagram. Functions are what the machine is supposed to do during its life cycle. The functional block diagram should display all the sequential and parallel functions, iterative loops, and external inputs to the system. The external inputs cross the boundary of the system as inputs to the system functions. Similarly, there can be system output as well. The system functions have to interact with the external environments and sources while functioning.

For example, system control inputs can be external to the system boundary. System-generated waste functions need to be displayed so that they can be managed properly, such as heat and sound management for the system. Similarly, there can be internally generated system functions that will interact with other system elements. For example, the sound generated by a machine while cutting will interact with the enclosures and create resonance for the system. Sound and heat management are big design challenges for machines such as forging presses and hammer presses. Functional analysis should include a functional flow diagram, functional block diagram, energy flow diagram, data flow diagram, and functional behavior diagram. The functional flow diagram displays function sequencing and parallelism of functionalities. The block diagram also displays iteration loops for the functions. The behavior diagram displays several items, such as sequencing, iterative loops, external and internal interfaces, external inputs and outputs, control inputs and outputs, etc.

19.7   Advantages of Functional Decomposition

- The functional block diagram helps the team assimilate and identify customer requirements for the machine system as a whole. This step of the process can identify the superficial functions that are not required.

- The functional block diagram is also a good tool to divide the total requirements into functions for sub-systems and components.

- This step also helps to identify the focus areas of design and development, including management priorities.

- It can help to simplify the process by combining several functions into one wherever possible. This also identifies the functions that can be done concurrently or that have to be accomplished in sequence.

- The functional diagram can also be used to create a timeline and critical path for the project. This analysis can be extended to manufacturing activities or manufacturing process improvements.

- The functional block diagram and analysis help the team identify several possible paths for activities. This will allow the team to select the best possible path to carry out for any activity.

- Functional decomposition helps the team desegregate activities into manageable functions that each sub-system or component can do.

- Functional analysis also helps divide and understand sub-functions, such as performance, reliability and quality, manufacturing, assembly, etc. It streamlines the total requirements into sub-requirements for each sub-system.

- The block diagram also helps the team identify the relationships and links among activities, i.e., serial events or parallel events. It is a system function characterization process.

- The tasks of the functional flow down, i.e., functional analysis, activity block diagram development, functional architecture development, functional decomposition, timeline management, and CPM development, will help manage customer requirements in a very efficient and productive way and will also help the company to achieve higher customer satisfaction.

- Since this flow-down analysis does not show how to accomplish these functions, the team can select alternate paths to carry them out. It only shows the functional relationships and links.

- The functional block diagram can be used to develop a timeline diagram for all functions.

19.8   Summary

- Functional block development, functional analysis, etc. do not show the timeline for each function of the system. They are tools deployed to break the total system functions into sub-system and component functions that will support customer requirements. Customers do not really care about the timing of functions, but they need the functional performance of the machine.

- The total task, i.e., functional decomposition, functional block diagram, functional architecture, trade-off analysis, could be called a system functional behavior development task. Keep in mind that this is a recommended process for complex system development and the detailed process might not be necessary to develop a simple machine.

- For a very complex system, such as flexible manufacturing systems or automated systems, functional analysis is almost a must and a very good stepping stone between the target requirements matrix and system design phase. The importance of such an intermediate step is not obvious among machine tool builders, and it is very often ignored since this process takes some time to develop.

- The functional decomposition process helps the team to identify the requirements that were not considered in the target requirement matrix. In order to establish proper relationships among activities, functional decomposition of target requirements is necessary as it identifies the functions of each sub-system and component.

- Each function of the system can be analyzed in details to identify alternative ways of achieving the same functions. The functions can be integrated logically and efficiently to eliminate unnecessary functions for the system. This helps to develop alternate systems and possible functional architecture, which is then converted into alternative system designs.

- Functional decomposition and a functional block diagram are necessary to identify the constraints and conflicting functional requirements for the system. A complex system has many constraints, and it also has a large number of requirements. Functional analysis will help the team put these two aspects together to find a common feasible solution.

- Functional analysis consists of several tasks: functional decomposition, allocation of functional requirements into sub-systems, development of functional interfaces and boundaries, development of functional architecture, etc.

- System functional requirements can be categorized into several important tasks, such as performance requirements, reliability requirements, quality requirements, maintainability requirements, add-on requirements, cost requirements, etc.

- Functional architecture can be converted into a digital or physical form. It can also be used to develop alternate digital architectures, which then can be used to develop a physical prototype.

- A functional decomposition process can be applied to various types of machining systems, which are conglomerations of sub-systems that act together to give rise to total systems. Functional decomposition or functional analysis is a part of the system engineering process.

- The functional block diagram can also be used to divide development tasks into several blocks that can be handled by separate teams to gain higher productivity and reduce the project timeline.

- The system engineering process and methodology can be tailored for a specific project. Every system has its own constraints and requirements that need to be managed properly to satisfy customer requirements. Eventually, this will enhance customer satisfaction. When system complexity grows into an unmanageable situation, system engineering process can help to manage the situation.

- The team has to assess the complexity of any project. For a simple task or design, application of system engineering could seem to be unnecessary and time-consuming. This will elongate the project time and increase the project cost. So, the team has to assess the project complexities and apply the SE process accordingly. For a very large and complex project, SE process adaptation is almost a must for success.

- System engineering standards and handbooks can be consulted for guidance. SE standards started in 1969, and they guide the team in identifying important steps and procedures to be followed for a complex project. The standards provide guidelines only as to what should be done and not how it should be done. As always, military SE standard is very rigorous/extensive and also very hard to follow, but the commercially available SE standards, such as IEEE-1220, can be used for machining systems without much difficulty.

- The SE process helps to manage the cost and timeline of a project. Hence, it increases productivity and efficiency of product development. For machine tool projects, development cost is a major issue, and project time and cost constraints must be satisfied for the product to be profitable.

- Documentation at each step of the system engineering process is necessary for future reference and consultation when a decision turns out to have a negative result.

19.9 References and Bibliography

Blanchard and Fabrycky, *System Engineering & Analysis*, Prentice-Hall, NY.

Boppe, C.W, 1980–81, Class Notes for System Engineering, MIT Course, ESD. 33J, Cambridge, MA.

DeMarco, T., 1978, *Structured Analysis and System Specification*, Yourdan, NY.

Elm, J., 2013, "The Value of System Engineering," System Engineering, May 13.

Flood, R.L., 1988, Dealing with Complexity, Plenum, NY.

Freidenfelds, J., 1981, *Capacity Expansion: Analysis of Simple Models with Applications*, Elsevier North Holland, NY.

Grady J O., 1995, *System Engineering Planning and Enterprise Identity*, CRC Press, Ann Arbor, MI.

INCOSE Systems Engineering Handbook, 1998.

Kaneshige, K., and Izawa, K., editors, 1965, "Proceedings IFAC Tokyo Symposium on System Engineering for Control System Design."

Maier, M.W., and Rechtin E., 2000, The Art of Systems Architecting, CRC Press LLC, Boca Raton, FL.

MIL-STD-499A, 1974, Military Standard, System Engineering Management, May 1.

Nahmias, S., 2008, *Production and Operations Management*, fourth edition, McGraw-Hill Erwin Companies, NY.

Porter, M.E., 1990, The Competitive Advantage of Nations, The Free Press, NY.

Sterman, J.D., "Learning in and About Complex Systems," *System Dynamic Review* 10(2–3), 291–330.

Sterman, J.D., 1981, Systems Analysis and Management, Petrocelli Books, Inc.

Ulrich, T.K., and Eppinger, S.D., 2011, *Product Design and Development*, McGraw-Hill Companies, NY.

19.10   Review Questions

- Develop a target requirement matrix for a machining center. Starting from this target requirement, conduct a functional decomposition. Conduct a functional analysis. Develop a functional architecture for the machining center as well.

- Can we convert a functional architecture into a digital and physical prototype? What are the advantages of using a digital prototype for prove-out vs. a physical prototype? What are the disadvantages of using a digital prototype for functional prove-out? What are the pros and cons of this SE approach for a large and complex machine development project? What are the risks associated with accepting this method?

- What are the salient features of functional decomposition and functional analysis process? Can system engineering principles be applied to revive a machine that did not sell well? Is system engineering necessary to revive the machine tool industry in the USA? What are the pros and cons of such a revival? Does an integrated approach have an advantage over a conventional compartmentalized process for the machine tool industry?

- What are the advantages and disadvantages of using functional decomposition for a very simple legacy-based machine tool project? Does the functional decomposition approach help to design a machine? If so, how? Does it increase productivity and improve efficiency?

- Can we use an SE approach for a brand-new, technology-based machining project? Does an SE approach depend on the complexity of new technology?

C H A P T E R 20

20   Failure Modes and Effects Analysis

20.1   Introduction

Every system fails. Statistically speaking, eliminating failure from a system is not possible. Failure is inherent to every system, but the failure rate or failure mechanism can be controlled or managed to extend the design life of equipment. The identification of the failure modes of any system has been discussed briefly in a previous chapter, but it has not been dealt with in detail. This chapter will focus on the development and design of the failure modes and effects analysis (FMEA) process for machine tool applications. I know of very few machine tool companies, including overseas companies, that really use this powerful method for machine tool development and maintenance. It enhances the design life of any machine and eventually reduces the life cycle of the machine. It also creates a defense against machine failures. Obviously, it enhances customer satisfaction and creates a better perspective on the machine. This process can be used for any type of machine or equipment. It is almost a universal process that must be applied before a machine is shipped to customers. This process is very commonly used in the aerospace and automobile industries, where any failure is taken very seriously.

An FMEA is a disciplined and systemic process that consists of several activities that are geared toward identifying and recognizing the potential failure criteria and modes of any product or process. It also identifies the corrective actions to be taken to reduce or eliminate the chances of such failure modes occurring during machine usage. Hence, an FMEA process helps the team take preventive action that will extend the life of a machine. The corrective actions help to reduce production downtime and manage failures. This process also helps the team refine and optimize the design. The idea is to identify the weaknesses in the design and how they lead to failure. Moreover, this process is very much team-oriented and requires the effective participation and coordination of team members for its success. The process and results have to be documented properly, as this fault-finding is very document-intensive. The FMEA process can be applied to both design and process. Hence, it consists of the following steps:

- Understand the design thoroughly
- Recognize potential failure modes of the design or process
- Evaluate each potential failure mode
- Identify the design or process steps to be taken to avert the failures

- Document each step of the process

Hence, an FMEA procedure allows the team to think about their design and how it can be improved to create a defect-free product. It is true that all the causes of potential defects cannot be completely eradicated, but attempting to do so will definitely make a design or process as close to a defect-free design as possible. Such procedures can help the team establish a discipline that will potentially reduce the chance of failures. This process can be applied before a design is implemented for manufacturing or anytime a failure occurs. This helps the team identify the root cause of failures and how they can be averted in the future. FMEA is very effective for machinery failures, but it has not been extensively explored in the machine tool industry. This process is suitable for any design, business, or manufacturing process, i.e., DFMEA or PFMEA.

20.2   Basic Purposes of FMEA Procedure

The purposes of such a process in machine tool design and manufacturing are numerous:

- FMEA helps to make the product safer and more reliable.

- FMEA helps reduce product life-cycle cost by eliminating most failures.

- FMEA helps to refine the manufacturing process, reducing the manufacturing cost and time.

- FMEA helps to reduce product development cost and duration.

- FMEA optimizes the design against failures or refines the design to a great extent.

- FMEA documents the proceedings so that responsible team members can take action.

- FMEA helps develop an action plan for the potential failures of the product.

- FMEA manages and mitigates the risk of failure.

- FMEA helps to identify and prioritize the severity of potential failures and failure modes.

- FMEA lists the definitive actions to be taken and the engineers responsible for taking the action

- FMEA identifies the weaknesses of a design or process

- FMEA identifies the potential defects of a design or process that could be a safety hazard or cause fatalities.

- FMEA identifies the critical factors of a design and can become an input for system specifications

- FMEA is a controlled document required to keep track of the necessary actions to be taken, who is responsible for implementing them, and when they should be completed.

- FMEA is a time-limited, disciplined process

If applied properly and implemented before a failure happens, FMEA has been found to eliminate most common failures of a design. It has been applied very effectively in the automobile and aerospace industries, and this proven methodology should be used in the machine tool industry as a common procedure during the design and manufacturing of the products. Time is the essence with this process. It is a time-bound process, i.e., potential failure modes and action to be taken to avert such failures must be completed before the failures occur. Sometimes a similar procedure can be applied to determine the root cause of a failure in design or manufacturing.

To get the most out of the FMEA process, it must be applied before the design is converted into a product. It is a highly iterative process. After an action has been taken to eliminate a failure mode, its effects are evaluated to see if it has properly eliminated the failure and failure modes. In some remote cases, an action to avert one such failure might cause other failures that were not anticipated before the action was implemented. FMEA also helps to manage the cost of a design or process. It is far costlier to incorporate changes after the fact since more changes are required to correct a design after the failure has surfaced. FMEA is a continuous process for any design and process. It to change the future design or manufacturing activities to reduce cost and time and enhance the reliability of a machine.

20.3 Types of FMEA Processes

There are two basic FMEA processes that are commonly used by the aerospace and automobile industries: design FMEA, or DFMEA, and process FMEA, or PFME. DFMEA can be used during system-, sub-system, or component-level design stages. On the other hand, PFMEA can be used for any process, such as manufacturing or business processes. PFMEA can also be used in assembly, manufacturing, and reliability assessment processes. DFMEA can be used throughout the system design process, i.e., the left side of the "V." The benefits of using DFMEA are as follows:

- Identifies inputs to the system requirements and design process

- Helps identify design weaknesses and critical failure modes due to component or sub-system interactions

- Helps the team identify critical failure modes and their severity

- Helps the team identify and prioritize failure modes so that actions can be taken within a reasonable amount of time and minimal incurred costs for implementation

- Helps mitigate risk associated with the design elements

- Helps create design verification methods to prove out the failure modes

- Helps refine and optimize any design, including the elimination of redundant systems

- Helps create and compare design alternatives

- Helps design the machine, which reduces manufacturing time and cost

- Helps the team to identify the critical failure modes and reduce the risk of failures

- Helps create a prioritized list of failures according to their severity and consequences

- Enhances customer satisfaction due to the reduction of failures, which reduces life-cycle cost and production downtime

- Helps the team optimize the design and track the action items and their suitability

- Separates all potential machine failures into critical and non-critical failures

- Helps the team catch design faults before they lead to unwarranted fatalities or consequences

The advantages and benefits of PFMEA are very similar to those of DFMEA. The benefits of PFMEA are as follows:

- Identifies the critical process failure modes during product manufacturing

- Helps identify bottlenecks in manufacturing and reduce manufacturing delays that would hamper production

- Helps identify manufacturing assembly causes of failure that affect the production rate or quality of a product

- Helps identify hazardous manufacturing conditions and reduce the risk of fatalities during assembly or machining parts

- Helps document manufacturing process abnormalities and problems

- Helps create a list of activities that need attention to enhance the production rate

- Helps develop a manufacturing plan that ensures the reliability and quality of a product

- Identifies operator safety concerns and reduces the risk of fatalities during assembly or manufacturing operations

- Helps design products with a minimal number of potential manufacturing risks and concerns

Design FMEAs are also used to critically review the safety aspects of a machine, and the process is always carried out before the design is released for manufacturing or production runs. It can be considered an audit plan for manufacturability or for the production suitability of any product. Design FMEAs are used to eliminate potential harmful failure modes that can endanger operators' lives during operation. It reveals design deficiencies and helps address them before the machine is produced for sale. Also, this process documents critical failure modes and their severities. Since the process is carried out before the machine is manufactured, the focus is on potential issues and not the actual occurrences of the failures.

The PFMEA is primarily intended for manufacturing and assembly operations. Similar to the DFMEA, the PFMEA focuses on potential failure modes caused by manufacturing or assembly operations. Process deficiencies are identified, and positive actions are taken to eliminate them from manufacturing operations. It helps identify the process control required during manufacturing so that manufacturing issues do not affect the product's failures. Hence, it helps to develop a manufacturing control plan that mitigates the risk of failure due to manufacturing and assembly operations. Actually, PFMEA substantiates the DFMEA by controlling manufacturing activities. It ensures that the product is produced as per the intended directions of the design group. Sometimes PFMEA identifies the requirements of quality checks during manufacturing to keep manufacturing deficiencies under control. In essence, DFMEA and PFMEA are complementary to each other, and they work together to make a product near perfect.

At the end of the DFMEA procedure, several outcomes are possible. First and foremost, potential failure modes and their severities are determined and listed. The outcome also contains potential critical characteristics. The severity determines the priority of actions to be taken. Then a list of design actions is created to reduce the severity or rate the occurrence of failures. This list also updates the system-level requirements that affect the design. DFMEA outputs should also clearly define the design verification and validation plan to confirm the elimination of any failure mode. Test verification also helps the design community enact design changes if required. Hence, the output from the DFMEA procedure helps the team refine the design, if necessary, to reduce or eliminate design weaknesses. This DFMEA report is a living document that needs to be updated when a design is updated or revised to change the performance or reliability of any design. To make this very productive and useful, proper documentation is a necessity, and it must contain potential failure modes, severity, occurrence rates, actions to be taken, and the person responsible for taking the action.

Similarly, the PFMEA output also contains potential failure modes and effects, severity, occurrence rates, operator safety characteristics, and safety hazards for the manufacturing system. The output also depicts the control plan that is required to control manufacturing process activities and eliminate potential failures or special characteristics. PFNEA is all about the process refinements required to eliminate or reduce process failures. PFMEA should address concerns during manufacturing or assembly. Operator safety, ergonomics, and health considerations should also be taken into account, and process and control documents created accordingly. Actually, the procedures adopted for DFMEA or PFMEA are very similar except the first is for design activities and the next one is about process or manufacturing activities. Both serve similar objectives, e.g., eliminating potential failure causes from the product.

20.4　FMEA Process Details

FMEA actions are taken by responsible engineers or any designated person within a stipulated time. FMEA is a team event, and its results depend on the team's efforts. Hence, it is critical that the team of responsible engineers must have a thorough understanding of and familiarity with the subject, context, and environment for the design or process. The members have to understand the criticality of the design or process. The responsible person or team must also understand the design process, manufacturing activities, assembly details, service and maintenance requirements, operator safety and health requirements, etc. The team members will have the responsibility and authority to pull resources from all other activities to resolve the issue. Normally, one engineer or team is tasked with one potential failure concern.

Each team should also have a subject member expert (SME) to deal with the issue. The whole process belongs to a team, and a team approach is very critical to the success of this process. The activity should also include manufacturers, suppliers, vendors, purchasing engineers, etc. if required. If the problem is about a purchased item, the responsible team member must coordinate with the supplier to resolve the issue. The whole process is initiated after an FMEA document has been finalized. In some cases, suppliers conduct their own FMEAs to resolve the issues and pass them off to the responsible engineers.

As mentioned earlier, an FMEA is a living document and needs to be updated when an issue is resolved or added to the list. The list must be current and correct. The team leader is supposed to lead this process. The end users, suppliers, and design team will benefit from such FMEA actions. Moreover, a DFMEA is required every time a part is changed or revised or a borrowed part is added to the system. Any system change necessitates reconsideration of the FMEA action to make sure that the system behavior has not changed to create another concern. Interactions among system elements are changed when a part is added or removed from the system. For example, adding a new loading device to a machine changes the manufacturing process to a great extent. Such a situation needs a reevaluation of previous PFMEA actions.

As mentioned earlier, DFMEA must be initiated and completed before the design is released for manufacturing or production. Similarly, PFMEA should be started and completed before manufacturing begins. The PFMEA report is also a living document and needs updating whenever a new process element is added or removed in manufacturing. PFMEA should be carried out whenever a new fixture is added, a new tooling system is incorporated, another machine is added to the manufacturing system, etc. Again, both the DFMEA and PFMEA must be initiated before any change is incorporated. The process must be reviewed and discussed before it is changed. FMEA action is pre-event process and not a post-event one. The time required for any FMEA depends on the extent of the actions to be taken. There is no fixed time for completing any action. The time and associated costs vary from situation to situation, but they need to be controlled as the actions start. In some cases, if a machine has been designed and produced before, previous experiences should be consulted before any new FMEA is started.

The FMEA process team must be cross-functional to provide different opinions about any design or process. Healthy team interaction is very important, and conflicts must be managed properly and in a timely way. Healthy discussion must be promoted, and unnecessary interactions must be discouraged. Each team member must be knowledgeable about the system and have past experience with the potential failures of similar systems. Each member has to understand the functional performance of the system and its working environment. The team also should have some members who are experts in the FMEA process or are at least familiar with it. For example, the FMEA team for a machining center might include a design engineer, supplier, tooling engineer, manufacturing representative, service engineer, purchasing engineer, assembly engineer, quality engineer, etc.

The team should not consist of more 7–10 people since too many cooks spoil the broth. The team should be manageable and cohesive, and it has to have a leader who has the authority to take action and carry the team to a resolution within a reasonable amount of time. The leader maintains a team environment where a full discussion about an issue is possible. The team leader also determines and develops team logistics and is responsible for publishing the team findings and end results. The team leader is fully responsible for the team's actions and behavior. The team should be geared toward producing results within the shortest possible time. The team leader has the following primary responsibilities:

- Form a cross-functional team
- Develop a plan for team functionalities, including team meetings and agenda
- Keep and update team discussions
- Resolve and manage any conflict
- Pull resources from outside the team whenever required
- Summarize the results and make decisions

The FMEA methodology consists of the four following main actions:

- Form of a team
- Develop a DFMEA document for the system, sub-system, or component
- Develop a PFMEA document for the system, sub-system, or component
- Publish the results and maintain the document

These four steps need information to be completed, and the information has to be fed to the team to start the action. It comes from various sources at the beginning of the brainstorming sessions and is managed by the team leader. The team has to understand the design requirements and the design itself. Before a critical review of the failure modes begins, the design requirements for the machine and its sub-systems need to be critically reviewed and understood. In particular, performance and reliability requirements must be understood thoroughly. The time and budget constraints need to be reviewed as well. The output of the design FMEA will be as follows:

- Potential, critical, and significant characteristics of the design
- Design revisions
- Revisions of target requirements

The critical inputs for the PFMEA include manufacturing data for the process, legacy process data, quality data, operator safety data, machinery data, assembly data, process flow data, and manufacturing constraint data. The outputs of the PFMEA process can be numerous depending on the situation and its criticality. In general, the outputs are critically significant events data, a process control plan, a quality control plan, recommended action to avoid failure modes, etc. The FMEA process is initiated for the following reasons:

- A new product is being designed, i.e., a new machine with new technology has to be designed.

- A legacy design is being revised, i.e., an existing turning machine is taken up for updating the system, such as by adding more power or changing the machine configuration.

- Units or sub-assemblies are added to an existing system, i.e., an indexing unit with more tool stations, a new spindle system, or a tailstock is being added to an existing machine as an upgrade.

- A failure is investigated, i.e., a critical failure has occurred during machine operation.

- The system environment has changed, i.e., a stand-alone machine is being added to a flexible manufacturing system (FMS).

- A machine application or process is changed, i.e., part material is changed.

- The quality or reliability of a system has changed, i.e., performance degradation.

Team coordination and management are very critical to the success of any FMEA process. Team motivation and structure are also very crucial. The team must be very cohesive and binding, and team members must be geared toward finding a solution to anticipated problems. Before an FMEA process is conducted, every team member has to understand the design or manufacturing process thoroughly. The system configuration has to be understood. The design requirements and constraints must be discussed and understood properly as well, and the design or manufacturing process must be stable and viable. Finally, the design or manufacturing team must be eager to change the design or process upon the recommendations of the FMEA team.

In order to analyze a system, a process flow diagram or functional diagram is generated first to understand the boundary or process flow of the system. For this diagram, blocks are used to represent the functions of a system. The diagram also shows the linkages between sub-systems and how each unit functions in relationship to the others. There is no standard for generating a function diagram. Normally, they are drawn hierarchically, i.e., from system to sub-systems to components as per the function. For system analysis, the system should be the starting point, whereas for a sub-system, e.g., an indexing or spindle unit, start with the sub-system and flow downwards. Gradually, the system functions are broken into lower sub-functions until the end is reached. For each step, inputs and outputs must be identified and mentioned in the block diagram. The boundaries for each block have to be developed and represented properly. Let us explain this process using the index turret system of a CNC turning machine, Fig 20.1.

12-Station Indexing Unit

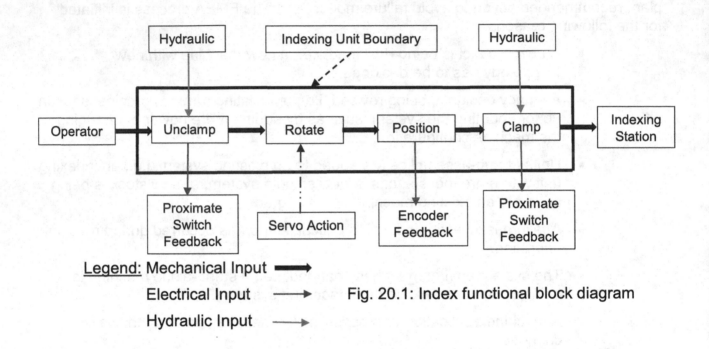

Fig. 20.1: Index functional block diagram

The process flow block diagram for the indexing unit is also shown in Fig.20.2 below. Such a diagram can be used when a new process is being started or designed. For effective and productive PFMEA, a process flow diagram is almost a necessity to understand the process steps required. The details of the block diagram, what should be included and excluded, can be ascertained by the team members, i.e., the scope of the FMEA project.

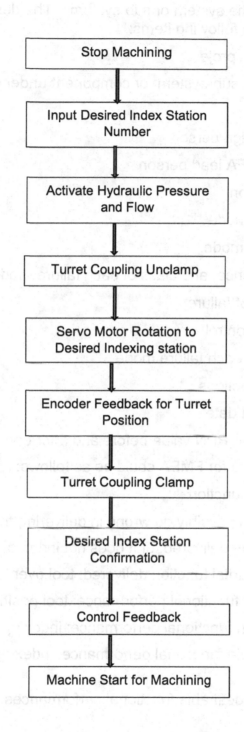

```
┌─────────────────────┐
│   Stop Machining    │
└─────────────────────┘
          │
          ▼
┌─────────────────────┐
│ Input Desired Index │
│   Station Number    │
└─────────────────────┘
          │
          ▼
┌─────────────────────┐
│ Activate Hydraulic  │
│  Pressure and Flow  │
└─────────────────────┘
          │
          ▼
┌─────────────────────┐
│ Turret Coupling     │
│     Unclamp         │
└─────────────────────┘
          │
          ▼
┌─────────────────────┐
│ Servo Motor Rotation│
│ to Desired Indexing │
│      station        │
└─────────────────────┘
          │
          ▼
┌─────────────────────┐
│ Encoder Feedback for│
│  Turret Position    │
└─────────────────────┘
          │
          ▼
┌─────────────────────┐
│ Turret Coupling     │
│      Clamp          │
└─────────────────────┘
          │
          ▼
┌─────────────────────┐
│ Desired Index       │
│ Station Confirmation│
└─────────────────────┘
          │
          ▼
┌─────────────────────┐
│  Control Feedback   │
└─────────────────────┘
          │
          ▼
┌─────────────────────┐
│ Machine Start for   │
│     Machining       │
└─────────────────────┘
```

Fig. 20.2: Process flow block diagram for indexing function

Inputs for the DFMEA or PFMEA can come from engineering drawings, legacy data for quality and reliability, previous failure patterns and modes, legacy failure data, quality function deployment (QFD) data, service and warranty data, customer service log data, customer service data, process flow diagrams, etc. The input source depends on the nature of concerns for the system or sub-systems. The design FMEA format/table should include the following items:

- Title of the FMEA project
- Name of system, sub-system, or component under consideration
- Machine model
- Responsible design person
- Responsible FMEA lead person
- Date of completion
- Potential failure mode
- Effects of failure mode
- Severity, occurrence, and class of each failure mode
- Potential cause of failure
- Current design control
- RPN number for each failure mode
- Recommended actions
- Action taken with date
- Action results with RPN value before and after corrective action taken

The eight-step procedure for FMEA could be as follows:

- Step 1: Select a function
- Step 2: What can possibly go wrong in delivering the function
  - No function delivered: unit does not index at all
  - Over or partial function delivered: tool over indexing/overshoot
  - Degraded functional performance: tool positioning gradual error
  - Off-and-on functional performance: indexing failures at random
  - Undesirable functional performance: indexing stops in between stations
  - Wrong/undesirable functional performances: wrong indexing positions
- Step 3: Effects of functional failure, severity, and occurrence rate of functional failure
- Step 4: Determine potential causes of each failure mode
- Step 5: Determine present design control
- Step 6: What should be done to avoid this failure:

- o Design change
- o Modify the design
- o Add more control for detection
- o Identify future actions
- Step 7: Identify responsible engineer and completion date
- Step 8: Document the findings and process details

A DFMEA process block diagram could be as follows:

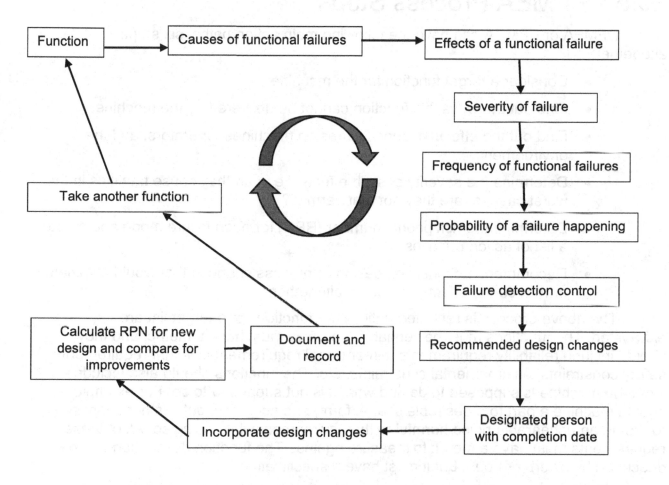

Fig. 20.3: DFMEA process flow block diagram for a new machine

The process shown in Fig.20.3 starts with the target list of functional requirements for the machine. The next step is to identify the potential failure modes and criteria for this function. Then the potential effects for the failure modes are determined. For this function, the criticality of failure, i.e., severity, probability, and occurrence rate, is determined. The severity (S) rating of any potential failure leads to the priority of failures. The severity, occurrence rate, and detection ratings are given a number between 1 and 10 to create a comparative list of priorities. A rating of 10 means the highest value possible, i.e., a failure mode that leads to a fatality gets a 10 to show that it is most important to eliminate this failure. The occurrence rating is denoted by "O." The product of S and O, i.e., S x O, is also used to understand the priority of fixing the design. A recommendation for design changes depends on this product, S x O.

The next step is to identify the detection rating (D) of any failure mode, i.e., how easily a failure mode can be detected before any damage occurs. This needs a physical verification of the suggested design change. The product of S, D, and O is called the risk priority number (RPN), i.e., RPN = S x D x O. These numbers are all relative, and they are arrived at after brainstorming among team members.

20.5   FMEA Process Steps

DFMEA or PFMEA can have the following main action paths, six steps altogether:

- Consider a target function for the machine.

- Find out the ways this function cannot be delivered by the machine.

- Find out the effects of such failures on machines, operators, and the environment.

- Determine the severity of such effects, i.e., can they cause fatalities in the worst case, or are they not that harmful?

- Determine the risk priority number (RPN) for each failure mode and create a list of action priorities.

- Recommend actions, i.e., design or process changes that would eliminate or reduce the effects or failures altogether.

The above process is repeated until all the functions of a target list are exhausted. The functions must be verifiable, i.e., they must have a metric, and they should include reliability requirements, dimensional requirements, weight requirements, safety constraints, environmental constraints, etc. The functions should also include what the machine is supposed to do and what it is not supposed to do. For example, it should machine a part to a desirable shape, form, and accuracy, but it is not supposed to create an unbearable noise harmful to the environment while doing so. All of these requirements must have a metric to measure against. The function description can be described by a verb or noun, but it must have a specification.

As mentioned before, each function will have a failure mode or modes, such as no function, over-reacted function, partial or degraded functions, intermittent functions, or undesirable function. For an indexing unit, we can describe these characteristics as follows:

- No Function: tool does not index at all.

- Partial Function: tool can only index in two stations out the 12

- Degraded Function: tool indexing takes four seconds, which is more than the specified two seconds for indexing.

- Over Function: tool rotates in 0.1 seconds, which does not allow other units to position and creates a crash condition.

- Intermittent Function: when commanded, the indexing unit sometimes rotates or sometimes fail to rotate at random without any pattern.

- Unintended Function: The tool has to rotate forward or in reverse to reduce the rotation time. The unintended function is that the tool rotates in the forward direction when it is supposed to move in reverse, i.e., the operator did not expect such movement.

Each failure mode must be attached to a specific function that causes the failure mode. In specific strength-based failure modes, a hardware-based approach could also be used. In the hardware-based approach, the focus is on the failure of hardware, housing, components, etc. The examples of hardware-based failure modes are soiled, corroded, broken, deformed, cracked, aged, etc. It is easier to connect the failure with the design revisions. The examples of function-based failure modes are:

- Tool does not stop at the desired position

- Tool is not clamped properly before machine starts machining

- Tool rotates while machining a part

- Tool rotates until the halfway point and then gets stuck

- Indexing unit does not accept control commands

For the PFMEA, the focus should be on the following aspects:

- Operators and maintenance personnel

- Environmental factors

- Customer usage

- Loading and unloading factors

- Interfacing with other devices

- Safety and hazards

- Government safety regulations

- OSHA regulations

Once the prioritized list of actions is created using RPN values, the goal is to eliminate the failure modes totally or partially. If the RPN value is very high and the failure can affect the health and hazards of the operators and factory personnel, government and OSHA regulations, performance, reliability, etc. the failures or cause of failures must be eradicated or eliminated completely from the design. The design should ensure such failures never happen during usage. The necessary design or process change can eliminate such failures. Moreover, the design should be verified against the failure after the recommended action has been incorporated to confirm the positive effects of the recommended actions.

As has been mentioned before, the potential causes of failures show the design weaknesses that need to be eliminated. The design weaknesses lead to potential failures, and the failure modes are also sign of weaknesses. If the housing of an indexing unit is not strong enough to withstand the inertia and metal-cutting forces, it will eventually fail. The cause of any failure leads to the failure mode. The weak spots in a design fail first and then the failure progresses to the rest of the structure. In order to find the potential cause of the failure, team members have to brainstorm to determine the root cause of the failure modes. The answers to the following questions might lead them to the cause of failures:

- Does the unit vibrate or resonate to cause the failure?
- What are the environmental reasons that might lead the unit to failure?
- Is there a strength mismatch between the interacting elements of the design?
- Does every component in a design satisfy the engineering standards?
- Is the unit capable of delivering performance or function within the specified range of tolerances?
- Is the design unsuitable for delivering the desirable functional performance?
- Is the design too complicated for the situation to cause failures?
- Are the inputs causing the failures?
- Is the operator causing the machine to fail?
- Is the cause of failure external or internal to the machine?

The direct failure path must be determined first. Then secondary or tertiary causes can be identified. The failure paths must be chalked out first, and then the quickest failure path mode is selected. Sometimes the nature of the failure can provide clues as to why it occurred. A fishbone diagram can be drawn to find the possible failure modes of the design. If any design failure is due to several reasons, they should be considered together. The design change suggestion should eliminate all of these grouped reasons without bringing in other ones. Unless the causes occur simultaneously, they should not be grouped together. The causes must be linked together. The root-cause analysis of any failure mode might help determine the potential causes.

For example, tool breakage might be caused for several reasons: the tool holder is loose, the tool is too weak for the cutting, the tool is vibrating due to its speed, the tool mounting is wrong, etc. They can all occur together or work together to give rise to a tool failure. So, vibration, tool strength, and loose mounting can all interact to cause tool breakage. They should be grouped together if, when they act together, they cause failures (AND conditions). Otherwise, they are OR conditions of failures and should be listed separately. For indexing to fail, several functions or interactions are required, such as the servo motor must act, the gears must not be broken, the interlinking drive shaft must not be broken, etc. The tool indexing will fail if any of these actions do not happen. So, they cannot be combined into one cause for this indexing failure. These are OR conditions and not AND conditions. To get power in the spindle unit, the spindle motor must rotate, and the belt must be in proper condition. These causes can be grouped together (AND conditions).

Many design faults can be attributed to improper component material, and the root cause of any part failure could be due to material strength. Improper machining could be another cause of part weakness. The discussion on root causes should start with the failures with the highest severity ratings, i.e., failures that cause bodily harm, fatalities, health hazards, etc. Another option is to take up the items that have the highest S x O values, i.e., the product of severity and occurrence. These failures must have the highest priority for elimination. The root causes for these failures must be determined using root-cause analysis (RCA). A cause and effect diagram, or fishbone diagram, can be constructed to identify the root causes of such failures. The first-level cause might not be the root cause of the failure.

For example, if the spindle failure is due to bearing seizure (first-level cause) and the bearing seizure is due to lubrication failure (second-level cause), then the root cause is the lubrication failure. The determination of the root cause is very important since the design has to be changed to eliminate them. If the lubrication was proper but the bearings failed due to overloading conditions, then the bearings have to be sized properly to eliminate the bearing failure. Hence, root-cause analysis is required to find the proper design actions that should be taken to remove the failure cause from the design. If a shaft bends under load, the shaft material might not be the root cause. It is quite possible that the shaft's diameter is not sized properly to withstand the external loading conditions. Hence, the shaft diameter must be sized up, and material change might not help the situation.

The next topic is how to determine the occurrence rating. For an existing design in use, the occurrence of failures can be easily determined, but for new designs, it is very difficult to find the occurrence rating (O) of a failure. In such cases, if the severity is high for the failure, it is customary to have a high occurrence rating as well. One thing should be remembered: DFMEA or PFMEA is all about potential failures. Hence, the analysis is pretty much directional. All the failures might not actually happen during usage, but there is a chance of them happening, or a probability of occurrence. Hence, the values are relative and not absolute. If the occurrence rating of a failure is known, the comparative values might be used as well. The occurrence rating might also depend on how the failures are detected or identified. If the detection is manual, identification might not always be possible.

If the identification is through electronic control, it is quite possible to identify the failures almost all the time when a failure occurs. If the control device is not sophisticated or accurate enough to identify the failure, the occurrence rating could be wrong, too. Hence, in order to find the occurrence rating, detection control methods must be explored properly and in detail. The occurrence rating (O) is given a scale of 1 to 10, where O = 1 means that the occurrence of such failure is very rare and O = 10 means the failure occurs very often. The machine bed cracking is very rare and could be given an $O = 1$, whereas the spindle bearing fails very often, and its occurrence rating could be as high as 9. Even if the values are comparative by nature, values must be given to get a proper priority list for design or process action. In some standards, the criticality (C) of any failure is determined by multiplying severity and occurrence, i.e., $C = S \times O$. If the criticality of a failure is very high, it must be addressed with design changes first. S denotes the consequences of a failure mode, and O denotes the frequency of the causes.

Once the priority FMEA table is completed, the next step is to recommend the action for design revisions or changes. Actions have to be taken to eliminate the cause of failures. The design is the best point to start from. It has to ensure that failure must not happen during usage or manufacturing. Recommended suggestions must include the failures with very high values of criticality or the failures that lead to multiple failure conditions. The idea is to reduce the proneness of failure in the design. The design must reduce the chance or risk of failure, i.e., it has to be robust enough based on external or internal conditions of usage. The design should promote robustness, reliability, safety, and ease of manufacturing.

20.6   Detection Control and Rating

For machine failures, measures must be taken to identify failure conditions ahead of actual occurrence. This means that the design must include measures to identify failure conditions or prevent them from happening. The design must also accurately identify the types of failures and prevent failures before their occurrence. This is called detection control for the design. It is only done for DFMEA; PFMEA does not include such control steps. The design steps recommended by DFMEA only detect failures but do not avoid them. In some machines, design control elements such as torque limiters, switches, diagnostics, etc. are put in place to identify failures and reduce their effects. There are two types of control methods that can be used in the design:

- Active Detection and Control Methods: In these methods, devices and controls are installed in a design to identify failures. They are post-operative methods. The design control is activated after the failure has occurred. These methods must be accurately tested and confirmed before the machine is released for production. If the spindle belts are slipping, the spindle revolution will be reduced during acceleration and deceleration. This property can be used to denote belt failure through the spindle encoder.

  Similarly, axes encoders must identify when the slide fails to move after a command is given to do so. This post-operative method only lets operators know about failures after they have happened. These methods must be tested and validated during prototype evaluation, and the validation or verification procedures must include confirmation of such failures. The position of proximity switches or relays must be tested to ensure slide movement.

- Preventive Detection Control Methods: These are pre-operative methods to prevent failures. For example, the machine can only be turned on in automatic mode when the main door switch is activated, i.e., the door is confirmed to be closed. This is done to ensure that parts do not fly out of the workholding devices and hurt the operator. Another example is electrical cabinets that cannot be opened until the machine is shut off. Such design elements ensure that operator mistakes do not happen. The design and installation of such devices can be taken up after design reviews, tests, customer feedback, design specifications for safety and hazard control, etc.

The detection effectivity has to be confirmed by validation tests. In order to add rating values for detection control of any design, one has to find out whether tests have been conducted and failure detection has been confirmed or not. Depending on the requirement for these tests, detection methods are given a rating, called the detection rating of a design. Moreover, control devices do deteriorate from time to time. Hence, a detection rating might change depending on the outcome of these tests for the machine. It is not a constant number and needs to be evaluated every so often. Hence, if the design is released for production without any need for testing to confirm a detection method, it gets a very high rating. The confirmation of the detection method and its sensitivity depends on the type of test and the test methodology. If the test is not designed to accurately identify the effectivity of the detection control, the design might seem very effective against failure mode detection when, in reality, it might not have proper detection control. Before any such tests are designed, failure modes have to be determined. The team has to decide what a failure really means. For some machines, a slight crack in the housing might not constitute a failure until and unless the crack has propagated and met with other cracks to dislodge any material out of the housing. The failure mode and its criticality to the functional performance of the design must be determined before the test is designed to detect the failure.

Depending on whether a test for the confirmation of a failure mode has been done or not, a detection rating for the design is established for that particular failure mode. The rating depends on the type of failure, failure mode, and cause of failures and how the failure has been detected. A rating value is given to denote how effective the detection control is at preventing a failure with or without a validation test. The verification of the detection control method of any design is the most effective element to determine the detection control rating for a failure mode, i.e., can the failure mode be detected or not? If a design seems to have the proper detection control method to identify a failure and does not have any test validation, it is given the lowest rating, and if a test has been done to verify a failure mode, it is given the lowest rating, say, a 1 on a scale of 1 to 10. The detection rating also depends on the unit design as well. If the design includes effective detection control methods so that the team thinks that any sort of testing, confirmation, or validation is not required before the design is released for production, it gets the highest rating of 10. This means that there is a good chance that failure will not be detected since it has not been confirmed through testing.

If the detection method is not sufficient or needs lots of academic evaluation, the design gets a very high rating, such as a 9 or 10. Again, prevention of any failure is better than a cure. Hence, confirmation of failure mode by testing is the best option, but it is time-consuming and costly. A testing confirmation of any failure that will disrupt production activities is the best way to avert failures. The design gets a very high rating of, say,10 when it has enough theoretical detection control to prevent a machine failure or reduce the functionality of the machine and when such failure mode detection has not been verified through testing. Any function with a very high detection rating means it needs attention or verification for critical failure modes. A detection rating of 1 or 2 means that the method is highly reliable in detecting a failure mode before a design is released for production. One has to be extremely cautious when giving a design this such a rating, i.e., the lower the rating, the higher the chance of failures being undetected during usage. Comparatively, this low rating should rarely be used in any design as it is extremely difficult to find all the failure modes.

Once the failure modes, causes, and severity of each failure function, occurrence rating, detection rating, etc. have been identified, a combined rating, called the RPN (risk priority number), for the design is calculated:

RPN = criticality x detection = severity x occurrence x detection= S x O x D

All the functions under consideration need to have this combined rating to understand which ones need the attention most to avoid a failure. These numbers provide a comparative value among the functions with possible failure modes and causes. The functions or failures are ranked accordingly depending on these combined RPN numbers. The idea is to reduce the RPN values for each function or failure mode by incorporating suggested design or process changes. Sometimes this priority ranking fails to purify a design or process. In many cases, a company might have some priorities of action based on previous experiences. Most of the time, severity gets the highest priority out of S, O, and D, and then the occurrence rating, or frequency of failure, gets the next priority. The detection rating is not given that much priority in machine tools.

For machine tools, severity and criticality must be given the highest priority for suggesting any design changes. In general, the focus is geared toward reducing the RPN values among all the functionalities of a machine. Lowering the RPN values, especially the severity and criticality, will enhance the reliability and robustness of a machine and will eventually help to increase customer satisfaction, which is the ultimate objective of any design. Keeping in mind that design changes at the end are very costly and time-consuming, these rankings could be done as soon as any design is completed. Unit-wise detection is more effective than the whole system failure ranking. Reduction in severity and criticality ratings will also help to reduce the overall RPN ranking of any design.

Various steps can be taken to reduce the severity and criticality of failure modes. To do this, design changes are normally recommended. Such changes could include material changes, component machining changes, geometry or tolerance changes, etc. Similarly, to reduce process-related severity, flow, operator training, and environmental changes could be recommended. The ultimate objective is to eliminate the effects of a failure mode that could affect the functional performance of the machine. If too much deformation is the cause of any failure of functional performance, the dimensions or material strength can be changed. In order to change the frequency of failures, the process or design can be changed.

In other words, design or process change might be the only way to reduce the severity or criticality of any failure mode. Root-cause analysis might help to understand the cause of any failure. In order to prevent the effects of failure, controls can be put in place to prevent that failure from occurring. Any extent of damage can only be controlled by putting control mechanisms into the design, such as limit switches, relays, proximity switches, torque limiters, diagnostic tools, etc. Detection equipment can also help the detection level of any design. Control devices should be properly installed in a machine to enhance the detection of failure ahead of time so that corrective actions can be taken in the operation. Such devices will make it easier to detect a failure before it gets out of control.

20.7   Action Steps for FMEA Process

The last step of the FMEA process is to suggest actions to be taken by the core design or manufacturing team. The suggestions could include the items described below:

- First and foremost, the responsibility of the FMEA is to suggest any modifications required to avoid failure modes. This is the primary responsibility. If a hydraulic system seems to be unreliable for any machine, an electric system could be suggested.

- In order to enhance the reliability of a system, redundancy must be reduced. The unnecessary ornaments of the design must be reduced.

- System linking or interaction among sub-systems must be made simple or eliminated wherever necessary by combining performances. If spindle vibration is reducing tool life, the spindle design must be changed to increase tool life.

- On the other hand, the team might suggest adding a parallel system to the main events so that the machine can still perform in case of the failure of a critical sub-system. This enhances the reliability and productivity of any system. The added parallel system must be compatible with other elements of the system. For a critical axes end limit switch, an additional switch or safety stops could be provided to prevent severe damage to the axis elements in case the primary switch fails to register runaways.

- Enhance proper and detailed diagnostics of failures.

- Add sufficient detection devices and controls to identify the occurrence of failures before they actually occur. For example, the vibration level of a machine might suggest that bearings need replacement. If the noise level of a machine goes up, noise or vibration levels must be automatically displayed to warn the operator about the degradation.

- The team might suggest adding safety devices to make the machine safer for the operator and maintenance personnel. Parallel safety devices for critical actions must be provided to reduce the risk of fatalities. For example, the machine cannot be operated in automatic mode when the doors are open. If the doors open during operation, the machine must cease operating automatically.

- Design changes such as material changes, shape changes, machining changes, etc. can be suggested to make the design more reliable. For example, steel could be suggested for cast components to reduce deformation or chance of breakage.

- Automatic gauging can be added to reduce errors during the manual checking of parts.

- Instead of manual loading or unloading, robots or loading devices can be added to the system.

- For very unsafe operation, an unmanned system can be used. For example, forging machines must have automatic loading and unloading devices to reduce manual intervention.

- To reduce abrupt failure, components can be made stronger by changing material or their dimensions.

- To enhance the detection rating of a system, the proper tests can be used.

The recommended action might be accompanied by the engineer who is going to interact with the design or manufacturing team to make sure that suggestions are taken care of. The responsible person and date of completion must be mentioned in the FMEA report. Every recommended action must have an engineer and date of completion assigned to it. It should be kept in mind that DFMEA or PFMEA is not an open-loop or one-time event. Continuous updates must be made to the document whenever a failure has been avoided or added to the system. It is a living document and needs updating from time to time. The FMEA report must be carefully reviewed to make sure that changes are explained in detail and correctly. The feasibility of such changes must also be kept in mind. The suggestions should also include the reasons for change and its benefits. For groundbreaking changes, the FMEA team should consult the core teams before suggesting changes. Cost and time implications must be very well thought out. Suggestions also become very critical when changes must be incorporated in machines that are working in customers' shops.

20.8   Common FMEA Concerns

The FMEA process is sometimes very difficult and convoluted. The team might not come to a resolution on some issues since all the team members are not convinced about the severity or criticality of a failure mode. Sometimes the suggested changes are not practical for the design. The changes might be cost-prohibitive, or they might bring other failure modes into the design. Changes might bring in unreliability or reduce productivity. Many FMEA procedures are never completed to the fullest extent because of these difficulties. Like anything else, there are issues or concerns in any FMEA process:

- Recommended actions are not feasible or are impractical for the design. For example, robotic loading and unloading might not be feasible for a machine due to space limitations or cost overruns.

- Most of the time, FMEA design actions remain unfulfilled because the engineers responsible for them move to other locations or leave the company.

- Suggested changes do not really solve the issues: severity, detection control, or criticality. Suggestions are not complete or are vague. Suggestions are not properly prioritized.

- Most problems come from the team's interaction during the FMEA process. Team members act against each other and do not cooperate to come to a decision about criticality or severity. Team conflict is a major issue. If the team is not cohesive and disciplined, the FMEA process will fail. The team has to be highly motivated to solve problems.

- The FMEA process must be a team event and not an individual event. The team leader is a coordinator and cannot dictate what the team does all the time. It is not the responsibility of the team leader to come up with the necessary changes; the team as a whole has to agree on the solution to any problem.

- The FMEA process is not carried out by a multidisciplinary team. Team structure is an issue. To solve a material issue, a metallurgist must be On the team.

- The team does not stay within the boundaries of its responsibilities. DFMEA is about the design, and PFMEA is about the process. To solve a material flow issue or bottleneck issue in the manufacturing process, the team should not expand its responsibilities. The team should be focused on its Jobs and responsibilities.

- FMEAs must be done before the design or manufacturing processes are firmed up. If the design has already been released for production, design FMEAS are less effective, and it is very costly to incorporate changes. FMEAs must take care of potential issues for any design or process.

- The suggested changes should not reflect bias, i.e., all the failure modes have the same or worst possible rating. No design should be bad for performance. Severity or criticality ratings must be discussed among team members and arrived at by consensus.

- The analysis of any failure must be based on the data. It should consider warranty data, customer feedback, and operator feedback to make sure that the failure mode is real.

- The FMEA table or template must be complete and signed by all team members.

- The FMEA process should not be used for management or business decision failures

20.9   DFMEA Design and Implementation Steps

The FMEA template has to be designed to include the following information:

- System designation and details: part number details, revision

- Machine model details: model number with revisions

- Team details: design and manufacturing process team details with responsibilities, design lead and manufacturing lead

- FMEA dates: start date, issue date, completion date, revision date

- FMEA ID details: number, revision number, location

- FMEA team details: team members, designation, department details

The functional description should be very specific and must be expressed in very clear terms. The potential failure modes should start with the design, process, and requirements. Customer feedback is also required to understand the priority. Sometimes legal or business requirements can also be considered. The failure description must not contain words that create confusion among team members. It is better to start with a verb: "RPM variation is more than 10 RPM at the highest speed," "Single station index time is more than two seconds," etc. The requirements must have a metric to measure against. For example, "Spindle power is not enough for machining cast iron" is too ambiguous to start with. A functional flow diagram can be used to describe a function.

In order to find the potential failure modes, the team has to ask the following questions:

- Can the design deliver the intended performance?

- Is the system not delivering the functions as intended? If so, what are the reasons?

- For assumed potential failure modes, how can they be proven to be real?

- Can failure modes be validated analytically or by using tests?

- Does the team understand the requirements of the design?

- Do the team members understand the manufacturing process for the design?

- Is the operating sequence for the design known with data?

- Has there been legacy data for the design failures?

- Are the environmental conditions known in detail?

- Is there any add-on required for the design in the future?

- Is the operator trained for the machine?

- Can the operator understand the failure modes?
- Is a potential failure mode critical to the machine's performance?

Potential failure modes must be studied and verified before design changes are suggested. The failure modes must be described in actual terms with some analytical data and should not be absolutely imaginary. Next is the task of identifying the effects of all the failures. In order to define and understand the potential effects of failures, the team has to ask the following questions:

- Is the failure going to hurt the operator? If so, how?
- Is the failure going to hurt the maintenance personnel? If so, how?
- How is the failure going to affect the connected sub-assemblies?
- Are the failures going to affect the functioning of the machines?
- Are the effects harmful to the operator?
- Can the failures affect the functioning of nearby machines?
- What are the financial consequences?
- What are the effects on purchasing and spare parts?

In a nutshell, the effects of failure modes on the functioning of machines or other sub-systems must be determined. If the failure is very localized and does not affect the other connected units, the severity could be less. The potential consequences of any failure on the components, sub-assemblies, operators, other machines, environments, customers, safety hazards, etc. must be discussed and documented for further processing. For example, a spindle failure might have its effects on the tooling systems, part holding units, operator safety, and indexing unit of a machine. Since any sub-assembly of a machine works in unison with many other components, a failure in any unit affects the functioning of others. A slide crash might affect the tooling system, indexing unit, spindle housing, sheet metal enclosures, etc., so the effects of any failure mode on neighboring systems must also be considered in detail. Some failure modes might affect environments in a negative way.

In some cases, customer feedback about similar machines could be of some help in finding the potential effects of failures. For example, excessive vibration, excessive noise, excessive temperature. poor surface finish, wrong indexing position, axes positioning errors, etc. could be the result of potential failure modes. The team has to find out which effects are due to which failures. Customers normally notice a change in machine performance during usage. Talking to maintenance personnel about these effects and the ways they overcame these problems could give some clues about the machine failures and their effects.

The risk priority number (RPN) is a measure of risk taken when a design is released for production. The RPN has three factors: severity (S), occurrence rate (O), and detection control (D). An effective design change is required to lower the severity number. The severity denotes the harmful effects of a failure and is given a number between 10 and 1. The problem is in the determination of the severity number for a failure mode. The team has to reach a logical conclusion for the severity number of any failure mode effect since this could be a highly debated topic. The highest value of severity rating of 10 is given to the effects that affect the safety of the operator or maintenance personnel. On the other hand, the lowest value of 1 is given when any failure has no effect at all. In between these values, the team can design a comparative number for all other effects, such as very high (8), high (6), and low (4). These numbers are comparative and are assigned to each effect of a failure mode. For example, the effect of a hydraulic pump failure could lead to chuck jaw failure, which causes the part to come out of the chuck and hit the operator. This event could be fatal, and such injury effects are given a value of 10. Safety hazards or injury potential effects are always given 9 or higher. On the other hand, paint peeling off the sheet metal has no effect on the functioning of the machine, and the severity number could be 1.

These severity numbers could break down a design into categories or design characteristics, such as potential critical characteristics and potential significant characteristics. All other effects are deemed as having no special characteristics. A very high value of severity, such as 8 or higher, is denoted by potential critical characteristics, and middle value of severity (4 to 7) could be defined as a potentially significant characteristic. Obviously, the potentially critical characteristic failure mode/causes must be eliminated before the design is released for production. Any severity value above 4 or 5 needs special attention. An FMEA process is initiated for such severe failure modes. Root-cause analysis (RCA) is initiated for any failure modes that are given values between 5 and 10. Eventually, the FMEA process results in a design change procedure before the design is finalized and released for manufacturing. The high severity rating normally denotes a design weakness that needs to be eliminated. To avert chuck failures due to loss of hydraulic power, a safety lock in the chuck cylinder must be designed and incorporated into the design.

The weakness or faults in a machine due to the manufacturing process are not normally included in the DFMEA process since the PFMEA process takes care of that. Only design weaknesses are considered in the DFMEA process. For design FMEA process, it is assumed that there is a manufacturing process control plan in existence and the product is produced within engineering standards and practices. PFMEA could identify the design elements that cause manufacturing variations. In such cases, the PFMEA should identify these causes.

The PFMEA could identify some manufacturing process control that would eliminate such variations in the manufacturing, i.e., orientation effects, assembly variations due to operator mistakes, process capability issues, etc. If these manufacturing variations are due to design deficiencies, it could be covered in the DFMEA, or the PFMEA could refer such failure modes/causes to the design core team, and DFMEA could be initiated immediately. The DFMEA task could be treated as a design improvement process as well. Any design will have some inherent weaknesses, and the primary purposes of initiating a design FMEA is to eliminate such weaknesses. This should also make the design robust and enhance customer satisfaction.

A robust design procedure also makes a product more reliable and highly productive. DFMEA should not be considered a fault-finding process. On the other hand, DFMEA could be considered an improvement step for the product before the design is released for manufacturing and line production. Sometimes DFMEA and PFMEA can be initiated together. If the manufacturing process faults are due to some design issues such as wrong mold drafts, restricted surface finish requirements, assembly tolerances, torque specifications, etc., PFMEA can identify such deficiencies, and DFMEA is initiated to take care of them in the design. Some very common failure causes might be due to:

- Wrong material specifications
- Wrong analysis procedure used for design strength analysis
- High stress concentration in a component
- Insufficient amount of lubrication for the machine slides
- Poor maintenance procedures recommended
- Excessive heat generation
- Incorrect stress and strength analysis
- Poor geometric tolerances
- Improper hardness of component specified

The failure mechanisms for the above design faults could be as follows:

- Rupture and breakage
- Creeping and yielding
- Thermal distortion
- Thermal and loading cycle fatigue
- Excessive wear and tear
- Abrasion and corrosion
- Excessive deformation
- Crash damage
- Very high maintenance cost

Once the severity of effects due to failures has been identified, the occurrence rate for such failure modes must be considered. The occurrence of a failure mode is the probability that a failure will happen during usage, i.e., the likelihood of a failure mode occurring. Again, the likelihood of such an occurrence is denoted by a comparative number between 1 and 10. In order to find the occurrence rating number, the team can discuss the following topics:

- Legacy data for similar designs from customers
- Carryover design concerns passed on to the present design
- Present design characteristics and concerns
- New technology concerns

- Non-conventional design or brand-new design
- Success or failure of similar designs in the competition's machines
- Design complexity
- Environmental conditions
- Engineering analysis and simulation procedures
- Radically different design

When the failure probability or likelihood is imminent and almost unavoidable, the occurrence rating is given a value of 9 or 10. The failure rates are present in most of the product. For example, failure of a relay due to excessive heat is found in almost in every machine produced, and it is given a rating of 9 or 10. If the component fails repeatedly for all products, the rating could be 6 to 8. If the failure is very unlikely or never experienced, it could get a rating of 1, the lowest number on the scale. The probability of failures should depend on the actual failures occurring in the design in the field or during testing of the new product. For example, if a spindle lubrication failure is found to occur in 30% of products, it could be given a value of 9 to 10.

Next is the topic of design control detection, i.e., how easy or difficult it is to detect a failure mode. The design control factor is also given a rating, which is used in the estimation of RPN values of failure modes. The value depends on how easy it is to detect failures, i.e., can the design be released for production without any control or lots of tests required to identify the failure modes. The control procedures that can be used to identify a failure mode are field tests, durability tests, design audits, quality inspections, analyses, and laboratory testing. These tests are used as design verifications or control tests. Such tests are meant to reduce the occurrences of failures and determine the nature of failures and their modes. These control tests are required to identify the occurrence rate and eventually suggest design changes to eliminate such failure modes from the design. They evaluate a design and can be considered design control tests. The idea of design control is to prevent a failure mode from happening in a machine during actual operation. Once the tests identify the correct failure mode, the cause for such failures needs to be evaluated. The primary purpose of applying a design control method is to identify design weaknesses, and corrective action needs to be taken. This is a detection procedure applied to the design.

Hence, design control is basically an analytical tool or a physical test to identify a failure and its mode. If it is absolutely impossible to ascertain a failure mode using a possible design control, it is given a rating of 10, which means that there is an absolute uncertainty to creating this failure mode using design control methods. Consequently, the design control cannot create the failure mechanism, for example, casting failures due to casting impurities. It is extremely difficult to have a design control that can identify consistently and within reasonable testing or analysis time. On the other end of the scale, if the design control can absolutely find the failure and its mode with certainty, it is given a rating of 1. For example, the failure of a door switch can be found very easily and with certainty using a reasonable design control test. This scale of 1 to 10 can be subdivided depending on the machine type and complexity.

The detection control is given a rating of 5 when there is a moderate chance of detecting a failure and its mode with reasonable accuracy and time using a simple design control test. For example, lubrication failure in the axes slide can be very quickly and accurately determined by doing a simple test. If a design is very mature or a legacy design with no history of failure, it can be released for production without much design control testing or analysis. Another common case of machine failure is due to heavy and unusual loading conditions due to material hardness or structures. There is no reasonable physical test that could possibly consider the external loading conditions for any machine. If the force used is beyond RWUP loading, it is very difficult to identify a failure mode using a normally accepted design control test.

The risk priority number (RPN) could vary between 1 and 1,000. The RPN is a measure of design risk. The lower number is comparatively better for any machine. It should also be kept in mind that high-severity failures must be taken care of irrespective of the RPN values since they denote hazardous conditions for the machine and severe failures could jeopardize operator safety and health. Before a design change is recommended, the team has to consider the failures that have the highest severity or occurrence rating. Hence, the criticality of a failure is the most important factor for issuing a design change notification.

The RPN characterization of the failures for any machine helps the team to recommend design changes for failures with the highest criticality valued or higher RPN values. Highest-ranked concerns have the highest priority for design changes. The recommended changes will reduce the severity or criticality of the failures for a machine. An unsafe machine must not leave the work premises without a change to reduce the safety risk. Time and cost are not considered more important than the safety of any machine. For example, the machine must have safety shrouding irrespective of cost and time to design and assemble them in a machine. The design control might include tolerance design control, design of experiments, durability testing, laboratory bogey testing, field testing, simulation, and analysis to name a few. Design complexity and technology are the most important factors in devising design control tests. Unproven technology might need more controls to prove the design and make it failure-free. Unprecedented or unforeseen failure might not need any design controls since physical or analytical tests cannot be designed using unreasonable inputs and environmental conditions. The design controls must be productive and effective as well.

The documentation of the initiation, process, and proceedings of any FMEA process should receive the highest level of attention. The recommendations must be documented so that a responsible person or engineer can follow through and solve the issue within a given time. The FMEA process is not endless. Once corrective actions have been taken and implemented, the newly revised ranking of the design based on severity, occurrence, and detection must be evaluated, and improvements must be recorded for future action. The FMEA table must be complete. The new and old values for RPN numbers must be noted. The success stories must be advertised for recognition and the failures must be re-evaluated to make the product a success.

<u>Example of DFMEA</u>

- **Item/Function:** Axes Movement: 30 inches of movements with 5 HP servo motor @ 2,000 RPM and 1,000 complete motions per day

- **Potential Failure Modes:**
  - Axes not moving/stopped when commanded through control (Mode #1)
  - Does not move by the desired amount (Mode #2)

- **Potential Effects of Failure:**
  - Mode #1 Effects:
    - Cannot machine any part (9) – Effect #1
    - Cannot remove the job from chuck (8) – Effect #2
    - Tool stuck in the part (4) – Effect #3
    - Tool is broken (4) – Effect #4

- **Severity (S):** 9 (worst of all effects)

- **Failure Class:** Potential Critical Characteristic for "Cannot machine any part": both Effect #1 and Effect #2

- **Occurrence Rating (O):** Effect #1: 6 (Found to happen 4 times in 200 hours)

  Effect #2: 3 (Found to happen once in 200 hours)

- **Current Design Control (D):** 4 for Effect #1: Intermittent failure; easy to find

  for Effect #1: Intermittent failure; hard to find

- **RPN:**
  - **156 for Effect #1 (higher priority for action)**
  - **135 for Effect #2**

- **Action Items:** Effect #1

- **Responsible Person: JM**

- **Date of Completion: December 2019**

20.10   PFMEA Design and Implementation Steps

The FMEA process is very well recorded in SAE standards, and most of this information has been taken from SAE Recommendation Practice J1739, which is written basically for automobiles and heavy trucks. These standards have been customized for machines in this section. The previous section was also written using the recommendations made by SAE J1739. The template for PFMEA is very similar to that for DFMEA.

In this section, the PFMEA process will be discussed very briefly. First of all, a process template has to be designed for the PFMEA as well. The design can vary from company to company depending on their manufacturing process control and methods adopted for production. The header of the PFMEA process is similar to the DFMEA template, as outlined below:

Header design:

- System designation and details: part number details, revision
- Machine model details: model number with revisions
- Core team details: design and manufacturing process team details with responsibilities, design lead and manufacturing lead
- Process responsibility: name of the engineer
- PFMEA dates: start date, issue date, completion date, revision date
- PFMEA ID details: number, revision number, location
- PFMEA team details: team members, designation, department details
- PFMEA number: company data

The process elements need to be described in detail for the FMEA. The team has to understand why the PFMEA is being initiated. For a complicated manufacturing process, the issues could be subdivided as per the manufacturing sections. The process flow must be charted and understood thoroughly. The next action is to analyze the process flow in detail. The process flow chart should include sources of variation, process identification, and a visual identification of the process and process characteristics. For example, process details for a welding joint could be as follows:

Sources of Variation: Material strength

Process Identification: WELD-001

Operation Flow ID: ⟶ | WLD001 | ⟶

Process Characteristics: Electric Welding

Hence, the whole process for the welding section, i.e., material flow, process flow, process layout, etc., must be described in detail. For each process, the sources of variation must be recorded as well. The data for the PFMEA process comes from the operation flow block diagram and the process characteristics, which represent what is being accomplished in that section of the process.

The process characteristic could mean the hardness of the weld, the length of the weld, the type of welding, the equipment used, the dimension of the weld, the start point and end point, etc. Most of this data comes from the engineering part drawing. Process characteristics define the method, equipment, operators, and quality controls used to manufacture a part as per the engineering drawing. In order to understand the links between processes, a process characteristic matrix table can be created. This table connects each process characteristic with others through an operation or operations. The table also shows the dependency of each operation on others. For further details on this table, SAE J1739 should be consulted. An example is as follows:

Product Characteristics	WELD001	WELD002	WELD003
Material Inspection	x		
Material Clamped in Position		x	
Stitch Welding Operation		x	x

In PFMEA, process functions include process requirements and product quality requirements. For example, welding two parts together by stitch welding is the description of the process. Welding requirements are given in the part drawing. The product quality requirements denote the hardness of the metals before and after welding, length of welding, quality of welding, etc. The quality specifications also come from the engineering part drawing. The potential process failure mode is defined as the way the designated or accepted process can fail to meet the requirements and specifications, i.e., how the process requirements are not met. If the blank material is too hard and quality inspection fail to record that, welding strength will be affected. The failure of the inspection process is a potential failure mode that could result in the failure of obtaining the specified welding strength. The insufficient voltage availability for the welding machine could cause the welding failure. So, the process or operation comes first and then the consequences of operation. The DFMEA must be studied in detail to understand the product requirements, and the process has to be designed according to the product requirements as specified in the engineering drawing.

In PFMEA, the team has to ask the following pertinent questions to understand process deviations or failures:

- What are the causes of product rejection due to process faults?

- What should the inspection process be to maintain quality?

- What are the sources of variations or deviations from requirements?

- What are the requirements for the welder and the welding equipment?

- What are the critical requirements of the part as per the design specifications?

The process failure modes can be described as follows:

- Quality Inspection: improper or insufficient Quality inspection controls and procedure to separate bad parts from good ones

- Manufacturing: deformation, dimensions, roughness, shape, form, etc.

- Assembly: deformation, wrong parts, part missing, etc.

- Incoming material inspection: material specifications, quality check, etc.

- Material Flow: part delivery fault, mishandling, etc.

The team needs to list all the faults in a particular process that could generate failure modes. The team has to understand why the process failed to meet the specifications, and it has to understand why a process output could be unacceptable for the next operation. The potential failure modes can be numerous, and some are listed below:

- Wrong chamfer
- Wrong radius
- Cracked part
- Bent part
- Excessive deformation
- Poor surface finish
- Loose part in assembly
- Part damage due to rough handling
- Wrong assembly
- Wrong orientation
- Unfinished operation
- Improper hardness
- Unevenness

The DFMEA process takes care of the design weaknesses. The PFMEA concentrates on the potential process faults that can affect the intended design to deviate from specifications. The focus is on the process inputs and outputs and not on the design weaknesses. The PFMEA assumes that design is not the problem but that the process variation is the issue at hand.

Here are some examples of the effects of a potential process failure:

- Unit does not function at all as intended
- Unit has intermittent function
- Unit does not look good due to poor paint finish
- Noise during operation
- Interference
- Assembly jammed
- Unit cannot be assembled with other parts in the next assembly station
- The operator's safety and health are affected
- Unsafe operation
- Environmentally non-compliant operations

Severity criticality and detection ratings are almost similar to those in the DFMEA process. Severity denotes the seriousness or gravity of any process failure mode. The seriousness of each failure mode effect needs to be evaluated, and a rating has to be assigned to it. The most serious effects need to be taken care of first. The ranking of effects starts with the effect and its severity. If any process failure affects the operator's safety or health, that is the highest form of effect, and it gets the highest ranking of 10 on a scale of 1 to 10. If a process failure could occur without any warning, it gets the highest possible severity ranking. If any wrong assembly creates a machine tool crash and has the possibility of hurting any operator, that would be considered a dangerous process. If the failure mode does not have any effect on the success of an operation, it gets the lowest ranking of 1. Smoke, harmful rays, and very bright light created during the welding process might create a hazardous condition for the operator. This failure effect must be eliminated effectively. Various levels of effects can be established, and severity ratings can be given to all of them.

For example, slower than normal material flow might affect the production rate in a day. This could get a severity rating of 5–6 and be called a "moderate" severity operation. Similar to the DFMEA process, potential failure characteristics could be given different classifications, such as critical (severity 9–10), significant (severity 6–8), high-impact (severity 4–5), operator safety (severity 9–10), or none. The cause of failure is defined as the reason for a process failure. Causes for each process failure mode must be determined and listed accordingly. Root-cause analysis can be initiated to identify the actual cause of a process failure. In order to remove the causes of process failures, they must be described and detailed properly. Depending on the severity rating and occurrence (frequency of happening) rating, the criticality ranking of a failure mode is determined. Process malfunctions or errors should be mentioned and should include general failure terms such as operator error, wrong inputs, material defects, etc. Below are examples of root causes of process failures:

- Wrong tooling for the material being machined
- Blunt tools
- Part too heavy for the loading robot
- Wrong orientation
- Too much depth of cut
- Feed rate too high
- Wrong current setting for the welding machine
- Wrong programming methodology
- End limit switch failure
- Conveyor not working

The sources of manufacturing variations must be identified to understand the mechanism of failures. The inputs to the process must be as per the specifications for the process. Next is the task of finding the occurrence rating. In order to do this, one might consider the process capability index, product failure rate, and process variability data. For process capability data, the manufacturing process must be stable, and errors due to assignable causes must be eliminated. Statistical data such as statistical process control (SPC) data, can be collected for a process to determine its stability. If statistical data for a brand-new process is not available, proper experience and engineering-based judgment should be used. A simulation engineering process could be used to determine the cause of failure. Failure history or legacy data for the similar process could also be used to determine the occurrence rating of a failure mode. It should be remembered that failure modes with the highest severity ratings must be taken care of irrespective of the occurrence rating.

Similar to the DFMEA occurrence rating scaling process, the process failure occurrence rating table should be generated. The scale is between 1 and 10. If the chance of failure is very high and is unavoidable, i.e., the failure rate is 50% or more, the occurrence rating is given a value of 10. If the chance of failure is very remote or highly unlikely to occur (one failure in, say, 10,000 machines), it gets a ranking of 1. In between 1 and 10, several suitable categories can be generated with different occurrence ratings. For example, an occurrence rating of 4–5 could mean failure rates of 1 in 250 products manufactured. These are comparative values. Isolated or occasional failure modes could be given a ranking of 2.

The process control methods and ranking are almost similar to those explained in the DFMEA process above. The process control methods could consist of the following:

- Prevention of process control failures, modes, and occurrence of failure rates
- Detection of failure mechanisms and process controls to eliminate the failures
- Detection of failure modes and causes
- Eliminate redundancy in the process
- Combine process elements to reduce the risk of failure
- Develop preventive process control plans

The detection control methods are designed to enhance process failure detection at an early stage or before it occurs. The process control methods could include process audits before and after implementation, automatic process checking, quality inspection and statistical analysis, validation tests, and verification. The process audit could include process ability, identification of redundancies and ambiguous controls, process capability studies, etc. Process control checking could be done using SPC, gauging, process bottleneck checking, etc. The inspection could be in-process automatic checking or out-of-line checking, routine inspection, end-of-line reliability testing, sound verification testing, in-process parameter checking and evaluation, etc.

Hence, detection of a failure mode means the reliability or chances of detection of failure. How easily the failures can be detected or if detection is not possible at all is the focus of detection control. Can the current control detect the failure mode? What are the chances that the proposed failure detection control will be detected, and how easily can it be done? Depending on the situation, the detection rating is assigned to a failure mode. The scale is again between 1 and 10. If the detection is very hard to do, it gets a value of 10, whereas if the detection is very easy to do or automatic, the rating is 1. For hard-to-find issues, more control is required, and without a properly proven process control in place, the process should not be used for continuous runs or production. For in-process control using gauge or laser devices in machining, gauges must be frequently calibrated and properly maintained, and gauge accuracy must be checked very frequently. The process control must not allow operators to bypass the manual gauge usage. The gauging could be made automatic.

The RPN value of a process failure is also similar to that discussed for the DFMEA process. RPN = S x O X D = criticality x detection rating. These values, between 1 and 1,000, are used to rank the failure modes. Corrective actions must be taken with very high RPN values or failure modes that give rise to safety and health concerns for the operators or other personnel. Process failure modes with very high severity ratings must be eliminated before the process is established to produce machines for customers.

The next step is to recommend actions for high-severity process failure modes. The corrective actions must be recommended for very highly ranked process failure modes. The suggested action must reduce the severity of failure or frequency of occurrence. The detection rating should be reduced as well. The process failure mode corrections are treated differently from DFMEA suggested actions. Nevertheless, if manufacturing process failure is due to stringent design requirements, a design concern should be raised. If the process is not capable of producing parts within design requirements, the design should be looked into since process change is always very costly if new equipment or new quality inspection is required to produce the part or machine as a whole. The design should promote ease of manufacturing and assembly. If design requirements are complicated, the manufacturing process also gets complicated. A simple design is easy to manufacture and assemble. If the design cannot be reproduced with ease, the cost and time for production go up, forcing the product to become noncompetitive in the marketplace. In my opinion, products fail very often due to poor design and much less often due to a poor manufacturing process if manufacturing satisfies the design requirement.

People buy a product for its capabilities and not due to the manufacturing process. The design determines the functionality of a product. For a PFMEA to be very productive and effective, suggested process changes must be easy to implement with minimal cost and time. The process should be followed up after control is implemented to ensure the effectivity of changes. For failures with ranking above 9, a design change is not an option but a necessity to make the process safer and more productive. So, after an action has been implemented in an existing process, an audit must be done by the PFMEA team to ensure the change has made a positive difference for the product and environment. The severity, occurrence, and detection rating must be re-evaluated to determine improvements over the current process.

20.11　Summary

- The success of an FMEA process depends mostly on the team cross-functional structure. The team should be very cohesive, and members should be highly motivated to make the process a grand success. Conflicts will occur among team members, but they should not be avoided but resolved. Team knowledge about the FMEA process and about the product and process are also very important for the success of an FMEA project. The team leader has the responsibility of leading the team through thick and thin. The team also needs the full support of management to make things happen through the FMEA process.

- The FMEA process starts with the machine requirements, customer feedback, competitive data, environmental regulation data, legacy machine data, etc. To get the most out of the FMEA process, it must be applied before the design is converted into a product. The FMEA process is highly iterative. After an action has been taken to eliminate a failure mode, the effects are evaluated to see if it has been eliminated properly. In some rare cases, an action to avert one such failure might cause other failures that were not anticipated before the action was implemented.

- An FMEA should be done when a new design or a new process is created. The potential characteristics must be determined and resolved before the design is released for production or a new process is started. The process must be stable before a PFMEA is initiated. Similarly, the design must be mature before DFMEA is conducted.

- There are two basic FMEA processes that are commonly used by the aerospace and automobile industries. These are design FMEA, or DFMEA, and process FMEA, or PFMEA. The DFMEA can be used during system-, sub-system-, or component-level design stages. On the other hand, PFMEA can be used for any process, such as manufacturing or business processes. PFMEA can also be used in assembly, manufacturing, and reliability assessment processes.

- Function diagrams, hardware block diagrams, or reliability block diagrams should be constructed to understand the design or process. Function diagrams come from functional analysis. Functional block diagrams divide the design or process into smaller chunks so that it becomes easier to understand and handle its functionalities.

- The team has to decide what should be included in an FMEA investigation and what should be excluded. The boundary definition is very important. Otherwise, the process becomes very complicated and winding, and the focus on the real problem is lost.

- The team has to design the FMEA header and FMEA table for the issue before anything is started. This process should start with the function and what is wrong with it. The causes and effects of these malfunctions must be determined at the outset of the process.

- The severity of the effects, criticality, and detection rating must be decided by team consensus. This is the most difficult task for the team when trying to determine the solution to a problem. Recommended actions depend on the severity or criticality numbers.

- The risk priority number (RPN) is a measure of risk taken when a design is released for production. The RPN has three factors: severity (S), occurrence rate (O), and detection control (D). An effective design change is required to lower the severity number. The severity denotes the harmful effects of a failure and is denoted by a number between 10 and 1. The problem is determining the severity number for a failure mode. The team has to reach a logical conclusion for the severity number of any failure mode effect since this could be a highly debated topic.

- The FMEA process is very well recorded in SAE standards, and most of this information has been taken from SAE Recommendation Practice J1739, which is written basically for automobiles and heavy trucks. These standards have been customized for machines in this chapter.

- The RPN value of a process failure is also similar to that discussed for the DFMEA process. RPN = S x O x D = criticality x detection rating. These values, between 1 and 1,000, are used to rank the failure modes accordingly. Corrective actions must be taken when there are very high RPN values or failure modes that give rise to safety and health concerns for operators or other personnel. Process failure modes with very high severity ratings must be eliminated before the process is established.

20.12   References and Bibliography

Boothroyd, G., 1992, *Assembly Automation and Product Design*, Marcel Dekker, NY.

Dewhurst, P., and Boothroyd, G., 1987, "Design for assembly in action," *Assembly Engineering*, January.

Dewhurst, P., and Boothroyd, G., 1989, *Product Design for Assembly*, Boothroyd Dewhurst, Inc., Wakefield, RI.

Ford Design Institute, 1998, Lecture Notes on Reliability and Robustness, Ford Motor Company, Dearborn, MI.

Lucie-Smith, Edward, 1982, *A History of Industrial Design*, Van Nostrand Reinhold Company, NY.

McGrath, M.E., 1995, *Product Strategy for High-Technology Companies*, McGraw-Hill, NY.

Nahmias, S., 2008, *Production and Operations Management*, fourth edition, McGraw-Hill Erwin Companies, NY.

SAE J1739 Standard, Society of Automobile Engineers, USA.

Ulrich, T.K., and Eppinger, S.D., 2011, *Product Design and Development*, McGraw-Hill Companies, NY.

20.13   Review Questions

- Can a DFMEA or PFMEA be applied successfully to machines? If so, why hasn't this process been used extensively in the machine tool industry? Does the complexity of an issue or a machine matter when applying such a process? Which one, the DFMEA or PFMEA, is more useful for machines?

- Is the team formation and structure important for the success of an FMEA process? Who should be included, and who should not be? Does the team need to be cross-functional? Is diversity important among team members? What are the primary responsibilities of a team leader? How should conflicts be resolved? How can a team be dysfunctional? Is a knowledge of the design or process important for all team members?

- How can the RPN number for a design failure issue be determined? Why is severity so important? Can severity or criticality be left alone before a design is released for production? How can control faults in a machine design be detected? What is more important: criticality or severity? Can we separate criticality from severity?

- A CNC turning machine frequently has an indexing issue. Create a hypothetical DFMEA and PFMEA process to resolve this issue.

SUBJECT INDEX

Topics	Page Number

AUTHOR INDEX

Authors	Page Number
Galbraith, J.R.	104
Garvin, D.A.	321
Grady J. O.	85,127
Green, P.	321
Griffin, A.	
Groeneveld, P.	164
Gupta, S.K.	443
Gupta, Y.	321
Haggart, V.A.	35,85
Hatley, D.J.	192
Hauser, J.R.	164
Hayes, R.H.	85
Hays, C.V.	247
Hein, L.	104
Hill, T.J.	35,127
Hippel, V.E.	220
Hopp, W.J.	35,104,164
Hunter, S.J.	320
Hunter, S.J.	353
Hunter, W.G.	353
Hunter, W.G.,	320
INCOSE Systems Engineering Handbook	568
Industrial Design, NY	414
ISO Standards	541
Izawa, K.	85
Jones, D.T.	85
Kankuro, K.	85
Kezner, K.	480
Kidder, T.	35,105,127
Kinnear, T.C.	164,192
Kleinschmidt, E.J.	164
Krugman, P.	35,105
Kumar, S.	321
Kumar, V.	192
Lenhard, A.P.	164
Lester, R.K.	85
Lewis, E.E.	353
Lucie-Smith, E.	414,606
Machine Tool Safety Standards	541
Maier, M.W.	85
Manne, A.S.	85
McGrath, J.E.	219
McGrath, M.E.	164,192
McKim, R.H.	219

The first edition of ***System** Engineering **for Machinery** Development* book has been written by a professional machine tool design engineer for students and practicing professionals who are trying to design machineries for higher productivity and efficiency and to satisfy customer requirements. This book is written by a practitioner for other practitioners. Every machine designer needs to follow the system engineering tools and methodologies to design globally competitive machines to enhance production, robustness of machine and reduce the life cycle cost of the machine. The contents of this book are primarily geared towards making the task of designing machine systems which would satisfy the users. I have worked relentlessly to identify the system engineering principles which would help the design professionals to design cost effective robust machines. The generalized system engineering tools have been customized for machine design and development for real world applications in machine tool industry. Hope readers of this book find the contents very useful for their day-to-day work.

Printed in the United States
By Bookmasters